Obesity and Gynecology

Obesity and Gynecology

Second Edition

Edited by

Tahir A. Mahmood
Victoria Hospital, Kirkcaldy, United Kingdom
University of St. Andrews, St. Andrews, United Kingdom

Sabaratnam Arulkumaran
Division of Obstetrics and Gynaecology, St. George's University of London, London, United Kingdom

Frank A. Chervenak
Lenox Hill Hospital, New York, NY, United States
Zucker School of Medicine at Hofstra/Northwell, Hempstead, NY, United States

ELSEVIER

Elsevier
Radarweg 29, PO Box 211, 1000 AE Amsterdam, Netherlands
The Boulevard, Langford Lane, Kidlington, Oxford OX5 1GB, United Kingdom
50 Hampshire Street, 5th Floor, Cambridge, MA 02139, United States

Copyright © 2020 Elsevier Inc. All rights reserved.

No part of this publication may be reproduced or transmitted in any form or by any means, electronic or mechanical, including photocopying, recording, or any information storage and retrieval system, without permission in writing from the publisher. Details on how to seek permission, further information about the Publisher's permissions policies and our arrangements with organizations such as the Copyright Clearance Center and the Copyright Licensing Agency, can be found at our website: www.elsevier.com/permissions.

This book and the individual contributions contained in it are protected under copyright by the Publisher (other than as may be noted herein).

Notices
Knowledge and best practice in this field are constantly changing. As new research and experience broaden our understanding, changes in research methods, professional practices, or medical treatment may become necessary.

Practitioners and researchers must always rely on their own experience and knowledge in evaluating and using any information, methods, compounds, or experiments described herein. In using such information or methods they should be mindful of their own safety and the safety of others, including parties for whom they have a professional responsibility.

To the fullest extent of the law, neither the Publisher nor the authors, contributors, or editors, assume any liability for any injury and/or damage to persons or property as a matter of products liability, negligence or otherwise, or from any use or operation of any methods, products, instructions, or ideas contained in the material herein.

British Library Cataloguing-in-Publication Data
A catalogue record for this book is available from the British Library

Library of Congress Cataloging-in-Publication Data
A catalog record for this book is available from the Library of Congress

ISBN: 978-0-12-817919-2

For Information on all Elsevier publications
visit our website at https://www.elsevier.com/books-and-journals

Publisher: Stacy Masucci
Acquisitions Editor: Tari Broderick
Editorial Project Manager: Sara Pianavilla
Production Project Manager: Sreejith Viswanathan
Cover Designer: Matthew Limbert

Typeset by MPS Limited, Chennai, India

Dedication

To Aasia, Gayatri and Judy
For their tolerance of our enthusiasm, support in times of happiness,
succor in times of challenge, and endless love
Tahir, Arul and Frank

Contents

List of contributors xiii
About the editors xvii
Preface—obesity in gynaecology xix

Section 1
Obesity and adolescence

1. Obesity and the onset of adolescence

Zana Bumbuliene, Gabriele Tridenti and Anastasia Vatopoulou

Obesity in childhood and adolescence: definition 3
Obesity in childhood and adolescence: incidence 3
Obesity in children and adolescents: etiology 4
Obesity and puberty: pathophysiology 7
Obesity and pubertal development 8
Obesity in childhood and adolescence: clinical manifestation 9
Obesity in children and adolescents: counseling 9
Obesity in children and adolescents: management 11
References 11

2. Obesity in adolescence

Gail Busby and Mourad W. Seif

Introduction 15
Prevalence of childhood obesity: a global perspective 15
Obesity and the pubertal transition 16
Factors affecting childhood and adolescent obesity 16
Adolescent obesity—adverse outcomes 16
Polycystic ovarian syndrome in adolescence 17
Obstetric outcomes in obese adolescents 18
Management principles 19
Management of polycystic ovarian syndrome in adolescence 20
Psychological morbidity 21
Conclusion 21
References 21

3. Obesity in polycystic ovary syndrome and infertility

Ioannis E. Messinis, Christina I. Messini and Konstantinos Dafopoulos

Introduction 23
Obesity and infertility—possible mechanisms 23
Diet, lifestyle changes, and bariatric surgery 26
Clomiphene citrate 27
Aromatase inhibitors 28
Follicle-stimulating hormone 28
Insulin sensitizers 29
In vitro fertilization 29
References 30

Section 2
Contraception

4. Obesity and sexual health

Sharon Cameron and Michelle Cooper

Introduction 37
Obesity and sexual behavior 37
Obesity and sexual function 38
Obesity and sexual health outcomes 39
Conclusion 40
References 40

vii

5. Obesity and contraception

Sujeetha Damodaran and Krishnan Swaminathan

Introduction	43
Risks of obesity in pregnancy	43
Classification of obesity based on body mass index	43
Potential concerns with obesity and contraception	43
Obesity and contraceptive efficacy	44
Evidence (or lack of) for contraceptive efficacy in overweight or obese women	45
Safety of hormonal contraceptives in obese women	46
Obesity, contraception, and cardiovascular disease	49
Obesity, contraception, and venous thromboembolism	49
Obesity, contraception, and cancer	50
Contraceptive issues after bariatric surgery	50
Intrauterine contraceptive devices in obese women	51
Sterilization procedures in obese women	51
References	52

6. Contraceptive choices for women before and after bariatric surgery

Agnieszka Jurga-Karwacka and Johannes Bitzer

Introduction—Bariatric Surgery	57
Long acting contraceptives	58
Progesterone-only injection	61
Oral hormonal contraception [56]	62
Progestogen-only pill	62
Safety and health benefits	62
Contraceptive patch and ring	63
Barrier method	63
Emergency contraception	63
References	63

7. Long-term contraceptive care in obese and superobese women

Johannes Bitzer

Introduction	67
Combined hormonal contraceptives	67
Copper intrauterine device	68
Levonorgestrel-containing intrauterine systems	69
Etonogestrel-releasing implant	70
References	72

Section 3
Male and female Infertility

8. Obesity and hirsutism

Mostafa Metwally

Introduction	77
Obesity and ovarian function	77
Obesity and androgen production	77
Hirsutism	77
Obesity and polycystic ovarian syndrome	78
The role of adrenal androgens in obese women with hirsutism	78
Management of hirsutism associated with obesity	78
Treatment	79
Conclusion	81
References	81

9. Obesity and female infertility

Suresh Kini, Mythili Ramalingam and Tahir A. Mahmood

Introduction	83
Epidemiology	83
Pathophysiological basis of infertility in obese women	83
The clinical effects of obesity on female infertility	85
Challenges of managing obese women	87
Treatment options	87
Conclusion	89
References	89

10. Obesity and recurrent miscarriage

Andrew C. Pearson and Tahir A. Mahmood

Introduction	91
Obesity and miscarriage	91
Obesity and recurrent miscarriage	92
Polycystic ovarian syndrome	92
Ovarian dysfunction	92
Endometrial changes in obesity	93
Immunological factors	93
Male obesity and recurrent miscarriage	93
Management	93
Conclusion	95
References	95

11. Obesity and assisted reproduction

Mark Hamilton and Abha Maheshwari

Introduction	97
Prevalence of obesity in the assisted reproduction sector	97
Evidence of reduced fertility in the obese	97
Specific issues relating to assisted reproduction treatment	98
Rationale for the use of assisted reproduction	98
Practical management of obese women undertaking assisted reproduction treatment	99
Clinical procedures	99
Effect of obesity on the results of assisted reproduction treatment	100
Safety issues for mothers and offspring	101
Ethical issues relevant to access to services	102
Conclusion	103
References	103

12. Obesity and sexual dysfunction in men

Darius A. Paduch and Laurent Vaucher

Physiology of sexual function	105
Sexual dysfunction and obesity-related comorbidities	110
Multidisciplinary approach to treatment	113
References	116

13. Male obesity—impact on semen quality

Vanessa Kay and Sarah Martins da Silva

Introduction	119
Impact on semen quality	119
Combined semen parameters	120
Etiological theories	121
Treatment	123
Conclusion	124
References	124

14. Evidence-based assisted reproduction in obese women

Brenda F Narice and Mostafa Metwally

Introduction	127
Impaired ovarian folliculogenesis	127
Altered endometrial receptivity	128
Obesity and in vitro fertilization	128
Obesity and frozen-thawed embryo transfer	129
Obesity and intrauterine insemination	129
Conclusion	129
References	130

15. Obesity, bariatric surgery, and male reproductive function

Man-wa Lui, Jyothis George and Richard A. Anderson

Introduction	135
Bariatric surgical techniques	135
Pathophysiology in obesity	136
Obesity and spermatogenesis	138
Obesity and Sertoli cell function	139
Obesity and erectile dysfunction	139
Transgenerational epigenetic effects	139
Practical considerations	139
References	140

16. Medical interventions to improve outcomes in infertile obese women planning for pregnancy

Vikram Talaulikar

Introduction	143
Impact of high body mass index on fertility and pregnancy	143
Lifestyle interventions to improve outcomes in infertile obese women planning for pregnancy	143
Dietary interventions	144
Diet	144
Role of exercise	144
Weight-loss medications and fertility outcomes	145
Metformin	145
Sibutramine	147
Orlistat	147
Liraglutide	147
Barriers to weight loss	148
Conclusion	148
References	149

17. Surgical interventions to improve fertility potential in obese men and women

Joseph Chervenak and Frank A. Chervenak

Introduction	151
Obesity and fertility	151
Nonsurgical management of obesity	152
Bariatric surgery as a weight loss measure	152
Types of bariatric surgery	152
The impact of bariatric surgery on fertility	153
Bariatric surgery and polycystic ovarian syndrome	154
The potential of bariatric surgery for a negative impact on fertility	154
Pregnancy after bariatric surgery	154
Assisted reproduction after bariatric surgery	154
Obesity in the male	155
Conclusion	155
References	155

Section 4
General Gynaecology

18. Obesity and gynecology ultrasound

Kiran Vanza, Mathew Leonardi and George Condous

Key points	159
Introduction	159
Pelvic ultrasound	159
Ultrasound settings	163
Ergonomic considerations	163
Clinical applications	164
Conclusion	168
Acknowledgments	168
References	168

19. Obesity and menstrual disorders

Jane J. Reavey, W. Colin Duncan, Savita Brito-Mutunayagam, Rebecca M. Reynolds and Hilary O.D. Critchley

Obesity: the problem	171
Abnormal uterine bleeding—Causes	171
PALM-COEIN Classification	171
Malignancy and hyperplasia	172
Ovulatory dysfunction	172
The endometrium	173
Polycystic ovary syndrome	174
Obesity in the absence of polycystic ovary	174
Summary	175
Acknowledgments	175
References	175

20. Incontinence and pelvic organ prolapse in the obese woman

Clare F. Jordan and Douglas G. Tincello

Introduction	179
Incidence and prevalence	179
Normal bladder function and causes of incontinence in women	179
Urodynamic stress incontinence	180
Detrusor overactivity	181
Obesity and urinary incontinence	181
Obesity and fecal incontinence	182
Obesity and prolapse	182
Weight loss and the effects upon continence and prolapse	183
Continence and prolapse surgery in the obese woman	184
Conclusion	184
References	185

21. Urinary and fecal incontinence in obese women

Vasilios Pergialiotis and Stergios K. Doumouchtsis

Introduction—epidemiology	189
Pathophysiology of incontinence in the obese population	190
Outcomes of incontinence procedures in obese women	191
Incontinence symptoms following weight loss	192
Conclusion	192
References	192

22. Role of obesity in cancer in women

Joanna M. Cain

Overview	195
Epidemiologic evidence for links between obesity and cancer	195
Cancers unique to or more common in women	196
Potential mechanisms for oncogenesis	196
Clinical implications for prevention and treatment of cancer in obese patients	197
Summary and ongoing needs for research	198
References	198

23. Obesity and breast cancer

Chiara Benedetto, Emilie Marion Canuto and Fulvio Borella

Epidemiology	201
Pathogenetic mechanisms	202
Diagnosis	204
Therapy	204
Prognosis	206
References	206

24. Obesity and female malignancies

Ketankumar B. Gajjar and Mahmood I. Shafi

Introduction	209
Epidemiology	209
Mechanisms relating obesity to female malignancies	211
Effect of obesity on management of female malignancies	212
Future directions	213
References	213

25. Challenges in gynecological surgery in obese women

Chu Lim and Tahir A. Mahmood

Introduction	217
Indications for surgery	217
Risk of obese women undergoing surgery	218
Physiological changes in the obese patients	218
Preoperative evaluation	218
Equipment and general considerations	219
Anesthetic challenges	219
Thromboprophylaxis	219
Intraoperative challenges	219
Open abdominal surgery	220
Postoperative issues	220
Medicolegal implication	221
Conclusion	221
References	221

26. Laparoscopic and robotic surgery in obese women

Manou Manpreet Kaur and Thomas Ind

Introduction	223
Physiological changes in (obese) surgical patient	226
Benefits of minimally invasive surgery	227
Alternatives for class III (morbidly) obese patients	230
Cost-effectiveness	231
Complications	231
Preoperative preparation	232
Intraoperative considerations	234
Postoperative considerations	238
Conclusion	239
References	239

27. Obesity and venous thromboembolism

Julia Czuprynska and Roopen Arya

Introduction	245
The interplay between obesity and venous thromboembolism risk	245
Hormonal contraception	246
Guideline recommendations	248
Hormone replacement therapy	249
Assisted conception	251
Gynecological surgery	251
Gynecological cancer	252
Conclusion	252
References	252

28. Obesity and cardiovascular disease in reproductive health

Isioma Okolo and Tahir A. Mahmood

Introduction	255
Risk assessment in obese individuals	255
The significance of adipose distribution	256
Pathogenesis of visceral obesity	256
Lipid metabolism in nonpregnant obese women	257
Cardiac adaptation to obesity	257
Lipid metabolism in pregnant obese women	257
Gynecology practice	258
Maternal obesity and in utero programing for cardiovascular disease	259
Interventions to address obesity in reproductive health	260
Interventions to improve outcomes in pregnancy	261
Conclusions	261
References	261

29. Female obesity and osteoporosis

Rashda Bano and Tahir A. Mahmood

Introduction	265
Relationship between fat and bone: epidemiologic and clinical observations	266
Adipocyte hormones	267
Obesity of the bone	269
Clinical and diagnostic implication of the concept-obesity of bone	269
Treatment implications of the concept-obesity of the bone	269
Conclusion	270
Conflict of interest	271
References	271
Future research	272

30. Obesity, menopause, and hormone replacement therapy

Marta Caretto, Andrea Giannini, Tommaso Simoncini and Andrea R. Genazzani

Introduction	273
The menopausal obesity: role of estrogens	273
Obesity, lifestyle intervention, and hormone replacement therapy	274
Emerging menopausal therapies	277
Conclusion	278
Conflict of interest	278
References	278

31. Obesity and chronic pelvic pain

I-Ferne Tan and Andrew W. Horne

Introduction	281
Obesity and pain physiology	281
The genetics of obesity and chronic pain	283
The psychological impact of obesity and chronic pelvic pain	284
The impact of obesity on the assessment of chronic pelvic pain	284
The impact of obesity on the treatment of women with chronic pelvic pain	285
The impact of obesity on the surgical management of women with chronic pelvic pain	285
Obesity and endometriosis	286
Obesity and adenomyosis	287
Obesity and abdominal myofascial pain syndrome	287
Obesity and nongynecological causes of chronic pelvic pain	287
Conclusion	287
References	288

32. Obesity and clinical psychosomatic women's health

Mira Lal and Abhilash H.L. Sarhadi

Introduction	293
Section 1: menstrual problems and obesity	296
Section 2: infertility/subfertility and clinical psychosomatic aspects	300
Section 3: physical, mental, and sexual violence with obesity in migrants	302
Section 4: obesity and severe pelvic/perineal dysfunction	305
Section 5: psychosomatic impact of gynecological tumors in the obese	305
Conclusions	307
References	307

33. Obesity and psychosexual disorders

Ernesto González-Mesa

Biological and psychological mechanism	313
Polycystic Ovarian syndrome	315
Impaired body image	316
References	316

34. Professionally responsible clinical management of obese patients before and during pregnancy

Frank A. Chervenak and Laurence B. McCullough

Introduction	319
Professional ethics in obstetrics	319
Professional ethics in clinical practice with obese patients before and during pregnancy	320
Conclusion	322
References	322

Index 323

List of contributors

Richard A. Anderson, MRC Centre for Reproductive Health, The Queen's Medical Research Institute, University of Edinburgh, Edinburgh, United Kingdom

Roopen Arya, King's College Hospital, London, United Kingdom

Rashda Bano, Obstetrics and Gynaecology, Royal Infirmary of Edinburgh, Edinburgh, United Kingdom

Chiara Benedetto, Department of Surgical Sciences, Sant'Anna Hospital, University of Torino, Torino, Italy

Johannes Bitzer, Department of Obstetrics and Gynecology, University Hospital of Basel, Basel, Switzerland; Post Graduate Diploma of Advanced Studies in Sexual Medicine, University of Basel, Basel, Switzerland

Fulvio Borella, Department of Surgical Sciences, Sant'Anna Hospital, University of Torino, Torino, Italy

Savita Brito-Mutunayagam, MRC Centre for Reproductive Health, The Queen's Medical Research Institute, The University of Edinburgh, Edinburgh, United Kingdom

Zana Bumbuliene, Clinic of Obstetrics & Gynecology, Institute of Clinical Medicine, Faculty of Medicine, Vilnius University, Vilnius, Lithuania

Gail Busby, Division of Gynaecology, St. Mary's Hospital, Manchester University Hospitals NHS Foundation Trust, Manchester, United Kingdom

Joanna M. Cain, Department of Obstetrics and Gynecology, University of Massachusetts Medical School, Worcester, MA, United States

Sharon Cameron, Consultant Gynaecologist, NHS Lothian, University of Edinburgh, Edinburgh, Scotland; Sexual Health Services, NHS Lothian, University of Edinburgh, Edinburgh, Scotland

Emilie Marion Canuto, Department of Surgical Sciences, Sant'Anna Hospital, University of Torino, Torino, Italy

Marta Caretto, Division of Obstetrics and Gynecology, Department of Clinical and Experimental Medicine, University of Pisa, Pisa, Italy

Frank A. Chervenak, Department of Obstetrics and Gynecology, Zucker School of Medicine at Hofstra/Northwell, Lenox Hill Hospital, New York, NY, United States

Joseph Chervenak, Obstetrics and Gynecology, New York Presbyterian/Weill Cornell, New York, NY, United States

George Condous, Acute Gynaecology, Early Pregnancy and Advanced Endoscopy Surgery Unit, Nepean Hospital, Kingswood, New South Wales, Australia; Sydney Medical School Nepean, The University of Sydney, Sydney, Australia

Michelle Cooper, Obstetrics & Gynaecology, NHS Lothian, University of Edinburgh, Edinburgh, Scotland

Hilary O.D. Critchley, MRC Centre for Reproductive Health, The Queen's Medical Research Institute, The University of Edinburgh, Edinburgh, United Kingdom

Julia Czuprynska, King's College Hospital, London, United Kingdom

Konstantinos Dafopoulos, Department of Obstetrics and Gynaecology, Faculty of Medicine, School of Health Sciences, University of Thessaly, Larissa, Greece

Sujeetha Damodaran, KMCH Institute of Allied Health Sciences, Coimbatore, India

Sarah Martins da Silva, Reproductive and Developmental Biology, School of Medicine, Ninewells Hospital and Medical School, University of Dundee, Dundee, United Kingdom

Stergios K. Doumouchtsis, Laboratory of Experimental Surgery and Surgical Research N.S. Christeas, National and Kapodistrian University of Athens, Athens, Greece; Department of Obstetrics and Gynaecology, Epsom and St Helier University Hospitals NHS Trust, London, United Kingdom; St George's, University of London, London, United Kingdom; American University of the Caribbean School of Medicine, Coral Gables, FL, United States

W. Colin Duncan, MRC Centre for Reproductive Health, The Queen's Medical Research Institute, The University of Edinburgh, Edinburgh, United Kingdom

Ketankumar B. Gajjar, Department of Gynaecological Oncology, Nottingham University Hospitals NHS Trust, Nottingham, United Kingdom

Andrea R. Genazzani, Division of Obstetrics and Gynecology, Department of Clinical and Experimental Medicine, University of Pisa, Pisa, Italy

Jyothis George, MRC Centre for Reproductive Health, The Queen's Medical Research Institute, University of Edinburgh, Edinburgh, United Kingdom; Boehringer Ingelheim, Frankfurt, Germany

Andrea Giannini, Division of Obstetrics and Gynecology, Department of Clinical and Experimental Medicine, University of Pisa, Pisa, Italy

Ernesto González-Mesa, Obstetrics and Gynecology, Malaga University School of Medicine, Malaga, Spain

Mark Hamilton, University of Aberdeen, Aberdeen, United Kingdom

Andrew W. Horne, MRC Centre for Reproductive Health, University of Edinburgh, Edinburgh, United Kingdom

Thomas Ind, The Royal Marsden Hospital, London, United Kingdom

Clare F. Jordan, University Hospitals of Leicester NHS Trust, Leicester, United Kingdom

Agnieszka Jurga-Karwacka, Department of Gynecology and Gynecological Oncology, University Hospital Basel, Basel, Switzerland

Manou Manpreet Kaur, The Royal Marsden Hospital, London, United Kingdom

Vanessa Kay, Assisted Conception Unit, Ninewells Hospital, Dundee, United Kingdom

Suresh Kini, Assisted Conception Unit, Department of Obstetrics and Gynaecology, Ninewells Hospital, Dundee, United Kingdom

Mira Lal, St James's University Hospital, Leeds, United Kingdom; The Dudley Group NHS Foundation Trust, Dudley, United Kingdom

Mathew Leonardi, Acute Gynaecology, Early Pregnancy and Advanced Endoscopy Surgery Unit, Nepean Hospital, Kingswood, New South Wales, Australia; Sydney Medical School Nepean, The University of Sydney, Sydney, Australia

Chu Lim, Obstetrics and Gynaecology, Victoria Hospital, Kirkcaldy, United Kingdom

Man-wa Lui, Department of Obstetrics and Gynaecology, Queen Mary Hospital, The University of Hong Kong, Hong Kong, P.R. China

Abha Maheshwari, Reproductive Medicine, NHS Grampian, Aberdeen, United Kingdom

Tahir A. Mahmood, Department of Obstetrics and Gynaecology, Victoria Hospital, Kirkcaldy, United Kingdom

Laurence B. McCullough, Department of Obstetrics and Gynecology, Zucker School of Medicine at Hofstra/Northwell, Lenox Hill Hospital, New York, NY, United States

Christina I. Messini, Department of Obstetrics and Gynaecology, Faculty of Medicine, School of Health Sciences, University of Thessaly, Larissa, Greece

Ioannis E. Messinis, Department of Obstetrics and Gynaecology, Faculty of Medicine, School of Health Sciences, University of Thessaly, Larissa, Greece

Mostafa Metwally, Academic Unit of Reproductive and Developmental Medicine, The University of Sheffield and Sheffield Teaching Hospitals, The Jessop Wing, Sheffield, United Kingdom; Consultant in Reproductive Medicine and Surgery, Sheffield Teaching Hospitals, University of Sheffield, Sheffield, United Kingdom

Brenda F Narice, Academic Unit of Reproductive and Developmental Medicine, The University of Sheffield and Sheffield Teaching Hospitals, The Jessop Wing, Sheffield, United Kingdom

Isioma Okolo, Obstetrics & Gynaecology, NHS Lothian, Edinburgh, United Kingdom

Darius A. Paduch, Consulting Research Services, Inc, Red Bank, NJ, United States; Department of Urology, The Smith Institute for Urology, Northwell Health, New Hyde Park, NY, United States; Clinique de Genolier, Genolier, Switzerland

Andrew C. Pearson, Department of Obstetrics and Gynaecology, Victoria Hospital, Kirkcaldy, United Kingdom

Vasilios Pergialiotis, Laboratory of Experimental Surgery and Surgical Research N.S. Christeas, National and Kapodistrian University of Athens, Athens, Greece; 3rd Department of Obstetrics and Gynecology, Attikon Hospital, National and Kapodistrian University of Athens, Athens, Greece

Mythili Ramalingam, Assisted Conception Unit, Department of Obstetrics and Gynaecology, Ninewells Hospital, Dundee, United Kingdom

Jane J. Reavey, MRC Centre for Reproductive Health, The Queen's Medical Research Institute, The University of Edinburgh, Edinburgh, United Kingdom

Rebecca M. Reynolds, University/BHF Centre for Cardiovascular Science, The Queen's Medical Research Institute, The University of Edinburgh, Edinburgh, United Kingdom

Abhilash H.L. Sarhadi, Independent Scholar, Stourbridge, United Kingdom

Mourad W. Seif, Division of Gynaecology, St. Mary's Hospital, Manchester University Hospitals NHS Foundation Trust, Manchester, United Kingdom; Academic Unit of Obstetric and Gynaecology, University of Manchester at St. Mary's Hospital, Manchester, United Kingdom

Mahmood I. Shafi, Nuffield Health, Cambridge, United Kingdom

Tommaso Simoncini, Division of Obstetrics and Gynecology, Department of Clinical and Experimental Medicine, University of Pisa, Pisa, Italy

Krishnan Swaminathan, Kovai Medical Center & Hospital, Coimbatore, India

Vikram Talaulikar, Reproductive Medicine Unit, University College London Hospital, London, United Kingdom

I-Ferne Tan, Department of Obstetrics and Gynaecology, Nepean Hospital, Sydney, Australia

Douglas G. Tincello, University Hospitals of Leicester NHS Trust, Leicester, United Kingdom; Department of Health Sciences, University of Leicester, Leicester, United Kingdom

Gabriele Tridenti, Department of Obstetrics & Gynecology, Santa Maria Nuova Hospital, Reggio Emilia, Italy

Kiran Vanza, Acute Gynaecology, Early Pregnancy and Advanced Endoscopy Surgery Unit, Nepean Hospital, Kingswood, New South Wales, Australia; Sydney Medical School Nepean, The University of Sydney, Sydney, Australia

Anastasia Vatopoulou, Department of Obstetrics and Gynecology, Aristotle University of Thessaloniki, Thessaloniki, Greece

Laurent Vaucher, Department of Urology, The Smith Institute for Urology, Northwell Health, New Hyde Park, NY, United States; Clinique de Genolier, Genolier, Switzerland

About the editors

Tahir A. Mahmood, CBE, MD, FRCPI, FFRSH, MBA, FACOG, FRCPE, FEBCOG, FRCOG

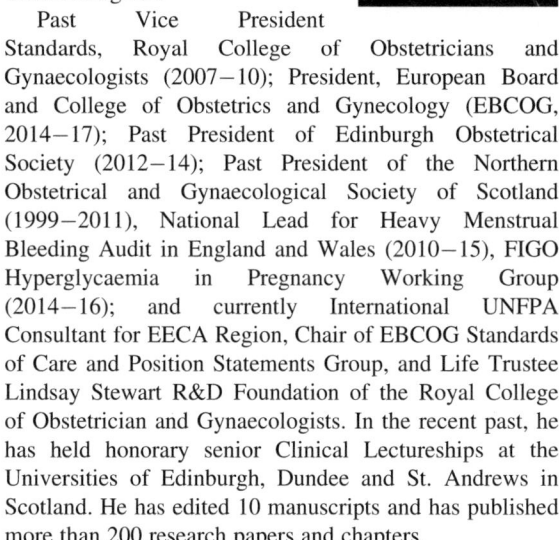

Consultant Gynecologist and Clinical Director Women, Children and Clinical Services Directorate, Victoria Hospital, NHS Fife, Kirkcaldy, Scotland, United Kingdom

Past Vice President Standards, Royal College of Obstetricians and Gynaecologists (2007–10); President, European Board and College of Obstetrics and Gynecology (EBCOG, 2014–17); Past President of Edinburgh Obstetrical Society (2012–14); Past President of the Northern Obstetrical and Gynaecological Society of Scotland (1999–2011), National Lead for Heavy Menstrual Bleeding Audit in England and Wales (2010–15), FIGO Hyperglycaemia in Pregnancy Working Group (2014–16); and currently International UNFPA Consultant for EECA Region, Chair of EBCOG Standards of Care and Position Statements Group, and Life Trustee Lindsay Stewart R&D Foundation of the Royal College of Obstetrician and Gynaecologists. In the recent past, he has held honorary senior Clinical Lectureships at the Universities of Edinburgh, Dundee and St. Andrews in Scotland. He has edited 10 manuscripts and has published more than 200 research papers and chapters.

He was appointed as Commander of the Order of the British Empire (CBE) in the New Year's Honours list (2012) by Her Majesty Queen Elizabeth, the second.

Sir Sabaratnam Arulkumaran, PhD, DSc, FRCOG, FRCS, FACOG, Emeritus Professor of Obstetrics and Gynaecology (O&G) of St. George's University, London; Foundation Professor of O&G, University of Nicosia, Visiting Professor, Institute of Global Health Innovation, Imperial College, London; Past President of the Royal College of Obstetricians and Gynaecologists (2007–10); President: International Federation of Obstetrics and Gynaecology (FIGO) (2012–15); Head, Dept. of Obstetrics and Gynaecology, St George's University Hospital, London, United Kingdom.

Frank A. Chervenak, MD, MMM currently serves as Chair of Obstetrics and Gynecology, Lenox Hill Hospital; Chair of Obstetrics and Gynecology and Associate Dean of International Medicine, Zucker School of Medicine at Hofstra/Northwell. He has published 327 papers in peer-reviewed literature and has coauthored or coedited 38 textbooks. Research interests include ultrasound and ethics in obstetrics and gynecology and physician leadership. He has been named a member of the National Academy of Medicine of the National Academies. Dr. Chervenak has served on the Board of Governors of the American Institute in Ultrasound and Medicine and the Society of Perinatal Obstetricians. He has served as President of the World Association of Perinatal Medicine, International Fetal Medicine in Surgery Society, the New York Perinatal Society and the New York Academy of Medicine Section of Obstetrics and Gynecology, and the New York Obstetrical Society. Currently, Dr. Chervenak serves as President of the International Society of the Fetus as a Patient, Vice President of the International Academy of Perinatal Medicine, and Codirector of the Ian Donald Inter-University School of Medical and Ultrasound. He has been awarded doctor honoris causa from 11 international universities. He has been admitted as a fellow ad eundem of the Royal College of Obstetricians and Gynaecologists of Great Britain and as a Foreign Member of the Russian Academy of Sciences. He has been named an Honorary Member of the Mexican Academy of Pediatrics and a "Knight of Medicine" by the University of Georgia.

Preface—obesity in gynaecology

The obesity epidemic has recently emerged as one of the greatest challenges in the provision of obstetric and gynecologic care in both developed and developing parts of the world. To this end the editors have assembled a comprehensive textbook embracing all aspects of obstetric and gynecologic care that are impacted by this increasingly prevalent condition. The implications of obesity in childhood and adolescence resonate into adulthood and throughout the whole life cycle of a woman. This theme runs through our book which has been divided into two volumes.

The first volume comprises 34 chapters dealing with *obesity in obstetrics*. The genetic, molecular, and psychological bases of obesity are explored. Many and varied clinical topics on adolescent pregnancy, preconception, and antepartum care are discussed, and the important role of ultrasound in early and late pregnancy is explained. The book includes helpful tools such as algorithms for the provision of antepartum care and an overview of complications of pregnancy specific to and affected by obesity in pregnancy. Management of specific complications and comorbidities including hypertension, preeclampsia, venous thromboembolism, hyperglycemia, and metabolic syndrome in pregnancy is presented. The important association with diabetes is elucidated including strategies for universal screening, management, including metformin usage, insulin resistance, placental dysfunction, and fetal growth disorders. The developmental priming of risk of later disease with the goal of developing strategies to prevent future morbidities is included. Labor and delivery care present particular challenges that are addressed in detail including induction, cesarean delivery, risk of stillbirth, sepsis, and the especially important issue of maternal mortality. Interventions to improve the care of obese women during pregnancy including modification of eating behavior, lifestyle changes, antiobesity drugs, and bariatric surgery are presented. Developments in minimal access surgery techniques are now allowing an increasing number of bariatric procedures being performed for obese women planning pregnancy. Our understanding of the impact of these interventions on the practice of obstetrics practice is evolving all the time and the challenges for the care of post–bariatric surgery women in pregnancy have been fully explored. This volume concludes with an important discussion of the impact of obese maternal patients on the provision of maternity services and on their future quality of life.

The second volume comprises 34 chapters dealing with *obesity in gynaecology*. It begins with a detailed discussion of issues related to adolescence including polycystic ovarian syndrome, sexual behavior, contraception, and hirsutism. Several chapters discuss infertility issues including recurrent pregnancy loss, assisted reproduction, sexual dysfunction, male obesity, and its effect on semen quality as well as potential therapies. Obesity can complicate imaging modalities and diagnostic approaches to improve diagnostic accuracy are presented. The causative and complicating role of obesity in myriad disorders is presented including menstrual disorders, urinary and fecal incontinence, gynecologic and breast cancer, surgical complications, venous thromboembolism, cardiovascular disease, osteoporosis, menopause, chronic pelvic pain, and psychosexual disorders. This volume also provides a compressive review of newer surgical approaches such as robotic surgery which is allowing challenging gynecological procedures to be performed for women with complex comorbidities and obesity. The positive impact of bariatric surgery in gynecological conditions has also been explored in each chapter. This volume concludes with the important topic of ethical and professional aspects of the care of women with obesity.

The editors are grateful to the international authors who have given of their time to contribute to these two volumes. We believe the result is of value to any physician or health-care provider who delivers obstetric or gynecologic services to enable them to provide optimal care for this ever-increasing subset of patients that we all encounter on a daily basis.

Tahir A. Mahmood, Sabaratnam Arulkumaran and
Frank A. Chervenak

Section 1

Obesity and adolescence

Chapter 1

Obesity and the onset of adolescence

Zana Bumbuliene[1], Gabriele Tridenti[2] and Anastasia Vatopoulou[3]

[1]Clinic of Obstetrics & Gynecology, Institute of Clinical Medicine, Faculty of Medicine, Vilnius University, Vilnius, Lithuania, [2]Department of Obstetrics & Gynecology, Santa Maria Nuova Hospital, Reggio Emilia, Italy, [3]Department of Obstetrics and Gynecology, Aristotle University of Thessaloniki, Thessaloniki, Greece

Obesity in childhood and adolescence: definition

Being overweight and obese is characterized by varying degrees of excess of body fat, or adiposity. Universally, diagnosis rests with the calculation of body mass index (BMI), attainable by dividing the body weight in kilograms by the height in meters squared (kg/m^2). In the adult, adiposity is clearly categorized. According to the World Health Organization (WHO), a BMI in adults of 25–30 kg/m^2 defines being overweight, whereas obesity is classified by stages or grades—Grade 1: BMI 30–34.9 kg/m^2, Grade 2: BMI 35–39.9 kg/m^2, and Grade 3: BMI ≥ 40 kg/m^2 [1]. In children and adolescents, obesity has not been as well defined as in adults and, therefore, is not a perfect measurement. Even if the alternatives of measuring waist/hip ratio, using dual X-ray absorptiometry or the assessment of body fat and skinfold thickness, might be more precise diagnostic tools, the evaluation of BMI, in the context of age- and sex-specific growth charts, is much more user-friendly and generally utilized worldwide; at present, the BMI is still considered the gold standard diagnostic for obesity/being overweight in childhood and adolescence. In the young and very young the BMI changes with age; therefore, BMI percentile charts are necessary to improve its diagnostic reliability (Fig. 1.1) [2].

Even by using a BMI percentile chart, different definitions of being overweight and obese exist. Cutoff points for being overweight are BMI 85th–95th percentiles for the Center of Disease Control (CDC) or at BMI 85th–97th percentiles for WHO, while obesity is defined as BMI greater or equal to the 95th percentile by CDC or as greater or equal to the 97th percentile by the WHO [3]. Other definitions, proposed by the International Obesity Task Force, the National Child Measurement Program, and the Scottish Intercollegiate Guideline Network (SIGN), are listed in Table 1.1. It is worth mentioning the SIGN categorizes "severe obesity" when the BMI is greater than 99.6th percentile.

Obesity in childhood and adolescence: incidence

According to 450 national surveys from 144 different countries in 2010, 43 million preschool children under the age of 5 years were overweight or obese (35 million in developing countries) and 92 million were at risk of being overweight [4]. Obesity among children, adolescents, and adults is set to be one of the most important public health concerns of the 21st century. More than 60% of children who are overweight before puberty will become overweight young adults, with earlier appearance of noncommunicable diseases and obesity-related health conditions, such as type 2 diabetes, hypertension, and cardiovascular disease [5]. According to other studies, 50%–77% of obese adolescents will become obese adults, at the risk of cardiovascular diseases, diabetes, and cancer [6]. Since the last three decades the incidence of obesity in childhood and adolescence has been a growing epidemic, with a rise of more than a half of overweight and a doubling of obesity [7]. All around the world, 1 in 10 young people aged 5–17 years is overweight or obese, and most of them live in developing countries, with bigger increasing rates than in the developed part of the world [8]. The worldwide prevalence of childhood overweight and obesity increased from 4.2% in 1990 to 6.7% in 2010, with 8.5% in Africa and 4.9% in Asia [9]. According to the WHO European Region, more and more youngsters are affected in Europe too, with generally higher prevalence in Southern European countries, and still show growing trends. Data collected by WHO Europe from school-age children from 36 European countries over a lot of 41 have shown a prevalence of overweight/obesity ranging from 5%

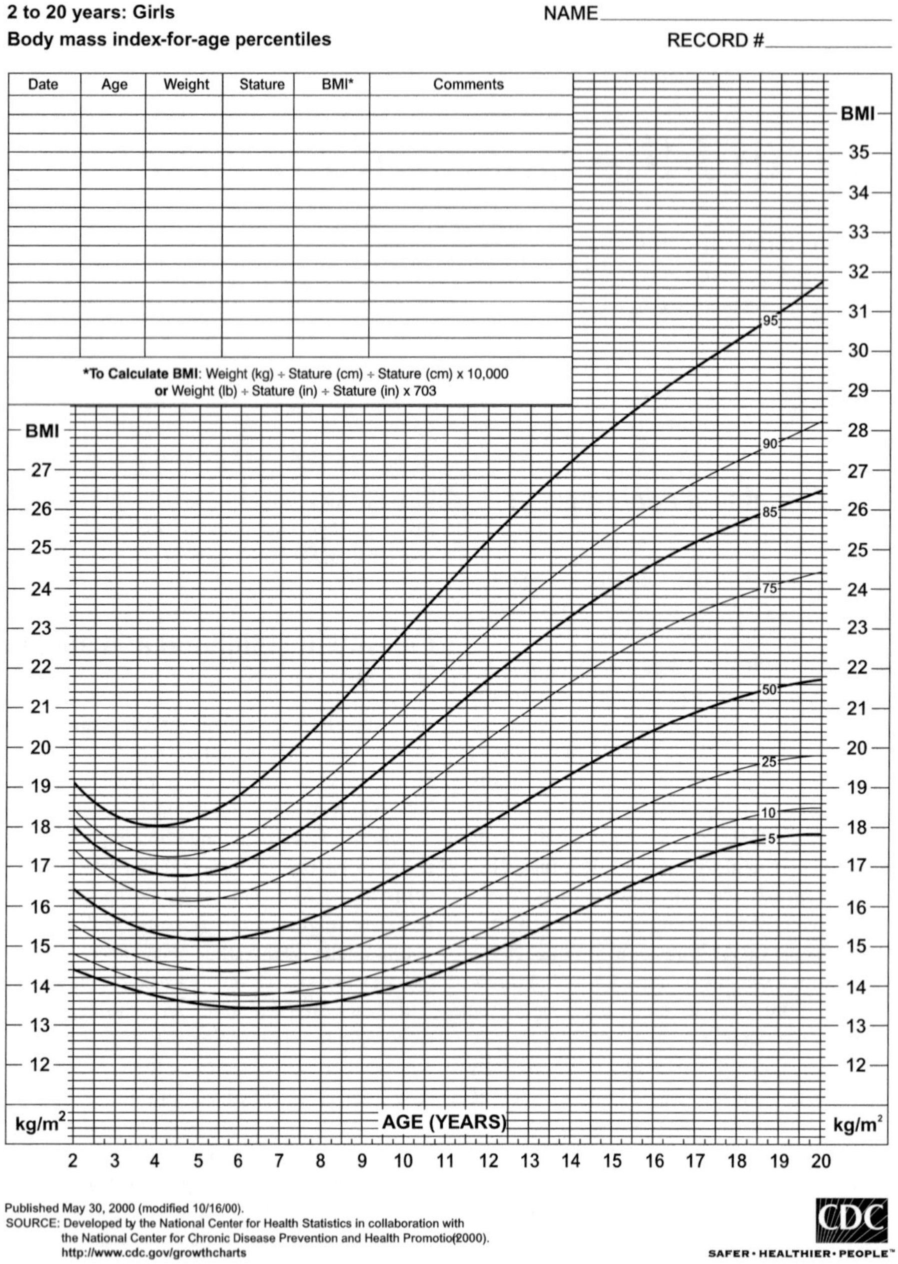

FIGURE 1.1 BMI percentile curves for girls. *BMI*, Body mass index.

to more than 25%, with great variability among countries and a still growing incidence in more than half of them. A general greater proportion of overweight/obesity was found in boys than in girls, as shown in Figs. 1.2 and 1.3 [10]. Outside Europe, 30% of North American children and adolescents are overweight or obese, with the highest rates among minorities and low-income families [2].

Obesity in children and adolescents: etiology

Obesity is a complex multifactorial condition that involves both genetic and nongenetic factors, with environmental, cultural, lifestyle, and behavioral influences. The main determinants of the overweight state in youth

TABLE 1.1 Childhood and adolescent obesity definitions as related to body mass index [2].

Definitions of childhood obesity	CDC	WHO	IOTF	NCMP	SIGN
Overweight	85th–95th	85th–97th	91st	>85th	>91st
Obesity	>95th	>97th	99th	>95th	>98th
Severe obesity					>99.6th

CDC, Center for Disease Control; IOTF, International Obesity Task Force; NCMP, National Child Measurement Program; SIGN, Scottish Intercollegiate Guideline Network; WHO, World Health Organization.

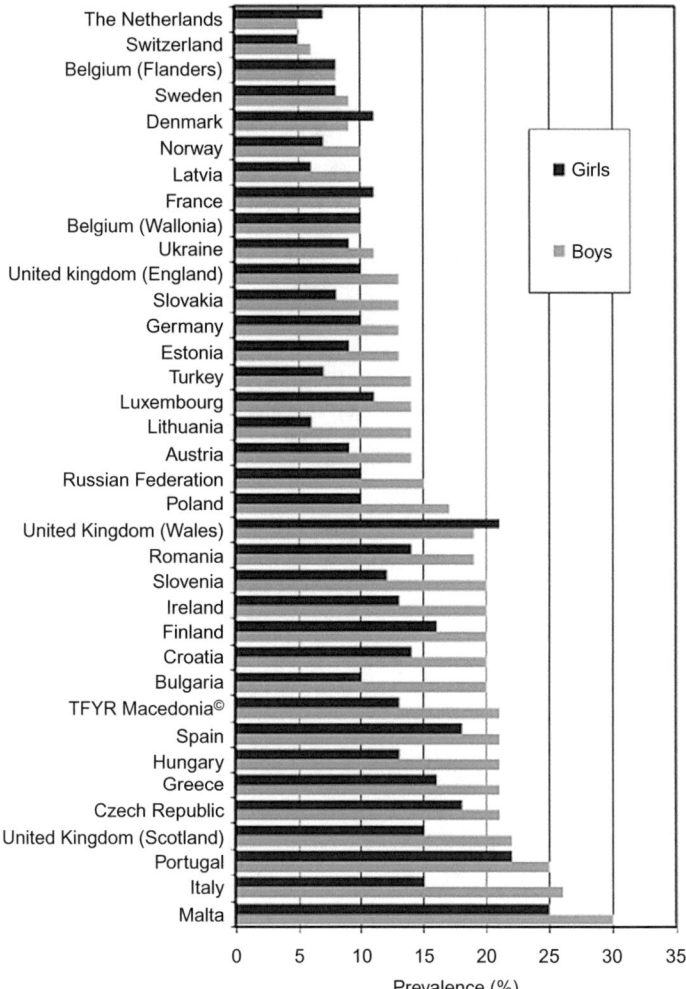

FIGURE 1.2 Prevalence of overweight (including obesity) among 11-year-olds in 36 countries and areas of the WHO European Region, 2005/2006. Source: *Health Behaviour in School-aged Children.*

^aTFYR Macedonia = the former Yugoslav Republic of Macedonia

are the lack of physical activity and unhealthy eating habits, resulting in excess energy intake getting stored in fat tissue. Socioeconomic status, race/ethnicity, media and marketing, and the physical environment may also play a role, but the association between socioeconomic factors and childhood obesity is heterogeneous across different countries [11]. A complex interaction between the obesogenic environment and the individual predisposition to

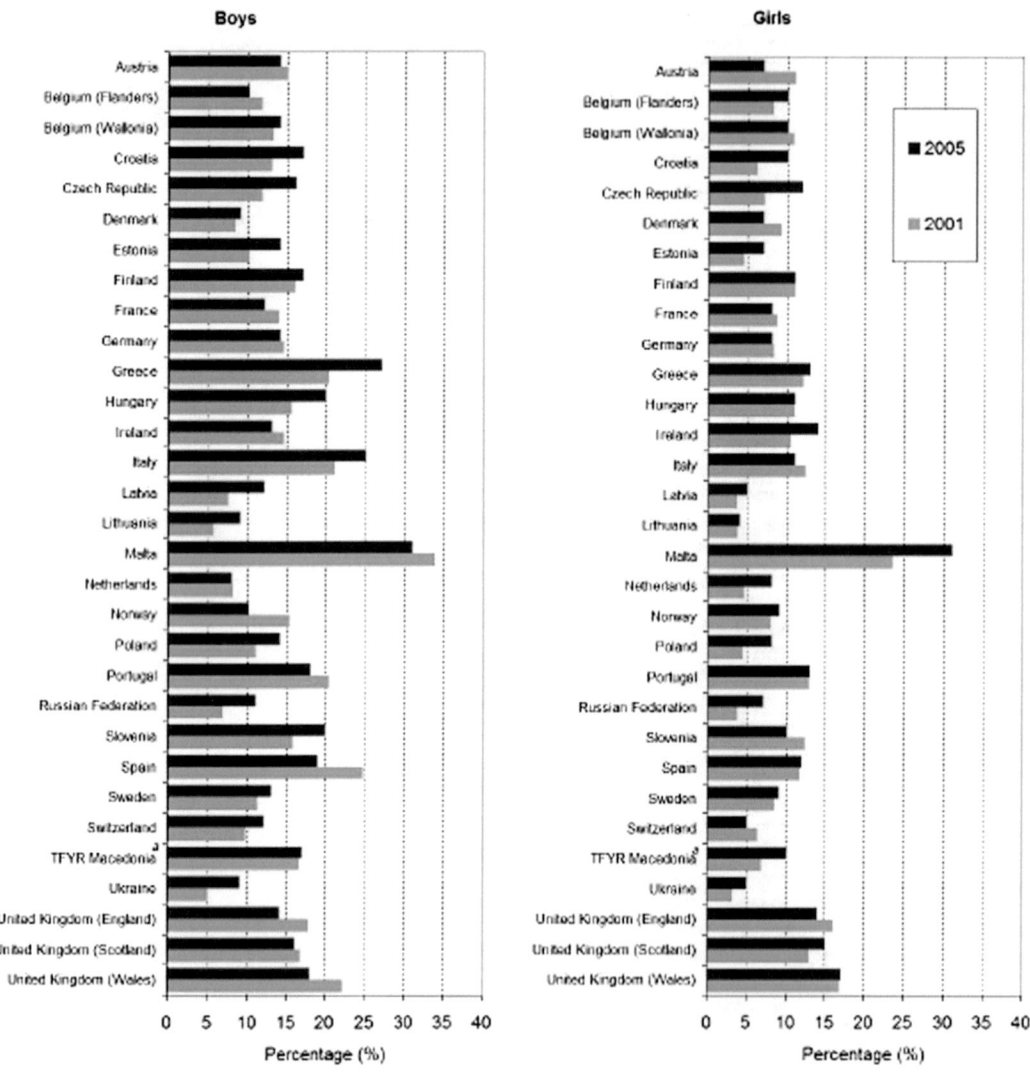

*TFYR Macedonia = the former Yugoslav Republic of Macedonia.

FIGURE 1.3 Prevalence of overweight (including obesity) among 13-year-olds in 2001 and 2005 in 31 countries and areas of the WHO European Region. Source: *Health Behaviour in School-aged Children*.

adiposity occurs, involving a great number of hormones that are mainly secreted by the gut, such as the appetite-stimulating ghrelin and the anorexigenic peptide YY, the pancreatic polypeptide, glucagon, and others. In 2014 Güngör, from Louisiana State University, United States, published the subsequent comprehensive list of possible etiological factors of obesity in childhood and adolescence [5]:

1. ***Genetic variations***: Genetic variations with rare genetic defects of leptin secretion and more frequent genetic syndromes causing obesity, such as Prader–Willi syndrome.

2. ***Epigenetics***: Epigenetics with in utero factors acting on deoxyribonucleic acid methylation which induce heritable changes in obesity expression. Further research is needed to support this statement.

3. ***Endocrine diseases***: Endocrine diseases mainly hypothyroidism, growth hormone (GH) deficiency, and cortisol excess.

4. ***Central nervous system diseases***: Mostly congenital or acquired hypothalamic pathologies (infiltrative diseases, tumors, or after treatment sequelae) that alter the hypothalamic regions in the charge of hunger and satiety.

5. *Intrauterine exposures*: Intrauterine exposures to gestational diabetes or extreme maternal adiposity; birth weight, with increased risk of childhood and adolescence obesity in both macrosomic and small for gestational age babies.
6. *BMI rebound*: An early postinfancy increase of BMI, occurring before 5.5 years of age, is a risk factor for the development of obesity in later ages.
7. *Diet*: Family food choices with high energy intake in early infancy and excessive consumption of sweetened soft drinks in childhood. Poor eating habits play a major role in the onset of early obesity, such as inadequate intake of vegetables and fruits, skipping breakfast, "eating out" frequently, "emotional (comfort) eating," fast food with high-calorie content. A higher protein intake in infancy and early childhood and a higher consumption of free sugars later in life may foster early obesity [11].
8. *Low-energy expenditure*: Due to poor physical activity and/or too much time spent in sedentary activities (e.g., television or other screen viewing).
9. *Sleep*: Shorter sleep duration in infancy and childhood may play a role.
10. *Infections*: Microbial infections and the composition of the gut flora might be associated with obesity but further evidence is requested.
11. *Iatrogenic*: With increased adiposity in children and teens due to:
    a. cranial irradiation or surgery-induced hypothalamic injury,
    b. psychotropic medication (e.g., olanzapine and risperidone),
    c. chemotherapeutics, and
    d. hormonal contraception (e.g., depot medroxyprogesterone acetate).
12. *Ethnic origin*: Being overweight is more frequent in Hispanic and South Asian children and adolescents.
13. *Country of birth*: Children from developing countries, often born underweight, are at higher risk of obesity with overnutrition.
14. *Residence in urban versus rural areas*: With a higher incidence of obese children in urban areas all over the world.
15. *Socioeconomic level*: With higher rates of obesity among children of the lowest socioeconomic groups living in high-income countries.

A subtype of endocrine-disrupting chemicals, called obesogens, may play a role in the onset of puberty, but further research is necessary [12].

Obesity and puberty: pathophysiology

A strong relationship exists between nutrition and pubertal development. In fact, an adequate nutritional status is a requisite for the central onset of puberty, being the key permissive factor for normal timing and tempo of the various pubertal steps [13]. Chronic malnutrition delays the onset of puberty. This has been indirectly demonstrated by the decreased age of menarche that has occurred in developed countries since the 19th century following the improvement in nutrition, hygiene, and general health [14]. At puberty, some degree of weight gain is physiological, with an increase in BMI and subcutaneous adiposity. Fat tissue may act as a metabolic trigger of central pubertal initiation, so that obesity may be associated with the premature activation of the gonadotropin-releasing hormone (GnRH) pulse generator [13]. Due to the rising epidemic of childhood obesity, evidence now exists that the increased height and BMI prior to puberty result in an earlier onset of puberty [15]. Further studies have confirmed that the nutritional status plays an important role in growth and body weight regulation, and it is now well recognized that excess adiposity in childhood can affect the processes of growth and puberty: obese children frequently show a tall stature for their age, associated with an accelerated epiphyseal growth plate maturation and early puberty [11]. Rapid weight gain in early life is linked to advanced puberty in both sexes, but mainly in girls [15]. Early puberty, premature adrenarche, and accelerated linear growth (with potentially impaired final height), occurring in obese children because of their increased subcutaneous body fat mass, are due to peculiar hormonal secretions secondary to being overweight [11]:

- *Increased leptin levels.* Mainly secreted by white adipose tissue, high leptin levels are present in obese children, stimulating both GH secretion and GH response to GH-releasing hormone with a resulting increase of linear growth. Leptin can also act directly on skeletal growth centers, by inducing both proliferation and differentiation of chondrocytes [16]. By acting on leptin receptors in the hypothalamus and in the gonadotrope cells of the anterior pituitary, leptin plays a role in pubertal development by directly and dose-dependently stimulating GnRH secretion in the arcuate hypothalamic neurons and also by fostering the release of LH and FSH from the anterior pituitary. Furthermore, leptin decreases hypothalamic neuropeptide Y levels, thus removing its inhibitory action on GnRH secretion, and shows a paracrine effect on the gonads. Finally, leptin dose-dependently stimulates adrenal 17alpha-hydroxylase and 17,20-lyase, with higher adrenal androgen levels that are involved in accelerated growth of obese girls [11]. Marked weight gain and obesity may, therefore, induce premature adrenarche [17].
- *Increased androgen levels* can favor precocious pubertal development with both peripheral and central

action on the hypothalamic−pituitary axis, enhancing pubertal rise of GnRH secretion [18].
- *Reduced sex hormone−binding globulin (SHBG) levels*, linked to peripheral obesity, increase the availability of sex steroids, among which is estradiol, which can induce premature thelarche [19].
- *Increased adipose tissue aromatization* of androgens into estrogens, which contribute to both accelerated growth rate and precocious puberty. Enlarged adipose tissue, acting as an endocrine organ, can release higher levels of sex steroid and adipokines that have a role in linear growth [13].
- *Increased Insulin levels*: Even if Insulin acts a negative feedback on GH secretion from the pituitary, its elevated levels in obesity, due to insulin resistance, highly stimulate the insulin growth factor-1 (IGF-1) receptor. In addition, insulin resistance may suppress insulin growth factor−binding protein 1 and 2, resulting in enhanced IGF-1 bioavailability and increased linear growth [20]. Hyperinsulinemia, with obesity-associated insulin resistance, may also stimulate the onset and progression of puberty by fostering pubertal weight gain and growth, as well as by increasing LH-stimulated ovarian and adrenal steroidogenesis [21].

In summary, many factors may be involved in early pubertal development in obese and overweight girls. Besides genetic factors, elevated BMI plays a major role, mediated by leptin, kisspeptin, and modified sex steroids bioavailability. Possible environmental exposures to endocrine-disrupting chemicals and epigenetic modifications must also be taken into account [22]. Obesity is connected to high IGF-1 concentrations, insulin resistance, increased adrenal androgen levels, excessive adrenarche, elevated leptin, and low SHBGs. All these factors promote the early activation of GnRH pulse generator, altering the timing of puberty.

Obesity and pubertal development

During adolescence, young people between 10 and 19 years of age experience physical, cognitive, and psychosocial maturation. The physical transition is known as puberty. Puberty is a developmental process during which a child becomes a young adult, characterized by the secretion of gonadal hormones and the development of secondary sexual characteristics that lead to sexual maturation and reproductive capability. The most visible changes during puberty in girls are growth in stature, development of the breasts and pubic hair growth, changes in body composition, and the menarche.

Puberty generally proceed in a predictable pattern, with some variation in the timing of onset, sequence, and tempo. The first sign of puberty begins from 8 to 13 years for girls. A major determinant of pubertal timing is genetic. Other factors that influence pubertal development are race, general health, nutrition, and environment effects [23,24].

A critical body weight or percent body fat is the primary determinant of the development of secondary sexual characteristics. It was proposed that the onset of the girls' growth spurt and menarche requires a critical weight of 47.8 kg, and that increased body fat can lead to an early height spurt start age and menarche age in puberty [25,26]. In girls over 16 years of age, the annual increase in BMI is associated with increase in fat mass [27]. An ongoing increase in body fat depends on the individual's nutritional status.

Puberty starts as the result of increase in the pulsatile secretion of the GnRH from the hypothalamus and the suppression of inhibitors of GnRH secretion. Leptin appears to be one of several factors that influence the activity of the GnRH pulse generator. Leptin is secreted in pulses that are positively correlated to gonadotropins, estradiol, and thyrotropin [28]. Higher serum leptin concentrations in girls are associated with increased body fat and an earlier onset of puberty [29]. However, leptin levels decrease with increasing Tanner stages of puberty, and there is increased sensitivity to leptin [30].

Obesity is often accompanied by elevated synthesis of androgens. Total testosterone is fourfold elevated in prepubertal 7- to 9-year-old girls with obesity, and 1.75-fold elevated in pubertal 10- to 12-year-old obese girls. Obesity is also associated with approximately 40% elevation of dehydroepiandrosterone sulfate levels [31]. BMI positively correlates with free testosterone index. Compared with normal weight controls, mean free testosterone in Tanner stage-matched girls with obesity is elevated two- to ninefold, depending on pubertal stage [32].

Puberty is associated with significant changes in body weight and alterations in body composition. The initial manifestation of secondary sexual characteristics predicts body morphology and composition. Girls with breast development as the first sign of puberty in comparison with girls who have pubic hair development first have an earlier age of menarche and greater BMI throughout puberty and as adults [33].

Girls who mature earlier were approximately twice as likely to be overweight as compared with those maturing at an average age, and early maturation is associated with greater adult adiposity [34]. Also earlier menarche (before 12 years of age) as compared with later menarche is associated with higher BMI during adulthood [35].

Current systematic review and metaanalysis examined the association between obesity and puberty timing based on scientific evidence, suggesting that obesity contributes to early onset of puberty in girls, including age at when puberty occurred [36]. The likelihood of persistence of childhood obesity into adulthood is related to age, parental obesity, and severity of obesity. The increase in pediatric obesity is alarming because of its association with health and as a critical early risk factor for adult morbidity and mortality.

The early onset of puberty in girls is a concern because it can cause significant health risks, and it is associated with higher rates of obesity, cardiovascular disease, menstrual irregularities, dysfunctional uterine bleeding, polycystic ovary syndrome (PCOS), and metabolic syndrome. Obesity can increase the risk of anovulation and is associated with reduced fertility [37,38].

Obesity in childhood and adolescence: clinical manifestation

Almost all systems of the human body are affected by being overweight and obese. These conditions also have an adverse effect on the quality of life of young people and are linked to various psychological and behavioral problems. As obesity poses several risks to the adolescent and young women, accurate identification, appropriate counseling, and treatment are essential. The clinical evaluation of the obese adolescent aims to identify comorbidities and causes that can be treated. (Table 1.2) Adolescent obesity is associated with serious medical problems, including cardiovascular disease, colon cancer, and diabetes mellitus, and has psychosocial consequences such as higher body dissatisfaction and lower educational attainment [39]. Overweight adolescents often experience significant low self-esteem and depression [40].

Problems specific to gynecology include earlier sexual maturation and reproductive dysfunction. Alterations in menstruation due to chronic anovulation and PCOS are common. Other gynecological problems include dysmenorrhea, risky sexual behavior and inefficient use of contraception, bone density abnormalities, macromastia, and an increased risk of breast and endometrial cancer. Obese adolescents are at greater risk of pregnancy and perinatal complications, such as preeclampsia, gestational hypertension, gestational diabetes mellitus, primary cesarean delivery, and induction of labor [41]. Newborns of obese teenagers are also more likely to have complications, including prematurity, small for gestational age weight, macrosomia, meconium aspiration, respiratory distress, and stillbirth to mention a few [42].

Obesity is a proinflammatory condition rising the risk of several chronic diseases in the adult, ranging from hypertension to dyslipidemia, diabetes, cardiovascular diseases, asthma, cancers, and many others, as reported in Table 1.3 [9]. Furthermore, adiposity in children and adolescents is a recognized predictor of the metabolic syndrome in both adolescence and adulthood. Obese children are fivefold more likely to have a diminished health-related quality of life, with a number of years of life lost proportional to the degree of their obesity and higher risk of dramatic psychological consequences [7]. Stigmatization, altered cognitive performances, low self-esteem, frustration, and emotional disorders were described, with positive correlation between depression and adolescent obesity [44].

Obesity in children and adolescents: counseling

The history should include the age of onset of obesity to exclude the presence of syndromes associated with obesity. Dietary habits (e.g., fast-food eating and foods with high caloric and low nutritional value) and assessment of eating patterns (e.g., timing and frequency of meals) should be done. Activity history is important with the evaluation of time spent in activities, physical education, or screen time. Medications that are associated with obesity include antiepileptics, steroids, and psychoactive drugs (risperidone). A review of the systems might indicate hypothyroidism, Cushing's syndrome, Albrights' hereditary osteodystrophy syndrome, hypothalamic congenital, or acquired diseases (previous surgery, autonomic dysfunction, and rapid onset of obesity). Family history of cardiovascular disease and diabetes is important. Finally, the obese adolescent should be asked about psychosocial

TABLE 1.2 Health consequences of obesity in adolescence [43].

Gynecologic/obstetric	Psychologic	General health
Faster growth	Psychosocial difficulties	Dyslipidemia
Early menarche	Eating disorders	Hypertension
Macromastia	School phobia	Hepatic steatosis
Polycystic ovary Syndrome	Depression	Abnormal glucose metabolism
Gestational hypertension	Chronic fatigue	Cholelithiasis
Preeclampsia		Sleep apnea
Gestational diabetes		Orthopedic disorders
Primary cesarean delivery		
Induction of labor		

TABLE 1.3 Comorbidities and complications of childhood obesity [5].

Endocrine
1. Glucose metabolism
- Insulin resistance
- Prediabetes (impaired fasting glucose/impaired glucose tolerance)
- Type 2 diabetes mellitus
- Metabolic syndrome
2. Growth- and puberty-related issue
   Girls
- Hyperandrogenism/polycystic ovarian syndrome
- Earlier menarche
   Boys
- Later pubertal onset
- Pseudo-micropenis (hidden penis)
- Reduced circulating androgens
3. Thyroid function aberrations

Cardiovascular
- Hypertension
- Dyslipidemia
- Other cardiovascular risks
- Adult coronary heart disease

Gastrointestinal
- Nonalcoholic fatty liver disease
- Steatohepatitis
- Cholestasis/cholelithiasis

Pulmonary
- Asthma
- Obstructive sleep apnea
- Obesity hypoventilation syndrome (Pickwickian syndrome)

Orthopedic
- Coxa vara
- SCFE
- Tibia vara (Blount disease)
- Fractures
- Legg–Calve–Perthes disease

Neurologic
- Idiopathic intracranial hypertension (pseudotumor cerebri)

Dermatologic
- Acanthosis nigricans
- Intertrigo
- Furunculosis

Psychosocial
- Low self-esteem
- Depression

SCFE, Slipped capital femoral epiphysis.

The aim of clinical examination is to look for dysmorphic features, short stature, and distribution of fat. If abdominal fat distribution is of male type, it may be associated with metabolic syndrome. Excessive fat in the interscapular area or in the face and neck is suggestive of Cushing's syndrome. Measurements of waist or hip circumference are not particularly helpful. Inspection of the skin for hirsutism or acanthosis nigricans could suggest hyperandrogenemia, hyperinsulinemia, or metabolic syndrome. Measurement of blood pressure should be done with the appropriate cuff. A blood pressure reading of ≥ 130/80 mmHg in adolescents 13 years and older on at least three occasions warrants further investigation. Other signs such as nystagmus, gait abnormalities, and cognitive or developmental delay can be present.

Pelvic examination is not necessary in younger patients but should be done in sexually active adolescents with appropriate testing for sexually transmitted infections (STIs). Great care and discretion should be exercised in keeping with the recommendations of Executive Committees for examining adolescents [45]. The nursing staff should be helpful and not critical.

Laboratory studies are not standardized but may be indicated to check for dyslipidemia, diabetes, and liver function tests for fatty liver disease. If hypertension is found, the patient should be referred accordingly for further investigation. Other tests are done if indicated for PCOS, vitamin D deficiency, Cushing's syndrome, and hypothyroidism, although mild elevations of TSH are considered normal [46].

Imaging if indicated can be done with radiographs for bone deformities and bone age. Ultrasound is helpful to examine the liver, gallbladder, and to assess uterine maturity and ovarian morphology.

Obese girls are more likely to have the thelarche as the first sign of puberty instead of the pubarche [47]. The timing of the menarche is affected less by obesity, although, as mentioned previously, it tends to be a few months earlier. Clinical consequences that are relevant to their reproductive health is that they become more vulnerable to provocation (teasing) [48], and their behavior is more advanced for their age, with the risks of early initiation of intercourse [49]. Health providers have to be aware of these issues and advise accordingly about abstinence or appropriate contraception and use of condoms. It is important to ask the adolescent about their relationships with their school friends and enquire about teasing or bullying at school, and how they respond to that. If this proves to be a significant problem, appropriate psychological referral may be needed [50].

Menstrual abnormalities are particularly more common in obese adolescents in the form of amenorrhea/oligomenorrhea, heavy periods, and sometimes irregular bleeding [41]. They are caused either by anovulation

issues such as mood changes, sleep disturbance, sense of loneliness and hopelessness, presence of friends, teasing at school, and smoking or substance abuse. Many teenagers think that smoking reduces weight gain and this must be addressed to.

(immaturity of the hypothalamic–pituitary–ovarian axis, and PCOS) or excessive estrogen production by the abundant fatty tissue that stimulates the endometrium and disrupts the normal feedback mechanisms [51]. The physician should exclude local causes and investigate the endometrium by imaging or even by endometrial sampling if the problem is not controlled with treatment, since 2–3 years of menstrual irregularity may result in endometrial hyperplasia and cancer even in this age group [41,52].

Large breasts (macromastia) are more common in obese adolescents and can be disfiguring and extremely distressing to the adolescent because of shoulder pain and difficulties in finding clothes and participating in activities [53]. The clinician should make sure that macromastia is not caused by tumor and counsel patient about cosmetic surgery.

Because of the complexity of intervening somatic and psychological comorbidities, referral and close collaboration with other disciplines may be necessary.

Obesity in children and adolescents: management

Principles of obesity treatment include the management of associated conditions, support of a long-term behavior, and dietary change coupled with increased physical activity. Sedentary behaviors and screen time should be decreased and good sleep ensured. Long-term maintenance should be planned carefully. In selected cases, pharmacotherapy and bariatric surgery may be applicable [54].

Menstrual irregularities in obese adolescents can be managed with the cyclic use of progestagens, hormonal contraceptives, or insertion of a levonorgestrel-releasing intrauterine system (LNG-IUS) [55]. In cases of severe dysmenorrhea, nonsteroidal antiinflammatory drugs are an option as they reduce the amount of bleeding. Chronic anovulation can lead to endometrial hyperplasia and cancer, and it is paramount that regular withdrawal bleeding is induced at least four times a year and the patients and family monitor menstrual patterns closely [56].

Hormonal contraception can be considered in obese teenagers who are sexually active. There is not much data on contraception in obese adolescents, and data are derived from the adult population. In the medical eligibility criteria for contraceptive use, obesity is in category 3 for administration of the combined pill as the risks of thrombosis increase. Combined hormonal contraception has less efficacy in obesity with an odds ratio of 1.65 (95% CI 1.09–2.50) of becoming pregnant compared to controls with normal weight [57] and special precautions such as delaying unprotected intercourse after the 10th day of pill initiation or continuous administration rather than cyclic appear safer. The vaginal ring appears to have more stable hormone levels, while the contraceptive efficacy of the patch in obese adolescents is unreliable [58]. Postcoital contraception with levonorgestrel is unreliable and ulipristal or timely insertion of an intrauterine device can be used instead [59]. Long-acting reversible contraception in the form of implants or LNG-IUS appears to be a good choice with added endometrial protection in cases of anovulatory bleeding.

Breast reduction surgery for macromastia is increasingly considered by adolescents [60], but a cautious attitude should be adopted regarding the age when the procedure should be done and appreciation of the goals surgery can achieve [61]. In general, adolescents are satisfied with the results, although complications and scarring do occur.

References

[1] Nicolai P, Lupian JH, Wolf A. An integrative approach to obesity. In: Rankel D, editor. Integrative medicine. 3rd ed. Philadelphia, PA: WB Saunders (Elsevier); 2012. p. 364–75.

[2] Thyson N, Frank M. Childhood and adolescent obesity definitions as related to BMI, evaluation and management options. Best Pract Res Clin Obstet Gynaecol 2018;48:158–64.

[3] Shields M, Tremblay MS. Canadian childhood obesity estimates based on WHO, IOTF and CDC cut-points. Int J Pediatr Obes 2010;5(3):265–73.

[4] de Onis M, Blossner M, Borghi E. Global prevalence and trends of overweight and obesity among preschool children. Am J Clin Nutr 2010;92:1257–64.

[5] Güngör NK. Overweight and obesity in children and adolescents. J Clin Res Pediatr Endocrinol 2014;6(3):129–43.

[6] Green Paper. Promoting healthy diets and physical activity: a European dimension for the prevention of overweight, obesity and chronic diseases. *Brussels*: *European Commission*; 2005 (COM (2005) 607 find).

[7] In-Iw S, Biro F. Adolescent women and obesity. J Pediatr Adolesc Gynecol 2011;24:58.

[8] WHO—Regional Office for Europe. Adolescent obesity and related behaviours: trends and inequalities in the WHO European Region, 2002–2014. Inchley J, Currie D, Jewell J, Breda J, Barnekow V, editors. Copenhagen; 2017.

[9] Elizondo-Montemayor D, Hernandéz-Escobar C, Lara-Torre E, Nieblas B, Gomez-Camona M. Gynecologic and obstetrics consequences of obesity in adolescent girls. J Pediatr Adolesc Gynecol 2017;30:156–68.

[10] European Environment and Health Information System. Prevalence of overweight and obesity in children and adolescents. Fact Sheet 2.3. CODE: RPG2_Hous_E2; 2009. <www.euro.who.int/ENHIS>; <http://ec.europa.eu/health/ph_determinants/life_-style/nutrition/documents/nutrition_gp_en.pdf>.

[11] Shalitin S, Kiess W. Putative effects of obesity on linear growth and puberty. Horm Res Pediatr 2017;88:101–10.

[12] Janesick AS, Blumberg B. Obesogens: an emerging threat to public health. Am J Obstet Gynecol 2016;214(5):559–65.

[13] Burt Solorzano CM, McCartney CR. Obesity and the pubertal transition in girls and boys. Reproduction 2010;140(3):399–410.

[14] Kaplowitz P. Link between body fat and the timing of puberty. Pediatrics 2008;121(Suppl. 3):S208–17.

[15] Unni JC. Onset of Puberty in Relation to obesity. Indian Pediatr 2016;53(5):379–80.

[16] Maor G, Rochwerger M, Segev Y, Phillip M. Leptin acts as a growth factor on the chondrocytes of skeletal growth centers. J Bone Mineral Res 2002;17:1034–43.

[17] Cizza G, Dorn LD, Lotsikas A, Sereika S, Rotenstein D, Chrousos GP. Circulating plasma leptin and IGF-1 levels in girls with premature adrenarche: potential implications of a preliminary study. Horm Metab Res 2001;33:138–43.

[18] Blank SK, McCartney CR, Chhabra S, Helm KD, Eagleson CA, Chang RJ, et al. Modulation of gonadotropin-releasing hormone pulse generator sensitivity to progesterone inhibition in hyperandrogenic adolescent girls—implications for regulation of pubertal maturation. J Clin Endocrinol Metab 2009;94(7):2360–6. Available from: https://doi.org/10.1210/jc.2008-2606 Epub 2009 Apr 7.

[19] McCartney CR, Blank SK, Prendergast KA, Chhabra S, Eagleson CA, Helm KD, et al. Obesity and sex steroid changes across puberty: evidence for marked hyperandrogenemia in pre- and early pubertal obese girls. J Clin Endocrinol Metab 2007;92 (2):430–6 Epub 2006 Nov 21.

[20] Luque RM, Kinerman RD. Impact of obesity on the growth hormone axis: evidence for a direct inhibitory effect of hyperinsulinemia on pituitary function. Endocrinology 2006;147:2754–63.

[21] Poretsky L, Cataldo NA, Rosenwaks Z, Giudice LC. The insulin-related ovarian regulation system in health and disease. Endocr Rev 1999;20:535–82.

[22] Biro FM, Kiess W. Contemporary trends in onset and completion of puberty, gain in height and adiposity. Endocr Dev 2016;29: 122–33. Available from: https://doi.org/10.1159/000438881 Epub 2015 Dec 17.

[23] Boynton-Jarrett R, Harville EW. A prospective study of childhood social hardships and age at menarche. Ann Epidemiol 2012; 22:731.

[24] Sun Y, Mensah FK, Azzopardi P, et al. Childhood social disadvantage and pubertal timing: a national birth cohort from Australia. Pediatrics 2017;139.

[25] Frisch RE, Revelle R. Height and weight at menarche and a hypothesis of critical body weights and adolescent events. Science 1970;169:397–9.

[26] Frisch RE, Revelle R, Cook S. Components of weight at menarche and the initiation of the adolescent growth spurt in girls: estimated total water, lean body weight and fat. Hum Biol 1973;45:469.

[27] Maynard LM, Wisemandle W, Roche AF, et al. Childhood body composition in relation to body mass index. Pediatrics 2001; 107:344.

[28] Sinha MK, Sturis J, Ohannesian J, et al. Ultradian oscillations of leptin secretion in humans. Biochem Biophys Res Commun 1996;228(3):733–8.

[29] Matkovic V, Ilich JZ, Skugor M, et al. Leptin is inversely related to age at menarche in human females. J Clin Endocrinol Metab 1997;82:3239.

[30] Hassink SG, Sheslow DV, de Lancey E, et al. Serum leptin in children with obesity: relationship to gender and development. Pediatrics 1996;98:201.

[31] Reinehr T, de Sousa G, Roth CL, Andler W. Androgens before and after weight loss in obese children. J Clin Endocrinol Metab 2005;90(10):5588–95.

[32] McCartney CR, Blank SK, Prendergast KA, et al. Obesity and sex steroid changes across puberty: evidence for marked hyperandrogenemia in pre- and early pubertal obese girls. J Clin Endocrinol Metab 2007;92(2):430–6.

[33] Biro FM, Lucky AW, Simbartl LA, et al. Pubertal maturation in girls and the relationship to anthropometric changes: pathways through puberty. J Pediatr 2003;142:643.

[34] Biro FM, McMahon RP, Striegel-Moore R, et al. Impact of timing of pubertal maturation on growth in black and white female adolescents: The National Heart, Lung and Blood Institute Growth and Health Study. J Pediatr 2001;138:636.

[35] Rosenfield RL, Lipton RB, Drum ML. Thelarche, pubarche, and menarche attainment in children with normal and elevated body mass index. Pediatrics 2009;123:84.

[36] Li W, Liu Q, Deng X, et al. Association between obesity and puberty timing: a systematic review and meta-analysis. Int J Environ Res Public Health 2017;14(10):1266.

[37] Burt Solorzano CM, McCartney CR, Blank SK, et al. Hyperandrogenaemia in adolescent girls: origins of abnormal gonadotropin-releasing hormone secretion. BJOG 2010;117(2):143–9.

[38] Shayya R, Chang RJ. Reproductive endocrinology of adolescent polycystic ovary syndrome. BJOG 2010;117(2):150–5.

[39] Huh D, Stice E, Shaw H, Boutelle K. Female overweight and obesity in adolescence: developmental trends and ethnic differences in prevalence, incidence and remission. J Youth Adolesc 2012; 41(1):76–85.

[40] Strauss RS. Childhood obesity and self-esteem. Pediatrics 2000;105:e15.

[41] Sukalich S, Mingione M, Glantz C. Obstetric outcomes in overweight and obese patients. Am J Ostet Gynecol 2006;195:851–5.

[42] Isgren AR, Kjathe P, Bloomberg M. Adverse neonatal outcomes in overweight and obese adolescents compared with normal weight adolescents and low risk adults. J Pediatr Adolesc Gynecol 2019;32:139–45.

[43] Tonkin RS, Sacks D. Obesity management in adolescence: Clinical recommendations. Paediatr Child Health 1998;3(6): 395–8.

[44] Boutelle KN, Hannan P, Fulkerson JA, Crow SJ, Stice E. Obesity as a prospective predictor of depression in adolescent females. Health Psychol 2010;29(3):293–8.

[45] American College of Obstetricians and Gynecologists. Committee Opinion Number 714, Obesity in adolescents. Obstet Gynecol 2017;130:e127–40.

[46] Reinehr T, de Sousa G, Andler W. Hyperthyrotropinemia in obese children is reversible after weight loss and is not related to lipids. J Clin Endocrinol Metab 2006;91(8):3088.

[47] Biro FM, Lucky AW, Simbartl LA, Barton BA, et al. Pubertal maturation in girls and the relationship to anthropometric changes: pathways through puberty. J Pediatr 2003;142:643–6.

[48] Randall-Arell JL, Utley R. The adolescent female's lived-experience of obesity. Qual Rep 2014;19(45):1–15 <http://www.nova.edu/ssss/QR/QR19/randallarell45.pdf>.

[49] Ratcliff MB, et al. Risk-taking behaviors of adolescents with extreme obesity: normative or not? Pediatrics 2011;127(5): 827–34.

[50] Varkula L, Heinberg L. Assessment of overweight children and adolescents. In: Heinberg L, Thompson K, editors. Obesity in youth. 1st ed. Washington, DC: American Psychological Association; 2009. p. 137.

[51] Wood PL, Bauman D. Gynaecological issues affecting the obese adolescent. Best Pract Res Clin Obstet Gynaecol 2015;29: 453—65.

[52] Stovall DW, Anderson RJ. Endometrial adenocarcinoma in teenagers. Adol Ped Gynecol 1989;2(3):157—9.

[53] Elizondo-Montemayor L, Hermamdez-Escobar C, Lata-Torre E, et al. Gynecologic and obstetric consequences of obesity in adolescent girls. J Pediatr Adolesc Gynecol 2017;30:156—68.

[54] Steinback KS, Lister NB, Gow ML, Baur LA. Treatment of adolescent obesity. Nat Rev Endocr 2018;14:331—44.

[55] Ju H, Jones M, Mishra G. A U-shaped relationship between body mass index and dysmenorrhea: a longitudinal study. PLoS One 2015;10:1.

[56] De Silva N. Abnormal uterine bleeding in adolescents. <https://www.uptodate.com/contents/abnormal-uterine-bleeding-inadolescentsmanagement?search=abnormal%20uterine%20bleeding&topicRef=116&source=related_link> [accessed 04.06.19].

[57] Gallo MF, Lopez LM, Grimes DA, Carayon F, Schulz KF, Helmerhorst FM. Combination contraceptives: effects on weight. Cochrane Database Syst Rev 2014;(1):CD003987. https://doi.org/10.1002/14651858.CD003987.pub5. Review. PMID:24477630.

[58] Dinger J, Cronin M, Möhner S, et al. Oral contraceptive effectiveness according to body mass index, weight, age, and other factors. Am J Obstet Gynecol 2009;201:e1.

[59] Glasier A, Cameron S, Blithe D, et al. Can we identify women at risk of pregnancy despite using emergency contraception? Data from randomized trials of ulipristal acetate and levonorgestrel. Contraception 2011;84:363.

[60] Xue AS, Wolfswinkel EM, Weathers WM, et al. Breast reduction in adolescents: indication, timing, and a review of the literature. J Pediatr Adolesc Gynecol 2013;26:228.

[61] [No authors listed]. Committee Opinion No. 662: breast and labial surgery in adolescents. Obstet Gynecol 2016;127:e138—40.

Chapter 2

Obesity in adolescence

Gail Busby[1] and Mourad W. Seif[1,2]

[1]Division of Gynaecology, St. Mary's Hospital, Manchester University Hospitals NHS Foundation Trust, Manchester, United Kingdom, [2]Academic Unit of Obstetric and Gynaecology, University of Manchester at St. Mary's Hospital, Manchester, United Kingdom

Introduction

Obesity in childhood and adolescence has major negative health impacts extending to adulthood. In addition to negative consequences that occur later in life, childhood and adolescent obesity confers increased risk of adverse outcomes, including asthma, increased risk of fractures, hypertension, early markers of cardiovascular disease, insulin resistance, and other endocrine abnormalities and psychological effects.

The incidence of childhood and adolescent obesity is increasing worldwide, both in developed and in developing countries. In recognition of the severity of this modern epidemic, the World Health Organization (WHO) published population-based strategies to control it [1]. Similarly, the Royal College of Obstetricians and Gynaecologists consider the prevention of childhood obesity to be a priority, due to the implications of obesity on reproductive, obstetric, and gynecological health [2].

There is good evidence that adolescent obesity leads to adult obesity. A cohort of 8834 American adolescents was followed-up until their adulthood, and it was found that a significant proportion of obese adolescents became severely obese by their early 30s. Among the individuals who were obese in adolescence, 37.1% men and 51.3% women became severely obese adults. Severe obesity was the highest among black women. In contrast, across all sex and racial/ethnic groups, less than 5% of adolescents who were at a normal weight became severely obese in adulthood [3].

Prevalence of childhood obesity: a global perspective

Worldwide, the prevalence of childhood obesity has been increasing over recent decades and increased from 4.2% in 1990 to 6.7% in 2010. This trend is expected to continue and reach a prevalence of 9.1% in 2020.

In 2010 based on the WHO criteria for weight (>2SD above median), the estimate worldwide for preschool children aged from birth to 5 years, who were overweight and obese, was 43 million. The prevalence of overweight and obesity in developed countries is about double than that in developing countries (11.7% and 6.1%, respectively); however, the majority of affected children live in developing countries (35 million) [4].

Although the increasing prevalence of childhood obesity is an international phenomenon, there are significant variations in prevalence throughout the world, with the highest rates being seen in Eastern Europe (levels >25%) and the lowest rates being found in Asia (levels <1%).

The prevalence of childhood obesity is also higher in Western and Southern Europe than that in Northern Europe. The prevalence rates are approximately double in Mediterranean nations than those of Northern European countries [5].

At a country level, the correlation between childhood obesity and ethnicity varies. In the United States the average weight of a child has risen by more than 5 kg within three decades, to a point where a third of the country's children are overweight or obese [6]. Obesity rates of both genders are the highest in Mexican Americans (31%), followed by non-Hispanic Blacks (20%), non-Hispanic Whites (15%), and Asian Americans (11%) [7]. In the United Kingdom, children of Bangladeshi or Black ethnicity were significantly associated with rapid weight gain in childhood [8].

In the United Kingdom the prevalence of childhood obesity has continued to rise as reported by the survey of the National Child Measurement Programme (NCMP) that included children of both sexes between 4–5 and 10–11 years of age. This is the case for both boys and girls and across both age groups. The NCMP data suggest that mean body mass index (BMI) has increased by around one BMI centile from the 2007/08 survey to the 2009/10 survey. Obesity prevalence among children living in the most deprived areas was roughly twice than that of the children living in the least deprived areas [9].

In 2009 the Health Survey for England reported that 31% of boys and 28% of girls aged 2–15 were classed as either overweight or obese; however, the mean BMI was higher among girls than among boys (difference of 0.2 kg/m^2). This difference was greatest among older children aged 12–15 where it ranged between 0.3 and 0.9 kg/m^2 [10]. Similar figures were reported in the same survey nearly 10 years later in 2018 with 31% boys and 27% girls overweight or obese. The prevalence of obesity increased in age, 16% boys and 23% girls being overweight or obese aged 2–4, rising to 36% boys and 37% girls aged 13–15. Worryingly, in the obese range, 4% boys and 9% girls were obese at ages 2–4 years, and this rose to 22% boys and 21% girls aged 13–15 [11]. Fig. 2.1 shows the aggregated prevalence of children at risk for overweight or children classified as overweight or obese in developed and developing countries and globally.

Obesity and the pubertal transition

The age of onset of puberty in girls has decreased over the past decades. Data collected from 1940 to 1994 support the contention that the larche and menarche are occurring earlier in the US girls. This apparent trend has coincided with the increase in the prevalence of obesity. It is unknown whether the early changes of puberty in obese girls are related to neuroendocrine maturation as, for example, estrogens from any source can result in the development of breast tissues [12]. Adipose tissues contain aromatase that can convert adrenal androgen precursors to estrogens. There also may be an obesity-related decrease in the hepatic metabolism of estrogens. Finally, peri-pubertal obesity is associated with insulin-induced reductions in sex hormone–binding globulins (SHBG), thereby increasing the bioavailability of sex steroids, including estrogens.

Unlike their female counterparts, pubertal development in obese boys may be delayed. The reasons for this are unclear, but increased aromatization of androgens to estrogens in adipose tissue with feedback inhibition of gonadotropin secretion may be involved.

Factors affecting childhood and adolescent obesity

The causes of obesity are complex and multifactorial. Increases in the amount of calorie-dense foods eaten and increases in the screen time (television, computer, and video games) with a simultaneous decrease in the amount of physical activity undertaken by children have been cited as reasons for the current epidemic [5].

Weight gain in early childhood (between 3 and 5 years) has been shown to be impacted upon by biological, early life, and social factors. Prepregnancy maternal overweight, and maternal and paternal overweight status at age 3 were all independently associated with more rapid weight gain in the child. Maternal smoking during pregnancy and postnatal exposure of the child through passive smoking was also independently associated with an increased risk of more rapid weight gain. Bangladeshi or black ethnicity and lone child status were also significant [12].

Adolescent obesity—adverse outcomes

General

Obese adolescents are more likely to develop pathologies such as diabetes mellitus that may be an insidious

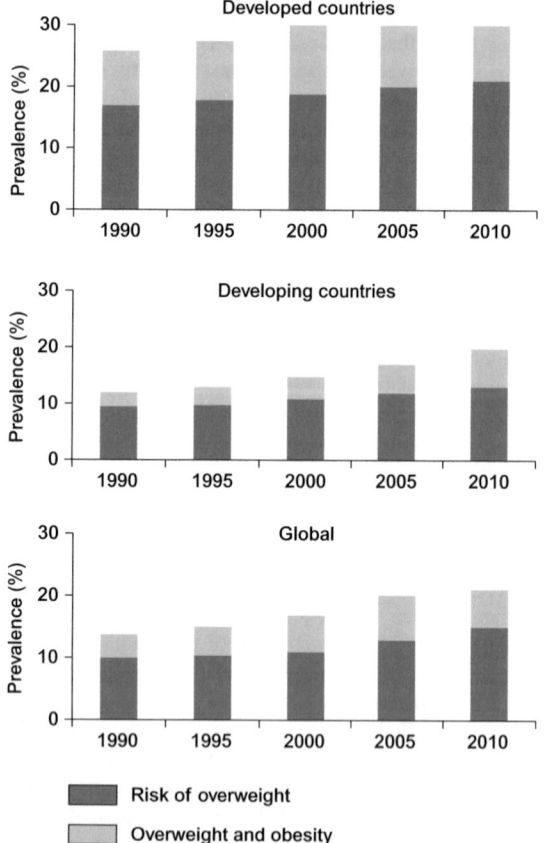

FIGURE 2.1 Aggregated prevalence of children at risk of overweight or of children classified as overweight or obese in developing and developed countries and globally. *Adapted from De Onis M, Blossner M, Borghi E. Global prevalence and trends of overweight and obesity among preschool children. Am J Clin Nutr 2010;92:1257–1264.*

presentation. Obese adolescents have a threefold-increased risk of developing hypertension due to sodium retention, increased sympathetic tone, or increased angiotensin system activity, but adolescents are usually asymptomatic and identified only via surveillance. Various pathologies are also related to childhood and adolescent obesity such as hyperlipidemia in the pattern of increased triglycerides, low-density lipoprotein, and decreased high-density lipoprotein (HDL).

Obstructive sleep apnea is four to six times more common in obese children and results in hypertension, left ventricular remodeling, daytime sleepiness, hyperactivity, restlessness, and inactivity.

Becoming obese significantly increases the risk of developing asthma. Overweight is significantly and independently associated with increased C-reactive protein concentration and other inflammatory indices, with repercussions on endothelial function.

Orthopedic complications such as musculoskeletal discomfort, impaired mobility, lower extremity asymmetry, and fractures are more common in childhood obesity. These problems further discourage physical activity, thereby exacerbating the underlying problem of obesity.

Obesity leads to various gastrointestinal diseases such as gastroesophageal reflux, nonalcoholic fatty liver disease, and cholelithiasis. Neurological disorders, in particular, the prevalence of benign intracranial hypertension, increases with increasing BMI [13]. Table 2.1 details the multiple adverse outcomes of childhood obesity.

Psychological effects

Child/adolescent obesity is independently associated with internalizing (emotional) difficulties, after adjusting for confounding variables such as gender and family income [14]. Psychological disturbances associated with childhood obesity include negative self-esteem, withdrawal from peer interaction, anxiety, depression, and suicide.

Nearly half of the obese adolescents report moderate-to-severe depressive symptoms and one-third report anxiety. Obese children are more likely to experience psychological or psychiatric problems than nonobese children. Girls are at greater risk than boys, especially concerning self-esteem [13].

In a study of nearly 1000 adolescents aged 12–18, it was found that elevated BMI was directly associated with depression at a 1-year follow-up. Social networking mapping studies indicate that overweight children have fewer secluded relationships than their normal-weight peers who have many relationships within a central network of children.

Teasing by an overweight adolescent's peers has been established to directly correlate with the child's suicidal ideation and number of attempts at suicide [15].

The metabolic syndrome

The metabolic syndrome is a constellation of cardiovascular risk factors associated with insulin resistance. These include glucose intolerance, dyslipidemia, hypertension, and central obesity.

There is no consensus regarding the definition of metabolic syndrome in children and adolescents; however, the International Diabetes Federation has proposed criteria depending on age groups [16].

The presence of the metabolic syndrome in adolescents in the United States has increased from 4.2% to 6.4% over the past two decades and is significantly higher in Hispanic and White youths. Nearly one-third of the overweight/obese adolescents meet the criteria for metabolic syndrome.

Childhood adiposity is a good predictor of metabolic syndrome in adulthood, and hence, a greater risk of cardiovascular disease, especially if there is a family history of type-2 diabetes. Metabolic syndrome is also more likely in adults who experienced a rapid increase in adipose tissue in childhood [17].

Polycystic ovarian syndrome in adolescence

Polycystic ovarian syndrome (PCOS) is the most common endocrinological problem in adult women. This was often unrecognized in adolescence. Although most adolescents and adults with PCOS are obese, only 20% of obese women have PCOS [18].

PCOS may present with any, or a combination of obesity, menstrual abnormalities, hirsutism, acanthosis nigricans, acne, hair loss, or premature adrenarche. PCOS is associated with insulin resistance and the metabolic

TABLE 2.1 Adverse outcomes of childhood obesity.

- Cardiovascular disease
- Insulin resistance and type-2 diabetes
- PCOS
- Dyslipidemia
- Hypertension
- Psychological and social morbidity
- Asthma
- Orthopedic—hips, ankles
- Breathing problems and sleep apnea
- Fatty liver disease
- Persistence of obesity into adulthood

PCOS, Polycystic ovarian syndrome.

syndrome, and subsequent adult morbidities may include infertility and cardiovascular disease.

In a study of 71 PCOS and 94 healthy adolescent girls, Fulghesu et al. [19] showed that the incidence of altered lipid profiles was not different between both groups of adolescents but instead was related to anthropometric characteristics (BMI, waist measurement, and waist-to-hip ratio). The differences, which were statistically significant between the groups, were hirsutism and grogens, including total testosterone levels and hyperinsulinemia. This suggests that PCOS confers no additional risk over obesity to dyslipidemia in adolescence.

There have been several different criteria established for the diagnosis of PCOS, including the NIH consensus in 1990 [20], the Rotterdam criteria in 2003 [21], and the criteria by Androgen Excess and PCOS Society in 2006 [22]. However, none of the definitions fits all cases and so the exact prevalence is difficult to define precisely.

The diagnostic label of PCOS in adolescence implies an increased risk for infertility, dysfunctional uterine bleeding, endometrial cancer, obesity, type-2 diabetes, dyslipidemia, hypertension, and possible cardiovascular disease; therefore diagnostic accuracy is important. The use of the adult diagnostic criteria in adolescence may not be appropriate, as the mean ovarian volume is higher in young women, hirsutism is uncommon and not related to PCOS, and acne is similarly not related to PCOS [23]. Finally, adolescents frequently have menstrual irregularity, thus making the definitive diagnosis difficult. The recently published guideline for the management of patients with PCOS has endorsed the Rotterdam criteria for diagnosis of PCOS in adults but has stated that in adolescence, both oligo-anovulation and hyperandrogenism must be present, and unlike in adults, ultrasound is not recommended for diagnosis [24].

The diagnosis of PCOS in adolescence may be delayed for the abovementioned reasons, as well as that physiological adolescent anovulation may mimic or mask PCOS. There are, however, risk factors that may assist in the identification of adolescents at risk for PCOS [25]. Fig. 2.2 shows the risk factors for adolescent PCOS.

Obstetric outcomes in obese adolescents

In one recent study, adolescent obesity has been shown to be independently associated with lifetime nulliparity. This study used self-reported weight and height at high school, and weight and height were measured in adulthood. The lifetime nulliparity percentage in this population of 3154 women was 16.7% in comparison with a nulliparity rate of 30.9 in women with a recalled adolescent BMI ≥ 30. The nulliparity rate increased with increasing BMI across all ranges [26].

The complications associated with teenage pregnancy include an increased incidence of very low birthweight babies (defined as birthweight <1500 g). Young teenagers aged 11–15 years have nearly double (4% vs 2%) the risk as women aged 20–22 years. In addition, this young age group more frequently delivered babies of birthweight between 1500 and 2500 g. In this study of 16,512 consecutive nulliparous women, the incidences of stillbirth and index values of fetal condition at birth were not significantly different between age groups [27].

Younger maternal age has also been shown to be associated with increased risk of fetal death and anemia during pregnancy. The risk of preeclampsia, cesarean section, instrumental vaginal delivery, and postpartum hemorrhage is lower in teenage pregnancy [28].

Obesity confers an increased risk of adverse fetal and pregnancy outcomes, including gestational diabetes, fetal macrosomia, and delivery by cesarean section and preeclampsia.

The combination of obesity and teenage pregnancy results in an increased risk of developing preeclampsia

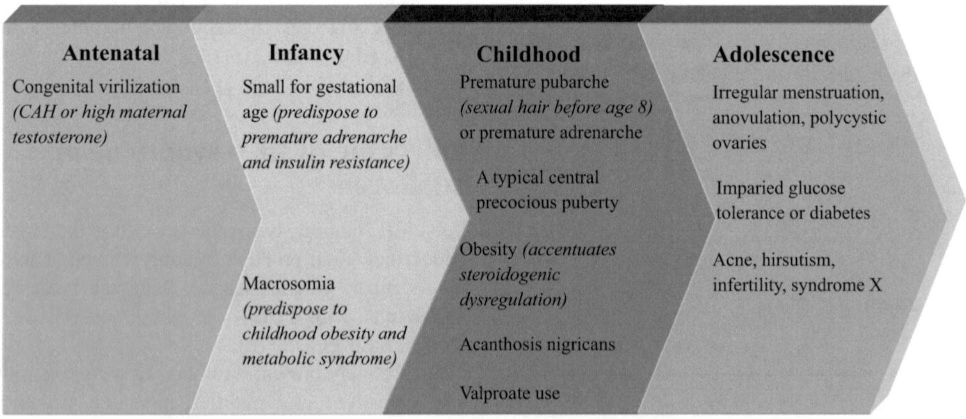

Antenatal
Congenital virilization *(CAH or high maternal testosterone)*

Infancy
Small for gestational age *(predispose to premature adrenarche and insulin resistance)*

Macrosomia *(predispose to childhood obesity and metabolic syndrome)*

Childhood
Premature pubarche *(sexual hair before age 8)* or premature adrenarche

A typical central precocious puberty

Obesity *(accentuates steroidogenic dysregulation)*

Acanthosis nigricans

Valproate use

Adolescence
Irregular menstruation, anovulation, polycystic ovaries

Imparied glucose tolerance or diabetes

Acne, hirsutism, infertility, syndrome X

FIGURE 2.2 Risk factors for adolescent PCOS. *PCOS*, Polycystic ovarian syndrome. *Adapted from Yii MF, Lim CED, Luo X, Wong WSF, Cheng ACL, Zhan X. Polycystic ovarian syndrome in adolescence. Gynecol Endocrinol 2009;25(10):634–9.*

and eclampsia. In a retrospective cohort study of 290,807 women in Florida, the United States, extremely obese (BMI ≥40) girls aged 16−17 years had the highest rate (13.2%) of preeclampsia and eclampsia compared with all the other age groups. The lowest rate of preeclampsia and eclampsia occurred in nonobese women aged 20−24 years (4.0%).

Extremely obese teenagers had a 71% increased risk of preeclampsia and eclampsia compared with extremely obese women aged 20−24 years [29].

The impact of childhood and adolescent obesity on adult health

Several studies have shown that childhood and adolescent overweight and obesity (based on BMI for age) have been associated with premature mortality at the adult stage of life [30]. In addition to an increased risk of premature mortality, childhood/adolescent obesity is also associated with an increased risk of later diabetes, stroke, coronary heart disease, and hypertension.

A retrospective study of 230,000 Norwegian adolescents was followed-up on average for 34.9 years. There were 9650 deaths within this time. Mean age of death was 40 years for men and 43 years for women. The relative risk of death from endocrine, nutritional and metabolic diseases, and diseases of the circulatory system was elevated in the two highest BMI groups of both men and women. The relative risks of death from diseases of the respiratory system and ill-defined causes were increased in the highest BMI group of both sexes [31].

In case of diseases of the circulatory system in men, ischemic heart disease was the dominant cause of death, whereas in women the death was due to cerebrovascular disease. In both men and women, the risk of death from ischemic heart disease was increased in the two highest BMI categories. Similar findings were reported in a Danish study that showed a linear association between BMI in childhood (7−13 years) and both fatal and nonfatal coronary heart disease events in adulthood (age 25 or older) [32].

The risk of death from diabetes was increased in men and women in the two highest BMI categories. There was an increased risk of sudden death for both sexes and of death from chronic lower respiratory diseases in men in the highest BMI category. There was no association between BMI and mortality from mental and behavioral disorders.

Management principles

Prevention

Prevention programs should address parental weight status and smoking habits, both of which are modifiable risk factors. Reductions in weight preconception may result in an additional reduction in risk. Parents should be encouraged to integrate healthy lifestyle behaviors into the family unit, thereby decreasing the risk of obesity in their children and improving their own physical and mental health.

The long-term impact of maternal weight loss was elegantly demonstrated by the reported impressive reduction in childhood/adolescent obesity in children of mothers who underwent bariatric surgery. An accompanying improvement in cardiometabolic risk factors also occurred that was sustained into adolescence and adulthood [33].

Obesity prevention programs in kindergartens and schools based on exercise therapy and dietary intervention have failed in preventing childhood obesity. It is widely accepted that changing the environment (e.g., banning all sugary drinks in favor of water) is more effective in preventing obesity. This may suggest that public health restrictions on advertising and marketing of sweetened drinks are a meaningful approach to the fight against childhood obesity.

Lifestyle interventions

A Cochrane review [34] regarding treatment in obese children included 64 randomized controlled trials with 5230 participants. Metaanalysis indicated a reduction in overweight at 6 and 12 months in response to lifestyle interventions. Dietary modification and exercise programs are effective in the treatment of already-obese children. Parental involvement was a major predictor of success.

Recommended family lifestyle changes:

- Be physically active for 1 h/day (moderate to vigorous intensity).
- Reduce screen time (sedentary behavior such as watching TV, using computers, and playing computer games to not more than 2 h/day).
- Encourage low-energy snacks (e.g., fruit, raw vegetables, and a plain biscuit).
- Avoid/cut down on high-energy foods such as crisps, chips, chocolate, sweets.
- Avoid grazing and keep food to mealtimes and small snacks.
- Avoid sugary juice.
- Parents should be positive about a healthy family lifestyle.

Drugs

The Cochrane review also concluded that consideration should be given to the use of drugs such as orlistat or sibutramine as an adjunct to lifestyle interventions in obese adolescents. This must, of course, be balanced with the risk of adverse events. Sibutramine is no longer available on the European market due to its side effects such as arterial and pulmonary hypertension.

Surgery

Gastric banding in obese adolescents is an effective intervention leading to substantial and durable reduction in obesity and to better health. This treatment requires long-term supportive follow-up. There is also a significant possibility of requiring revisional procedures [33].

Management of polycystic ovarian syndrome in adolescence

In adolescents with PCOS the target goals of therapy include protection of endometrial health, improvement in cosmetic appearance (e.g., hirsutism and acne), and reduction in weight and obesity-related metabolic complications.

Lifestyle interventions, including weight loss, are the first line of treatment of adolescents with PCOS. The study by Lass et al. [35] showed a significant decrease in the prevalence of metabolic syndrome and insulin resistance in adolescents with successful weight loss. Furthermore, testosterone concentrations, free testosterone index, luteinizing hormone (LH) levels, and LH/follicle-stimulating hormone (FSH) ratio decreased significantly. The prevalence of amenorrhea and/or oligomenorrhea decreased significantly in the weight-loss group.

Overall, the evidence regarding both the assessment and the management of PCOS was described as generally low-to-moderate quality. Of note, there was a clinical consensus recommendation that health professionals should be aware that in PCOS, there is a high prevalence of moderate-to-severe anxiety and depressive symptoms in adults, and a likely increased prevalence in adolescents. In addition, in an adolescent with irregular menstrual cycles, the value and optimal timing of assessment and diagnosis of PCOS should be discussed with the patient, taking into account diagnostic challenges at this life stage and psychosocial and cultural factors, keeping in mind that ultrasound is not necessary or recommended for diagnosis. Weight, height, and ideally waist circumference should be measured and BMI calculated and adolescent and ethnic-specific BMI and waist circumference categories need to be considered when optimizing lifestyle interventions and weight.

Lifestyle modification with calorie restriction and increased exercise should be considered as first-line treatment of the comorbidities of obesity and insulin resistance. A balanced diet is recommended to reduce dietary energy [36].

A mean weight loss of 6.5% has been reported to restore menstrual irregularity in obese adolescents. Both low-fat and low-carbohydrate diets resulted in weight loss [37].

Lifestyle modifications and weight loss, while being very effective in treating the hormonal abnormalities of PCOS, are difficult to sustain. In one study of adolescents with PCOS, 30% of the subjects enrolled into a study that included intensive lifestyle modification dropped out, whereas 40% attended less than 50% of the sessions and demonstrated no weight change [38]. Therefore, in most cases, pharmacological therapy for PCOS becomes necessary.

Combined oral contraceptive pills

Combined oral contraceptive pills (COCPs) are among the primary treatment options for adolescents with PCOS. COCPs improve symptoms via several mechanisms. Estrogens increase the production of SHBG, resulting in a decrease in circulating androgens, as well as their bioavailability. Progestins protect the endometrium against hyperplasia induced by unopposed estrogen stimulation. Some progestins such as drospirenone and cyproterone acetate have been proven to have antiandrogenic effects and therefore may be of added benefit in PCOS [39]. COCPs also suppress FSH and LH, resulting in reduced ovarian stimulation and androgen production.

None of these actions, however, affect insulin resistance in PCOS, and their use may actually be associated with long-term metabolic derangements such as glucose intolerance, abnormal lipid profiles, and cardiovascular diseases. It has been shown that in adolescents with PCOS the use of COCPs containing desogestrel or cyproterone as progestin was associated with decreased insulin sensitivity and increased total, low-density lipoprotein and HDL cholesterol and with variable changes in triglycerides [40].

The use of the COCP does have other benefits in this population, such as contraception in sexually active adolescents.

Insulin sensitizers

These drugs act to reduce insulin levels (metformin) and increase insulin sensitivity (metformin and thiazolidinediones), thus treating the metabolic comorbidities associated with PCOS and obesity.

Metformin increases insulin sensitivity by the liver, increases peripheral glucose uptake, decreases fatty acid oxidation, and decreases glucose absorption from the gut. Metformin therapy in adolescents seems to be associated with several benefits, including an improvement in glucose tolerance, a decrease in testosterone levels [41], and an improvement in menstrual cyclicity; some report between 90% and 100% resumption of menses in adolescents [42]. These studies were all relatively small, and, in most, significant weight loss occurred, making the data difficult to interpret the effect of metformin independent of weight loss. This has not yet been substantiated by metaanalysis of randomized controlled trials.

The dose of metformin required is approximately 1.5—2 g/day, and even larger doses up to 2.55 g/day have been reported. The dose should be titrated upward over a period of 1 month due to gastrointestinal side effects.

A metaanalysis of metformin versus COCPs in adolescents revealed that they were similar in effects on hirsutism. Metformin was superior for weight reduction and associated with decreased dysglycemia. COCPs were superior for menstrual regulation. This metaanalysis included only four relatively small studies, so results should be interpreted with caution, but, however, do suggest that therapy should be tailored to patients' individual needs [43].

Thiazolidinediones act as insulin sensitizers through their activation of the nuclear receptor peroxisome proliferator—activated receptors γ, leading to increased production of insulin-sensitive adipocytes and increased glucose uptake in these cells, increased secretion of adiponectin and decreased secretion of proinflammatory cytokines. The thiazolidinedione, pioglitazone, has been shown to ameliorate the signs and symptoms of PCOS in a cohort of women who failed a previous trial of metformin [44]. These medications have not been well studied in adolescence and remain off-label in this age group due to lack of evidence on efficacy and safety.

Psychological morbidity

An increased risk of mental health disease has been described in women and adolescents with PCOS. Significantly, higher levels of psychological distress, impaired quality of life, and increased hostility/irritability have been described among adolescents with PCOS compared with controls [45].

Screening and management of mental health conditions in adolescents with PCOS should therefore form an essential part of their management [46].

Conclusion

Obesity in childhood and adolescence is a global epidemic with health implications. There are geographic and ethnic variations in the prevalence of adolescent obesity, but it is an international problem.

All efforts should be made to control this epidemic and therefore to avoid or reduce the health-care burden of chronic illness on health-care systems as well as to reduce the personal cost, including premature mortality.

In particular, a clear focus on lifestyle and prevention measures should be adopted for all age groups from preconception to early childhood, adolescence, and adulthood.

References

[1] WHO. Population-based prevention strategies for childhood obesity: report of a who forum and technical meeting. Geneva: WHO Press; 2009.
[2] RCOG, 2011. Consensus views arising from the 53rd study group: obesity and reproductive health. RCOG.
[3] The NS, Suchindran C, North KE, Popkin BM, Gordon-Larsen P. Association of adolescent obesity with risk of severe obesity in adulthood. JAMA 2010;304(18):2042—7.
[4] De Onis M, Blossner M, Borghi E. Global prevalence and trends of overweight and obesity among preschool children. Am J Clin Nutr 2010;92:1257—64.
[5] Ben-Sefer E, Ben-Natan M, Ehrenfeld M. Childhood obesity: current literature, policy and implications for practice. Int Nurs Rev 2009;56(2):166—73.
[6] Lobstein T, Jackson-Leach R, Moodie ML. Child and adolescent obesity: part of a bigger picture. Lancet 2015;385(9986):2510—20.
[7] Sorof JM, Lai D, Turner J. Overweight, ethnicity and the prevalence of hypertension in school-aged children. Pediatrics 2004;113 (3):475—82.
[8] Griffiths LJ, Hawkins SS, Cole TJ, Dezateux C. Risk factors for rapid weight gain in preschool children: findings from a UK-wide prospective study Millenium Cohort Study Child Health Group Int J Obes 2010;34:624—32.
[9] National Obesity Observatory. National Child Measurement Programme. Changes in children's body mass index between 2006/7 and 2009/10. NOO. 2011.
[10] The Health and Social Care Information Centre. Statistics on Obesity, Physical Activity and Diet. England; 2011.
[11] Health Survey for England 2018: Overweight and Obesity in adults and children. <https://digital.nhs.uk/data-and-information/publications/statistical/health-survey-for-england/2018>.
[12] Burt Solorzano CM, McCartney CR. Obesity and the pubertal transition in girls and boys. Reproduction 2010;140:399—410.
[13] Stewart L. Childhood obesity. Medicine 2010;39(1):42—4.
[14] Bruni V, Dei M, Peruzzi E, Seravalli V. The anorectic and obese adolescent. Best Pract Res Clin Obstet Gynaecol 2010;24:243—58.
[15] Tiffin PA, Arnott B, Moore HJ, Summerbell CD. Modelling the relationship between obesity and mental health in children and adolescents: findings from the Health Survey for England 2007. Child Adolesc Psychiatry Ment Health 2001;5:31.
[16] Sinha A, Kling SA. Review of adolescent obesity: prevalence, etiology and treatment. Obes Surg 2009;19:113—20.
[17] Zimmet P, Alberti KG, Kaufman F. The metabolic syndrome in children and adolescents — an IDF consensus report. Pediatr Diabetes 2007;8:299.
   In-Iw S, Biro FM. Adolescent women and obesity. J Pediatr Adolesc Gynecol 2011;24:58—61.
[18] Alihu MH, Luke S, Kristensen S, Alio A, Salihu HM. Joint effect of obesity and teenage pregnancy on the risk of preeclampsia: a population-based study. J Adolesc Health 2010;46:77—82.

[19] Fulghesu A, Magnini R, Portoghese E, Angioni S, Minerba L, Melis GB. Obesity-related lipid profile and altered insulin incretion in adolescents with polycystic ovary syndrome. J Adolesc Health 2010;46:474–81.

[20] Zawadzki JK, Dunaif A. Diagnostic criteria for polycystic ovary syndrome: towards a rational approach. In: Dunaif A, editor. Polycystic ovary syndrome. Boston, MA: Blackwell Scientific; 1995. p. 377–84.

[21] Rotterdam. ESHRE/ASRM-Sponsored PCOS Consensus Workshop Group. Revised 2003 consensus on diagnostic criteria and long-term health risks related to polycystic ovary syndrome. Fertil Steril 2004;81:19–25.

[22] Azziz R, Carmina E, Dewailly D. The Androgen Excess and PCOS society criteria for the polycystic ovary syndrome: the complete taskforce report. Fertil Steril 2009;91:456–88.

[23] Hickey M, Doherty DA, Atkinson H. Clinical, ultrasound and biochemical features of polycystic ovary syndrome in adolescents: implications for diagnosis. Hum Reprod 2011;26(6):1469–77.

[24] Teede HJ, Misso ML, Costello MF, et al. Recommendations from the international evidence-based guideline for the assessment and management of polycystic ovary syndrome. Fertil Steril 2018;110:364–79.

[25] Yii MF, Lim CED, Luo X, Wong WSF, Cheng ACL, Zhan X. Polycystic ovarian syndrome in adolescence. Gynecol Endocrinol 2009;25(10):634–9.

[26] Polotsky AJ, Hailpern SM, Skurnick JH. Association of adolescent obesity and lifetime nulliparity – the Study of Women's Health Across the Nation (SWAN). Fertil Steril 2010;93(6):2004–11.

[27] Satin AJ, Leveno KJ, Sherman ML, Reedy NJ, Lower TW, McIntire DD. Maternal youth and pregnancy outcomes: middle school versus high school age groups compared with women beyond the teen years. Am J Obstet Gynecol 1994;171:184–7.

[28] de Vienne CM, Creveuil C, Dreyfus M. Does young maternal age increase the risk of adverse obstetric, fetal and neonatal outcomes: a cohort study. Eur J Obstet Gynecol Rep Biol 2009;147:151–6.

[29] Alihu MH, Luke S, Kristensen S, Alio A, Salihu HM. Joint effect of obesity and teenage pregnancy on the risk of preeclampsia: a population-based study. J Adol Health 2010;46:77–82.

[30] Reilly JJ, Kelly J. Long-term impact of overweight and obesity in childhood and adolescence on morbidity and premature mortality in adulthood: systematic review. Int J Obes 2011;35:891–8.

[31] Bjorge T, Engeland A, Tverdal A, Davey Smith G. Body mass index in adolescence in relation to cause-specific mortality: a follow-up of 230,000 Norwegian adolescents. Am J Epidemiol 2008;168:30–7.

[32] Baker JL, Olsen LW, Sorensen TIA. Childhood body-mass index and the risk of coronary heart disease in adulthood. N Engl J Med 2007;357:2329–37.

[33] Reinehr T, Wabitsch M. Childhood obesity. Curr Opin Lipidol 2011;22:21–35.

[34] Oude Luttikhuis H, Baur L, Jansen H. Interventions for treating obesity in children. Cochrane Database Syst Rev 2009;1 CD001872.

[35] Lass N, Kleber M, Winkel K, Wunsch R, Reinehr T. Effect of lifestyle intervention on features of polycystic ovarian syndrome, metabolic syndrome, and intima-media thickness in obese adolescent girls. J Clin Endocrinol Metab 2011;96(11):3533–40.

[36] Rosenfield RL. The diagnosis of polycystic ovary syndrome in adolescents. Pediatrics 2015;136:1154.

[37] Marzouk TM, Sayed Ahmed WA. Effect on dietary weight loss on menstrual regularity in obese young adult women with polycystic ovary syndrome. J Pediatr Adolesc Gynecol 2015;24:161–5.

[38] Hoeger K, Davidson K, Kochman L, Cherry T, Kopin L, Guzick DS. The impact of metformin, oral contraceptives, and lifestyle modification on polycystic ovary syndrome in obese adolescent women in two randomised, placebo-controlled clinical trials. J Clin Endocrinol Metab 2008;93:4299–306.

[39] Franks S, Layton A, Glasier A. Cyproterone acetate/ethinyl estradiol for acne and hirsutism: time to revise prescribing policy. Hum Reprod 2008;23:231–2.

[40] Mastorakos G, Koliopoulos C, Deligeoroglou E, Diamanti-Kandarakis E, Creatsas G. Effects of two forms of combined oral contraceptives on carbohydrate metabolism in adolescents with polycystic ovary syndrome. Fertil Steril 2006;85:420–7.

[41] Arslanian SA, Lewy V, Danadian K, Saad R. Metformin therapy in obese adolescents with polycystic ovary syndrome and impaired glucose tolerance: amelioration of exaggerated adrenal response to adrenocorticotropin with reduction of insulinaemia/insulin resistance. Clin Endocrinol Metab 2002;87:1555–9.

[42] De Leo V, Musacchio MC, Morgante G, Piomboni P, Petraglia F. Metformin treatment is effective in obese teenage girls with PCOS. Hum Reprod 2006;21:2252–6.

[43] Al Khalifah RA, Florez ID, Dennis B, et al. Metformin or oral contraceptives for adolescents with polycystic ovarian syndrome: a meta-analysis. Pediatrics 2016;137(5).

[44] Gluck CJ, Moreira A, Goldenberg N, Sieve L, Wang P. Pioglitazone and metformin in obese women with polycystic ovary syndrome not optimally responsive to metformin. Hum Reprod 2003;18:1618–25.

[45] Giudi J, Gambineri A, Zanotti L, et al. Psychological aspects of hyperandrogenic states in the late adolescent and young women. Clin Endocrinol (Oxf) 2015;83:872–8.

[46] Javed A, Chelvakumar G, Bonny AE. Polycystic ovary syndrome in adolescents: a review of past year evidence. Curr Opin Obstet Gynecol 2016;28:373–80.

Chapter 3

Obesity in polycystic ovary syndrome and infertility

Ioannis E. Messinis, Christina I. Messini and Konstantinos Dafopoulos

Department of Obstetrics and Gynaecology, Faculty of Medicine, School of Health Sciences, University of Thessaly, Larissa, Greece

Introduction

The National Institutes of Health (NIH) diagnostic criteria [1] for the polycystic ovary syndrome (PCOS) are the presence of hyperandrogenism and chronic oligo-anovulation, with the exclusion of other causes of hyperandrogenism such as adult-onset congenital adrenal hyperplasia, hyperprolactinemia, and androgen-secreting neoplasms. The European Society of Human Reproduction and Embryology/American Society for Reproductive Medicine (ESHRE/ASRM)-sponsored PCOS consensus workshop in Rotterdam [2] concluded that ultrasound morphology of the ovaries should be included in the diagnostic criteria and that at least two of the following criteria are sufficient for the diagnosis: oligo-anovulation, clinical and/or biochemical signs of hyperandrogenism, and polycystic ovaries on ultrasound, while other causes of hyperandrogenism should be excluded. The PCOS is the most common endocrinopathy in women. According to the NIH criteria, the prevalence of PCOS ranges between 6% and 10%, and with utilization of the ESHRE/ASRM consensus criteria it is as high as 15% [3]. The most recent definition issued by "the Androgen Excess and PCOS Society (AES 2006)" describes hyperandrogenism as the main criterion associated with either oligo-anovulation or polycystic morphology of the ovaries or both [4]. The prevalence of PCOS tends to be higher with the Rotterdam criteria than with the AES and NIH definitions [5].

The main cause of infertility in women with PCOS is anovulation. The PCOS represents 75% of all anovulatory disorders causing infertility, and 90% of women with oligomenorrhea [6]. On the other hand, on average 79% of women with PCOS have oligomenorrhea [4]. The incidence of oligomenorrhea and the other manifestations are variable in different phenotypes. Women with PCOS are commonly (35%−80%) overweight (body mass index, BMI, above 25 kg/m^2) or obese (BMI above 30 kg/m^2) [7,8]. The range depends on the setting of the study and the ethnic characteristics of the patients. Women in the United States with PCOS have a higher BMI than their European counterparts. This may explain the increase in the incidence of the PCOS in the US population, which parallels the increase in obesity [9]. It has also been shown that women with PCOS may be more likely to exhibit an abdominal and/or visceral pattern of fat distribution [10], although the distribution of fat to abdominal regions is not probably related to insulin resistance in such women [11]. On the other hand, PCOS (defined by NIH criteria) was found in nearly 35% of morbidly obese women from Spain undergoing bariatric surgery [12], compared with only 6.5% in unselected female blood donors from Spain, and studied by the same investigators [13]. However, obesity is not included in the diagnostic criteria for PCOS.

Obesity may intensify the severity of the phenotypic characteristics of the PCOS, including disturbed menstrual cycle [14]. In particular, a higher prevalence of anovulation has been found in obese as compared to nonobese women with PCOS [15]. Recent experimental data in rodents have shown that high-fat high-sugar diet induces PCOS [16]. The extent to which these data can explain a possible role for obesity at a critical time in the development of PCOS has not been investigated. Obesity seems to have a negative effect on spontaneous and induced ovulation in PCOS.

Obesity and infertility—possible mechanisms

Hyperandrogenism

Ovarian hyperandrogenism in PCOS may arrest the development of antral follicles. Increased androgen secretion originates from intrinsic amplified steroidogenetic

Obesity and Gynecology. DOI: https://doi.org/10.1016/B978-0-12-817919-2.00003-6
© 2020 Elsevier Inc. All rights reserved.

capacity of theca cells, due to 17α-hydroxylase/17,20-lyase (CYP17a1), HSD3B2, and side-chain cleavage enzyme (CYP11A1) activities. Besides, endocrine mechanisms may contribute, including luteinizing hormone (LH) hypersecretion that stimulates thecal androgen secretion, relative follicle-stimulating hormone (FSH) insufficiency resulting in reduced aromatase activity and hyperinsulinemia synergizing with LH for thecal androgen production. Intraovarian mechanisms, such as anti-Müllerian hormone (AMH) inhibition of FSH and subsequent inhibition of aromatase activity, may further deteriorate hyperandrogenism. In addition, obesity amplifies the stimulating action of LH on the theca cells [17]. Finally, ovarian hyperandrogenism in PCOS may arrest folliculogenesis through inhibition of granulosa cells proliferation and maturation, secretion of estrogen and progesterone, action of aromatase, and increase of 5α-reductase activity [18,19].

The severity of hyperandrogenism seems to be amplified in obese women with PCOS. It has been shown that obese women with PCOS have higher total and free T levels as compared to nonobese PCOS [15,20]. In fact, the increase of body weight and fat tissues, especially in the form of abdominal obesity, is associated with an abnormality of sex-steroid balance. This is mainly due to the reduction of sex hormone−binding globulin (SHBG) levels in circulation, resulting in an increased fraction of free androgens in blood. Reduced SHBG synthesis in liver originates from hyperinsulinemia that compensates for insulin resistance associated with obesity. Although hyperinsulinemia is associated with PCOS, it is clear that obese women with PCOS exhibit a higher degree of insulin resistance and hyperinsulinemia [21]. Finally, increased androgens in obese women with PCOS further contribute to the inhibition of SHBG secretion. It is obvious that obesity may deteriorate hyperandrogenism in women with PCOS, a mechanism that is involved in anovulatory infertility.

Hypersecretion of luteinizing hormone

In anovulatory women with PCOS, 75% have high LH levels, while in up to 94% an elevated LH/FSH ratio may be found [22]. Anovulation and lack of progesterone [23] and hyperandrogenemia followed by progesterone negative feedback effect attenuation [24,25] are the main mechanisms for inappropriate LH secretion. This is characterized by accelerated LH pulse frequency and amplitude and elevated LH response to gonadotrophin-releasing hormone (GnRH). However, obesity in PCOS is associated with blunted LH secretion, acting at the pituitary and not at the hypothalamic level. The overall quantity of GnRH secreted and the LH pulse frequency is not affected by BMI [26]. The mechanisms that may mediate the negative effect of BMI on LH secretion may be hyperinsulinemia, since insulin infusions decrease basal and GnRH-induced LH secretion [27]. Leptin may also be important as data have shown an inverse correlation between leptin and LH levels and LH pulse amplitude; furthermore, a decrease in leptin levels and an increase in LH pulse amplitude were found following short-term caloric restriction [28,29]. According to the ceiling hypothesis, high LH levels in circulation may lead to premature luteinization and inhibit the proliferation of granulosa cells resulting in anovulation [30]. Although, in lean women with PCOS, elevated LH is a significant mechanism leading to hyperandrogenemia and anovulation, in obese women with PCOS probably this is not the case.

Hyperinsulinemia

As reported previously, in obese women with PCOS, insulin resistance and hyperinsulinemia are higher than in lean women with PCOS. High insulin levels in circulation may be mainly related to anovulation in women with PCOS. Hyperinsulinemia may cause premature maturation of granulosa cells, because they respond prematurely to LH (small follicles of 4 mm), which is in contrast to the normal response that occurs when follicles reach the 10 mm diameter [31]. Premature exposure of granulosa cells to LH inhibits their proliferation and further development. In normal folliculogenesis, when granulosa cells can respond to LH at about 10 mm, they undergo two more cell divisions to reach the preovulatory size (20−25 mm). In the anovulatory PCOS, granulosa cells responsive to LH in follicles as small as 4 mm in diameter undergo two more cell divisions reaching a maximum size of around 8−10 mm [32]. Furthermore, high insulin levels amplify LH-stimulated androgen secretion from theca cells through LH receptor's upregulation [33]. The role of hyperinsulinemia is significant in deteriorating fertility in obese women with PCOS since when such women lose weight and subsequently become ovulatory, they also have a reduction in insulin resistance and central adiposity [34].

Adipokines

In obesity, many genes were dysregulated in adipocytes of obese compared with nonobese individuals [35]. In omental fat of obese women with PCOS, there was a different expression pattern in genes compared to obese non-PCOS women [36]. Adipokines secreted by adipose tissues may mediate the deleterious effects of obesity upon fertility in women with PCOS. These substances include leptin, adiponectin, interleukin-6 (IL-6), plasminogen activator inhibitor-1 (PAI-1), resistin, and tumor necrosis factor-α (TNF-α).

Leptin

Leptin is a 16 kDa protein that is secreted almost exclusively by the adipocytes and is produced by the obese (*ob*) gene. It may serve as a link between fat tissues and the brain, since by acting at the level of the hypothalamus, leptin decreases food intake and increases energy expenditure [37]. Besides, it may have a role in reproductive function, exerting effects upon the hypothalamic—pituitary—ovarian axis at central and gonadal levels. Leptin receptors have been demonstrated in the hypothalamus and pituitary as well as in theca cells, granulosa cells, oocytes, endometrial cells, and preimplantation embryos [14]. In obesity, circulating leptin levels are high due to leptin resistance. Furthermore, in women with PCOS, increased leptin levels in circulation compared to weight-matched controls have been found in some but not all studies. Leptin levels have been found to be positively correlated with insulin resistance in such women, although some contradictory findings have been reported. Similarly, in some but not all studies, following treatment with insulin-sensitizing agents, leptin concentrations in blood had decreased. Also, levels of soluble leptin receptor have been found to be lower in obese than in lean women with PCOS, but they increased significantly after treatment with metformin [38].

Leptin may affect reproductive function at many levels. Physiologically, in women, by central action, leptin seems to be important for the hypothalamic—pituitary function and puberty. However, in obese women with PCOS as compared to obese controls, abnormalities in the relationship between leptin and LH secretory characteristics have been found [39]. At the level of the ovary, leptin was found to modulate basal and FSH-stimulated steroidogenesis in cultured human lutein granulosa cells, with high concentrations suppressing the secretion of estradiol and progesterone [40]. In vivo and in vitro experiments in animals have shown that high levels of leptin, representing hyperleptinemia of obesity, may inhibit folliculogenesis [41,42]. Leptin may have a role in regulation of embryo implantation and endometrial receptivity, and it has been suggested that obesity-related perturbations of the leptin system can possibly interfere with embryo implantation, therefore causing infertility [14]. In conclusion, it seems that obesity may further intensify hyperleptinemia in PCOS women, deteriorating the reproductive function further.

Adiponectin

Adiponectin, a 30 kDa protein, is the most abundant serum adipokine, secreted exclusively by the adipose tissue. Adiponectin, in contrast to leptin, is downregulated in obesity and may have both antiinflammatory and insulin-sensitizing effects. A metaanalysis showed that women with PCOS have lower levels of adiponectin, independently of BMI [43]. Because adiponectin has direct insulin-sensitizing effects, decreased levels of adiponectin in PCOS women could, in addition to obesity, contribute to systemic insulin resistance and hyperinsulinemia, thereby declining fertility. Besides, a direct role of low levels of adiponectin on folliculogenesis is possible. It was found that human theca cells express adiponectin and adiponectin receptors (AdipoR1 and AdipoR2), while granulosa cells express AdipoR1 and AdipoR2, but not adiponectin. Human recombinant adiponectin increased IGF-I-induced P and E2 production by human granulosa cells through an increase in the IGF-I-induced p450 aromatase protein level [44].

Interleukin-6

IL-6 is an inflammatory cytokine, and approximately 30% of circulating levels are derived from adipose tissue. Circulating IL-6 levels increase in obesity and they are associated with increased insulin resistance. In rats, intracerebroventricular injection of IL-6 inhibits LH secretion [45], although in another study this effect was not replicated [46]. In vitro in rat, IL-6 has been observed to prevent LH-triggered ovulation, inhibit LH-/FSH-induced estradiol production [47], and in human granulose tumor cells to suppress aromatase activity [48]. Furthermore, women with PCOS had elevated serum and follicular IL-6 levels when compared with non-PCOS controls stimulated all for in vitro fertilization (IVF) [49]. It seems that IL-6, in the high levels seen in obese women, may contribute to impaired fertility in women with PCOS.

Plasminogen activator inhibitor type-1

PAI-1 is a regulator of blood-fibrinolytic activity and is mainly produced by white adipose tissue and visceral fat. Circulating PAI-1 levels increase in obesity and correlate with the elements of the metabolic syndrome [50]. Unlike in normal-weight subjects, overweight/obese patients with PCOS had higher PAI-1 levels than BMI-matched controls [51]. PAI-1 has been associated with miscarriage in women with PCOS [52,53].

Resistin

Resistin, a 12.5 kDa polypeptide, is a member of the cysteine-rich secretary proteins called "resistin-like molecules" or "found in inflammatory zone" and is secreted by adipocytes. Resistin is associated with insulin resistance in mice [54]. However, there was no difference of plasma resistin levels between PCOS and control women with or without insulin resistance [55] and probably resistin may not be implicated in infertility in PCOS women.

Tumor necrosis factor-α

TNF-α is synthesized in adipose tissues by adipocytes and other cells in the tissue matrix [50]. Blood levels and adipocyte production of TNF-α correlate with BMI and hyperinsulinemia, and TNF-α impairs insulin action by inhibiting insulin signaling. TNF-α may affect several levels of the reproductive axis: inhibition of gonadotrophin secretion, ovulation, steroidogenesis, corpus luteum regression, and endometrial development [50]. A metaanalysis involving nine studies of circulatory TNF-α levels revealed no statistically significant differences between PCOS and controls [56]. Obese women with PCOS probably have an additional factor, impairing fertility at multiple levels.

Ghrelin

Ghrelin is a 28-amino-acid peptide hormone produced mainly by the stomach and is the endogenous ligand for the growth hormone (GH) secretagogue receptor type 1a. Ghrelin stimulates the secretion of GH, prolactin, and adrenocorticotrophic hormone from pituitary and also increases appetite, promotes food intake, and regulates energy balance via hypothalamic action. Evidence has been provided that ghrelin may affect reproductive function in animals and humans. Plasma ghrelin concentrations have been shown to be lower in obese when compared with normal subjects. In obese women with PCOS, lower ghrelin levels have been found than those expected based on their obesity [57]. Furthermore, obese women with PCOS showed a negative correlation between ghrelin and insulin resistance, while regardless of the presence of PCOS, a marked negative correlation existed between ghrelin and androstenedione levels [57]. Therefore it is possible that in obese women with PCOS, ghrelin may contribute to modification of factors such as insulin resistance and androgens mediating a negative effect on fertility.

Other biomarkers

A number of other substances produced by the fat tissue or by other organs have also been investigated as biomarkers in PCOS, which may or may not be associated with obesity. These include visfatin, vaspin, apelin, retinol-binding protein 4, kisspeptin, copeptin, irisin, and zonulin [58]. However, for some of these substances, data are limited, and further research is needed.

Impact of obesity on infertility treatment

Treatment of anovulatory infertility in women with PCOS involves various modalities of ovulation induction [59]. Although the basis of such treatment is the administration of different drugs, in obese women diet and lifestyle changes are considered the first-line approach [60]. In the case of noncompliance, various treatments or interventions, including clomiphene citrate, gonadotrophins, insulin sensitizers, and laparoscopic ovarian drilling (LOD), are applied. Nevertheless, the treatment outcome during ovulation induction in PCOS may be influenced by the excessive body fat. Of late, bariatric surgery has also gained ground in recent years in the case of infertile women with excessive obesity.

Diet
Lifestyle
Bariatric surgery

Lifestyle modifications are based on diet and exercise and aim at the restoration of the disturbed reproductive function. Weight-loss programs applied to obese patients with PCOS result in the improvement of the abnormal biochemical and hormonal parameters. Especially, a reduction in serum-free testosterone and insulin concentrations and an increase in SHBG values have been reported, while in more than 50% of the cases regular ovulation and menstruation are reestablished [60]. It has been suggested that even 5%–10% decrease in body weight of overweight women with PCOS can be effective [34,61–63]. In such cases a 30% decrease in visceral fat has been estimated [64]. A steady decrease of intraabdominal fat is associated with restoration of ovulation [65]. Although energy-restricted diet is the key factor, information regarding the specific type of exercise that is more effective is limited. It has been suggested that the addition of aerobic resistance exercise to an energy-restricted diet did not further improve reproductive outcomes [66]. In contrast the combination of hypocaloric diet and sibutramine, an oral anorexiant, showed a better effect at 6 months on the weight loss and also led to reduction in androgens levels and insulin resistance in women with PCOS than the diet alone [67]. However, sibutramine has been withdrawn from the market in the majority of the European countries and therefore it is not recommended. Evidence from studies suggests that diets with reduced glycemic load may provide a better control of hyperinsulinemia and the metabolic consequences as well as menstrual cyclicity [68]. However, a recent metaanalysis has not provided clear evidence that lifestyle intervention can have an impact on glucose tolerance, although it improves free androgen index, but there were no studies investigating the effect of this intervention on live birth, miscarriages, or menstrual regularity [69].

It is clear that in the majority of the women, a period of 3–6 months is required to lose 5%–10% of body weight. This might be a hindrance to the women who are very anxious to get pregnant quickly. On the other hand, it is not known if the reduction of weight, via caloric restriction or pharmacological intervention during the periconceptional period, has a negative impact on the conceptus or it is more reasonable to postpone conception until the end of the effort to lose weight. However, taking into account that obesity can adversely affect human reproduction by increasing perinatal and maternal risks, it is advisable for the women to reduce their weight before attempting to conceive [70]. Obese women during pregnancy carry a greater risk for congenital anomalies, miscarriage, gestational diabetes, and hypertension either after spontaneous or after assisted conception [71,72]. An increased miscarriage rate was seen in a retrospective analysis of data in obese women (BMI > 28 kg/m^2) undergoing ovulation induction when compared with normal controls [73]. Similarly, a mixed population of obese women undergoing IVF/ICSI (intracytoplasmic sperm injection) treatment showed an increased miscarriage rate as compared to controls with normal weight [74], while a recent metaanalysis showed that women with a BMI of >25 kg/m^2 had a higher miscarriage rate regardless of the mode of conception [75]. It is evident from this information that weight reduction prior to any intervention in obese women with PCOS would be an advisable approach for the improvement of the treatment outcome. National guidelines in the United Kingdom advise for a weight loss to a BMI of <30 kg/m^2 prior to the start of any treatment for ovulation induction [76].

Bariatric surgery is now widely used for the treatment of human obesity. Recent evidence has demonstrated a marked improvement in fertility in women with PCOS following bariatric surgery [77]. In particular, a significant improvement in reproductive characteristics of PCOS has been reported after 12 months of the operation [78]. Although bariatric surgery may have a positive impact on pregnancy complications related to obesity, new risks may arise related to nutritional deficiencies, anemia, and changes in maternal glucose metabolism [79]. Therefore women undergoing bariatric surgery are advised not to become pregnant for 1−1.5 years after surgery as there is an increased risk of delivering small-for-gestational-age babies [80]. Bariatric surgery is recommended for the women with a BMI of ≥ 35 kg/m^2 [81].

Clomiphene citrate

Clomiphene citrate is an antiestrogenic compound that belongs to the selective estrogen receptor modulators. By binding to the estrogen receptors, clomiphene blocks the negative feedback effect of estrogens on the central nervous system and this leads to an increased secretion of GnRH and gonadotrophins from the hypothalamus and the pituitary, respectively. When this drug is administered orally to women for a few days in the early follicular phase of the cycle, it creates an intercycle type of FSH rise that leads to follicle recruitment selection [82]. The selected dominant follicle then secretes estrogens, while the sequence of hormonal events is similar to that in the normal menstrual cycle resulting in the occurrence of an endogenous LH surge at mid-cycle.

Clomiphene is used as first-line treatment in anovulatory infertile women with PCOS [83]. The protocol for ovulation induction involves the administration of clomiphene at a starting dose of 50 mg/day immediately after a spontaneous period or withdrawal bleeding induced by the administration of progesterone. Detailed analysis of the literature shows that clomiphene treatment leads to ovulation in 70%−86% of the women, while the pregnancy rate is lower (34%−43%) [84]. Nevertheless, in properly selected patients, cumulative pregnancy rates as high as 63% at 6 months and 97% at 10 months have been reported [85,86]. In cases of clomiphene failure or clomiphene resistance, a second-line treatment is used.

Clomiphene resistance is attributed to several hormonal and clinical characteristics of the women. For example, free androgen index, BMI, age, and cycle abnormalities play an important role [87,88]. In a multivariate prediction model, it was shown that decreased insulin sensitivity, hyperandrogenemia, and obesity are associated with reduced response to clomiphene treatment in PCOS [89]. Especially, women with less-reduced insulin sensitivity had a higher possibility of ovulating on clomiphene treatment [90], while obesity had a negative impact on the treatment outcome with clomiphene [91]. Obese women with PCOS respond less well to clomiphene, and the chance of ovulation is reduced particularly in women with amenorrhea as compared to those with oligomenorrhea [88]. Such women may need higher dosages of clomiphene even up to 250 mg/day, although the evidence is limited to retrospective data [92]. It has been reported that patients who are expected to ovulate and become pregnant on clomiphene have significantly lower BMI than those who remain anovulatory during treatment [90]. A recent study has shown that in women with a BMI of ≥ 23 kg/m^2, clomiphene-resistant PCOS occurred more frequently than that in lean women [93]. It has been proposed that leptin produced by the adipose tissue is a more direct index of ovarian dysfunction in PCOS that can predict women remaining anovulatory during treatment with clomiphene citrate [89]. Clinical data have demonstrated a negative impact of high serum AMH levels in PCOS on ovarian response to clomiphene treatment [94]. The relationship between AMH and BMI in PCOS is rather obscure because, on the one hand, AMH levels correlate negatively with BMI [95], and on the other hand, high AMH levels in overweight/obese women with PCOS decrease after diet but not after exercise possibly via the reduction in free testosterone levels [96]. Furthermore, bariatric surgery normalizes high AMH levels in obese women with or without PCOS [97].

Aromatase inhibitors

Aromatase inhibitors are drugs that are given orally to women suffering from breast cancer. These compounds inhibit the action of the enzyme aromatase, which converts androgens into estrogens. Consequently, the reduced production of estrogens and the reduced circulating levels of these steroids lead to the attenuation of the negative feedback and the increase in the secretion of gonadotrophins from the pituitary. For these reasons, aromatase inhibitors can be used for ovulation induction in PCOS. Letrozole is one of the third-generation aromatase inhibitors used more extensively than others for the treatment of infertility. Based on evidence derived from a metaanalysis, letrozole is equally effective compared with clomiphene in inducing ovulation in naïve women with PCOS [98]. Nevertheless, recent evidence has demonstrated that letrozole is superior to clomiphene in terms of cumulative live birth and is recommended as first-line treatment for ovulation induction in anovulatory women with PCOS [99–101]. It should be noted, however, that letrozole is still considered an "off-label" medication for infertility treatment due to possible teratogenic effects in pregnancy, although this has been debated [83,102]. It is advisable, therefore, to discuss with the patients the possible risks and benefits. Regarding the role of BMI in the treatment outcome during ovulation induction with letrozole as compared to clomiphene in women with PCOS, a higher cumulative live-birth rate was found with letrozole [99]. Nevertheless, no clear evidence was provided that relative efficacy differed according to BMI tertile, although in extremely obese women (BMI $>$ 39.4 kg/m^2), lower cumulative live-birth rate was found as compared to women with a BMI of $\leq$ 30.3 kg/m^2 with either of the two drugs [99].

Follicle-stimulating hormone

In the early 1980s, low-dose protocols of human menopausal gonadotrophin or FSH were introduced for ovulation induction in PCOS as a second-line treatment in women with clomiphene failure or resistance. Low dosages were adopted in order to stimulate single follicle maturation and thus to prevent multiple pregnancies and the ovarian hyperstimulation syndrome (OHSS). Two protocols, the step-up and the step-down, are in use. In the *step-up protocol* the starting dose was initially 75 IU FSH/day, but it was subsequently reduced to 50 IU [103]. A long period of 2 weeks on the starting dose has been adopted with an increase by 25 IU every week if there is no ovarian response. In the *step-down protocol* the starting dose was initially 150 IU, but it is now 100 IU, a few days later, following the selection of a dominant follicle, reducing down to 75 IU and then to 50 IU [104,105]. With the step-up protocol, monofollicular development is achieved in about 70% of the cycles, while there is a low rate of multiple pregnancies ($\sim$6%) and the OHSS ($<$1%) [106]. Both protocols are equally effective in terms of pregnancy rate, although with the step-up protocol higher monofollicular development and lower hyperstimulation are achieved [105]. The treatment is monitored only by ultrasound scans of the ovaries, while estradiol measurement is not required. When clomiphene and FSH were considered consecutive treatments in women with no other cause of infertility, a cumulative pregnancy rate at 12 months of 91% and a live-birth rate at 24 months of 71% were reported [107,108]. No difference in clinical outcome has been found between urinary-derived gonadotrophins and recombinant FSH preparations in anovulatory women with PCOS [109].

The effectiveness of the treatment highly depends on various parameters, including BMI. An earlier study in PCOS women with clomiphene resistance has shown that during ovulation induction with low-dose gonadotrophins, moderate obesity was associated with a lower ovulation rate and a higher miscarriage rate, although the proportion of women who had at least one pregnancy was similar with that of women with normal BMI [73]. In the same study, significantly higher doses of gonadotrophins were required in the group of the obese than in the group of the lean women, a finding that was confirmed in a subsequent study [110]. A more recent study, including women with PCOS and increased BMI but $<$35 kg/m^2, showed that obesity was associated with higher insulin resistance and free androgen index and a higher number of immature follicles [111]. Although a higher FSH threshold for ovarian stimulation was noted and a greater total dose of gonadotrophins with a longer duration of stimulation was needed, careful monitoring was necessary due to the increased risk of overresponse [111]. In the context of the step-up protocol, efforts have been made to calculate the individual FSH response dose based on various screening characteristics, including BMI [112]. However, due to a rather complicated prediction model based on a mathematical equation, such an approach has not been proven reliably effective. Similarly, the prediction of the individual effective dose of FSH using several screening characteristics such as BMI, clomiphene resistance or failure, free IGF-I, and FSH has also been attempted in the context of the step-down protocol [113]. Despite the initial optimism with both protocols, it was later shown that in the step-up protocol, the predicted FSH dose was higher than the observed response dose [114]. Similar to clomiphene, the ovarian response to gonadotrophins decreases with increasing AMH levels [115].

Laparoscopic ovarian drilling

LOD is used as a second-line treatment competing with FSH in women with PCOS and clomiphene resistance.

Retrospective data have shown high ovulation and pregnancy rates [116]. A previous metaanalysis has demonstrated no advantage of LOD over FSH regarding clinical pregnancy, live birth, and miscarriage rates except for a significantly lower multiple pregnancy rate [117]. A subsequent study has compared prospectively LOD with clomiphene as a first-line treatment in PCOS but showed no difference between the two treatment modalities regarding the cumulative pregnancy rate at 12 months [118]. There are now certain indications for LOD, such as in clomiphene-resistant patients, particularly those with persistently elevated LH, in the case of laparoscopic assessment of the pelvis for infertility problems and if the patients are unable to visit the hospital regularly. It should be emphasized, however, that LOD must not be used for nonfertility indications. Regarding the influence of obesity in the effectiveness of LOD, a retrospective study, including 200 patients with PCOS, who were treated unsuccessfully with clomiphene, showed that LOD applied to women with a BMI of ≥ 35 kg/m^2 induced significantly lower ovulation and pregnancy rates as compared to moderately overweight and normal-weight women [119]. Nevertheless, once ovulation was achieved, BMI had no influence on the conception rates in agreement with previous reports [116,119,120]. Several studies have shown a negative correlation between BMI and the response to various medical treatments, including clomiphene [92,121] and gonadotrophins [122]. However, LOD has been shown to sensitize clomiphene-resistant patients to this drug. According to a review article, parameters that may predict poor reproductive outcome following LOD include obesity (BMI > 25 kg/m^2), duration of infertility greater than 3 years, basal serum LH below 10 IU/L, marked biochemical hyperandrogenism, and serum AMH ≥ 7.7 ng/mL [123].

Ovarian drilling can be also performed by fertiloscopy, which provides as good results as laparoscopy, but experience is needed because the risk of complications is greater [124].

Insulin sensitizers

To alleviate insulin resistance in women with PCOS, insulin-sensitizing drugs are currently used. Metformin is the main representative and has been used more extensively than other compounds with similar actions. Although metformin is superior to placebo in inducing ovulation, it is not considered an ovulation-inducing agent, since it only increases the number of spontaneous ovulations in PCOS patients with oligomenorrhea from one ovulation to two ovulations every 5 months [125]. Prospective randomized trials have shown that metformin used for the treatment of anovulatory infertility in PCOS is inferior to clomiphene regarding live-birth rate, while the addition of metformin to clomiphene has no advantages over clomiphene alone [126,127]. Two metaanalyses have demonstrated that metformin in combination with clomiphene might be useful only in cases of clomiphene resistance before moving to the second-line treatment of low-dose FSH protocols [128,129].

It is known that about 40% of patients with PCOS are obese. Insulin resistance is one of the main characteristics of these women. Treatment with metformin would be expected to improve insulin sensitivity and the metabolic and reproductive functions. Several studies have demonstrated that obese but not lean women with PCOS may benefit from the treatment with metformin [130–133]. Nevertheless, metformin administration over a period of 6 months in the context of a diet and lifestyle changes program was not better than placebo regarding body-weight reduction [134]. A subsequent metaanalysis confirmed the initial observations that metformin treatment alone resulted in a greater reduction in BMI than placebo, but when added to a diet program, it did not show any advantage over placebo [135]. A more recent study, however, has shown that with metformin, the weight loss was greater than that with lifestyle changes, although there was a high dropout rate [136].

It is clear that when PCOS patients are treated in the context of an ovulation induction program, body weight and BMI are important. Although it would be expected metformin to be useful in obese women, it has been shown that women with a lower BMI as compared to those with a higher BMI had a higher possibility of becoming pregnant on this drug [137]. In that respect, metformin could be used as a first-line treatment in women with a BMI of <32 kg/m^2, since both the clinical pregnancy and live-birth rates did not differ from those achieved with clomiphene alone [138]. The most recent guidelines suggest that metformin can be used alone for ovulation induction, but it is less effective than other agents and if this drug is being used in women with a BMI of ≥ 30 kg/m^2, clomiphene could be added [101]. The same guidelines recommend the use of metformin in combination with clomiphene in clomiphene-resistant patients [101], although a combination of metformin with FSH might also be useful in such a group of patients [139].

In vitro fertilization

When the abovementioned treatment modalities fail to induce a pregnancy, women with PCOS are treated in the context of an IVF program even without any specific indication for this method. Although the number of oocytes retrieved is usually higher in PCOS as compared to controls, there is no significant difference in fertilization, pregnancy, and live-birth rates [140]. Furthermore, in

IVF/ICSI cycles, obesity and PCOS were found to independently decrease the size of the oocytes [141], also affecting egg quality and endometrial, receptivity [142]. Overall, in women with PCOS, IVF outcome is worse in those with high BMI as compared to women with normal or low BMI [143], although this issue needs further investigation [144]. The ideal protocol for ovarian stimulation in PCOS for IVF has not been identified. However, obese women with PCOS require higher amounts of FSH for ovarian stimulation [125]. On the other hand, the addition of metformin at any stage of the procedure had no benefits regarding pregnancy and live-birth rates [145,146], except for a significant reduction in the risk for the OHSS [146]. In any case, for ovarian stimulation for IVF, a GnRH-antagonist protocol combined with GnRH agonist triggering is preferred with the need to freeze all embryos [147]. Liraglutide, a glucagon-like peptide-1 receptor agonist, commonly used to reduce weight in obese patients, has been shown to increase pregnancy rate in obese women with PCOS after IVF and embryo transfer, following combined treatment with metformin for 12 weeks in a pilot randomized study [148].

References

[1] Zawadski JK, Dunaif A. Diagnostic criteria for polycystic ovary syndrome; towards a rational approach. In: Dunaif A, Givens JR, Haseltine F, editors. Polycystic ovary syndrome. Boston, MA: Blackwell Scientific; 1992. p. 377–84.

[2] The Rotterdam ESHRE/ASRM-Sponsored PCOS Consensus Workshop Group. Revised 2003 consensus on diagnostic criteria and long-term health risks related to polycystic ovary syndrome (PCOS). Hum Reprod 2004;19:41–7.

[3] The Amsterdam ESHRE/ASRM-Sponsored Third PCOS Consensus Workshop Group. Consensus on women's health aspects of polycystic ovary syndrome (PCOS). Hum Reprod 2012;27:14–24.

[4] Azziz R, Carmina E, Dewailly D, Diamanti-Kandarakis E, Escobar-Morreale HF, Futterweit W. The Androgen Excess and PCOS Society criteria for the polycystic ovary syndrome: the complete task force report. Fertil Steril 2009;91:456–88.

[5] Skiba MA, Islam RM, Bell RJ, Davis SR. Understanding variation in prevalence estimates of polycystic ovary syndrome: a systematic review and meta-analysis. Hum Reprod Update 2018;24:694–709.

[6] Homburg R. Polycystic ovary syndrome. Best Pract Res Clin Obstet Gynaecol 2008;22:261–74.

[7] Franks S. Polycystic ovary syndrome. Trends Endocrinol Metab 1989;1:60–3.

[8] Cupisti S, Kajaia N, Dittrich R, Duezenli H, Beckmann MW, Mueller A. Body mass index and ovarian function are associated with endocrine and metabolic abnormalities in women with hyperandrogenic syndrome. Eur J Endocrinol 2008;158:711–19.

[9] Mokdad AH, Ford ES, Bowman BA, Dietz WH, Vinicor F, Bales VS. Prevalence of obesity, diabetes, and obesity-related health risk factors, 2001. JAMA 2003;289:76–9.

[10] Carmina E, Bucchieri S, Esposito A, Del Puente A, Mansueto P, Orio F. Abdominal fat quantity and distribution in women with polycystic ovary syndrome and extent of its relation to insulin resistance. J Clin Endocrinol Metab 2007;92:2500–5.

[11] Manneråˊs-Holm L, Leonhardt H, Kullberg J, Jennische E, Odén A, Holm G. Adipose tissue has aberrant morphology and function in PCOS: enlarged adipocytes and low serum adiponectin, but not circulating sex steroids, are strongly associated with insulin resistance. J Clin Endocrinol Metab 2011;96:E304–11.

[12] Escobar-Morreale HF, Botella-Carretero JI, Alvarez-Blasco F, Sancho J, San Millán JL. The polycystic ovary syndrome associated with morbid obesity may resolve after weight loss induced by bariatric surgery. J Clin Endocrinol Metab 2005;90:6364–9.

[13] Asunción M, Calvo RM, San Millán JL, Sancho J, Avila S, Escobar-Morreale HF. A prospective study of the prevalence of the polycystic ovary syndrome in unselected Caucasian women from Spain. J Clin Endocrinol Metab 2000;85:2434–8.

[14] Brewer CJ, Balen AH. The adverse effects of obesity on conception and implantation. Reproduction 2010;140:347–64.

[15] Kiddy DS, Sharp PS, White DM, Scanlon MF, Mason HD, Bray CS. Differences in clinical and endocrine features between obese and non-obese subjects with polycystic ovary syndrome: an analysis of 263 consecutive cases. Clin Endocrinol (Oxf) 1990;32:213–20.

[16] Roberts JS, Perets RA, Sarfert KS, Bowman JJ, Ozark PA, Whitworth GB, et al. High-fat high-sugar diet induces polycystic ovary syndrome in a rodent model. Biol Reprod 2017; 10.1095/biolreprod.116.142786.

[17] Glueck CJ, Goldenberg N. Characteristics of obesity in polycystic ovary syndrome: etiology, treatment, and genetics. Metabolism 2019;92:108–20.

[18] Pradeep PK, Li X, Peegel H, Menon KM. Dihydrotestosterone inhibits granulosa cell proliferation by decreasing the cyclin D2 mRNA expression and cell cycle arrest at G1 phase. Endocrinology 2002;143:2930–5.

[19] Jakimiuk AJ, Weitsman SR, Magoffin DA. 5Alpha-reductase activity in women with polycystic ovary syndrome. J Clin Endocrinol Metab 1999;84:2414–18.

[20] Holte J, Bergh T, Gennarelli G, Wide L. The independent effects of polycystic ovary syndrome and obesity on serum concentrations of gonadotrophins and sex steroids in premenopausal women. Clin Endocrinol (Oxf) 1994;41:473–81.

[21] Gambineri A, Pelusi C, Vicennati V, Pagotto U, Pasquali R. Obesity and the polycystic ovary syndrome. Int J Obes Relat Metab Disord 2002;26:883–96.

[22] Taylor AE, McCourt B, Martin KA, Anderson EJ, Adams JM, Schoenfeld D. Determinants of abnormal gonadotropin secretion in clinically defined women with polycystic ovary syndrome. J Clin Endocrinol Metab 1997;82:2248–56.

[23] Dafopoulos K, Venetis C, Pournaras S, Kallitsaris A, Messinis IE. Ovarian control of pituitary sensitivity of luteinizing hormone secretion to gonadotropin-releasing hormone in women with the polycystic ovary syndrome. Fertil Steril 2009;92:1378–80.

[24] Eagleson CA, Gingrich MB, Pastor CL, Arora TK, Burt CM, Evans WS. Polycystic ovarian syndrome: evidence that flutamide restores sensitivity of the gonadotropin-releasing hormone pulse generator to inhibition by estradiol and progesterone. J Clin Endocrinol Metab 2000;85:4047–52.

[25] McCartney CR, Marshall JC. CLINICAL PRACTICE. Polycystic ovary syndrome. N Engl J Med 2016;375:54–64.

[26] Pagán YL, Srouji SS, Jimenez Y, Emerson A, Gill S, Hall JE. Inverse relationship between luteinizing hormone and body mass index in polycystic ovarian syndrome: investigation of hypothalamic and pituitary contributions. J Clin Endocrinol Metab 2006;91:1309–16.

[27] Lawson MA, Jain S, Sun S, Patel K, Malcolm PJ, Chang RJ. Evidence for insulin suppression of baseline luteinizing hormone in women with polycystic ovarian syndrome and normal women. J Clin Endocrinol Metab 2008;93:2089–96.

[28] Laughlin GA, Morales AJ, Yen SS. Serum leptin levels in women with polycystic ovary syndrome: the role of insulin resistance/hyperinsulinemia. J Clin Endocrinol Metab 1997;82:1692–6.

[29] Van Dam EW, Roelfsema F, Veldhuis JD, Helmerhorst FM, Frölich M, Meinders AE. Increase in daily LH secretion in response to short-term calorie restriction in obese women with PCOS. Am J Physiol Endocrinol Metab 2002;282:E865–72.

[30] Hillier SG. Current concepts of the roles of follicle stimulating hormone and luteinizing hormone in folliculogenesis. Hum Reprod 1994;9:188–91.

[31] Willis DS, Watson H, Mason HD, Galea R, Brincat M, Franks S. Premature response to luteinizing hormone of granulosa cells from anovulatory women with polycystic ovary syndrome: relevance to mechanism of anovulation. J Clin Endocrinol Metab 1998;83:3984–91.

[32] McNatty KP, Smith DM, Makris A, Osathanondh R, Ryan KJ. The microenvironment of the human antral follicle: interrelationships among the steroid levels in antral fluid, the population of granulosa cells, and the status of the oocyte in vivo and in vitro. J Clin Endocrinol Metab 1979;49:851–60.

[33] Poretsky L, Cataldo NA, Rosenwaks Z, Giudice LC. The insulin-related ovarian regulatory system in health and disease. Endocr Rev 1999;20:535–82.

[34] Clark AM, Ledger W, Galletly C, Tomlinson L, Blaney F, Wang X. Weight loss results in significant improvement in pregnancy and ovulation rates in anovulatory obese women. Hum Reprod 1995;10:2705–12.

[35] Lee YH, Nair S, Rousseau E, Allison DB, Page GP, Tataranni PA. Microarray profiling of isolated abdominal subcutaneous adipocytes from obese vs non-obese Pima Indians: increased expression of inflammation-related genes. Diabetologia 2005;48:1776–83.

[36] Cortón M, Botella-Carretero JI, Benguría A, Villuendas G, Zaballos A, San Millán JL. Differential gene expression profile in omental adipose tissue in women with polycystic ovary syndrome. J Clin Endocrinol Metab 2007;92:328–37.

[37] Messinis IE, Milingos SD. Leptin in human reproduction. Hum Reprod Update 1999;5:52–63.

[38] Liu RB, Liu Y, Lv LQ, Xiao W, Gong C, Yue JX. Effects of metformin treatment on soluble leptin receptor levels in women with polycystic ovary syndrome. Curr Med Sci 2019;39:609–14.

[39] Roelfsema F, Kok P, Veldhuis JD, Pijl H. Altered multihormone synchrony in obese patients with polycystic ovary syndrome. Metabolism 2011;60:1227–33.

[40] Karamouti M, Kollia P, Kallitsaris A, Vamvakopoulos N, Kollios G, Messinis IE. Modulating effect of leptin on basal and follicle stimulating hormone stimulated steroidogenesis in cultured human lutein granulosa cells. J Endocrinol Invest 2009;32:415–19.

[41] Duggal PS, Van Der Hoek KH, Milner CR, Ryan NK, Armstrong DT, Magoffin DA. The in vivo and in vitro effects of exogenous leptin on ovulation in the rat. Endocrinology 2000;141:1971–6.

[42] Srivastava PK, Krishna A. Increased circulating leptin level inhibits folliculogenesis in vespertilionid bat, Scotophilus heathii. Mol Cell Endocrinol 2011;337:24–35.

[43] Toulis KA, Goulis DG, Farmakiotis D, Georgopoulos NA, Katsikis I, Tarlatzis BC. Adiponectin levels in women with polycystic ovary syndrome: a systematic review and a meta-analysis. Hum Reprod Update 2009;15:297–307.

[44] Chabrolle C, Tosca L, Ramé C, Lecomte P, Royère D, Dupont J. Adiponectin increases insulin-like growth factor I-induced progesterone and estradiol secretion in human granulosa cells. Fertil Steril 2009;92:1988–96.

[45] Rivier C, Vale W. Cytokines act within the brain to inhibit luteinizing hormone secretion and ovulation in the rat. Endocrinology 1990;127:849–56.

[46] Watanobe H, Hayakawa Y. Hypothalamic interleukin-1 beta and tumor necrosis factor-alpha, but not interleukin-6, mediate the endotoxin-induced suppression of the reproductive axis in rats. Endocrinology 2003;144:4868–75.

[47] Mikuni M. Effect of interleukin-2 and interleukin-6 on ovary in the ovulatory period – establishment of the new ovarian perfusion system and influence of interleukins on ovulation rate and steroid secretion. Hokkaido Igaku Zasshi 1995;70:561–72.

[48] Deura I, Harada T, Taniguchi F, Iwabe T, Izawa M, Terakawa N. Reduction of estrogen production by interleukin-6 in a human granulosa tumor cell line may have implications for endometriosis-associated infertility. Fertil Steril 2005;83(Suppl. 1):1086–92.

[49] Amato G, Conte M, Mazziotti G, Lalli E, Vitolo G, Tucker AT. Serum and follicular fluid cytokines in polycystic ovary syndrome during stimulated cycles. Obstet Gynecol 2003;101:1177–82.

[50] Gosman GG, Katcher HI, Legro RS. Obesity and the role of gut and adipose hormones in female reproduction. Hum Reprod Update 2006;12:585–601.

[51] Koiou E, Tziomalos K, Dinas K, Katsikis I, Kandaraki EA, Tsourdi E. Plasma plasminogen activator inhibitor-1 levels in the different phenotypes of the polycystic ovary syndrome. Endocr J 2012;59:21–9.

[52] Glueck CJ, Wang P, Fontaine RN, Sieve-Smith L, Tracy T, Moore SK. Plasminogen activator inhibitor activity: an independent risk factor for the high miscarriage rate during pregnancy in women with polycystic ovary syndrome. Metabolism 1999;48:1589–95.

[53] Glueck CJ, Sieve L, Zhu B, Wang P. Plasminogen activator inhibitor activity, 4G5G polymorphism of the plasminogen activator inhibitor 1 gene, and first-trimester miscarriage in women with polycystic ovary syndrome. Metabolism 2006;55:345–52.

[54] Steppan CM, Bailey ST, Bhat S, Brown EJ, Banerjee RR, Wright CM. The hormone resistin links obesity to diabetes. Nature 2001;409:307–12.

[55] Zhang J, Zhou L, Tang L, Xu L. The plasma level and gene expression of resistin in polycystic ovary syndrome. Gynecol Endocrinol 2011;27:982–7.

[56] Escobar-Morreale HF, Luque-Ramírez M, González F. Circulating inflammatory markers in polycystic ovary syndrome: a systematic review and metaanalysis. Fertil Steril 2011;95:1048–58.

[57] Pagotto U, Gambineri A, Vicennati V, Heiman ML, Tschöp M, Pasquali R. Plasma ghrelin, obesity, and the polycystic ovary syndrome: correlation with insulin resistance and androgen levels. J Clin Endocrinol Metab 2002;87:5625–9.

[58] Polak K, Czyzyk A, Simoncini T, Meczekalski B. New markers of insulin resistance in polycystic ovary syndrome. J Endocrinol Invest 2017;40:1−8.

[59] Messinis IE. Ovulation induction: a mini review. Hum Reprod 2005;20:2688−97.

[60] Hoeger KM. Role of lifestyle modification in the management of polycystic ovary syndrome. Best Pract Res Clin Endocrinol Metab 2006;20:293−310.

[61] Kiddy DS, Hamilton-Fairley D, Seppälä M, Koistinen R, James VH, Reed MJ. Diet-induced changes in sex hormone binding globulin and free testosterone in women with normal or polycystic ovaries: correlation with serum insulin and insulin-like growth factor-I. Clin Endocrinol (Oxf) 1989;31:757−63.

[62] Clark AM, Thornley B, Tomlinson L, Galletley C, Norman RJ. Weight loss in obese infertile women results in improvement in reproductive outcome for all forms of fertility treatment. Hum Reprod 1998;13:1502−5.

[63] van Dam EW, Roelfsema F, Veldhuis JD, Hogendoorn S, Westenberg J, Helmerhorst FM. Retention of estradiol negative feedback relationship to LH predicts ovulation in response to caloric restriction and weight loss in obese patients with polycystic ovary syndrome. Am J Physiol Endocrinol Metab 2004;286: E615−20.

[64] Després JP, Lemieux I, Prud'homme D. Treatment of obesity: need to focus on high risk abdominally obese patients. BMJ 2001;322:716−20.

[65] Kuchenbecker WK, Groen H, van Asselt SJ, Bolster JH, Zwerver J, Slart RH. In women with polycystic ovary syndrome and obesity, loss of intra-abdominal fat is associated with resumption of ovulation. Hum Reprod 2011;26:2505−12.

[66] Thomson RL, Buckley JD, Noakes M, Clifton PM, Norman RJ, Brinkworth GD. The effect of a hypocaloric diet with and without exercise training on body composition, cardiometabolic risk profile, and reproductive function in overweight and obese women with polycystic ovary syndrome. J Clin Endocrinol Metab 2008;93:3373−80.

[67] Florakis D, Diamanti-Kandarakis E, Katsikis I, Nassis GP, Karkanaki A, Georgopoulos N. Effect of hypocaloric diet plus sibutramine treatment on hormonal and metabolic features in overweight and obese women with polycystic ovary syndrome: a randomized, 24-week study. Int J Obes (Lond) 2008;32:692−9.

[68] Marsh KA, Steinbeck KS, Atkinson FS, Petocz P, Brand-Miller JC. Effect of a low glycemic index compared with a conventional healthy diet on polycystic ovary syndrome. Am J Clin Nutr 2010;92:83−92.

[69] Lim SS, Hutchison SK, Van Ryswyk E, Norman RJ, Teede HJ, Moran LJ. Lifestyle changes in women with polycystic ovary syndrome. Cochrane Database Syst Rev 2019;(3):CD007506 10.1002/14651858.CD007506.pub2.

[70] Nelson SM, Fleming RF. The preconceptual contraception paradigm: obesity and infertility. Hum Reprod 2007;22:912−15.

[71] Cedergren MI. Maternal morbid obesity and the risk of adverse pregnancy outcome. Obstet Gynecol 2004;103:219−24.

[72] Linné Y. Effects of obesity on women's reproduction and complications during pregnancy. Obes Rev 2004;5:137−43.

[73] Hamilton-Fairley D, Kiddy D, Watson H, Paterson C, Franks S. Association of moderate obesity with a poor pregnancy outcome in women with polycystic ovary syndrome treated with low dose gonadotrophin. Br J Obstet Gynaecol 1992;99:128−31.

[74] Fedorcsák P, Dale PO, Storeng R, Ertzeid G, Bjercke S, Oldereid N. Impact of overweight and underweight on assisted reproduction treatment. Hum Reprod 2004;19:2523−8.

[75] Metwally M, Ong KJ, Ledger WL, Li TC. Does high body mass index increase the risk of miscarriage after spontaneous and assisted conception? A meta-analysis of the evidence. Fertil Steril 2008;90:714−26.

[76] National Institute for Clinical Excellence (NICE)/National Collaborating Centre for Women's and Children's Health. Fertility: assessment and treatment for people with fertility problems. London: RCOG Press; 2004.

[77] Casimiro I, Sam S, Brady MJ. Endocrine implications of bariatric surgery: a review on the intersection between incretins, bone, and sex hormones. Physiol Rep 2019;7(10):e14111.

[78] Skubleny D, Switzer NJ, Gill RS, Dykstra M, Shi X, Sagle MA, et al. The impact of bariatric surgery on polycystic ovary syndrome: a systematic review and meta-analysis. Obes Surg 2016;26:169−76.

[79] Falcone V, Stopp T, Feichtinger M, Kiss H, Eppel W, Husslein PW, et al. **Pregnancy** after bariatric surgery: a narrative literature review and discussion of impact on **pregnancy** management and outcome. BMC Pregnancy Childbirth 2018;18:507 10.1186/s12884-018-2124-3.

[80] Johansson K, Stephansson O, Neovius M. Outcomes of pregnancy after bariatric surgery. N Engl J Med 2015;372:2267.

[81] Balen AH, Morley LC, Misso M, Franks S, Legro RS, Wijeyaratne CN, et al. The management of anovulatory infertility in women with polycystic ovary syndrome: an analysis of the evidence to support the development of global WHO guidance. Hum Reprod Update 2016;22:687−708.

[82] Adashi EY. Clomiphene citrate-initiated ovulation: a clinical update. Semin Reprod Endocrinol 1986;4:255−76.

[83] Thessaloniki ESHRE/ASRM-Sponsored PCOS Consensus Workshop Group. Consensus on infertility treatment related to polycystic ovary syndrome. Hum Reprod 2008;23:462−77.

[84] Messinis IE. Clomiphene citrate. In: Tarlatzis B, editor. Ovulation induction. Paris: Elsevier; 2002. p. 87−97.

[85] Hammond MG, Halme JK, Talbert LM. Factors affecting the pregnancy rate in clomiphene citrate induction of ovulation. Obstet Gynecol 1983;62:196−202.

[86] Kousta E, White DM, Franks S. Modern use of clomiphene citrate in induction of ovulation. Hum Reprod Update 1997;3:359−65.

[87] Imani B, Eijkemans MJ, te Velde ER, Habbema JD, Fauser BC. Predictors of patients remaining anovulatory during clomiphene citrate induction of ovulation in normogonadotropic oligoamenorrheic infertility. J Clin Endocrinol Metab 1998;83:2361−5.

[88] Imani B, Eijkemans MJ, te Velde ER, Habbema JD, Fauser BC. A nomogram to predict the probability of live birth after clomiphene citrate induction of ovulation in normogonadotropic oligoamenorrheic infertility. Fertil Steril 2002;77:91−7.

[89] Imani B, Eijkemans MJ, de Jong FH, Payne NN, Bouchard P, Giudice LC. Free androgen index and leptin are the most prominent endocrine predictors of ovarian response during clomiphene citrate induction of ovulation in normogonadotropic oligoamenorrheic infertility. J Clin Endocrinol Metab 2000;85:676−82.

[90] Palomba S, Falbo A, Orio F, Tolino A, Zullo A. Efficacy predictors for metformin and clomiphene citrate treatment in anovulatory

infertile patients with polycystic ovary syndrome. Fertil Steril 2009;91:2557—67.
[91] Al-Azemi M, Omu FE, Omu AE. The effect of obesity on the outcome of infertility management in women with polycystic ovary syndrome. Arch Gynecol Obstet 2004;270:205—10.
[92] Dickey RP, Taylor SN, Curole DN, Rye PH, Lu PY, Pyrzak R. Relationship of clomiphene dose and patient weight to successful treatment. Hum Reprod 1997;12:449—53.
[93] Sachdeva G, Gainder S, Suri V, Sachdeva N, Chopra S. Obese and non-obese polycystic ovarian syndrome: comparison of clinical, metabolic, hormonal parameters, and their differential response to clomiphene. Indian J Endocrinol Metab 2019;23:257—62.
[94] Mahran A, Abdelmeged A, El-Adawy AR, Eissa MK, Shaw RW, Amer SA. The predictive value of circulating anti-Müllerian hormone in women with polycystic ovarian syndrome receiving clomiphene citrate: a prospective observational study. J Clin Endocrinol Metab 2013;98:4170—5.
[95] Sova H, Unkila-Kallio L, Tiitinen A, Hippeläinen M, Perheentupa A, Tinkanen H, et al. Hormone profiling, including anti-Müllerian hormone (AMH), for the diagnosis of polycystic ovary syndrome (PCOS) and characterization of PCOS phenotypes. Gynecol Endocrinol 2019;35:595—600.
[96] Nybacka Å, Carlström K, Fabri F, Hellström PM, Hirschberg AL. Serum anti-Müllerian hormone in response to dietary management and/or physical exercise in overweight/obese women with polycystic ovary syndrome: secondary analysis of a randomized controlled trial. Fertil Steril 2013;100:1096—102.
[97] Chiofalo F, Ciuoli C, Formichi C, Selmi F, Forleo R, Neri O, et al. Bariatric surgery reduces serum anti-Mullerian hormone levels in obese women with and without polycystic ovarian syndrome. Obes Surg 2017;27:1750—4.
[98] Requena A, Herrero J, Landeras J, Navarro E, Neyro JL, Salvador C. Use of letrozole in assisted reproduction: a systematic review and meta-analysis. Hum Reprod Update 2008;14:571—82.
[99] Legro RS, Brzyski RG, Diamond MP, Coutifaris C, Schlaff WD, Casson P, , et al.NICHD Reproductive Medicine Network Letrozole versus clomiphene for infertility in the polycystic ovary syndrome. N Engl J Med 2014;371:119—29.
[100] Franik S, Eltrop SM, Kremer JA, Kiesel L, Farquhar C. Aromatase inhibitors (letrozole) for subfertile women with polycystic ovary syndrome. Cochrane Database Syst Rev 2018;5:CD010287.
[101] Teede HJ, Misso ML, Costello MF, Dokras A, Laven J, Moran L, , et al.International PCOS Network Recommendations from the international evidence-based guideline for the assessment and management of polycystic ovary syndrome. Hum Reprod 2018;33:1602—18.
[102] Elizur SE, Tulandi T. Drugs in infertility and fetal safety. Fertil Steril 2008;89:1595—602.
[103] White DM, Polson DW, Kiddy D, Sagle P, Watson H, Gilling-Smith C. Induction of ovulation with low-dose gonadotropins in polycystic ovary syndrome: an analysis of 109 pregnancies in 225 women. J Clin Endocrinol Metab 1996;81:3821—4.
[104] Fauser BC, Donderwinkel P, Schoot DC. The step-down principle in gonadotrophin treatment and the role of GnRH analogues. Baillieres Clin Obstet Gynaecol 1993;7:309—30.
[105] Christin-Maitre S, Hugues JN, Recombinant FSH Study Group. A comparative randomized multicentric study comparing the step-up versus step-down protocol in polycystic ovary syndrome. Hum Reprod 2003;18:1626—31.
[106] Homburg R, Howles CM. Low-dose FSH therapy for anovulatory infertility associated with polycystic ovary syndrome: rationale, results, reflections and refinements. Hum Reprod Update 1999;5:493—9.
[107] Messinis IE, Milingos SD. Current and future status of ovulation induction in polycystic ovary syndrome. Hum Reprod Update 1997;3:235—53.
[108] Eijkemans MJ, Imani B, Mulders AG, Habbema JD, Fauser BC. High singleton live birth rate following classical ovulation induction in normogonadotrophic anovulatory infertility (WHO 2). Hum Reprod, 18. 2003. p. 2357—62.
[109] Weiss NS, Kostova E, Nahuis M, Mol BWJ, van der Veen F, van Wely M. Gonadotrophins for ovulation induction in women with polycystic ovary syndrome. Cochrane Database Syst Rev 2019;1:CD010290.
[110] Loh S, Wang JX, Matthews CD. The influence of body mass index, basal FSH and age on the response to gonadotrophin stimulation in non-polycystic ovarian syndrome patients. Hum Reprod 2002;17:1207—11.
[111] Balen AH, Dresner M, Scott EM, Drife JO. Should obese women with polycystic ovary syndrome receive treatment for infertility? BMJ 2006;332:434—5.
[112] Imani B, Eijkemans MJ, Faessen GH, Bouchard P, Giudice LC, Fauser BC. Prediction of the individual follicle-stimulating hormone threshold for gonadotropin induction of ovulation in normogonadotropic anovulatory infertility: an approach to increase safety and efficiency. Fertil Steril 2002;77:83—90.
[113] van Santbrink EJ, Eijkemans MJ, Macklon NS, Fauser BC. FSH response-dose can be predicted in ovulation induction for normogonadotropic anovulatory infertility. Eur J Endocrinol 2002;147:223—6.
[114] van Wely M, Fauser BC, Laven JS, Eijkemans MJ, van der Veen F. Validation of a prediction model for the follicle-stimulating hormone response dose in women with polycystic ovary syndrome. Fertil Steril 2006;86:1710—15.
[115] Amer SA, Mahran A, Abdelmaged A, El-Adawy AR, Eissa MK, Shaw RW. The influence of circulating anti-Müllerian hormone on ovarian responsiveness to ovulation induction with gonadotrophins in women with polycystic ovarian syndrome: a pilot study. Reprod Biol Endocrinol 2013;17(11):115.
[116] Li TC, Saravelos H, Chow MS, Chisabingo R, Cooke ID. Factors affecting the outcome of laparoscopic ovarian drilling for polycystic ovarian syndrome in women with anovulatory infertility. Br J Obstet Gynaecol 1998;105:338—44.
[117] Farquhar C, Lilford RJ, Marjoribanks J, Vandekerckhove P. Laparoscopic 'drilling' by diathermy or laser for ovulation induction in anovulatory polycystic ovary syndrome. Cochrane Database Syst Rev 2007;3:CD001122.
[118] Amer SA, Li TC, Metwally M, Emarh M, Ledger WL. Randomized controlled trial comparing laparoscopic ovarian diathermy with clomiphene citrate as a first-line method of ovulation induction in women with polycystic ovary syndrome. Hum Reprod 2009;24:219—25.
[119] Amer SA, Li TC, Ledger WL. Ovulation induction using laparoscopic ovarian drilling in women with polycystic ovarian syndrome: predictors of success. Hum Reprod 2004;19:1719—24.

[120] Gjønnaess H. Ovarian electrocautery in the treatment of women with polycystic ovary syndrome (PCOS). Factors affecting the results. Acta Obstet Gynecol Scand 1994;73:407–12.

[121] Lobo RA, Paul W, March CM, Granger L, Kletzky OA. Clomiphene and dexamethasone in women unresponsive to clomiphene alone. Obstet Gynecol 1982;60:497–501.

[122] Fedorcsák P, Dale PO, Storeng R, Tanbo T, Abyholm T. The impact of obesity and insulin resistance on the outcome of IVF or ICSI in women with polycystic ovarian syndrome. Hum Reprod 2001;16:1086–91.

[123] Abu Hashim H. Predictors of success of laparoscopic ovarian drilling in women with polycystic ovary syndrome: an evidence-based approach. Arch Gynecol Obstet 2015;291:11–18.

[124] Pouly JL, Krief M, Rabischong B, Brugnon F, Gremeau AS, Dejou L, et al. Ovarian drilling by fertiloscopy: feasibility, results and predictive values. Gynecol Obstet Fertil 2013;41:235–41.

[125] Harborne L, Fleming R, Lyall H, Norman J, Sattar N. Descriptive review of the evidence for the use of metformin in polycystic ovary syndrome. Lancet 2003;361:1894–901.

[126] Moll E, Bossuyt PM, Korevaar JC, Lambalk CB, van der Veen F. Effect of clomifene citrate plus metformin and clomifene citrate plus placebo on induction of ovulation in women with newly diagnosed polycystic ovary syndrome: randomised double blind clinical trial. BMJ 2006;332:1485.

[127] Legro RS, Barnhart HX, Schlaff WD, Carr BR, Diamond MP, Carson SA. Clomiphene, metformin, or both for infertility in the polycystic ovary syndrome. N Engl J Med 2007;356:551–66.

[128] Moll E, van der Veen F, van Wely M. The role of metformin in polycystic ovary syndrome: a systematic review. Hum Reprod Update 2007;13:523–7.

[129] Creanga AA, Bradley HM, McCormick C, Witkop CT. Use of metformin in polycystic ovary syndrome: a meta-analysis. Obstet Gynecol 2008;111:959–68.

[130] Pasquali R, Gambineri A, Aiscotti D, Vicennati V, Gagliardi L, Colitta D. Effect of long-term treatment with metformin added to hypocaloric diet on body composition, fat distribution, and androgen and insulin levels in abdominally obese women with and without the polycystic ovary syndrome. J Clin Endocrinol Metab 2000;85:2767–74.

[131] Chou KH, von Eye Corleta H, Capp E, Spritzer PM. Clinical, metabolic and endocrine parameters in response to metformin in obese women with polycystic ovary syndrome: a randomized, double-blind and placebo-controlled trial. Horm Metab Res 2003;35:86–91.

[132] Hoeger KM, Kochman L, Wixom N, Craig K, Miller RK, Guzick DS. A randomized, 48-week, placebo-controlled trial of intensive lifestyle modification and/or metformin therapy in overweight women with polycystic ovary syndrome: a pilot study. Fertil Steril 2004;82:421–9.

[133] Trolle B, Flyvbjerg A, Kesmodel U, Lauszus FF. Efficacy of metformin in obese and non-obese women with polycystic ovary syndrome: a randomized, double-blinded, placebo-controlled cross-over trial. Hum Reprod 2007;22:2967–73.

[134] Tang T, Glanville J, Hayden CJ, White D, Barth JH, Balen AH. Combined lifestyle modification and metformin in obese patients with polycystic ovary syndrome. A randomized, placebo-controlled, double-blind multicentre study. Hum Reprod 2006;21:80–9.

[135] Nieuwenhuis-Ruifrok AE, Kuchenbecker WK, Hoek A, Middleton P, Norman RJ. Insulin sensitizing drugs for weight loss in women of reproductive age who are overweight or obese: systematic review and meta-analysis. Hum Reprod Update 2009;15:57–68.

[136] Ladson G, Dodson WC, Sweet SD, Archibong AE, Kunselman AR, Demers LM. The effects of metformin with lifestyle therapy in polycystic ovary syndrome: a randomized double-blind study. Fertil Steril 2011;95:1059–66.

[137] Johnson NP, Bontekoe S, Stewart AW. Analysis of factors predicting success of metformin and clomiphene treatment for women with infertility owing to PCOS-related ovulation dysfunction in a randomised controlled trial. Aust N Z J Obstet Gynaecol 2011;51:252–6.

[138] Johnson N. Metformin is a reasonable first-line treatment option for non-obese women with infertility related to anovulatory polycystic ovary syndrome – a meta-analysis of randomised trials. Aust N Z J Obstet Gynaecol 2011;51:125–9.

[139] Bordewijk EM, Nahuis M, Costello MF, Van der Veen F, Tso LO, Mol BW, et al. Metformin during ovulation induction with gonadotrophins followed by timed intercourse or intrauterine insemination for subfertility associated with polycystic ovary syndrome. Cochrane Database Syst Rev 2017;1:CD009090.

[140] Heijnen EM, Eijkemans MJ, Hughes EG, Laven JS, Macklon NS, Fauser BC. A meta-analysis of outcomes of conventional IVF in women with polycystic ovary syndrome. Hum Reprod Update 2006;12:13–21.

[141] Marquard KL, Stephens SM, Jungheim ES, Ratts VS, Odem RR, Lanzendorf S. Polycystic ovary syndrome and maternal obesity affect oocyte size in in vitro fertilization/intracytoplasmic sperm injection cycles. Fertil Steril 2011;95:2146–9.

[142] Talmor A, Dunphy B. Female obesity and infertility. Best Pract Res Clin Obstet Gynaecol 2015;29:498–506.

[143] Provost MP, Acharya KS, Acharya CR, Yeh JS, Steward RG, Eaton JL, et al. Pregnancy outcomes decline with increasing body mass index: analysis of 239,127 fresh autologous in vitro fertilization cycles from the 2008-2010 Society for Assisted Reproductive Technology registry. Fertil Steril 2016;105:663–9.

[144] Sheng Y, Lu G, Liu J, Liang X, Ma Y, Zhang X, et al. Effect of body mass index on the outcomes of controlled ovarian hyperstimulation in Chinese women with polycystic ovary syndrome: a multicenter, prospective, observational study. J Assist Reprod Genet 2017;34:61–70.

[145] Tang T, Glanville J, Orsi N, Barth JH, Balen AH. The use of metformin for women with PCOS undergoing IVF treatment. Hum Reprod 2006;21:1416–25.

[146] Tso LO, Costello MF, Albuquerque LE, Andriolo RB, Freitas V. Metformin treatment before and during IVF or ICSI in women with polycystic ovary syndrome. Cochrane Database Syst Rev 2009;15:CD006105.

[147] Costello MF, Garad RM, Hart R, Homer H, Johnson L, Jordan C, et al. A review of second- and third-line infertility treatments and supporting evidence in women with polycystic ovary syndrome. Med Sci (Basel) 2019;7(7). pii: E75. doi:10.3390/medsci7070075.

[148] Salamun V, Jensterle M, Janez A, Vrtacnik Bokal E. Liraglutide increases IVF pregnancy rates in obese PCOS women with poor response to first-line reproductive treatments: a pilot randomized study. Eur J Endocrinol 2018;179(1):1–11.

Section 2

Contraception

Chapter 4

Obesity and sexual health

Sharon Cameron[1,2] and Michelle Cooper[3]

[1]Consultant Gynaecologist, NHS Lothian, University of Edinburgh, Edinburgh, Scotland, [2]Sexual Health Services, NHS Lothian, University of Edinburgh, Edinburgh, Scotland, [3]Obstetrics & Gynaecology, NHS Lothian, University of Edinburgh, Edinburgh, Scotland

Introduction

The increasing prevalence of obesity is a major public health concern. A systematic analysis of international data estimated that the number of individuals with obesity or overweight at 2.1 billion in 2013 (an increase from 921 million in 1980) [1]. In global terms, 36.9% of men and 38.0% of women have a body mass index (BMI) of 25 or above. Although there are rising trends in obesity in low-, middle-, and high-income settings, these are often more pronounced in developed countries [1].

Data from national surveys from the United States have shown a steady rise in the prevalence of obesity. In 2015/16 4 out of 10 adults in the United States had a BMI of 30 kg/m^2 or more [2]. Corresponding data from the same year in England show almost two-thirds of all adults had a BMI of 25 or greater, including 27% of men and 30% of women classified as obese [3]. In Scotland, 29% of adults had a BMI in the obese range in 2017 [4]. Projected trends estimate an additional 65 million more adults with obesity in the United States and 11 million adults in the United Kingdom by 2030 [5].

Data from Scotland have shown clearly that obesity is more prevalent in men and women from the most deprived areas compared to the least deprived areas—a gap of 35%–20%, respectively, in 2017 [4]. In the United States, there is a similar tendency of increasing prevalence of obesity among women (but not men) from lowest income groups [2].

Sexual health is an important part of overall health, well-being, and quality of life. While the association between obesity and physical illnesses is well established, there is now growing recognition of the negative impact that obesity can have on sexual health. This may be mediated directly through the physical and psychosocial effects or indirectly through concurrent comorbidities and can affect sexual behavior, sexual function, and sexual health outcomes.

Obesity and sexual behavior

Probably, the largest contribution of information to date regarding BMI and sexual behavior comes from the French national survey of sexual behaviors that was conducted in 2005/06 (Contexte de la Sexualite en France) [6]. This was a population-based survey of over 10,000 men and women, selected at random, who underwent lengthy telephone interviews about their sexual practices, contraception, and history of sexually transmitted infections (STIs). Respondents also provided data on their height and weight. Interestingly, in this study, only one-half of the female respondents who had a BMI that put them in the obese range actually considered themselves to be so. This lack of awareness of realizing oneself as being "very overweight" was even more marked among men, with only one-quarter of obese men considering themselves to be in this category. Compared to normal-weight respondents, those with obesity were less likely to have had more than one sexual partner in the previous year, and women (but not men) were more likely not to have had any sexual partner. However, there was no difference between the groups for individuals with sexual partners, in terms of frequency of sexual intercourse and the proportion who considered themselves to be "very satisfied" with their sexual life [6].

Women with obesity were likely to have sexual partners who also had obesity; a tendency that was less marked for men. This French study also showed that women with obesity were more likely to have met a sexual partner through the Internet than women of normal weight. The authors of the paper suggested that women with obesity might find it more difficult to attract a sexual partner and/or that they can establish a rapport with a potential partner while at the same time concealing their weight. This study also showed that in terms of the importance placed on sexuality for ones "personal life balance," there was a significant trend for women of higher BMI to

Obesity and Gynecology. DOI: https://doi.org/10.1016/B978-0-12-817919-2.00004-8
© 2020 Elsevier Inc. All rights reserved.

downgrade the importance of sexuality for their wellbeing. However, this was not the case for men. The authors also suggested that the gender differences observed in sexual activity of men and women with obesity may be due to one or a combination of psychological factors such as low self-esteem; social factors such as the stigma attached to being overweight (often greater for women than for men) or physical factors associated with obesity.

In contrast to the findings of this French study, data from a nationally representative database from the United States from 2002 (National Survey of Family Growth), which surveyed women of reproductive age about sexual behavior, showed no difference in the objective measures of sexual behavior such as the number of lifetime male partners or the frequency of sexual intercourse between women of normal and raised BMI [7]. One possibility is that this might suggest that being a woman with obesity in the United States is not associated with the same stigma as in France.

In general, however, for women in the Western countries, there is often a sociocultural association between slender physique and physical attractiveness. Studies have shown that individuals perceive their obesity as a serious psychosocial handicap. This can lead to some of the psychosocial manifestations of negative body image such as low self-esteem, which could lead to difficulty in initiating a healthy sexual relationship. In one small study of individuals with morbid obesity who lost weight, most subjects stated that they would prefer to be of normal weight and have a major physical handicap (such as being deaf, dyslexic, or blind) rather than have obesity again [8]. Moreover, mental health conditions such as anxiety and depression (which can independently affect sexual function) are more common in those above a normal BMI. One metaanalysis found that women with obesity had a 67% greater odds of developing depression than women of normal weight [9].

Some of the physical consequences associated with obesity that may limit sexual intercourse for women include musculoskeletal problems, urinary incontinence, and menorrhagia. In addition, women with obesity may also experience hyperandrogenism [10]. Thus the associated acne or hirsutism may also contribute to lowering a woman's confidence about her physical attractiveness.

It should be remembered, however, that most of the studies to date about sexual health outcomes of interest have been self-reported, including self-reporting of both height and weight. Previous research studies have noted that with reporting of weight there is a tendency for respondents to underestimate weight [7]. Since this appears to affect individuals of all BMI's equally, it is likely that this would lead to an underestimation of the association of obesity and the outcome of interest.

Furthermore, weight is a dynamic variable and so BMI at the time of survey may not be an accurate reflection of BMI at the time of the event of interest. There is also evidence that men and women may underreport sexual behavior. Using data from the National Health and Nutrition Examination Survey from the United States, which was conducted using computer-assisted self-interview (widely used to gather information on sensitive topics), 18% of the male and 28% of the female respondents who reported no lifetime sex partners ever tested positive for antibodies to Herpes simplex type 2 that was used as a serological marker of sexual exposure [11]. In this study, individuals who were overweight or with obesity reported fewer sex partners than individuals of normal weight, although this was not reflected in their HSV-2 status [11]. In addition to underreporting of sexual partners, it is also possible that in surveys there may be an overreporting of the number of sexual partners or frequency of intercourse, possibly due to social pressure, and this may be more likely to occur among men [7].

Obesity and sexual function

There is evidence that men with obesity are more likely to experience sexual dysfunction. In a survey of over 3000 men aged 40–79 from across Europe, impaired sexual function was reported more frequently by men with obesity than those of normal BMI [12]. In the previously mentioned French survey, men who were overweight or obese were more than twice as likely than normal-weight men to have experienced erectile dysfunction in the previous year [6]. This may be due to penile vascular impairment, as a consequence of the atherosclerosis associated with obesity [13,14]. Psychological considerations such as low self-esteem due to obesity may also play a role. There is evidence, however, that erectile dysfunction can improve with weight reduction. In one study, approximately one-third of men with obesity with erectile dysfunction experienced improved sexual function after weight loss and lifestyle changes [15]. In older men, obesity has also been associated with lower sexual satisfaction [16].

With regards sexual function among women, multivariate analysis of results from the most recent NATSAL found an association between obesity (and many of its associated medical comorbidities) and lower sexual function scores [17]. Although the nature of the survey did not allow for further exploration, the causes are likely to be multifactorial in origin. Experts acknowledge that female sexual dysfunction is often a more complex condition, because it has its origins in a number of factors, including relationship, emotional, psychological, medical, hormonal, reproductive, and aging [18].

Other studies have reported that women with obesity may suffer from a lack of libido and reduced satisfaction with sexual life [19]. A study that examined self-reported quality of sexual life among 500 individuals in the United States with obesity, using a validated measure of weight-related quality of life, showed that obesity was associated with a high prevalence of lack of sexual enjoyment, lack of sexual desire, difficulty with sexual performance, and avoidance of sexual encounters [19]. Furthermore, in this study the impairment in quality of sexual life was more marked for women than for men. Another study that used a validated sexual functioning questionnaire among men and women with obesity showed that scores for female participants were lower than those of men. In addition, scores for women with obesity were generally lower than scores reported for cancer survivors [20]. In women with known sexual dysfunction, one study also found that BMI strongly correlated with more abnormal female sex function index (FSFI) scores [21]. Of the sexual function parameters measured, arousal, lubrication, orgasm, and satisfaction were most affected by increasing BMI [21].

These effects may not be limited to heterosexual relationships. While there is limited evidence about female sexual dysfunction in same-sex couples, it is recognized that same-sex partnered females are more likely to have obesity [22].

Obesity and sexual health outcomes

Obesity may impact upon sexual health outcomes in several ways. First, girls who are heavier will attain secondary sexual characteristics and the menarche earlier than their normal-weight peers, potentially allowing for more reproductive years [23]. However, survey data from a cross-sectional database of women of reproductive age in the United States showed that there was no association between age at first intercourse and BMI [7]. Data from the French national survey of sexual behaviors [6] showed that women with obesity who were under 30 years of age were four times more likely than women of normal weight to report an unintended pregnancy or an abortion. This higher unintended pregnancy rate might seem surprising, since obesity is linked to anovulation and thus reduced fertility. However, the French study showed that women with obesity were less likely to use oral contraceptive pills and to attend contraceptive services in general and more likely to rely on less effective methods such as "withdrawal," which may partly explain the higher risk of unintended pregnancies. Obesity was also linked to nonuse of contraception in a systematic review of factors linked to adverse sexual health outcomes in women of reproductive age [24]. It is also possible that clinicians may be reluctant to prescribe combined hormonal contraception to obese women due to concerns about higher risk of venous thromboembolism [25]. It is also possible that reliance on less effective methods may reflect a difficulty in negotiating condom use with a sex partner, greater sexual risk-taking, or misconceptions regarding one's fertility. However, data from a study in the United States from 2002 (National Survey of Family Growth) showed no association between BMI and self-reported history of unintended pregnancy in the past 5 years, among more than 7600 women of reproductive age surveyed [26]. In addition, there was no significant difference between women of different BMI groups and current method of contraception nor on women's perceived fertility [26]. The authors of this study did point out, however, that women probably underreport a history of unintended pregnancy, just as history of abortion tends to be underreported in surveys [7].

In contrast, there is some evidence that risk-taking behaviors may be different between adolescent women of normal weight and those with extreme obesity. Data from the US survey of adolescents showed that young women with extreme obesity were less likely to have had sex than those who had been more likely to report having taken alcohol or drugs before the last sexual encounter (BMI ≥ 99th percentile for age and gender) [27]. In another cross-sectional study of sexually active female adolescents, increased BMI was linked to higher number of sexual partners and participation in "riskier" sexual practices [28]. However, differences in sexual behavior may not be the only explanation for the poorer sexual health outcomes observed.

One US study of postpartum women reported that among women who had been using contraception at the time of conception, women of raised BMI had almost twice the rate of unintended pregnancy compared to women of normal weight [29]. While an association between higher weight and contraceptive failure has been reported [30,31], it is hard to distinguish between method failure and user failure and so the effect of weight on failure rates has been suggested to be a reflection of issues of compliance [32]. A study that examined risk factors for failure of two types of oral emergency contraception (levonorgestrel and ulipristal acetate) found that failure rates were significantly higher among women with obesity and that this was more marked for women receiving levonorgestrel [33]. Pharmacokinetic studies have shown that the emergency contraceptive dose of levonorgestrel takes a longer time to reach steady-state levels in women with obesity and that this is not observed with ulipristal acetate [34]. The authors suggested that this could be a possible mechanism for reduced efficacy of emergency contraception in women with obesity.

Where is limited evidence regarding the effect of obesity on the incidence of STIs, an important finding from the French national survey of sexual behavior was that

TABLE 4.1 Factors associated with obesity that may impact negatively upon sexual health.

Factor	Contribution
Social	Stigma of obesity
Psychological	Reduced self-esteem
Physical	Musculoskeletal
	Menorrhagia (women)
	Urinary incontinence (women)
	Acne, hirsutism (women)
	Erectile dysfunction (men)
Behavior	Less likely to access contraceptive services (women)
	Less effective contraception (women)
	Less condom use (men)

among men in their late teens and 20s the odds of contracting an STI in the previous 5 years were more than 10 times greater for men with obesity than for men of normal weight [6]. There was no difference, however, between women of different BMI groups in the history of STI. Evidence for higher sexual risk-taking among men with obesity compared to normal-weight men has also been reported for men who have sex with men (MSM), with MSM who have raised BMI being more likely to have unprotected anal intercourse than normal-weight counterparts in one study [35]. The authors of this study suggested that this may be because these men might feel less able to negotiate safe sex and may be less well placed to be selective about sexual partners.

Conclusion

In addition to the physical problems caused by obesity, there is evidence that obesity negatively impacts on sexual health. In particular, there is the possibility that having obesity affects one's sexual risk-taking behavior. Clearly, however, the relative contribution of social (stigma of obesity), psychological (low self-esteem), and physical consequences of obesity upon sexual health is difficult to disentangle (Table 4.1). There is some evidence that women with raised BMI may be less likely to use effective contraception than normal-weight counterparts, thus placing themselves at a higher risk of unintended pregnancy. In addition to the personal distress that may be associated with an unintended pregnancy, clearly, there are significantly increased obstetric and neonatal risks for a woman of raised BMI when she embarks upon a pregnancy. There is evidence that emergency hormonal contraception may be less effective in women with obesity. Another area of concern is that men with obesity may be less likely to have protected sex and have a higher rate of STIs than men of normal BMI. In addition, for both sexes, there is evidence that obesity may affect the quality of sexual life and, in men, can give rise to sexual dysfunction via erectile difficulties.

From a public health perspective, we need to continue to pursue effective strategies (prevention and cure) to tackle obesity. Health promotion services should highlight how obesity affects not only one's physical health but also impacts one's sexual life. Health-care professionals should also be encouraged to have sensitive discussions around weight loss and be able to signpost individuals to appropriate services for weight reduction, since these may ultimately prove effective and result in an improved quality of both physical and sexual health.

References

[1] Ng M, Fleming T, Robinson M, et al. Global, regional, and national prevalence of overweight and obesity in children and adults during 1980-2013: a systematic analysis for the Global Burden of Disease Study 2013. Lancet 2014;384:766–81.

[2] Centers for Disease Control and Prevention. US data trends: overweight and obesity. Available from: <https://www.cdc.gov/obesity/data/index.html>. [accessed 14.04.20]

[3] Statistics on obesity, physical activity and diet, England. Available from: <https://digital.nhs.uk/data-and-information/publications/statistical/statistics-on-obesity-physical-activity-and-diet/statistics-on-obesity-physical-activity-and-diet-england-2019>; 2019 [accessed 14.04.20]

[4] Scottish Government. Available from: <https://digital.nhs.uk/data-and-information/publications/statistical/statistics-on-obesity-physical-activity-and-diet/statistics-on-obesity-physical-activity-and-diet-england-2019>; 2018.

[5] Wang YC, McPherson K, Marsh T, et al. Health and economic burden of the projected obesity trends in the USA and the UK. Lancet 2011;378:815–25.

[6] Bajos N, Wellings K, Laborde C, Moreau CCSF Group. Sexuality and obesity, a gender perspective: results from French national random probability survey of sexual behaviours. BMJ 2010;340:c2573.

[7] Kaneshiro B, Jensen JT, Carlson NE, Harvey SM, Nichols MD, Edelman AB. Body mass index and sexual behavior. Obstet Gynecol 2008;112(3):586−92.

[8] Rand CS, Macgregor AMC. Successful weight loss following obesity surgery and the perceived liability of morbid obesity. Int J Obes 1991;15:577−9.

[9] Luppino FS, de Wit LM, Bouvy PF, Stijnen T, Cuijpers P, Penninx BW, et al. Overweight, obesity, and depression: a systematic review and meta-analysis of longitudinal studies. Arch Gen Psychiatry 2010;67(3):220.

[10] Barber TM, McCarthy MI, Wass JA, Franks S. Obesity and polycystic ovary syndrome. Clin Endocrinol (Oxf) 2006;65(2): 137−145.

[11] Nagelkerke NJ, Bernsen RM, Sgaier SK, Jha P. Body mass index, sexual behaviour, and sexually transmitted infections: an analysis using the NHANES 1999-2000 data. BMC Public Health 2006;6:199.

[12] Han TS, Tajar A, O'Neill TW, Jiang M, Bartfai G, Boonen S, et al.EMAS Group Impaired quality of life and sexual function in overweight and obese men: the European Male Ageing Study. Eur J Endocrinol 2011;164(6):1003−11.

[13] Derby CA, Mohr BA, Goldstein I, Feldman HA, Johannes CB, mcKinlay JB. Modifiable risk factors and erectile dysfunction: can lifestyle modify risk? Urology 2000;56:302−6.

[14] Chung WS, Sohn JH, Park YY. Is obesity an underlying factor in erectile dysfunction? Eur Urol 1999;36:68−70.

[15] Esposito K, Giugliano F, Di Palo C, Giugliano G, Marfella R, D'Andrea F, et al. Effect of lifestyle changes on erectile dysfunction in obese men: a randomized controlled trial. JAMA 2004;291 (24):2978−84.

[16] Adolfsson B, Elofsson S, Rössner S, Undén AL. Are sexual dissatisfaction and sexual abuse associated with obesity? A population-based study. Obes Res 2004;12(10):1702−9.

[17] Polland AR, Davis M, Zeymo A, Iglesia CB. Association between comorbidities and female sexual dysfunction: findings from the third National Survey of Sexual Attitudes and Lifestyles (Natsal-3). Int Urogynecol J 2019;30(3):377−83.

[18] Goldbeck-Wood S. Commentary: Female sexual dysfunction is a real but complex problem. BMJ 2010;30:341.

[19] Kolotkin RL, Binks M, Crosby RD, Østbye T, Gress RE, Adams TD. Obesity and sexual quality of life. Obesity (Silver Spring) 2006;14(3):472−9.

[20] Ostbye T, Kolotkin RL, He H, Overcash F, Brouwer R, Binks M, et al. Sexual functioning in obese adults enrolling in a weight loss study. J Sex Marital Ther 2011;37(3):224−35.

[21] Esposito K, Ciotola M, Giugliano F, Bisogni C, Schisano B, Autorino R, et al. Association of body weight with sexual function in women. Int J Impot Res 2007;19(4):353.

[22] Blosnich JR, Hanmer J, Yu L, Matthews DD, Kavalieratos D. Health care use, health behaviors, and medical conditions among individuals in same-sex and opposite-sex partnerships: a cross-sectional observational analysis of the Medical Expenditures Panel Survey (MEPS), 2003−2011. Med Care. 2016;54(6):547.

[23] Cooper C, Kuh D, Egger P, Wadsworth M, Barker D. Childhood growth and age at menarche. Br J Obstet Gynecol 1996;103: 814−817.

[24] Edelman NL, de Visser RO, Mercer CH, McCabe L, Cassell JA. Targeting sexual health services in primary care: a systematic review of the psychosocial correlates of adverse sexual health outcomes reported in probability surveys of women of reproductive age. Preventive medicine. 2015;81:345−56.

[25] UK Medical Eligibility Criteria for Contraceptive Use (UKMEC). Faculty of sexual & reproductive healthcare. Available from: <https://www.fsrh.org/standards-and-guidance/documents/ukmec-2016/>; 2016.

[26] Kaneshiro B, Edelman A, Carlson N, Nichols M, Jensen J. The relationship between body mass index and unintended pregnancy: results from the 2002 National Survey of Family Growth. Contraception 2008;77(4):234−8.

[27] Ratcliff MB, Jenkins TM, Reiter-Purtill J, Noll JG, Zeller MH. Risk-taking behaviors of adolescents with extreme obesity. Paediatrics 2011;127(5):827−34.

[28] Gordon LP, Diaz A, Soghomonian C, Nucci-Sack AT, Weiss JM, Strickler HD, et al. Increased body mass index associated with increased risky sexual behaviors. J Pediatr Adolesc Gynecol 2016;29(1):42−7.

[29] Brunner Huber LR, Hogue CJ. The association between body weight, unintended pregnancy resulting in a livebirth, and contraception at the time of conception. Matern Child Health J 2005;9 (4):413−20.

[30] Vessey MP, Law;ess M, Yeates D, McPherson K. Progestogen-only oral contraception. Findings in a large prospective study with special reference to effectivenss. Brit J Fam Plann 1985;10: 117−121.

[31] Brunner Huber LR, Toth JL. Obesity and oral contraceptive failure: findings from the 2002 National Survey of Family growth. Am J Epidemiol 2007;166:1306−11.

[32] Westhoff CL, Torgal AH, Mayeda ER, Stanczyk FZ, Lerner PJ, Benn EKT, et al. Ovarian suppression in normal-weight and obese women during oral contraceptive use. Obstet Gynecol 2010;116: 275−283.

[33] Glasier A, Cameron ST, Blithe D, Scherrer B, Mathe H, Levy D, et al. Can we identify women at risk of pregnancy despite using emergency contraception? Data from randomized trials of ulipristal acetate and levonorgestrel. Contraception 2011;84(4): 363−367.

[34] Praditpan P, Hamouie A, Basaraba CN, Nandakumar R, Cremers S, Davis AR, et al. Pharmacokinetics of levonorgestrel and ulipristal acetate emergency contraception in women with normal and obese body mass index. Contraception 2017;95(5):464−9.

[35] Moskowitz DA, Seal DW. Revisiting obesity and condom use in men who have sex with men. Arch Sex Behav 2010;39(3):761−5.

Chapter 5

Obesity and contraception

Sujeetha Damodaran[1] and Krishnan Swaminathan[2]

[1]KMCH Institute of Allied Health Sciences, Coimbatore, India, [2]Kovai Medical Center & Hospital, Coimbatore, India

Introduction

Obesity continues to be a major public health concern across the globe. The prevalence of obesity has doubled over the past 30 years with 15% of women worldwide classified as obese as of 2014. More worrying is the data that shows more than 42 million children under the age of 5 years overweight as of 2013 [1]. Not surprisingly, the prevalence of obesity in pregnancy is rising exponentially, with obesity rates of around 32% reported in women between 20 and 39 years of age [2]. This is the age-group where apart from obesity, pregnancy can be complicated by diabetes and hypertension. These "three musketeers" are a devastating combination for women in pregnancy leading to a wide array of poor obstetric and neonatal outcomes. In addition, obesity in pregnancy also poses a serious challenge to the skills of the obstetrician, anesthetist, and midwives. Interestingly, studies have shown than obese women were more likely to have an unplanned pregnancy than women with a normal body mass index (BMI). Obese women reported less contraceptive usage, more contraceptive failure, and lower intake of preconceptional folic acid, which can greatly compromise prepregnancy and pregnancy care [3]. Prevention of unwanted pregnancy in obese women is, therefore, a major priority for health-care professionals. The aim of this chapter is to review the various contraceptive options available to the obese woman. Knowledge of such contraceptive options may lead to a reduction of unwanted pregnancies and support women in their quest for a well-planned pregnancy.

Risks of obesity in pregnancy

Maternal obesity is linked with a range of serious maternal and fetal outcomes (Table 5.1). Adverse maternal outcomes include increased risk of miscarriage [4], gestational diabetes with odds ratio (OR) of 6.28 (3.01–13.06) [5] and its consequences, pregnancy-associated hypertension (2.2- to 7.7-fold increase) [6], preterm and extremely preterm ($<$27 weeks) births [7], prolonged first stage of labor [8], higher rates of anesthetic complications with difficult intubation in nearly one-third of obese women [9], higher chances of operative deliveries [10], wound infections [11], longer hospital stay [12], and shorter duration of breastfeeding [13]. Maternal obesity also contributes to adverse fetal outcomes, including macrosomia [14], shoulder dystocia [15], infant mortality [16], and predisposition to obesity later in life [17]. A systematic review of reviews [18] confirmed the above mentioned findings, emphasizing the need for obese women to lose weight before they conceive and for health-care professionals to support obese women in this endeavor.

Classification of obesity based on body mass index

The WHO and the National Institute of Health classify obesity as given in Table 5.2.

It has to be pointed out that the arbitrary cutoffs are based on data derived from Whites, as ethnic-based data is currently unavailable.

Potential concerns with obesity and contraception

The main concerns with contraceptive methods in obese woman are as follows:

1. Historically, overweight and obese women have been excluded from trials in contraception, leading to lack of robust evidence regarding the safety and efficacy of contraceptive methods in such a population. Are contraceptive methods efficacious in the obese population?
2. "One dose fits all" has been the traditional practice with hormonal contraception. However, the effects of

TABLE 5.1 Effects of obesity on pregnancy outcomes.

Condition	Type of study	Effect[a]
GDM [5]	Metaanalysis	OR, 2.14 (1.82–2.53)[b]
		OR, 3.56 (3.05–4.21)[c]
		OR, 8.56 (5.07–16.04)[d]
PIH [6]	Metaanalysis	OR, 2.5 (2.1–3.0)[c]
		OR, 3.2 (2.6–4.0)[d]
C-section [7]	Population-based cohort study	RR, 2.6 (2.04–2.51)[c]
		RR, 3.38 (2.49–4.57)
Preeclampsia [5]	Metaanalysis	OR, 1.6 (1.1–2.25)[c]
		OR, 3.3 (2.4–4.5)[d]
Preeclampsia [8]	Retrospective cohort study	OR, 7.2 (4.7–11.2)[d]
Induction of labor [8]	Retrospective cohort study	OR, 1.8 (1.3–2.5)[d]
Postpartum hemorrhage [8]	Population-based cohort study	OR, 1.5 (1.3–1.7)[e]
Preterm delivery (<33 weeks) [8]	Population-based cohort study	OR, 2.0 (1.3–2.9)[e]
Stillbirth [9]	Systematic review and metaanalysis	OR, 1.47[b]
		RR, 2.07[e]
Stillbirth [10]	Population-based cohort study	OR, 2.8 (1.5–5.3)[e]
Neonatal death [10]	Population-based cohort study	OR, 2.6 (1.2–5.8)[e]

BMI, Body mass index; GDM, gestational diabetes mellitus; OR, odds ratio; PIH, pregnancy-induced hypertension; RR, relative risk.
[a]All OR and RR are compared to normal-weight pregnant women (BMI 18–25). Values in parentheses indicate 95% CI.
[b]BMI 25–30.
[c]BMI 30–35.
[d]BMI > 35.
[e]BMI > 30.
Source: Reproduced by permission of J Am Board Fam Med 2011;24(1):75–85.

TABLE 5.2 Classification of obesity.

Classification	BMI (kg/m^2)
Underweight	<18.5
Normal	≥ 18.5–24.9
Overweight	25–29.9
Class I obesity	30–34.9
Class II obesity	35–39.9
Class III obesity	≥ 40

BMI, Body mass index.

obesity on drug pharmacokinetics and pharmacogenetics, especially steroidal contraceptives, are poorly understood. Is there a higher risk of failure due to altered pharmacokinetics in obesity?

3. As a generalization, women tend to blame contraception for weight gain. This perceived weight gain is a leading cause of discontinuation of contraception at least in some parts of the world [19,20]. Does contraception in obesity promote further weight gain?
4. Obesity, per se, doubles the risk of venous thromboembolism (VTE) as compared with someone with a normal BMI [21]. There is always a concern as to whether oral contraceptives increase the risk of a potentially life-threatening complication such as VTE and other health hazards associated with obesity (diabetes, dyslipidemia, cardiovascular disease, hepatobiliary disease, and cancer). Are metabolic health and thromboembolic risk high in obese women using contraceptives?
5. Finally, procedure-dependent contraceptive methods [intrauterine devices (IUDs) and sterilization] are technically more challenging to perform in an obese woman than their normal BMI counterparts.

Obesity and contraceptive efficacy

Mechanisms by which obesity could potentially affect contraceptive efficacy

Obesity can have profound effects on different physiologic processes, including absorption, distribution, metabolism, and excretion of contraceptive drugs [22]. Higher cardiac output, increased gut perfusion, accelerated gastric emptying, and alterations in enterohepatic recirculation have been reported in obese human and animal models [23–29], factors that can potentially affect absorption of contraceptive drugs. Obesity is also associated with altered body composition with an increase in fat mass, which can affect the distribution of hydrophilic and lipophilic drugs [30]. Other physiological alterations in obesity that can have a potential impact in contraceptive drug metabolism and excretion include increased splanchnic and renal flow, fatty infiltration of liver, inflammatory cytokines, reduction in expression of biliary canalicular transporters, increased kidney size, and reduced urinary pH. The combined consequences of the previous alterations include a reduction in specific cytochrome P450 activities, altered biliary metabolism and enterohepatic circulation, increased renal clearance, and tubular secretion [31].

In spite of all the potential mechanisms by which obesity could affect contraceptive efficacy, there have been few studies to date that have investigated the pharmacokinetics of contraceptive steroids in obese women. In a study conducted to determine whether increased BMI affects the pharmacokinetics of oral contraceptives, the levenorgestrel (LNG) half-life in obese subjects was twice that of normal BMI subjects and the time taken to reach a steady state was doubled as well [32]. In another study conducted to compare the pharmacokinetics of oral contraceptives in obese and normal-weight women, obese women had a lower area under curve and lower maximum values for ethinyl estradiol than normal-weight women [33]. But the observed differences in pharmacokinetics did not translate into more ovarian follicular activity in obese oral contraceptive users. Finally, in a 26-week prospectively designed experimental study conducted to determine the incidence of ovulation and follicular development in different classes of obese women using depot medroxy progesterone acetate subcutaneous (DMPA-SC), median medroxy progesterone acetate (MPA) was consistently lowest among class III obese women but above the levels needed to inhibit ovulation [34]. Translation of these mechanistic findings into confirmed evidence for failure (i.e., pregnancy) has not been robustly studied except probably for emergency contraception with LNG. The risk of pregnancy was greater in obese women using LNG for emergency contraception compared to women with normal BMI (OR 4.41) [35]. It is, therefore, very clear that we need more clinical trials to understand the impact of obesity on drug pharmacokinetics and therapeutics that helps in counseling obese women in day-to-day clinical practice.

Evidence (or lack of) for contraceptive efficacy in overweight or obese women

Obesity increases metabolic rate, circulating volume, and absorption of contraceptive steroids by adipose tissue. Hence, women with obesity may take a longer time to achieve steady-state levels of contraceptives compared to women of normal BMI. Therefore there are logical concerns regarding the efficacy of contraceptives in obese women.

The current literature is inadequate to provide clear information to the overweight or obese woman as to the efficacy of contraception. There are multiple reasons for this lack of clarity. As previously mentioned, a number of studies excluded women over a certain weight/BMI cutoff, thereby greatly limiting the ability to draw reasonable conclusions. The limited numbers of studies that are available are too heterogeneous, assessing different delivery mechanisms (e.g., patch, injectables, and rings).

A Cochrane database review (2016) of 17 studies with a total of 63,813 women examined the effectiveness of hormonal contraceptives in preventing pregnancy among women who are overweight or obese versus women with normal or lower BMI or weight. Most studies in this review did not show a higher pregnancy risk in overweight or obese women. Two of the five combined oral contraceptive (COC) studies in this review found that BMI was associated with pregnancy but in different directions. Similar findings were noted in the five implant studies in this review. Analysis of data from other contraceptive methods such as depot medroxy progesterone acetate (DMPA), levonorgestrel IUD, and the two rod levonorgestrel implant, and the levonorgestrel implant showed no association of pregnancy with overweight or obesity [36]. A consensus opinion from the European Society of Contraception (2015) also did not generally find any robust evidence for decreased efficacy of different contraceptive methods in overweight or obese women but went on to conclude that progestin-only contraceptives and IUDs are effective in most women with obesity and have minimal or no metabolic effects. Combined hormonal contraceptives (CHCs) are also effective but in view of risks such as VTE in obese women (will be discussed later in this chapter), should be considered if other methods are not acceptable [37].

Contraceptives and weight gain: myth or truth?

Whether this is a myth or truth, many women and clinicians worldwide believe that an association exists between weight gain and oral contraceptives [38–42]. Young women, particularly, may be preoccupied with body image, and this fear of potential weight gain can deter already obese individuals and clinicians from initiating combination contraceptives. It can also lead to early discontinuation among users [43–45]. In a prospective nationwide study [46], 6 months after a new oral contraceptive prescription, only 68% of women were still continuing the medication. Of the women who discontinued, 46% of women did so because of side effects, predominantly perceived weight gain. More importantly, 80% of those who discontinued failed to adopt another method or adopted a less effective method, putting themselves at high risk of unintended pregnancy. It is, therefore, very clear that concerns about possible weight gain limit the use of a very effective method of contraception. However, a causal relationship between weight gain and combination contraceptives has not been clearly established. In addition, possible causal relationship between contraceptive use and weight gain is beset with a lot of confounding factors. Women tend to gain weight with time, a contemporaneous control group is needed, which may be ethically difficult to justify, contraceptives are complicated with plenty of formulations of varying strengths and regimes, duration of use is extremely variable and finally no consensus exists as to what constitutes excessive weight gain.

Potential mechanisms by which contraceptives can cause weight gain

In general, weight gain is due to one of the following factors: fluid retention, fat deposition, or muscle mass. Treatment with hormonal contraceptives may lead to a considerable activation of the renin—angiotensin—aldosterone system [47]. Fluid retention could, therefore, be induced by the mineralocorticoid activity of contraceptive steroids [48]. Experimentally, estrogens increase the size and number of subcutaneous adipocytes [49] that can be associated with increased subcutaneous fat in breasts, hips, and thighs [50]. In a prospective study to determine whether the use of low-dose estrogen oral contraceptives is associated with changes in weight, body composition, or fat distribution, there was no overall impact of low-dose oral contraceptive on any of the parameters previously mentioned. However, when weight gain did occur, it was related to the increase in body fat and not due to fluid retention or fat distribution [51]. Finally, the anabolic properties of COCs can have an effect on satiety and appetite that could result in increased food intake and weight gain.

COCs and weight gain

Weight gain is often considered a limiting side effect of using combination contraceptives (estrogen and progestin) by many clinicians and women worldwide [38,39,41]. Nevertheless, a causal relationship between combined oral contraceptives and weight gain has not been clearly established. In a Cochrane database review (2014) to evaluate the potential association between weight gain and combined contraceptive use, 49 studies were evaluated that had 85 weight change comparisons for 52 distinct contraceptive pairs or placebos. The review did not find any large association between combined oral contraceptive use and weight gain. Of these, four trials had a placebo group or no intervention group. Even in these trials, there was no evidence to support a causal association. The authors concluded that the available evidence was insufficient to confirm or refute a causal association but no large effect was evident [52].

In a longitudinal study to assess the long-term effects of COCs on body weight, COC use was not found to be a predictor for weight increase in the long term [53]. Interestingly, this long-term follow-up study showed a weight increase of 10.6 kg for women between 19 and 44 years of age. There are further studies describing weight change in women with increasing age [54—56] and it may be possible that the perceived weight gain with COC may be related to the natural changes in weight from a lifetime perspective. There have been other studies that show minimal or no weight increase with COC to suggest a causal relationship [57—60]. Based on the previous statements, there is little evidence for significant weight alterations in relation to COC pill use but this has to be confirmed with further long-term studies.

POCs and effect on weight

Progestogen only contraception (POCs) are ideally suited for women who have contraindications to or who are unable to tolerate estrogens. There has always been a concern of weight gain associated with progesterone preparations especially DMPA and LNG implants. In a study designed to compare women using either DMPA or an oral contraceptive pill and those who were not using hormonal contraception [61], weight gain was reported more with DMPA users (OR, 2.3). There is also some evidence that obese adolescent users of DMPA may gain more weight compared to those with normal BMI [62], and weight gain is an important reason for discontinuation of DMPA in this age-group [63]. LNG implants have also been implicated in weight gain [64].

In a Cochrane database review (2016) to evaluate the potential association between POCs and alterations in body weight, the authors could find only limited evidence for weight gain when using POCs [65]. In this review of 22 eligible studies that included a total of 11,450 women, mean weight gain at 6 or 12 months was less than 2 kg for most studies. The overall quality of evidence was low as most studies were nonrandomized and there was high loss to follow-up or early discontinuation. More weight gain was noted in 2 and 3 years but was similar for both groups, reinforcing our previous discussion on natural changes in weight over a period of time. Therefore appropriate counseling regarding the degree of weight gain associated with POCs based on current evidence will help to reduce discontinuation of POCs and unintended pregnancies.

Safety of hormonal contraceptives in obese women

Obesity is associated with a diverse array of health hazards, including hypertension [66,67], diabetes mellitus

TABLE 5.3 UKMEC criteria (2016) for contraceptive use—obesity and selected clinical conditions that are of particular relevance to obese women.

UKMEC	DEFINITION OF CATEGORY
Category 1	A condition for which there is no restriction for the use of the method
Category 2	A condition where the advantages of using the method generally outweigh the theoretical or proven risks
Category 3	A condition where the theoretical or proven risks usually outweigh the advantages of using the method. The provision of a method requires expert clinical judgement and/or referral to a specialist contraceptive provider, since use of the method is not usually recommended unless other more appropriate methods are not available or not acceptable
Category 4	A condition which represents an unacceptable health risk if the method is used

CONDITION	Cu-IUD	LNG-IUS	IMP	DMPA	POP	CHC
			I = Initiation, C = Continuation			
Obesity						
a) BMI ≥30–34 kg/m²	1	1	1	1	1	2
b) BMI ≥35 kg/m²	1	1	1	1	1	3
Smoking						
a) Age <35 years	1	1	1	1	1	2
b) Age ≥35 years						
(i) <15 cigarettes/day	1	1	1	1	1	3
(ii) ≥15 cigarettes/day	1	1	1	1	1	4
(iii) Stopped smoking <1 year	1	1	1	1	1	3
(iv) Stopped smoking ≥1 year	1	1	1	1	1	2
History of bariatric surgery						
a) With <30 kg/m² BMI	1	1	1	1	1	1
b) With ≥30–34 kg/m² BMI	1	1	1	1	1	2
c) With ≥35 kg/m² BMI	1	1	1	1	1	3

TABLE 5.3 (Continued).

CARDIOVASCULAR DISEASE (CVD)						
Multiple risk factors for CVD (such as smoking, diabetes, hypertension, obesity and dyslipidaemias)	1	2	2	3	2	3
Hypertension						
a) Adequately controlled hypertension	1	1	1	2	1	3
b) Consistently elevated BP levels (properly taken measurements)						
(i) Systolic >140–159 mmHg or diastolic >90–99 mmHg	1	1	1	1	1	3
(ii) Systolic ≥160 mmHg or diastolic ≥100 mmHg	1	1	1	2	1	4
c) Vascular disease	1	2	2	3	2	4
Diabetes						
a) History of gestational disease	1	1	1	1	1	1
b) Non-vascular disease						
(i) Non-insulin dependent	1	2	2	2	2	2
(ii) Insulin dependent	1	2	2	2	2	2
c) Nephropathy/retinopathy/neuropathy	1	2	2	2	2	3
d) Other vascular disease	1	2	2	2	2	3

CONDITION	Cu-IUD	LNG-IUS	IMP	DMPA	POP	CHC
	I = Initiation, C = Continuation					

Condition	Cu-IUD (I/C)	LNG-IUS (I/C)	IMP	DMPA	POP	CHC
Cervical cancer						
a) Awaiting treatment	I:4 / C:2	I:4 / C:2	2	2	1	2
b) Radical trachelectomy	3	3	2	2	1	2
Breast conditions						
a) Undiagnosed mass/breast symptoms	1	2	2	2	2	I:3 / C:2
b) Benign breast conditions	1	1	1	1	1	1
c) Family history of breast cancer	1	1	1	1	1	1
d) Carriers of known gene mutations associated with breast cancer (e.g. BRCA1/BRCA2)	1	2	2	2	2	3
e) Breast cancer						
(i) Current breast cancer	1	4	4	4	4	4
(ii) Past breast cancer	1	3	3	3	3	3
Endometrial cancer	I:4 / C:2	I:4 / C:2	1	1	1	1
Ovarian cancer	1	1	1	1	1	1

[68], dyslipidemia [69], heart disease [70], stroke [71], venous thrombosis [72,73], hepatobiliary disease [74], and cancer [75,76]. It is, therefore, imperative that an obese woman, who chooses a contraceptive method, does so without increasing her health risks. The dictum in medicine is "Do no harm," doing good always comes later. There should not be a situation where the "cure is worse than the problem!" Table 5.3 clearly outlines the criteria for contraceptive use in Obesity and selected clinical conditions of relevance to this Chapter.

Obesity, contraception, and cardiovascular disease

Potential mechanisms by which contraceptives can impact on cardiovascular disease involve its effects on vasculature and metabolic parameters (lipids and glucose). Current users of oral contraceptives had a moderately increased risk of hypertension, which decreased quickly with the cessation of the drug [77]. Oral contraceptives use, especially in smokers, may be associated with increased levels of fibrinogen and intravascular fibrin deposition [78], factors that have a role in arterial thrombotic diseases. In general, the estrogen component of the oral contraceptive pill causes modest increases in triglycerides, which is offset by increases in HDL and lowering of LDL cholesterol [79]. The androgenic progestogens (norgestrel and LNG) usually increase the serum LDL and lower serum HDL concentrations, but the newer progestogens such as desogestrel appear to be more favorable [80]. Oral contraceptives can also affect carbohydrate metabolism, mainly through the actions of progestogens. Studies have shown insulin resistance, glucose intolerance, and increased risk of diabetes mellitus, especially with POCs [81,82]. The big question is whether these statistically significant changes in fibrinogen, lipid, and glucose levels in hormonal contraceptive users are clinically relevant, especially in an obese population. Unfortunately, there are minimal data at present regarding the effects of hormonal contraception on the above mentioned parameters in the obese population, as this group has been traditionally excluded from most studies.

Some, but not all studies, report that oral contraceptives may be associated with an increased risk of myocardial infarction and stroke. In a study to assess whether current use of newer low-dose oral contraceptives increased the risk of myocardial infarction, only women who were heavy smokers (defined in this study as smoking ≥25 cigarettes/day) were at increased risk of myocardial infarction, with no evidence of increased risk in nonsmokers or light smokers [83]. However, a metaanalysis of 10 studies suggested an overall doubling of cardiovascular mortality, mainly driven by coronary heart disease in women using low-dose COC (less than 50 µg of ethinyl estradiol) [84]. Low-dose oral contraceptives may also be associated with a small increase in risk of ischemic stroke, but the absolute risk is very low [85]. Myocardial infarction and stroke are rare events in women of reproductive age-group; therefore even a doubling of this risk would still result in a very low attributable risk. In a systematic review (2016) evaluating the effects of CHC use in obese women (BMI ≥ 30) and the risk of acute myocardial infarction and stroke, results were inconclusive with conflicting evidence [86]. Currently, there is no evidence for any detrimental metabolic or cardiovascular effects of nonhormonal contraception, barrier methods, copper IUDs, and sterilization.

Obesity, contraception, and venous thromboembolism

Exogenous estrogens and obesity may increase blood coagulability with an increase in procoagulant factors (factors VII, VIII, XII, fibrogen). Therefore the risk of VTE increases with obesity and the use of oral contraceptives. The question is whether the risk is additive or multiplicative? In addition, there is substantial controversy surrounding the actual risk of VTE with different formulations of CHC.

Baseline risk of VTE in obese women range from 6 to 11/10,000 women-years [87,88]. In parallel, the risk of VTE with CHCs is much higher in users than nonusers (3–15/10,000 women-years in users vs 1–5/10,000 women-years in nonusers) [89]. With obesity and CHCs being independent risk factors for VTE, the logical question is to know the risks of VTE in an obese woman taking CHCs. The answer lies in the PILGRIM study in which a total of 968 women with one event of VTE with COC were compared to 874 women on COC with no personal history of VTE [90]. Severe obesity as defined by a BMI ≥ 35 was associated with a 3.46 increased risk of VTE, which is in line with similar results from previous studies [91].

Does the dose and formulation matter? While it is difficult to compare the effects of the dose of EE (ethinyl estradiol) since many preparations differ with respect to the progestin component, there is good evidence to link increased doses of estrogen to VTE risk. A number of studies have confirmed VTE associations with 50 µg EE formulations compared with sub 50 µg EE formulations [92,93]. There is some evidence to suggest that lowering the EE dose to <50 µg reduces the risk of VTE. A 2014 Cochrane systematic review suggested that the risk of VTE depends on both the progestogen used and the dose of EE [94]. Risk of VTE was similar in formulations of 30–35 µg EE with gestodene, desogestrel, cyproterone acetate (CPA), and drospirenone, but 50%−80% higher than similar dose of EE with levonorgestrel. The authors concluded that a sub 50 μg EE with levonorgestrel should be prescribed to reduce the risk of VTE with COCs.

Despite the higher risk of VTE in obese women using CHCs, it is still a rare event [93]. Therefore CHCs are safe for obese women who are otherwise healthy. However, the choice of CHC must be seriously reconsidered if there is a superimposition of additive risk factors such as smoking, diabetes, hypertension, older age, dyslipidemia, and overt arterial disease. In these circumstances, it would be unacceptable to defend the use of CHCs and it is better to think of alternate contraceptive methods [95,96].

Despite the concerns regarding certain progestogens contributing to VTE in COC users, currently there is no evidence for increased risk of VTE with POPs [97,98]. The safety of injectable progestins needs further investigations. Limited evidence from a systematic review (2016) suggested increased odds of VTE with the use of DMPA [98]. However, in a study comparing new DMPA users versus IUDs, there were lower D-dimer levels and longer peak time to thrombin generation in DMPA users suggesting a positive profile of DMPA against hypercoagulability [99]. Limited evidence shows conflicting results on the effects of patch/rings with VTE [100]. Overall, any potential risk likely represents a very small number of events at a population level.

Obesity, contraception, and cancer

Oral contraceptive use has been associated with increase in risk of cervical cancer [101], conflicting results with breast cancer [102−105], and decrease in risk of ovarian [106], uterine [104], and colorectal cancers [104]. Obesity, per se, has been associated with increased risk of cancer, especially breast, endometrial, ovarian, and colorectal [107,108]. It is, therefore, theoretically possible that oral contraceptive use in obese women may have significant effects on incident cancer.

In one of the longest follow-up study of 46,022 women on oral contraceptives observed for 44 years, most women do not seem to expose themselves to long-term cancer harms [109]. On balance, many women benefit from significant reductions in risks of certain types of cancer that seems to persist many years after stopping treatment. Ever use of oral contraceptives in this follow-up study was associated with a reduced incidence of colorectal, endometrial, ovarian, lymphatic, and hematopoietic cancers. An increased risk of breast and cervical cancers that was seen with current users appeared to be lost within 5 years of stopping oral contraception. While the data on this large inception cohort study was adjusted for potentially confounding factors such as age, smoking, parity, and social class, there was no data on BMI or lifestyle variables. It is not clear at this point whether obese women would benefit from oral contraceptive use as cancer prevention strategies [110], but it is reassuring to see the trends in ovarian and endometrial malignancies, for which obesity is an independent risk factor.

It is widely accepted that intrauterine contraceptive devices are not implicated in cancer [111,112]. There is no evidence to suggest that inert IUDs are associated with cervical cancer [113] nor has there been any difference in premalignant or malignant cervical pathologies between copper-containing IUDs and inert IUDs [114].

Contraceptive issues after bariatric surgery

The demand for bariatric surgery has greatly increased in recent times, as it is believed to be the most effective treatment method for the morbidly obese. The incidence of bariatric surgery in the United States increased by 800% between 1998 and 2005, predominantly accounted for by women of reproductive age-group [115]. Since the chances of fertility increase after bariatric surgery [116], such women are at higher risk of unintended pregnancies. The main concerns of such an unintended pregnancy in this group of women are the risks associated with maternal and fetal outcomes due to the nutritional effects of weight loss surgery [117,118]. The general consensus from major associations is that pregnancy should be avoided for 12−24 months after bariatric surgery, coinciding with the time when significant weight loss and postoperative complications occur [119].

There are legitimate concerns regarding the efficacy and safety of hormonal contraception in women who have undergone bariatric surgery. Data on pharmacokinetics, efficacy, and safety of oral contraceptive drugs after bariatric surgery was extremely limited to a couple of pharmacokinetic studies, some observational studies, and a single case report [120]. In a study evaluating the pharmacokinetics of two commonly used progestogens (norethisterone and LNG) in morbidly obese women after jejunoileal bypass compared to healthy controls, the mean plasma levels of both the progestogens were lower in the obese surgical patients at 1−8 hours after ingestion compared to controls [121]. Whether this translates into contraceptive failure is not clear, but the authors of this study recommend that low-dose progestogen-only minipills should not be used after jejunoileal bypass. This procedure is not performed nowadays, and there are no pharmacokinetic studies in more modern bypass procedures at this point of time. Three case reports on young women who had subdermal levonorgestrel (etonogestrel (ENG))—releasing implant (Implanon) prior to Roux-en-Y bypass surgery showed ENG concentrations sufficient to inhibit ovulation until 8 months after insertion and may therefore be a safe contraceptive choice for women undergoing bariatric surgery [122].

In terms of efficacy, there are concerns around malabsorption of oral contraceptives, especially in women undergoing malabsorptive procedures such as Roux-en-Y gastric bypass and biliopancreatic diversion with duodenal switch. In a prospective study of 40 women who underwent biliopancreatic diversion [123], 2 out of 9 patients, who postoperatively used oral contraceptives only, had their first pregnancy after surgery while still using the same contraception that they had used preoperatively. There does not appear to be any particular concern with restrictive procedures. In a study of 215 morbidly obese women who had agreed to be on contraception for 2 years after laparoscopic adjustable gastric banding [124], 7 women had unexpected pregnancies, but all these women were using unreliable methods such as periodic abstinence. There were no pregnancies observed in patients using oral contraception.

In terms of safety of oral contraceptives after bariatric surgery, evidence is limited to a single case report [125] of an 18-year-old lady who developed an acute ischemic stroke 4 months after a Roux-en-Y gastric bypass procedure. She was on oral contraceptive at the time of the event, but there were other confounding factors, including tobacco smoking and recreational drug use. Apart from the risk of cardiovascular disease, the other concern is the risk of VTE in postoperative users of contraceptive pills after bariatric surgery. In a survey of the members of American Society of Bariatric Surgery, the self-reported incidence of deep vein thrombosis and pulmonary embolism was 2.63% and 0.95%, respectively, even though routine prophylaxis was used by more than 95% of the members [126]. Finally, there is a theoretical concern to using DMPA in women who have undergone bariatric surgery. There is some evidence for increased bone turnover and reduced bone mineral density (BMD) in women who have undergone Roux-en-Y gastric bypass procedure [127,128]. DMPA use is associated with small and usually reversible changes in BMD [129], but at this point, it is not clear whether DMPA use further aggravates bone loss in women undergoing bariatric surgery.

To summarize, though there is no evidence for a significant decrease in oral contraceptive effectiveness post-—bariatric surgery, there are potential concerns regarding oral contraceptive efficacy in women undergoing malabsorptive procedures, more so if they have long-term diarrhea and/or vomiting. There does not appear to be an increased risk of cardiovascular disease in women using oral contraceptives post—bariatric surgery, but there are potential concerns regarding risk of VTE, especially if there is prolonged immobilization. The relationship between DMPA use and bone loss in post—bariatric surgery patients needs further investigation.

Intrauterine contraceptive devices in obese women

IUD insertion presents some challenges in the obese woman. It may be difficult to ascertain the size and direction of the uterus. In addition, visualization of the cervix may be difficult. Simple measures such as use of a larger speculum, placing a condom with the tip removed over the blades of the speculum, comfortable positioning of the patient, and use of ultrasound during the device insertion can overcome the abovementioned problems [130]. These difficulties are negated to a large extent by the long-term highly effective contraception with these devices. There is also evidence of some benefit with the LNG intrauterine system in selected obese women with abnormal uterine bleeding [131] with no huge dropout rates or complications compared to other BMI groups [132]. The European Society of Contraception for Obese women Statement (2015) suggests that IUD contraception is highly recommended in obese women with LNG-intrauterine system a safe and effective contraceptive method for obese women. In addition, this device is more beneficial than copper IUDs in those with heavy menstrual bleeding. To summarize, there are no specific contraindications to hormone-containing devices or copper IUDs in the obese woman, and such devices, apart from being effective contraceptives, may produce additional benefit in obese women with abnormal uterine bleeding or endometrial hyperplasia.

Sterilization procedures in obese women

Obesity can complicate tubal sterilization procedures [133–135]. In a large prospective multicenter cohort study of 9475 women who underwent interval laparoscopic sterilization, obesity was an independent predictor for one or more complications [134]. While there is evidence with a large international data set that obesity can be associated with higher incidence of surgical difficulties, technical failure rate, and longer surgical times compared to nonobese controls, none of the abovementioned issues led to serious consequences [133]. In addition, not all studies have shown a link between obesity and poor outcomes after laparoscopic sterilization procedures. In a retrospective study of 248 consecutive patients undergoing laparoscopic tubal sterilization [136], there were no differences in complications, mean operating times or blood loss between obese women and nonobese controls. However, morbidly obese women have their own set of problems, including theater requirements and hazards related to anesthetic procedures. In such cases, vasectomy for the woman's partner seems to be the best available option.

References

[1] World Health Organization. Global status report on noncommunicable diseases. Switzerland: WHO; 2014.

[2] Declerq E, Macdorman M, Cabral H, Stotland N. Prepregnancy body mass index and infant mortality in 38 U.S. States, 2012–2013. Obstet Gynecol 2016;127(2):279–87.

[3] McKeating A, O'Higgins A, Turner C, McMahon L, Sheehan SR, Turner MJ. The relationship between unplanned pregnancy and maternal body mass index 2009–2012. Eur J Contracept Reprod Health Care 2015;20(6):409–18.

[4] Poston L, Caleyachetty R, Cnattingius S, Corvalán C, Uauy R, Herring S, et al. Preconceptional and maternal obesity: epidemiology and health consequences. Lancet Diabetes Endocrinol 2016;4(12):1025–36.

[5] Gaillard R, Durmuş B, Hofman A, Mackenbach JP, Steegers EA, Jaddoe VW. Risk factors and outcomes of maternal obesity and excessive weight gain during pregnancy. Obesity (Silver Spring) 2013;21(5):1046–55.

[6] Li N, Liu E, Guo J, Pan L, Li B, Wang P, et al. Maternal prepregnancy body mass index and gestational weight gain on pregnancy outcomes. PLoS One 2013;8(12):e82310.

[7] Cnattingius S, Villamor E, Johansson S, Edstedt Bonamy AK, Persson M, Wikström AK, et al. Maternal obesity and risk of preterm delivery. JAMA 2013;309(22):2362–70.

[8] Carlhäll S, Källén K, Blomberg M. Maternal body mass index and duration of labor. Eur J Obstet Gynecol Reprod Biol 2013;171(1):49–53.

[9] Tan T, Sia AT. Anesthesia considerations in the obese gravida. Semin Perinatol 2011;35(6):350–5.

[10] Morken NH, Klungsøyr K, Magnus P, Skjærven R. Pre-pregnant body mass index, gestational weight gain and the risk of operative delivery. Acta Obstet Gynecol Scand 2013;92(7):809–15.

[11] Smid MC, Kearney MS, Stamilio DM. Extreme obesity and postcesarean wound complications in the maternal-fetal medicine unit cesarean registry. Am J Perinatol 2015;32(14):1336–41.

[12] Mamun AA, Callaway LK, O'Callaghan MJ, Williams GM, Najman JM, Alati R, et al. Associations of maternal prepregnancy obesity and excess pregnancy weight gains with adverse pregnancy outcomes and length of hospital stay. BMC Pregnancy Childbirth 2011;11:62.

[13] Amir LH, Donath S. A systematic review of maternal obesity and breastfeeding intention, initiation and duration. BMC Pregnancy Childbirth 2007;7:9.

[14] Dai RX, He XJ, Hu CL. Maternal pre-pregnancy obesity and the risk of macrosomia: a meta-analysis. Arch Gynecol Obstet 2018;297(1):139–45.

[15] Zhang C, Wu Y, Li S, Zhang D. Maternal prepregnancy obesity and the risk of shoulder dystocia: a meta-analysis. BJOG 2018;125(4):407–13.

[16] Meehan S, Beck CR, Mair-Jenkins J, Leonardi-Bee J, Puleston R. Maternal obesity and infant mortality: a meta-analysis. Pediatrics 2014;133(5):863–71.

[17] Leonard SA, Rasmussen KM, King JC, Abrams B. Trajectories of maternal weight from before pregnancy through postpartum and associations with childhood obesity. Am J Clin Nutr 2017;106(5):1295–301.

[18] Marchi J, Berg M, Dencker A, Olander EK, Begley C. Risks associated with obesity in pregnancy, for the mother and baby: a systematic review of reviews. Obes Rev 2015;16(8):621–38.

[19] Picardo CM, Nichols M, Edelman A, Jensen JT. Women's knowledge and sources of information on the risks and benefits of oral contraception. J Am Med Womens Assoc 2003;58(2):112–16.

[20] Hall KS, White KO, Rickert VI, Reame NK, Westhoff CL. An exploratory analysis of associations between eating disordered symptoms, perceived weight changes, and oral contraceptive discontinuation among young minority women. J Adolesc Health 2013;52.

[21] Eichinger S, Hron G, Bialonczyk C, Hirschl M, Minar E, Wagner O, et al. Overweight, obesity, and the risk of recurrent venous thromboembolism. Arch Intern Med 2008;168(15):1678–83.

[22] Edelman AB, Cherala G, Stanczyk FZ. Metabolism and pharmacokinetics of contraceptive steroids in obese women: a review. Contraception 2010;82(4):314–23.

[23] Wisén O, Hellström PM. Gastrointestinal motility in obesity. J Intern Med 1995;237(4):411–18.

[24] Dubois A. Obesity and gastric emptying. Gastroenterology 1983;84(4):875–6.

[25] Tosetti C, Corinaldesi R, Stanghellini V. Gastric emptying of solids in morbid obesity. Int J Obes Relat Metab Disord 1996;20(3):200–5.

[26] Wisén O, Johansson C. Gastrointestinal function in obesity: motility, secretion, and absorption following a liquid test meal. Metabolism 1992;41(4):390–5.

[27] Bolt HM. Interactions between clinically used drugs and oral contraceptives. Environ Health Perspect 1994;102(Suppl. 9):35–8.

[28] Dobrinska MR. Enterohepatic circulation of drugs. J Clin Pharmacol 1989;29(7):577–80.

[29] Geier A, Dietrich CG, Grote T. Characterization of organic anion transporter regulation, glutathione metabolism and bile formation in the obese Zucker rat. J Hepatol 2005;43(6):1021–30.

[30] Forbes GB, Welle SL. Lean body mass in obesity. Int J Obes 1983;7(2):99–107.

[31] Chagnac A, Weinstein T, Korzets A, Ramadan E, Hirsch J, Gafter U. Glomerular hemodynamics in severe obesity. Am J Physiol Renal Physiol 2000;278(5):F817–22.

[32] Edelman AB, Carlson NE, Cherala G. Impact of obesity on oral contraceptive pharmacokinetics and hypothalamic–pituitary–ovarian activity. Contraception 2009;80(2):119–27.

[33] Westhoff CL, Torgal AH, Mayeda ER, Pike MC, Stanczyk. Pharmacokinetics of a combined oral contraceptive in obese and normal-weight women. Contraception 2010;81(6):474–80.

[34] Segall-Gutierrez P, Taylor D, Liu X, Stanzcyk F, Azen S, Mishell DR. Follicular development and ovulation in extremely obese women receiving depo-medroxyprogesterone acetate subcutaneously. Contraception 2010;81(6):487–95.

[35] Glasier A, Cameron ST, Blithe D, Scherrer B, Mathe H, Levy D, et al. Can we identify women at risk of pregnancy despite using emergency contraception? Data from randomized trials of ulipristal acetate and levonorgestrel. Contraception 2011;84(4):363–7.

[36] Lopez LM, Bernholc A, Chen M, Grey TW, Otterness C, Westhoff C, et al. Hormonal contraceptives for contraception in overweight or obese women. Cochrane Database Syst Rev 2016;18(8):CD008452.

[37] Merki-Feld GS, Skouby S, Serfaty D, Lech M, Bitzer J, Crosignani PG, et al. European society of contraception statement on contraception in obese women. Eur J Contracept Reprod Health Care 2015;20(1):19−28.

[38] Turner R. Most British women use reliable contraceptive methods, but many fear health risks from use. Fam Plann Perspect 1994;26:183−4.

[39] Gaudet LM, Kives S, Hahn PM, Reid RL. What women believe about oral contraceptives and the effect of counseling. Contraception 2004;69(1):31−6.

[40] Emans SJ, Grace E, Woods ER, Smith DE, Kelin K, Merola J. Adolescents' compliance with the use of oral contraceptives. JAMA 1987;257:3377−81.

[41] Oddens BJ. Women's satisfaction with birth control: a population survey of physical and psychological effects of oral contraceptives, intrauterine devices, condoms, natural family planning, and sterilization among 1466 women. Contraception 1999;59(5):277−86.

[42] Le MG, Laveissiere MN, Pelissier C. Factors associated with weight gain in women using oral contraceptives: results of a French 2001 opinion poll survey conducted on 1665 women. Gynecol Obstet Fertil 2003;31:230−9.

[43] Rosenberg M. Weight change with oral contraceptive use and during the menstrual cycle. Results of daily measurements. Contraception 1998;58:345−9.

[44] Wysocki S. A survey of American women regarding the use of oral contraceptives and weight gain [abstract]. Int J Gynecol Obstet 2000;70:114.

[45] Rosenberg MJ, Waugh MS, Meehan TE. Use and misuse of oral contraceptives: risk indicators for poor pill taking and discontinuation. Contraception 1995;51:283−8.

[46] Rosenberg MJ, Waugh MS. Oral contraceptive discontinuation: a prospective evaluation of frequency and reasons. Am J Obstet Gynecol 1998;179(3 Pt 1):577−82.

[47] Kaulhausen H, Klingsiek L, Breuer H. Changes of the renin−angiotensin−aldosterone system under contraceptive steroids. Contribution to the etiology of hypertension under hormonal contraceptives. Fortschr Med 1976;94(33):1925−30.

[48] Corvol P, Elkik F, Feneant M, Oblin ME, Michaud A, Claire M. Effect of progesterone and progestogens on water and salt metabolism. In: Bardin CW, Milgrom E, Mauvais-Jarvis P, editors. Progesterone and progestogens. New York: Raven Press; 1983.

[49] Mattsson C, Olsson T. Estrogens and glucocorticoid hormones in adipose tissue metabolism. Curr Med Chem 2007;14(27):2918−24.

[50] Nelson AL. Combined oral contraceptives. In: Hatcher RA, Trussell J, Nelson AL, Cates W, Stewart FH, Kowal D, editors. Contraceptive technology. New York: Ardent Media, Inc; 2007. p. 193−270.

[51] Reubinoff BE, Grubstein A, Meirow D, Berry E, Schenker JG, Brzezinski A. Effects of low-dose estrogen oral contraceptives on weight, body composition, and fat distribution in young women. Fertil Steril 1995;63(3):516.

[52] Gallo MF, Lopez LM, Grimes DA, Carayon F, Schulz KF, Helmerhorst FM. Combination contraceptives: effects on weight. Cochrane Database Syst Rev 2014;29(1):CD003987.

[53] Lindh I, Ellström AA, Milsom I. The long-term influence of combined oral contraceptives on body weight. Hum Reprod 2011;26(7):1917−24.

[54] Sheehan TJ, DuBrava S, DeChello LM, Fang Z. Rates of weight change for black and white Americans over a twenty year period. Int J Obes Relat Metab Disord 2003;27(4):498−504.

[55] Nooyens AC, Visscher TL, Verschuren WM. Age, period and cohort effects on body weight and body mass index in adults: the Doetinchem Cohort Study. Public Health Nutr 2009;12(6):862−70.

[56] Flegal KM. Obesity, overweight, hypertension, and high blood cholesterol: the importance of age. Obes Res 2000;8(9):676−7.

[57] Gupta S. Weight gain on the combined pill − is it real? Hum Reprod Update 2000;6(5):427−31.

[58] Milsom I, Lete I, Bjertnaes A. Effects on cycle control and body-weight of the combined contraceptive ring, NuvaRing, versus an oral contraceptive containing 30 microg ethinyl estradiol and 3 mg drospirenone. Hum Reprod 2006;21(9):2304−11.

[59] Berenson AB, Rahman M. Changes in weight, total fat, percent body fat, and central-to-peripheral fat ratio associated with injectable and oral contraceptive use. Am J Obstet Gynecol 2009;200(3):329.e1−8.

[60] Beksinska ME, Smit JA, Kleinschmidt I, Milford C, Farley TM. Prospective study of weight change in new adolescent users of DMPA, NET-EN, COCs, nonusers and discontinuers of hormonal contraception. Contraception 2010;81(1):30−4.

[61] Berenson AB, Odom SD, Breitkopf CR, Rahman M. Physiologic and psychologic symptoms associated with use of injectable contraception and 20 μg oral contraceptive pills. Am J Obstet Gynecol 2008;199(4):351.e1−351.e12.

[62] Curtis KM, Ravi A, Gaffield ML. Progestogen-only contraceptive use in obese women. Contraception 2009;80(4):346−54.

[63] Bonny AE, Britto MT, Huang B, Succop P, Slap GB. Weight gain, adiposity, and eating behaviors among adolescent females on depot medroxyprogesterone acetate (DMPA). J Pediatr Adolesc Gynecol 2004;17(2):109−15.

[64] Sivin I. Risks and benefits, advantages and disadvantages of levonorgestrel-releasing contraceptive implants. Drug Saf 2003;26(5):303−35.

[65] Lopez LM, Ramesh S, Chen M, Edelman A, Otterness C, Trussell J, et al. Progestin-only contraceptives: effects on weight. Cochrane Database Syst Rev 2016;28(8):CD008815.

[66] Alpert MA, Hashimi MW. Obesity and the heart. Am J Med Sci 1993;306(2):117−23.

[67] Huang Z, Willett WC, Manson JE. Body weight, weight change, and risk for hypertension in women. Ann Intern Med 1998;128(2):81−8.

[68] Colditz GA, Willett WC, Rotnitzky A, Manson JE. Weight gain as a risk factor for clinical diabetes mellitus in women. Ann Intern Med 1995;122(7):481−6.

[69] Grundy SM, Barnett JP. Metabolic and health complications of obesity. Dis Mon 1990;36(12):641−731.

[70] Yusuf S, Hawken S, Ounpuu S. Effect of potentially modifiable risk factors associated with myocardial infarction in 52 countries (the INTERHEART study): case−control study. Lancet 2004;364(9438):937−52.

[71] Rexrode KM, Hennekens CH, Willett WC. A prospective study of body mass index, weight change, and risk of stroke in women. JAMA 1997;277(19):1539−45.

[72] Ageno W, Becattini C, Brighton T, Selby R, Kamphuisen PW. Cardiovascular risk factors and venous thromboembolism: a meta-analysis. Circulation 2008;117(1):93−102.

[73] Holst AG, Jensen G, Prescott E. Risk factors for venous thromboembolism: results from the Copenhagen City Heart Study. Circulation 2010;121(17):1896−903.

[74] Stampfer MJ, Maclure KM, Colditz GA, Manson JE, Willett WC. Risk of symptomatic gallstones in women with severe obesity. Am J Clin Nutr 1992;55(3):652−8.

[75] Deslypere JP. Obesity and cancer. Metabolism 1995;44(9 Suppl. 3): 24−7.

[76] Schapira DV, Clark RA, Wolff PA, Jarrett AR, Kumar NB, Aziz NM. Visceral obesity and breast cancer risk. Cancer 1994;74(2):632−9.

[77] Chasan-Taber L, Willett WC, Manson JE. Prospective study of oral contraceptives and hypertension among women in the United States. Circulation 1996;94(3):483.

[78] Scarabin PY, Vissac AM, Kirzin JM. Elevated plasma fibrinogen and increased fibrin turnover among healthy women who both smoke and use low-dose oral contraceptives − a preliminary report. Thromb Haemost 1999;82(3):1112.

[79] Kelsey B. Contraceptive for obese women: considerations. Nurse Pract 2010;35(3):24−31.

[80] Lobo RA, Skinner JB, Lippman JS, Cirillo SJ. Plasma lipids and desogestrel and ethinyl estradiol: a meta-analysis. Fertil Steril 1996;65(6):1100.

[81] Krauss RM, Burkman RT. The metabolic impact of oral contraceptives. Am J Obstet Gynecol 1992;167(4 Pt 2):1177.

[82] Kjos SL, Peters RK, Xiang A, Thomas D, Schaefer U, Buchanan TA. Contraception and the risk of type 2 diabetes mellitus in Latina women with prior gestational diabetes mellitus. JAMA 1998;280(6):533.

[83] Rosenberg L, Palmer JR, Rao RS, Shapiro S. Low-dose oral contraceptive use and the risk of myocardial infarction. Arch Intern Med 2001;161(8):1065.

[84] Baillargeon JP, McClish DK, Essah PA, Nestler JE. Association between the current use of low-dose oral contraceptives and cardiovascular arterial disease: a meta-analysis. J Clin Endocrinol Metab 2005;90(7):3863−70.

[85] Gillum LA, Mamidipudi SK, Johnston SC. Ischemic stroke risk with oral contraceptives: a meta-analysis. JAMA 2000;284(1):72.

[86] Horton LG, Simmons KB, Curtis KM. Combined hormonal contraceptive use among obese women and risk for cardiovascular events: a systematic review. Contraception 2016;94(6):590−604.

[87] Nightingale AL, Lawrenson RA, Simpson EL, Williams TJ, MacRae KD, Farmer RD. The effects of age, body mass index, smoking and general health on the risk of venous thromboembolism in users of combined oral contraceptives. Eur J Contracept Reprod Health Care 2000;5(4):265−74.

[88] Gronich N, Lavi I, Rennert G. Higher risk of venous thrombosis associated with drospirenone-containing oral contraceptives: a population-based cohort study. CMAJ 2011;183(18):E1319−25.

[89] Committee on Gynecologic Practice. ACOG Committee Opinion Number 540: risk of venous thromboembolism among users of drospirenone-containing oral contraceptive pills. Obstet Gynecol 2012;120(5):1239−42.

[90] Suchon P, Al Frouh F, Henneuse A, Ibrahim M, Brunet D, Barthet MC, et al. Risk factors for venous thromboembolism in women under combined oral contraceptive. The PILl Genetic RIsk Monitoring (PILGRIM) study. Thromb Haemost 2016; 115(1):135−42.

[91] Pomp ER, le Cessie S, Rosendaal FR, Doggen CJ. Risk of venous thrombosis: obesity and its joint effect with oral contraceptive use and prothrombotic mutations. Br J Haematol 2007;139(2):289−96.

[92] van Hylckama Vlieg A, Helmerhorst FM, Vandenbroucke JP, Doggen CJ, Rosendaal FR. The venous thrombotic risk of oral contraceptives, effects of oestrogen dose and progestogen type: results of the MEGA case-control study. BMJ 2009;339:b2921.

[93] Lidegaard Ø, Løkkegaard E, Svendsen AL, Agger C. Hormonal contraception and risk of venous thromboembolism: national follow-up study. BMJ 2009;339:b2890.

[94] de Bastos M, Stegeman BH, Rosendaal FR, Van Hylckama Vlieg A, Helmerhorst FM, Stijnen T, et al. Combined oral contraceptives: venous thrombosis. Cochrane Database Syst Rev 2014;3(3):CD010813.

[95] World Health Organization. Medical eligibility criteria for contraceptive use. 5th ed. Geneva: World Health Organization; 2015. WHO guidelines approved by the Guidelines Review Committee.

[96] Rocha ALL, Campos RR, Miranda MMS, Raspante LBP, Carneiro MM, Vieira CS, et al. Safety of hormonal contraception for obese women. Expert Opin Drug Saf 2017;16(12):1387−93.

[97] Mantha S, Karp R, Raghavan V, Terrin N, Bauer KA, Zwicker JI. Assessing the risk of venous thromboembolic events in women taking progestin-only contraception: a meta-analysis. BMJ 2012;345:e4944.

[98] Tepper NK, Whiteman MK, Marchbanks PA, James AH, Curtis KM. Progestin-only contraception and thromboembolism: a systematic review. Contraception 2016;94(6):678−700.

[99] Melhado-Kimura V, Bizzacchi JMA, Quaino SKP, Montalvao S, Bahamondes L, Fernandes A. Effect of the injectable contraceptive depot-medroxyprogesterone acetate on coagulation parameters in new users. J Obstet Gynaecol Res 2017;43(6):1054−60.

[100] Tepper NK, Dragoman MV, Gaffield ME, Curtis KM. Nonoral combined hormonal contraceptives and thromboembolism: a systematic review. Contraception 2017;95(2):130−9.

[101] Smith JS, Green J, Berrington de Gonzalez A. Cervical cancer and use of hormonal contraceptives: a systematic review. Lancet 2003;361(9364):1159.

[102] Hankinson SE, Colditz GA, Manson JE. Prospective study of oral contraceptive use and risk of breast cancer (Nurses' Health Study, United States). Cancer Causes Control 1997;8(1):65−72.

[103] Marchbanks PA, McDonald JA, Wilson HG. Oral contraceptives and the risk of breast cancer. N Engl J Med 2002;346(26):2025.

[104] Hannaford PC, Selvaraj S, Elliott AM, Angus V, Iversen L, Lee AJ. Cancer risk among users of oral contraceptives: cohort data from the Royal College of General Practitioner's oral contraception study. BMJ 2007;335(7621):651.

[105] Grabrick DM, Hartmann LC, Cerhan JR. Risk of breast cancer with oral contraceptive use in women with a family history of breast cancer. JAMA 2000;284(14):1791.

[106] Collaborative Group on Epidemiological Studies of Ovarian Cancer, Beral V, Doll R, Hermon C, Peto R, Reeves G. Ovarian

cancer and oral contraceptives: collaborative reanalysis of data from 45 epidemiological studies including 23,257 women with ovarian cancer and 87,303 controls. Lancet 2008;371(9609):303.
[107] Pan SY, Johnson KC, Ugnat AM, Wen SW, Mao Y, Canadian Cancer Registries Epidemiology Research Group. Association of obesity and cancer risk in Canada. Am J Epidemiol 2004;159 (3):259.
[108] Rapp K, Schroeder J, Klenk J. Obesity and incidence of cancer: a large cohort study of over 145,000 adults in Austria. Br J Cancer 2005;93(9):1062.
[109] Iversen L, Sivasubramaniam S, Lee AJ, Fielding S, Hannaford PC. Lifetime cancer risk and combined oral contraceptives: the Royal College of General Practitioners' oral contraception study. Am J Obstet Gynecol 2017;216(6):580.e1–9.
[110] Kwon JS, Lu KH. Cost-effectiveness analysis of endometrial cancer prevention strategies for obese women. Obstet Gynecol 2008;112(1):56–63.
[111] Misra JS, Engineer AD, Tandon P. Cytopathological changes in human cervix and endometrium following prolonged retention of copper-bearing intrauterine contraceptive devices. Diagn Cytopathol 1989;5:237–42.
[112] Rivera R, Best K. Current opinion: consensus statement on intrauterine contraception. Contraception 2002;65:385–8.
[113] Batar I. The Szontagh IUD and cervical carcinoma (results of a 10-year follow up study). Orv Hetil 1990;131:1871–4.
[114] Ganacharya S, Bhattoa HP, Batár I. Pre-malignant and malignant cervical pathologies among inert and copper-bearing intrauterine contraceptive device users: a 10-year follow-up study. Eur J Contracept Reprod Health Care 2006;11(2):89–97.
[115] Maggard MA, Yermilov I, Li Z. Pregnancy and fertility following bariatric surgery: a systematic review. JAMA 2008;300: 2286–96.
[116] Marceau P, Kaufman D, Biron S. Outcome of pregnancies after biliopancreatic diversion. Obes Surg 2004;14:318–24.
[117] Merhi ZO. Challenging oral contraception after weight loss by bariatric surgery. Gynecol Obstet Invest 2007;64:100–2.
[118] Gosman GG, King WC, Schrope B. Reproductive health of women electing bariatric surgery. Fertil Steril 2010;94:1426–31.
[119] Mechanick JI, Youdim A, Jones DB, Garvey WT, Hurley DL, McMahon MM, et al. American Association of Clinical Endocrinologists; Obesity Society; American Society for Metabolic & Bariatric Surgery Clinical practice guidelines for the perioperative nutritional, metabolic, and nonsurgical support of the bariatric surgery patient—2013 update: cosponsored by American Association of Clinical Endocrinologists, the Obesity Society, and American Society for Metabolic & Bariatric Surgery. Endocr Pract 2013;19.
[120] Paulen ME, Zapata LB, Cansino C, Curtis KM, Jamieson DJ. Contraceptive use among women with a history of bariatric surgery: a systematic review. Contraception 2010;82(1):86–94.

[121] Victor A, Odlind V, Kral JG. Oral contraceptive absorption and sex hormone binding globulins in obese women: effects of jejunoileal bypass. Gastroenterol Clin North Am 1987;16:483–91.
[122] Ciangura C, Corigliano N, Basdevant A. Etonorgestrel concentrations in morbidly obese women following Roux-en-Y gastric bypass surgery: three case reports. Contraception 2011;84 (6):649–51.
[123] Gerrits EG, Ceulemans R, van Hee R, Hendrickx L, Totté E. Contraceptive treatment after biliopancreatic diversion needs consensus. Obes Surg 2003;13:378–82.
[124] Weiss HG, Nehoda H, Labeck B, Hourmont K, Marth C, Aigner F. Pregnancies after adjustable gastric banding. Obes Surg 2001;11:303–6.
[125] Choi JY, Scarborough TK. Stroke and seizure following a recent laparoscopic Roux-en-Y gastric bypass. Obes Surg 2004;14: 857–60.
[126] Wu EC, Barba CA. Current practices in the prophylaxis of venous thromboembolism in bariatric surgery. Obes Surg 2000;10(1):7–13.
[127] Wucher H, Ciangura C, Poitou C, Czernichow S. Effects of weight loss on bone status after bariatric surgery: association between adipokines and bone markers. Obes Surg 2008;18:58–65.
[128] Coates PS, Fernstrom JD, Fernstrom MH, Schauer PR, Greenspan SL. Gastric bypass surgery for morbid obesity leads to an increase in bone turnover and a decrease in bone mass. J Clin Endocrinol Metab 2004;89:1061–5.
[129] Curtis KM, Martins SL. Progestogen-only contraception and bone mineral density: a systematic review. Contraception 2006;73:470–87.
[130] Grimes DA, Shields WC. Family planning for obese women: challenges and opportunities. Contraception 2005;72(1):1–4.
[131] Vilos GA, Marks J, Tureanu V, Abu-Rafea B, Vilos AG. The levonorgestrel intrauterine system is an effective treatment in selected obese women with abnormal uterine bleeding. J Minim Invasive Gynecol 2011;18(1):75–80.
[132] Saito-Tom LY, Soon RA, Harris SC, Salcedo J, Kaneshiro BE. Levonorgestrel intrauterine device use in overweight and obese women. Hawaii J Med Public Health 2015;74(11):369–74.
[133] Chi IC, Wilkens L. Interval tubal sterilization in obese women – an assessment of risks. Am J Obstet Gynecol 1985;152(3): 292–7.
[134] Jamieson DJ, Hillis SD, Duerr A, Marchbanks PA, Costello C, Peterson HB. Complications of interval laparoscopic tubal sterilization: findings from the United States collaborative review of sterilization. Obstet Gynecol 2000;96(6):997–1002.
[135] Chi IC, Wilkens LR, Reid SE. Prolonged hospital stay after laparoscopic sterilization. IPPF Med Bull 1984;18(4):3–4.
[136] Singh KB, Huddleston HT, Nandy I. Laparoscopic tubal sterilization in obese women: experience from a teaching institution. South Med J 1996;89(1):56–9.

Chapter 6

Contraceptive choices for women before and after bariatric surgery

Agnieszka Jurga-Karwacka[1] and Johannes Bitzer[2,3]

[1]*Department of Gynecology and Gynecological Oncology, University Hospital Basel, Basel, Switzerland,* [2]*Department of Obstetrics and Gynecology, University Hospital of Basel, Basel, Switzerland,* [3]*Post Graduate Diploma of Advanced Studies in Sexual Medicine, University of Basel, Basel, Switzerland*

Introduction—Bariatric Surgery

Bariatric surgery is the most effective treatment for morbid obesity [1]. Due to the consensus statement of the National Institutes of Health, following group of patients should be considered suitable candidates for bariatric surgery if they meet one of the criteria [2]:

- Body mass index (BMI) greater than 40;
- BMI of 30–40 plus one of the following obesity-associated comorbidities: severe diabetes mellitus, Pickwickian syndrome, obesity-related cardiomyopathy, severe sleep apnea, or osteoarthritis interfering with lifestyle.

Bariatric surgeries lead to weight loss and comorbidity improvement by the following mechanisms [3]:

1. gastric restriction: reduction of amount of food that can be consumed;
2. malabsorption: impaired food-digestion by smaller stomach and nutrients absorption by shortened intestine; or
3. combination of the both.

The most common bariatric procedures are [4]:

1. Adjustable gastric band: it is a most commonly performed and purely restrictive procedure: this is achieved by placing an inflatable band around the upper portion of stomach, thus creating a small stomach pouch above the band.
    a. vertical banded gastroplasty: sectioning off the cardia of the stomach by a longitudinal staple line and wrapping tight the outlet with a band or mesh [5];
    b. laparoscopic sleeve gastrectomy: removing approximately 80% of stomach;
    c. intragastric balloon; and
    d. endoluminal gastroplasty.

2. Biliopancreatic diversion (BPD) [6]
    a. jejunoileal bypass: dividing the jejunum near the ligament of Treitz and reconnecting it near the ileocecal valve, bypassing a long small bowel segment(b)—this procedure is no longer performed [7,8]
3. Roux-en-Y gastric bypass: it is the gold standard procedure done by creating a small stomach pouch by dividing the top of the stomach from the rest and connecting it with a bottom of divided small intestine, then connecting the top portion of the divided small intestine to the small intestine further down [9]
    a. BPD with duodenal switch: this procedure has two components: first, creating a smaller, tubular stomach pouch by removing a portion of stomach and then bypassing a large portion of the small intestine [8].

Most weight-loss surgeries today are performed using minimally invasive techniques (laparoscopic surgery) [10]. All the abovementioned procedures result in gut hormones changes that promote satiety and suppress hunger [4].

The majority of patients undergoing bariatric therapies (up to 80%) are women and most of them are in childbearing age (mean age 39 years) [7,11].

Reproductive and general health consequences of bariatric surgery:

1. Increased fertility in a short period of time postoperatively.
    Weight-reduction has a positive effect on sex hormone profiles and ovulation [12,13]. In a study of 24 amenorrhoic women with PCOS, normal menstrual cycles resumed after a mean of 3 months postoperatively and five women became pregnant after a weight-reduction of 57% [14].
2. In general, it has been shown that there is an improved maternal and fetal outcome during pregnancy and in

the postpartum period, compared to women with untreated obesity [11].

However, it has been reported that conceiving during the period of rapid weight-loss seen in the first 12−24 months following bariatric surgery is associated with higher rates of nutritional deficiencies and obstetric complications [13,15], such as a higher incidence of stillbirths in the first year after surgery [16].

3. Obese women who have undergone surgical treatment for obesity are advised not to conceive for the following 12−18 months [17] in order to ensure an optimal and a stable weight with optimal nutritional and vitamin status before the start of pregnancy [13].

Effective contraception is therefore of high importance for these women.

A range of personal factors, including long- and short-term requirements, future plans for pregnancy, sexual health risks, age, overall health, and use of other medications should be taken into consideration when choosing a method of contraception.

When counseling women who are planning to undergo a bariatric surgery, safety and efficacy of contraceptive methods for both, the pre− and post−weight loss period should be discussed.

The general advice of the UK Faculty for Sexual and Reproductive Health (FSRH) regarding contraception for post bariatric surgery patients is as follows:

"That long-acting reversible contraceptive (LARC) methods should be used for the first 2 years after surgery, as they are believed to be the most reliable ones (with failure rate <1 pregnancy per 100 women in a year) [17]. However, the evidence regarding use of contraception after bariatric surgery is very limited."

Long acting contraceptives

Copper intrauterine device

Different types of copper intrauterine device (Cu-IUD) with variable amount of copper content, frames, and duration of action (between 5 and 12 years) are available. The most frequently used are ParaGard, Flexi-T, Multiload, T-Safe, and GyneFix [18]. Their principle of action is by local inflammatory and spermicidal effect of copper ions. As this is a local action, it is expected not to be affected by BMI [18].

Efficacy:

Pearl Index (PI) for ideal use is 0.6 and for typical use 0.8 [19]. The real-life user failure rate is less than 1%. After prolonged continuous use, the cumulative pregnancy rate is 1.6% at 7 years and 2.2% at 8 and 12 years [18,20].

There are no studies comparing specifically the efficacy of Cu-IUD in obese women and women with normal BMI.

However, Gemzell-Danielsson et al. [21] have reported that the IUCD failure rate [for both Cu-IUD and levonorgestrel (LNG)-IUD] is less than one pregnancy per 100 woman-years independent of BMI. There are no data available on use of the Cu-IUD after bariatric surgery, but theoretically the effectiveness of the device should not be different due to its local mechanism of action [20].

Safety:

Although there are no studies in this area, but there are no theoretical reasons for Cu-IUD to cause any specific health risks in obese and/or postbariatric women—as recommended by the UK Medical Eligibility Criteria for Contraceptive Use (UKMEC) category 1: no restriction for use [17].

Cu-IUD does not increase the risk of cardiovascular diseases, including venous thormboembolic disease (VTE) and myocardial infarction. It has no metabolic impact on diabetes or metabolic syndrome. No weight increase has been reported [22].

The risk of complications during placement is minimal. It includes uterine perforation (0.1%) and pelvic inflammatory disease (PID; 0%−2%), which is thought to be not caused by the device itself but only by a preexisting vaginal and cervical infection [23].

The cumulative risk of IUD expulsion is 10% over 3 years of use [20]. There is no evidence of increased risk of expulsion or perforation in obese patients, although insertion may be sometimes difficult.

Side effects:

The usual side effects of Cu-IUD are hypermenorrhea, dysmenorrhea and lower abdominal pain, or discomfort [18,20].

Health benefits:

Cu-IUD has a protective effect on the endometrium, which in obese women is at higher risk of developing an endometrial carcinoma [24].

Contraindications [18,25]:

- severe uterine distortion (anatomic abnormalities, including uterus bicornuate, cervical stenosis, or fibroids severely distorting the uterine)
  - increased difficulty of insertion
  - increased risk of expulsion
- active pelvic infection (PID, endometritis, mucopurulent cervicitis, and pelvic tuberculosis)
  - increased risk of an infection exacerbation through a foreign body

  The IUD may be inserted in women who are at least 3 months after treatment for PID or puerperal/postabortion sepsis.
- known or suspected pregnancy
  - miscarriage
  - increased risk of septic abortion
- Wilson's disease or copper allergy

Although no adverse event related to copper allergy or Wilson's disease has ever been reported in a woman with a Cu-IUD, progesterone-only contraceptives are preferred for use in women with these conditions [26].

- unexplained abnormal uterine bleeding and preexisting pathology

Limitations of use:
Use of Cu-IUD may not be advisable if woman has any of the followings:

1. heavy menstrual bleeding
2. dysmenorrhea
3. endometriosis
4. women with high risk for sexually transmitted infections

Practical issues in obese women:
Insertion of the IUD can be challenging in severely obese women. Selecting a large speculum or placing the tip of a condom over the blades of the speculum may be helpful to optimize the visualization of the cervix. Transabdominal ultrasound may one to help to guide the insertion.

Summary for women before and after bariatric surgery:
There is no evidence that the efficacy, safety, and tolerability of Cu-IUDs are different from normal-weight women in obese patients before and after bariatric surgery due to the local mechanism of action.

Levonorgestrel containing intrauterine device

There are three types of LNG-intrauterine systems (IUS):

1. LNG 52 (Mirena) containing 52 mg of LNG with an average daily release of 20 μg LNG. Effective for at least 5 years.
2. LNG 14 (Jaydess) containing 13.5 mg of LNG with an average daily release of 6 μg LNG. Effective for 3 years.
3. LNG 20 (Kyleena) containing 19.5 mg of LNG with an average daily release of 9 μg LNG. Effective for 5 years.

The mechanism of action of LNG-IUS relies on local effects and not on systemic drug levels. The progestin secreted from the IUS causes thickening of cervical mucus and increase in expression of glycodelin A in endometrial glands, which inhibits binding of sperm to the egg cell [25,27,28]. Serum concentrations of progestin can partially inhibit the ovarian follicular development and lead to anovulation, but this is thought not to be the major contraceptive mechanism. It has been reported that at least 75% of women using an LNG-IUS have ovulatory cycles [27].

Efficacy:
PI for all the three IUS is around 0.2 and the cumulative pregnancy rate is 0.5%–1.1% after 5 years of continuous use with the LNG 20 IUD [19]. The 3-year cumulative pregnancy rate is 0.9% with the LNG 14 IUD. Available evidence suggests that LNG-IUS effectiveness is not reduced by higher body weight or BMI [29,30]. It seems also not to be affected by malabsorptive surgery [31,32].

A prospective cohort study reported no statistically significant difference in contraceptive failure rate during the first 2–3 years of use among intrauterine contraception (IUC) users (Cu-IUD or 52 mg LNG-IUS) who were of normal BMI ($n = 1584$), overweight BMI ($n = 1149$), or obese BMI ($n = 1467$). The overall IUC failure rate of less than one pregnancy per 100 woman-years did not vary by BMI [21].

Safety:
The use of LNG-IUS is not restricted by obesity alone, independent of the degree of obesity (UKMEC 1). The presence of coexisting risk factors for CVD (smoking, diabetes, hypertension, history of VTE) in addition to obesity puts LNG-IUS into UKMEC category 2, where the advantages of using this contraception method generally outweigh the theoretical or proven risks [17]. There are no reports about an increased risk of cardiovascular diseases. A Danish national registry-based cohort study [33] found that the risk of confirmed VTE was not increased in LNG-IUS users (adjusted relative risk 0.57; 95% CI 0.41–0.81). A metaanalysis of eight observational studies also showed no association between VTE risk and use of an LNG-IUS (adjusted relative risk 0.61; 95% CI 0.24–1.53) [34].

Due to contradictory study results, there are some concerns about the association between LNG-IUS use and breast cancer. Two large retrospective case–control studies of European women showed no increased risk of breast cancer in women using the LNG-IUS for contraception [35,36]. However, in two analyses of a large Finnish cohort, the risk of breast cancer, in particular of lobular and ductal cell cancers, was slightly increased (up to 1.3 times) in women using the LNG-IUS for heavy menstrual bleeding [17,18]. In the Danish cohort study, the authors found a relative increased risk of breast cancer of 1.21 (95% CI 1.11–1.33) [37]. Taking into account those results, a progestogen-only method is considered either not to be, or associated with a slightly increased risk for breast cancer [18,20]. However, according to FSRH, the risk of the LNG-IUS in current or active breast cancer patients is currently classified as category UKMEC 4 (unacceptable risk) and UKMEC 3 (risk outweigh benefits) in patients with a personal history of breast cancer in the last 5 years with no active disease [18].

In the general population, there are no significant differences in weight gain when hormonal and nonhormonal intrauterine methods are compared. There is also no evidence that LNG-IUS use has any impact on weight gain in women with higher BMI.

Side effects:

- Irregular bleeding: very common during the first 3–6 months

At 24 months follow-up, 50% of LNG 20 users have amenorrhea, 30% have oligoamenorrhea, and 11% have spotting. The pattern is similar with the LNG 14 IUD with lower rates of amenorrhea (13% vs 24% after 3 years) [18].
- Mood changes

  In the Danish cohort, LNG-IUS users had an increased risk of having to be prescribed antidepressants and a higher risk of being hospitalized with depression [38]. This may be of clinical importance in those women who suffer from preclinical or undiagnosed perimenopausal depression [39].
- Breast tenderness and acne (rarely)

Health benefits:
- Reduction in heavy menstrual bleeding and dysmenorrhea in patients without organic pathology or bleeding diathesis, including undergoing anticoagulation therapy. The efficacy regarding reduction of bleeding intensity in women with fibroids and adenomyosis is yet unclear and under investigation [25].
- Protection from PID, due to cervical mucus thickening, which acts as a barrier toward ascending infections [20,23].
- Treatment of endometriosis-related pain.
- Endometrial protection in perimenopausal and menopausal women using estrogen replacement.
- Risk reduction of endometrial cancer [18,25].
- Fewer painful crises in women with sickle cell disease.
- IUD can also be used in treatment of endometrial hyperplasia and cancer.

Contraindications:
- current or active breast cancer or personal history of breast cancer in the past 5 years with no active disease;
- active liver disease;
- other contraindications that are related to the intrauterine application and include severe deformity of the uterine cavity, acute sexually transmitted infections, unexplained vaginal bleeding, as well as known or suspected pregnancy.

Summary for women before and after bariatric surgery:
There is no evidence that the efficacy, safety, and tolerability of LNGIUDs are different in patients before and after bariatric surgery when compared with normal-weight women. There are no studies available as regards the systemic effect of the progestogen absorbed in utero among the abovementioned both study groups.

Progesterone-only implant

Hormonal implants are subdermal inserts that provide reliable long-time contraception [40] with very rare true contraceptives failures. The matrices consist of small flexible biologically degradable rods inserted under the skin of the upper arm. Following preparations are currently available globally:

Norplant II (Jadelle, Sinoplant) is composed of two flexible silicone rods, each containing 75 mg of LNG, and has duration of action of 5 years.

Implanon (Nexplanon) is an etonogestrel (ENG)-releasing hormonal implant. The single rod contains 68 mg ENG [3-keto-desogestrel (DSG)]. The duration of action after subdermal insertion is 3 years.

Effectiveness:

The PI is 0.05 [19]. In the US Contraceptive CHOICE project with 1168 individuals, the pregnancy rate at 4 years follow-up of ENG implant use was extremely low in all BMI groups of implant users [41]. In Europe, based on recent pharmacokinetic and clinical outcomes data, the FSRH Guidelines Development Group concluded that the ENG-releasing implant represents a highly effective contraceptive method with a licensed use-duration of 3 years in women of all BMIs [42].

In patients who underwent malabsorptive surgery, the subcutaneous (SC) application of steroid appears to be unaffected. Some evidence, level-3 data, suggest that although serum ENG concentration after surgery decreased with weight loss, it remained in the range considered to provide the contraceptive effect for at least 6 months [41]. Unfortunately, there are no long-term data available on long-term efficacy.

It is known that the serum level of ENG depends on body weight and decreases at a higher rate in women with obesity due the increased volume of distribution, effects on plasma protein binding, and altered clearance [43]. The absolute ENG plasma concentration required for suppression of ovulation suggested in the studies is 90 pg/mL but in reality it may be lower. In the study of prolonged ENG implant use (up to 5 years) in 237 women (25% overweight and 46% obese) that continued ENG implant use beyond 3 years, no pregnancies occurred during the period of prolonged use for all BMIs. Median levels of serum ENG remained above the suggested ovulation threshold of 90 pg/mL in all BMI classes after the fourth and the fifth year of implant use [44]. Clinical experience with Implanon in women weighing more than 80 kg is limited and plasma levels of ENG are lower in obese women and can be close to the concentration, which is necessary to effectively prevent ovulation [43]. There was a question as regards earlier replacement of the implant (2 years instead of 3 years) in women over 80 kg or at a higher BMI, but recent studies indicate that the efficacy is maintained over the indicated time of use independent of BMI.

Safety:

Obesity alone does not restrict the use of progestogen implant (IMP) (UKMEC 1) and even in the case of coexisting CVD-risk factors (smoking, diabetes, and hypertension), the advantages of using that method generally outweigh the theoretical or proven risks (UKMEC 2). Especially, there is no evidence that the implant increases the risk of thromboembolic events or myocardial infarction after bariatric surgery. In the National Danish Registry study, VTE incidence during the use of ENG implants (1.7 per 10,000 exposure years) did not significantly differ from the VTE incidence in nonusers of hormonal contraception [33]. Studies on hemostatic parameters show neither significant procoagulatory values nor increased blood pressure [45,46]. Further studies did not show negative effects on carbohydrate or lipid parameters used as markers of risk (fasting glucose, insulin, triglycerides, and HDL) [46]. The implant has little and clinically irrelevant impact on fasting glucose and insulin in obese women. In women with diabetes, HbA1c did not change in implant users neither did the daily insulin requirement [47,48]. There is no concern regarding bone mass loss [49].

There are not enough long-term studies regarding the risk of breast cancer in patients using SC implants. It has to be taken into consideration that obese women have a higher base risk for breast cancer, which theoretically could increase their sensitivity to the progestogen action on the breast. There is however until now no consensus about the impact of progestogen-only contraception on breast cancer risk in normal-weight women.

The question about whether the implant leads to an additional weight increase is not yet completely resolved. In a comparative 12-month study of three different progestin-only methods (including the ENG implant), a small increase in weight (2.1 kg) was observed in all of them in comparison to Cu-IUD [50].

Health benefits:

The main noncontraceptive benefit of progesterone-only implant is that it may help one to alleviate dysmenorrhea and ovulatory pain that are not associated with any identifiable pathological conditions [51]. While there is theoretically no reason why this would not be the case for women who are overweight or obese, this has not been specifically studied in women of different weight categories [29].

Practical issues:

Generally, the placement or removal of IMP should be unproblematic in women who are overweight or women with obesity. There are some theoretical concerns regarding difficult implant insertion in women with significant weight loss after surgery due to loose skin, while a correct subdermal placement of the implant is crucial in women of all BMIs.

Contraindications:

There are very few contraindications that the implant shares with other progesterone-only contraceptives: active or recent breast cancer, and active and chronic liver disease (except nodular hyperplasia). It has to be considered that same drugs (such as antiviral drugs) may affect the metabolism of ENG.

Summary for women before and after bariatric surgery:

There is no evidence that the efficacy, safety, and tolerability of progestogen implants are different in patients before and after bariatric surgery when compared with normal-weight women. Due to the fact that the mechanism of contraception is by the systemic ovulation inhibitory effect of the progestogens, there is a theoretical possibility that the efficacy of the device may be influenced by weight in general and rapid weight change (see before). Only few studies are available and until now there is no evidence of a reduced efficacy in women before and after bariatric surgery.

Progesterone-only injection

Depot medroxyprogesterone acetate (DMPA)—Depo-Provera is an injectable progestin-only contraceptive that provides reversible contraception.

Three progestogen-only injectable contraceptives are available:

- DMPA: intramuscular (DMPA-IM) and DMPA-SC, which are administered every 13 weeks
- Norethisterone (NET) enanthate: which is administered IM every 8 weeks

Efficiency:

The PI range is between 0.2 by perfect and 6 by typical use. There are limited data relating to DMPA use in women with obesity or after bariatric surgery, but the available evidence suggests that its effectiveness is not reduced [52—54].

Safety:

DMPA use is not contraindicated because of obesity alone (UKMEC 1), but its use should be restricted (UKMEC 3—the theoretical or proven risks usually outweigh the advantages of using the method) in the case of coexisting multiple risk factors for CVD (smoking, diabetes, and hypertension) [17]. The causal association between DMPA use and VTE cannot be evaluated because of limited evidence available.

DMPA use and bariatric surgery both appear to be associated with the loss of bone mineral density. However, there is no evidence for an additive effect of those two factors relating specifically to risk for osteoporosis or fracture. Since the clinical significance of this fact is unknown, postbariatric women should be made aware that there are other effective contraceptive methods that are not associated with reduced bone mineral density.

DMPA use is associated with some weight gain. Prior studies had demonstrated an association with weight gain in women under 18 years old with BMI ≥ 30 kg/m^2, and therefore, prescribing DMPA after bariatric surgery in adolescents should proceed with caution [55].

Health benefits:

- treatment of heavy menstrual bleeding, dysmenorrhea, and for pain associated with endometriosis [51];
- some protection against ovarian and endometrial cancers [25] but there are no studies taking into consideration patients' BMI.

Clinical tips:

In obese women, DMPA-IM injection may require deltoid administration or a use of a longer needle to reach the muscular layer in gluteal approach. Alternatively, DMPA-SC use can be considered.

Summary for women before and after bariatric surgery:

There is no evidence that the efficacy of progesterone-only injectables is different for normal-weight women when compared with patients before and after bariatric surgery. Due to the concerns regarding bone safety and the possible impact on weight gain progesterone-only injections would not be the first choice if intrauterine contraceptives or implants are available and no contraindications exist.

Oral hormonal contraception

Most available combined oral contraceptives (COCs) contain ethinylestradiol as the estrogenic component and one of various progestins in a sufficient dose [56]. Progestins are primarily responsible for the contraceptive effect that relies on suppressing the hypothalamic–pituitary–ovarian axis leading to inhibition of ovulation. Secondary contraceptive effects include thickening of cervical mucus and thinning of the endometrial lining [56]. The estrogen component contributes to ovulation suppression and control of irregular bleeding.

Progestogen-only pill

The progestogen-only pill (POP) contains only a progestin.

Efficiency:

PI of POP as well as COC is 9 by typical and 0.3 by perfect use [19]. The available evidence suggests that POP is not affected by body weight or BMI [17]. The findings of studies of COC effectiveness in relation to increased body weight/BMI vary. The most of high-quality studies suggest that the effectiveness of COC is not reduced in women with obesity, but the pharmacokinetics of steroid hormones appears to be changed in obese COC pill users compared to those with normal-weight. Therefore it could be potentially less effective when used imperfectly. However, there was no clear impact of high BMI on the end organ suppression by COC [57–59].

There are theoretical concerns about decreased OC absorption after bariatric procedures [60]. Postoperative complications such as long-term diarrhea and/or vomiting could probably further decrease OC effectiveness.

Theoretically, OCs are dissolved in the stomach and transformed by bacterial enzymes and the enzymes in the intestinal mucosa. The metabolized and nonmetabolized drugs are then absorbed through the intestinal mucosa and enter the portal circulation [61]. 17α-Ethinylestradiol, the estrogenic component of COC, undergoes first-pass metabolism caused by gut wall sulfation for at least 60% of volume of the content [62,63]. It may be assumed that absorption of oral contraceptives could be affected by malabsorptive and restrictive–malabsorptive character of bariatric procedures. However, the pharmacokinetic studies on that matter are inconclusive.

Victor et al. [64] investigated serum levels of two progestin-only pills, NET 3 mg and LNG 0.25 mg, in seven obese (BMI > 40 kg/m^2) women (20–44 years) after jejunoileal bypass. The reference group consisted of healthy normal-weight women. The mean plasma concentrations of both progestins were significantly lower in the operated patients at 1–8 hours after oral application. The authors concluded that the risk for contraceptive failure of POP is increased in patients with jejunoileal bypass, but the difference could be caused by the body weight and not bariatric procedure.

Another study [65] compared serum progesterone levels after COC in very obese female patients after bariatric surgery to those who did not undergo a surgical procedure and reported higher serum levels in operated women. Recent study of Ginstman et al. [66] showed no significant changes in the plasma concentrations of ENG in nine women using orally 75 μg DSG before and after Roux-en-Y bypass (measurements 8 weeks before, and 12 and 52 weeks after surgery).

However, in a retrospective chart review study, in which 276 women had documented the contraceptive method they use (COC 48.5%, sterilization 39%, IUD 6%, DMPA 3%, patch/ring 1.8%, barrier 1.4%, and vasectomy 0.4%), 16 pregnancies were identified in the first 18 months postoperatively. Although the contraception method used in this subgroup was documented only in three pregnancies, two of the women used COCs and one condom [67]. Another retrospective study by Vilallonga et al. [68] assessed the outcomes after laparoscopic Roux-en-Y gastric bypass in 19 patients aged under 18 years between 2 and 10 years post-bypass. Data were self-reported. Twelve of those patients reported using COC and two of them became pregnant (6 and 8 years after surgery).

Safety and health benefits

Use of COC as well as other combined hormonal contraceptives (CHC) leads to an increase in the cardiovascular and metabolic risk proportionally to the BMI. For this reason, CHC seems to be the least appropriate method of contraception for morbidly obese patients. The benefits regarding the protective effect on endometrial carcinoma and reduction of heavy menstrual bleeding do not in general outweigh the potential risks.

Due to unknown and potentially reduced efficacy of COC in postbariatric patients, as well as elevated risks associated with COC in obese women specifically, the use of OC is not advised in patients before and after bariatric surgery.

According to the UKMEC, obesity alone does not restrict the use of POP (UKMEC 1). Even when obesity is accompanied by other risk factors for CVD (e.g., smoking, diabetes, and hypertension), the use of POP is qualified as UKMEC 2. Study results show no association between POP use and increased risk of CVD [17].

There are however no specific studies investigating whether women with raised BMI using POP are at an increased risk of VTE and other cardiovascular events compared to normal-weight patients.

Contraceptive patch and ring

The transdermal patch and vaginal ring contain a combination of estrogen and progestin. Their mechanism of action is the same as COCs.

Both oral and nonoral CHC that are associated with an elevated risk for CVD in obese women are categorized as UKMEC 3. In addition, contraceptive patches may be less effective in patients with obesity, and caution is necessary.

Summary for women before and after bariatric surgery:

- CHC (oral, transdermal, and vaginal) carry an increased risk for cardiovascular complications in super obese patients before bariatric surgery [17]. After the surgery, oral contraceptives may have a reduced efficacy due to malabsorption. Nonoral CHC may be considered after weight stabilization and exclusion of other cardiovascular risk-increasing factors.
- Oral progestogen-only preparations do not seem to increase the cardiovascular risk on preoperative obese patients. Due to possible absorption problems, a reduction of efficacy will have to be expected.
- CHC and POPs associated with possible increase risks and reduced efficacy, therefore, are not suitable contraceptives for patients before and after bariatric surgery.

Barrier method

Barrier methods, including male and female condoms, cervical caps, and diaphragms, are an effective method of contraception if used correctly. Above that condoms offer an additional protection against sexually transmitted diseases.

Efficiency:
PI of condoms is 18 by typical and 2 by perfect use, PI of female diaphragm is 12 and 6, respectively [19]. The reliability of barrier methods should not be theoretically affected by obesity or bariatric procedures; however, there are no studies that make this direct comparison.

Clinical tips:
Weight loss or gain can alter the fit of a cap or diaphragm; so it would be a good practice to check regularly if it still fits during the period of rapid weight loss following a bariatric surgery.

Emergency contraception

Women with a history of bariatric surgery presenting for emergency contraception (EC) should be offered a Cu-IUD, assuming that there are no contraindications (see before) [17]. For cases, where the Cu-IUD is not acceptable, there is no evidence, which, if any, oral EC would be most effective option.

References

[1] Buchwald H. Consensus conference statement. Surg Obes Relat Dis 2005;1:371–81. Available from: https://doi.org/10.1016/j.soard.2005.04.002.

[2] Grundy SM, Barondess JA, Bellegie NJ, Fromm H, Greenway F, Halsted CH, et al. Gastrointestinal surgery for severe obesity. Ann Intern Med 1991;115:956–61. Available from: https://doi.org/10.7326/0003-4819-115-12-956.

[3] Santry HP. Trends in bariatric surgical procedures. JAMA 2005;294:1909. Available from: https://doi.org/10.1001/jama.294.15.1909.

[4] Ochner CN, Gibson C, Shanik M, Goel V, Geliebter A. Changes in neurohormonal gut peptides following bariatric surgery. Int J Obes 2011;35:153–66. Available from: https://doi.org/10.1038/ijo.2010.132.

[5] Balsiger B, Poggio J, Mai J, Kelly K, Sarr M. Ten and more years after vertical banded gastroplasty as primary operation for morbid obesity. J Gastrointest Surg 2000;4:598–605. Available from: https://doi.org/10.1016/S1091-255X(00)80108-0.

[6] Marceau P, Hould F-S, Simard S, Lebel S, Bourque R-A, Potvin M, et al. Biliopancreatic diversion with duodenal switch. World J Surg 1998;22:947–54. Available from: https://doi.org/10.1007/s002689900498.

[7] Belle SH, Berk PD, Courcoulas AP, Flum DR, Miles CW, Mitchell JE, et al. Safety and efficacy of bariatric surgery: Longitudinal Assessment of Bariatric Surgery. Surg Obes Relat Dis 2007;3:116–26. Available from: https://doi.org/10.1016/j.soard.2007.01.006.

[8] Bult MJF, van Dalen T, Muller AF. Surgical treatment of obesity. Eur J Endocrinol 2008;158:135–45. Available from: https://doi.org/10.1530/EJE-07-0145.

[9] Poves I, Cabrera M, Maristany C, Coma A, Ballesta-López C. Gastrointestinal quality of life after laparoscopic Roux-en-Y gastric bypass. Obes Surg 2006;16:19–23.

[10] Sundbom M. Laparoscopic revolution in bariatric surgery. World J Gastroenterol 2014;20:15135. Available from: https://doi.org/10.3748/wjg.v20.i41.15135.

[11] Maggard MA, Yermilov I, Li Z, Maglione M, Newberry S, Suttorp M, et al. Pregnancy and fertility following bariatric surgery. JAMA 2008;300:2286–96. Available from: https://doi.org/10.1001/jama.2008.641.

[12] Merhi ZO. Impact of bariatric surgery on female reproduction. Fertil Steril 2009;92:1501–8. Available from: https://doi.org/10.1016/j.fertnstert.2009.06.046.

[13] Kominiarek MA, Jungheim ES, Hoeger KM, Rogers AM, Kahan S, Kim JJ. American Society for Metabolic and Bariatric Surgery position statement on the impact of obesity and obesity treatment on fertility and fertility therapy endorsed by the American College of Obstetricians and Gynecologists and the Obesity Society. Surg Obes Relat Dis 2017;13:750–7. Available from: https://doi.org/10.1016/j.soard.2017.02.006.

[14] Eid GM, Cottam DR, Velcu LM, Mattar SG, Korytkowski MT, Gosman G, et al. Effective treatment of polycystic ovarian syndrome with Roux-en-Y gastric bypass. Surg Obes Relat Dis 2005;1:77–80. Available from: https://doi.org/10.1016/j.soard.2005.02.008.

[15] Gosman GG, King WC, Schrope B, Steffen KJ, Strain GW, Courcoulas AP, et al. Reproductive health of women electing bariatric surgery. Fertil Steril 2010;94:1426–31. Available from: https://doi.org/10.1016/j.fertnstert.2009.08.028.

[16] Johansson K, Cnattingius S, Näslund I, Roos N, Lagerros YT, Granath F, et al. Outcomes of pregnancy after bariatric surgery. Obstet Gynecol Surv 2015;70:375–7. Available from: https://doi.org/10.1097/01.ogx.0000466883.05862.27.

[17] FSRH guideline. (April 2019) overweight, obesity and contraception. BMJ Sex Reprod Health 2019;45:1–69. doi:10.1136/bmjsrh-2019-OOC.

[18] Faculty of Sexual and Reproductive Healthcare Clinical Guidance. Intrauterine contraception. 2015.

[19] Trussell J. Contraceptive failure in the United States. Contraception 2011;83:397–404. Available from: https://doi.org/10.1016/j.contraception.2011.01.021.

[20] Committee on Practice Bulletins—Gynecology. Practice bulletin no. 186. Obstet Gynecol 2017;130:e251–69. Available from: https://doi.org/10.1097/AOG.0000000000002400.

[21] Gemzell-Danielsson K, Apter D, Hauck B, Schmelter T, Rybowski S, Rosen K, et al. The effect of age, parity and body mass index on the efficacy, safety, placement and user satisfaction associated with two low-dose levonorgestrel intrauterine contraceptive systems: Subgroup Analyses of Data From a Phase III Trial. PLoS One 2015;10:e0135309. Available from: https://doi.org/10.1371/journal.pone.0135309.

[22] Vu Q, Micks E, McCoy E, Prager S. Efficacy and safety of long-acting reversible contraception in women with cardiovascular conditions. Am J Cardiol 2016;117:302–4. Available from: https://doi.org/10.1016/j.amjcard.2015.10.026.

[23] Hubacher D. Intrauterine devices & infection: review of the literature. Indian J Med Res 2014;140:53–7.

[24] Felix AS, Gaudet MM, La VC, Nagle CM, Shu XO, Weiderpass E, et al. Intrauterine devices and endometrial cancer risk: a pooled analysis of the Epidemiology of Endometrial Cancer Consortium. Int J Cancer 2015;136:E410–22. Available from: https://doi.org/10.1002/ijc.29229.

[25] NICE Clinical Guidelines N 30. (UK). NCC for W and CH. Long-acting reversible contraception: the effective and appropriate use of long-acting reversible contraception. 2005.

[26] Haimov-Kochman R, Ackerman Z, Anteby EY. The contraceptive choice for a Wilson's disease patient with chronic liver disease. Contraception 1997;56:241–4 doi:10.1016/S0010-7824(97)00141-8.

[27] Beatty MN, Blumenthal PD. The levonorgestrel-releasing intrauterine system: safety, efficacy, and patient acceptability. Ther Clin Risk Manag 2009;5:561–74.

[28] Bayer Ltd. Differentiation of Kyleena, Jaydess and Mirena intrauterine delivery systems (levonorgestrel). 2016.

[29] Merki-Feld GS, Skouby S, Serfaty D, Lech M, Bitzer J, Crosignani PG, et al. European Society of Contraception Statement on contraception in obese women. Eur J Contracept Reprod Health Care 2015;20:19–28. Available from: https://doi.org/10.3109/13625187.2014.960561.

[30] Saito-Tom L, Harris S, Soon R, Salcedo J, Kaneshiro B. Intrauterine device use in overweight and obese women. Contraception 2013;88:457–8. Available from: https://doi.org/10.1016/j.contraception.2013.05.105.

[31] Hillman JB, Miller RJ, Inge TH. Menstrual concerns and intrauterine contraception among adolescent bariatric surgery patients. J Women's Health 2011;20:533–8. Available from: https://doi.org/10.1089/jwh.2010.2462.

[32] Ciangura C, Corigliano N, Basdevant A, Mouly S, Declèves X, Touraine P, et al. Etonorgestrel concentrations in morbidly obese women following Roux-en-Y gastric bypass surgery: three case reports. Contraception 2011;84:649–51. Available from: https://doi.org/10.1016/j.contraception.2011.03.015.

[33] Lidegaard O, Nielsen LH, Skovlund CW, Lokkegaard E. Venous thrombosis in users of non-oral hormonal contraception: follow-up study, Denmark 2001-10. BMJ 2012;344:e2990. Available from: https://doi.org/10.1136/bmj.e2990.

[34] Mantha S, Karp R, Raghavan V, Terrin N, Bauer KA, Zwicker JI. Assessing the risk of venous thromboembolic events in women taking progestin-only contraception: a meta-analysis. BMJ 2012;345:e4944. Available from: https://doi.org/10.1136/bmj.e4944.

[35] Backman T, Rauramo I, Jaakkola K, Inki P, Vaahtera K, Launonen A, et al. Use of the levonorgestrel-releasing intrauterine system and breast cancer. Obstet Gynecol 2005;106:813–17. Available from: https://doi.org/10.1097/01.aog.0000178754.88912.b9.

[36] Dinger J, Bardenheuer K, Minh TD. Levonorgestrel-releasing and copper intrauterine devices and the risk of breast cancer. Contraception 2011;83:211–17. Available from: https://doi.org/10.1016/j.contraception.2010.11.009.

[37] Mørch LS, Skovlund CW, Hannaford PC, Iversen L, Fielding S, Lidegaard Ø. Contemporary hormonal contraception and the risk of breast cancer. N Engl J Med 2017;377:2228–39. Available from: https://doi.org/10.1056/NEJMoa1700732.

[38] Skovlund CW, Mørch LS, Kessing LV, Lidegaard Ø. Association of hormonal contraception with depression. JAMA Psychiatry 2016;73(11):1154–62. Available from: https://doi.org/10.1001/jamapsychiatry.2016.2387.

[39] Bitzer J, Rapkin A, Soares CN. Managing the risks of mood symptoms with LNG-IUS: a clinical perspective. Eur J Contracept Reprod Health Care 2018;23:321–5. Available from: https://doi.org/10.1080/13625187.2018.1521512.

[40] Darney P, Patel A, Rosen K, Shapiro LS, Kaunitz AM. Safety and efficacy of a single-rod etonogestrel implant (Implanon): results from 11 international clinical trials. Fertil Steril 2009;91:1646−53. Available from: https://doi.org/10.1016/j.fertnstert.2008.02.140.

[41] Birgisson NE, Zhao Q, Secura GM, Madden T, Peipert JF. Preventing unintended pregnancy: the contraceptive CHOICE project in review. J Women's Health 2015;24(5):349−53. Available from: https://doi.org/10.1089/jwh.2015.5191.

[42] National Collaborating Centre for Women's and Children's Health (Great Britain), National Institute for Health and Clinical Excellence (Great Britain), Royal College of Obstetricians and Gynaecologists (Great Britain). Long-acting reversible contraception: the effective and appropriate use of long-acting reversible contraception. RCOG Press; 2005.

[43] Díaz S, Pavez M, Moo-Young AJ, Bardin CW, Croxatto HB. Clinical trial with 3-keto-desogestrel subdermal implants. Contraception 1991;44:393−408. Available from: https://doi.org/10.1016/0010-7824(91)90030-J.

[44] Graesslin O, Korver T. The contraceptive efficacy of Implanon®: a review of clinical trials and marketing experience. Eur J Contracept Reprod Health Care 2008;13:4−12. Available from: https://doi.org/10.1080/13625180801942754.

[45] Edelman A, Jensen J. Obesity and hormonal contraception: safety and efficacy. Semin Reprod Med 2012;30:479−85. Available from: https://doi.org/10.1055/s-0032-1328876.

[46] Bender NM, Segall-Gutierrez P, Najera SOL, Stanczyk FZ, Montoro M, Mishell DR. Effects of progestin-only long-acting contraception on metabolic markers in obese women. Contraception 2013;88:418−25. Available from: https://doi.org/10.1016/j.contraception.2012.12.007.

[47] Vicente L, Mendonça D, Dingle M, Duarte R, Boavida JM. Etonogestrel implant in women with diabetes mellitus. Eur J Contracept Reprod Health Care 2008;13:387−95. Available from: https://doi.org/10.1080/13625180802382604.

[48] Chakhtoura Z, Canonico M, Gompel A, Scarabin P-Y, Plu-Bureau G. Progestogen-only contraceptives and the risk of acute myocardial infarction: a meta-analysis. J Clin Endocrinol Metab 2011;96:1169−74. Available from: https://doi.org/10.1210/jc.2010-2065.

[49] Curtis KM, Martins SL. Progestogen-only contraception and bone mineral density: a systematic review. Contraception 2006;73(5):470−87. Available from: https://doi.org/10.1016/j.contraception.2005.12.010.

[50] Rosenberg M. Weight change with oral contraceptive use and during the menstrual cycle. Contraception 1998;58:345−9. Available from: https://doi.org/10.1016/S0010-7824(98)00127-9.

[51] Bernardi M, Lazzeri L, Perelli F, Reis FM, Petraglia F. Dysmenorrhea and related disorders. F1000Res 2017;6:1645. Available from: https://doi.org/10.12688/f1000research.11682.1.

[52] Segall-Gutierrez P, Taylor D, Liu X, Stanzcyk F, Azen S, Mishell DR. Follicular development and ovulation in extremely obese women receiving depo-medroxyprogesterone acetate subcutaneously. Contraception 2010;81:487−95. Available from: https://doi.org/10.1016/j.contraception.2010.01.021.

[53] Jain J, Jakimiuk AJ, Bode FR, Ross D, Kaunitz AM. Contraceptive efficacy and safety of DMPA-SC. Contraception 2004;70:269−75. Available from: https://doi.org/10.1016/j.contraception.2004.06.011.

[54] World Health Organization. Multinational comparative clinical evaluation of two long-acting injectable contraceptive steroids: norethisterone oenanthate and medroxyprogesterone acetate. 1. Use-effectiveness. Contraception 1977;15:513−33. Available from: https://doi.org/10.1016/0010-7824(77)90102-0.

[55] Murthy A, Lam C. Depo-Provera (depot medroxyprogesterone acetate) use after bariatric surgery. Open Access J Contracept 2016;7:143−50. Available from: https://doi.org/10.2147/OAJC.S84097.

[56] Robinson JA, Burke AE. Obesity and hormonal contraceptive efficacy. Women's Health 2013;9(5):453−66. Available from: https://doi.org/10.2217/whe.13.41.

[57] Edelman AB, Cherala G, Munar MY, DuBois B, McInnis M, Stanczyk FZ, et al. Prolonged monitoring of ethinyl estradiol and levonorgestrel levels confirms an altered pharmacokinetic profile in obese oral contraceptives users. Contraception 2013;87:220−6. Available from: https://doi.org/10.1016/j.contraception.2012.10.008.

[58] Edelman AB, Carlson NE, Cherala G, Munar MY, Stouffer RL, Cameron JL, et al. Impact of obesity on oral contraceptive pharmacokinetics and hypothalamic−pituitary−ovarian activity. Contraception 2009;80:119−27. Available from: https://doi.org/10.1016/j.contraception.2009.04.011.

[59] Westhoff CL, Torgal AH, Mayeda ER, Pike MC, Stanczyk FZ. Pharmacokinetics of a combined oral contraceptive in obese and normal-weight women. Contraception 2010;81:474−80. Available from: https://doi.org/10.1016/j.contraception.2010.01.016.

[60] Schlatter J. Oral contraceptives after bariatric surgery. Obes Facts 2017;10:118−26. Available from: https://doi.org/10.1159/000449508.

[61] Edelman AB, Cherala G, Stanczyk FZ. Metabolism and pharmacokinetics of contraceptive steroids in obese women: a review. Contraception 2010;82:314−23. Available from: https://doi.org/10.1016/j.contraception.2010.04.016.

[62] Belle DJ, Callaghan JT, Gorski JC, Maya JF, Mousa O, Wrighton SA, et al. The effects of an oral contraceptive containing ethinyloestradiol and norgestrel on CYP3A activity. Br J Clin Pharmacol 2002;53:67−74. Available from: https://doi.org/10.1046/j.0306-5251.2001.01521.x.

[63] Back D, Breckenridge A, MacIver M, Orme M, Purba H, Rowe P, et al. The gut wall metabolism of ethinyloestradiol and its contribution to the pre-systemic metabolism of ethinyloestradiol in humans. Br J Clin Pharmacol 1982;13:325−30. Available from: https://doi.org/10.1111/j.1365-2125.1982.tb01382.x.

[64] Victor A, Odlind V, Kral JG. Oral contraceptive absorption and sex hormone binding globulins in obese women: effects of jejunoileal bypass. Gastroenterol Clin North Am 1987;16:483−91.

[65] Andersen AN, Lebech PE, Sørensen TI, Borggaard B. Sex hormone levels and intestinal absorption of estradiol and D-norgestrel in women following bypass surgery for morbid obesity. Int J Obes 1982;6:91−6.

[66] Ginstman C, Frisk J, Carlsson B, Ärlemalm A, Hägg S, Brynhildsen J. Plasma concentrations of etonogestrel in women using oral desogestrel before and after Roux-en-Y gastric bypass surgery: a pharmacokinetic study. BJOG An Int J Obstet Gynaecol 2019;126:486−92. Available from: https://doi.org/10.1111/1471-0528.15511.

[67] Paulen ME, Zapata LB, Cansino C, Curtis KM, Jamieson DJ. Contraceptive use among women with a history of bariatric surgery: a systematic review. Contraception 2010;82:86−94. Available from: https://doi.org/10.1016/j.contraception.2010.02.008.

[68] Vilallonga R, Himpens J, van de Vrande S. Long-term (7 years) follow-up of Roux-en-Y gastric bypass on obese adolescent patients (<18 years). Obes Facts 2016;9:91−100. Available from: https://doi.org/10.1159/000442758.

Chapter 7

Long-term contraceptive care in obese and superobese women

Johannes Bitzer

Department of Obstetrics and Gynecology, University Hospital of Basel, Basel, Switzerland

Introduction

Obesity is defined based on the body mass index (BMI, kg/m^2). BMI categories are defined by "The Centers for Disease Control (CDC) and Prevention and The World Health Organization (WHO)," which are as follows:

- Underweight <18.5 kg/m^2
- Normal 18.5–24.9 kg/m^2
- Overweight 25–29.9 kg/m^2
- Obese 30–39.9 kg/m^2 or Class I obesity 30–34.9 kg/m^2/Class II obesity 35–39.9 kg/m^2
- Very obese ≥40 kg/m^2 or otherwise referred to as severe, extreme, morbid, or Class III obesity [1]

There was a lack of evidence regarding contraceptive effectiveness and safety among overweight or obese women, until recently as they had been excluded in most of the clinical trials [2].

Obesity can lead to the following consequences, and they must be considered by the clinician and the woman when deciding upon the optimal method of contraception:

1. Increased risk of eclampsia, preterm birth, cesarean delivery, postpartum hemorrhage, and thromboembolic events during pregnancy and in the postpartum period [2].
    These include:
    a. Increase in risks of various cardiovascular, metabolic, and neoplastic diseases that may be either aggravated or reduced by contraceptive methods [3–7].
        i. Diabetes
        ii. Metabolic syndrome
        iii. Venous thromboembolism (VTE)
        iv. Myocardial infarction
        v. Breast cancer
        xvi. Endometrial carcinoma

2. Changes in the metabolism of sex steroids used in hormonal contraception.

Pharmacokinetic studies have shown that obesity may have an influence on half-life, clearance AUC (area under the curve), and time to reach steady state. This was, however, mainly related to the estrogen component, and there was no clear impact on end-organ suppression [8–10].

These obesity-related characteristics should be taken into account regarding the efficacy, health risks, side effects, and benefits of contraceptive methods [11].

Combined hormonal contraceptives

Combined hormonal contraceptives (CHC) seem the least appropriate methods when applying these criteria. Although CHC can prevent pregnancies effectively, these methods can also lead to a significant increase in the risk of the conditions mentioned previously, which are directly proportional to the incremental increase in BMI. The beneficial protective effect on endometrial carcinoma and reduction of heavy menstrual bleeding in general do not outweigh the risks.

In addition, it should be kept in mind that although effectiveness studies are controversial, pharmacokinetic studies as explained earlier give indirect evidence of lower plasma values that may have a negative impact on efficacy.

Therefore it can be stated that from safety (cardiovascular, metabolic, and malignancy risk) and efficacy perspective (no or very little impact on pharmacokinetics and independence of user compliance), copper intrauterine devices (IUDs) and long-acting progestogen-only contraceptives have the most favorable profile and should be considered the first choice [12].

Obesity and Gynecology. DOI: https://doi.org/10.1016/B978-0-12-817919-2.00007-3
© 2020 Elsevier Inc. All rights reserved.

Copper intrauterine device

There are many copper IUDs available with different frames and structures, different delivery rates of copper ions, and different duration of action. Their principal mode of action is through the local inflammatory and spermicidal effect of copper ions. As this is a local action, BMI per se is not expected to have any impact [13].

Different types of copper IUDs are available in different countries with different copper content, frames and last between 5 and 12 years. The most frequently used are ParaGard, Flexi T, Multiload, T-Safe, and GyneFix [13].

Efficacy

These devices are very effective and independent of user adherence.

Pearl Index (PI) for ideal use is 0.6 and for typical use 0.8. So the real-life user failure rate is less than 1%. After prolonged continuous use the cumulative pregnancy rate is 1.6% at 7 years and 2.2% at 8 and 12 years [13,14].

There are no studies indicating a reduced efficacy of copper IUDs in obese women compared to normal-weight women.

Health risks

No major health risks are reported except for rare complications in the context of use. The risk of complications during placement is minimal and includes uterine perforation (0.1%) and pelvic inflammatory disease (PID, 0%–2%). The cumulative risk of IUD expulsion is 10% over 3 years of use [14].

There is no evidence of increased risk of expulsion and perforation, although insertion maybe sometimes challenging.

Copper IUDs have no increased risk as regards cardiovascular diseases, including VTE and myocardial infarction. There are no metabolic impacts especially related to diabetes or metabolic syndrome. Furthermore, no weight increase has been reported [15].

The device by itself does not lead to increased risk of PID or infertility, but the vaginal and cervical infections act as the pathogenetic factors [16].

Side effects

The usual side effects are prolonged menstrual bleeding, dysmenorrhea, lower abdominal pain, or discomfort [13,14].

As obese women have a higher likelihood of heavy menstrual bleeding, there is a theoretical possibility that blood loss during menstruation may be higher than in normal-weight women, thus increasing the risk of iron deficiency. There are however no comparative data regarding the menstrual blood loss in different BMI groups. In clinical trials, there is so far no evidence that the side-effect profile would be any different. There are however no larger comparative studies either.

Additional health benefits

Obese women have a higher risk of endometrial carcinoma. Copper IUDs have a protective effect on the endometrium and thus provides an additional benefit [17].

Contraindications

- Severe uterine distortion—Anatomic abnormalities, including bicornuate uterus, cervical stenosis, or fibroids, severely distort the uterine cavity because of increased difficulty with insertion and increased risk of expulsion [11,13].
- Active pelvic infection—IUD insertion in women with active pelvic infection, including PID, endometritis, mucopurulent cervicitis, and pelvic tuberculosis, because the presence of a foreign body may impede the resolution of the infection. However, the IUD may be inserted in women who are at least 3 months posttreatment for PID or puerperal or postabortion sepsis.
- Known or suspected pregnancy—IUD insertion during pregnancy can lead to miscarriage and increases the risk of septic abortion.
- Wilson's disease or copper allergy—Although no adverse event related to copper allergy or Wilson's disease has ever been reported in a woman with a copper IUD, hormone-releasing IUDs are preferred for use in women with these conditions.
- Unexplained abnormal uterine bleeding—Evaluation of women with abnormal uterine bleeding should precede IUD placement since, after placement, the abnormal bleeding may be erroneously attributed to the IUD rather than the preexisting pathology.

Limitations of use

Following clinical conditions are not strictly contraindications, but the use of copper IUDs may be considered cautiously:

1. Heavy menstrual bleeding
2. Dysmenorrhea
3. Endometriosis
4. Women with high risk for sexually transmitted infections (STIs)

Practical issues in obese women

Visualizing the cervix and determining the size and direction of the uterus can be challenging in severely obese women during insertion of the device. Optimizing equipment by selecting a large speculum or removing the tip of a condom and placing it over the blades of the speculum can help with the visualization of the cervix. Ultrasound may be helpful to guide insertion as well.

Another limitation is the nonacceptance of intrauterine contraception (foreign body in the uterus) by the woman.

Levonorgestrel-containing intrauterine systems

There are three types of levonorgestrel-containing intrauterine systems (LNG-IUS) currently available:

1. LNG 52 (Mirena) containing 52 mg of LNG with an average daily release of 20 μg and LNG is effective for at least 5 years.
2. LNG 14 (Jaydess) containing 13.5 mg of LNG with an average daily release of 6 μg and LNG is effective for 3 years.
3. LNG 20 (Kyleena) containing 19.5 mg of LNG with an average daily release of 9 μg and LNG is effective for 5 years.

The progestin secreted by progestin-releasing IUDs thickens cervical mucus and also increases expression of glycodelin A in endometrial glands, which inhibit binding of sperm to the egg [11,18,19].

Serum concentrations of progestin can lead to partial inhibition of ovarian follicular development and ovulation. This is not, however, a major contraceptive mechanism; one study found that at least 75% of women using a levonorgestrel-releasing IUD had ovulatory cycles [18].

Efficacy

There is a very high efficacy for all three systems with PI around 0.2–0.33.

The cumulative pregnancy rate is 0.5–1.1 after 5 years of continuous use with the LNG 20 IUD. The 3-year cumulative pregnancy rate is 0.9 with the LNG 14 IUD.

There is no evidence of impaired contraceptive effectiveness in IUC users with obesity, either with the Cu IUD or the LNG-IUS [20,21].

A prospective cohort study reported no statistically significant difference in contraceptive failure rate during the first 2–3 years of use among IUC users (Cu-IUD or 52 mg LNG-IUS) who were of normal BMI ($n = 1584$), overweight BMI ($n = 1149$), or obese BMI ($n = 1467$) [59]. The overall IUC failure rate of less than one pregnancy per 100 woman-years did not vary by BMI [22].

Health risks

There is no increased risk of cardiovascular diseases.

A Danish national registry-based cohort study [23] found that the risk of confirmed VTE was not increased with the use of hormone-releasing IUDs [adjusted relative risk (aRR) 0.57; 95% CI 0.41–0.81].

A metaanalysis of eight observational studies also showed no association between VTE risk and the use of an LNG-IUS (aRR 0.61; 95% CI 0.24–1.53) [24].

The association between LNG-IUS use and breast cancer has been investigated. There is some controversy due to conflicting results. Two large retrospective case–control studies of European women showed no increased risk of breast cancer in women using the LNG-IUS for contraception [25,26].

However, two analyses of a large Finnish cohort suggest a small increased risk (up to 1.3 times) of breast cancer, in particular, lobular and ductal cell cancers, in women using the LNG-IUS for heavy menstrual bleeding [27,28]. In the Danish cohort study, the authors found a relative risk of 1.21 (95% CI 1.11–1.33) [29].

Taking into account the different results and the type of studies (observational studies), it can be concluded that the use of a progestogen-only method is either not, or to a minor degree (expressed as absolute risk), accompanied by an increased risk for breast cancer [13,14].

There are no data looking into the potentially different risk for obese women.

Taking into account the possible association between obesity and breast cancer risk, this remains an issue for future studies among progestogen-only users.

Side effects

- Irregular bleeding: The major side effect is irregular bleeding, which is very common during the first 3–6 months. At 24 months, 50% of LNG20 users have amenorrhea, 30% have oligomenorrhea, and 11% have spotting. The pattern is similar to the LNG 14 IUD with less amenorrhea (13 vs 24% after 3 years) [30].
- Mood changes have recently received special attention. In the Danish cohort, LNG-IUS users had an increased risk of having to be prescribed antidepressants and a higher risk of being hospitalized with depression [31].
- The impact of progestogens on the affective state of women is, however, complicated, and it seems that the negative impact is limited to a smaller group of vulnerable women. This may be of clinical importance in those

women who suffer from preclinical or undiagnosed perimenopausal depression [32].
- Other side effects are rare and include breast tenderness and acne.
- In the general population, there are no significant differences in weight gain when hormonal and nonhormonal intrauterine methods are compared, and there is no evidence to support a causal association between IUC use and weight gain.

Health benefits
- Reduction in heavy menstrual bleeding and dysmenorrhea in patients without organic pathology and bleeding due to bleeding diathesis, including anticoagulation therapy. The efficacy regarding reduction of bleeding intensity in women with fibroids and adenomyosis is yet unclear and under investigation [11].
- Protection from PID, due to cervical mucus thickening, which acts as a barrier toward ascending infections [14,16].
- Treatment of endometriosis-associated pain.
- Endometrial protection in premenopausal and menopausal women using estrogen hormone replacement and a concomitant reduction of the risk of endometrial cancer [11,13].
- Fewer painful crisis in women with sickle cell disease.
- IUD can also be used to treat endometrial hyperplasia and cancer.
- The LNG-IUS is licensed for use in the management of idiopathic menorrhagia. Amenorrhea can be expected up to 45% in women 6 months after insertion and up to 50% after 12 months. Continuation rates are highest in women aged 39–48 years [12].

Safe prescribing

The Faculty of Sexual and Reproductive Healthcare has published the UK Medical Eligibility Criteria for Contraceptive Use, indicating that the risk of the LNG-IUS is currently category 4 (unacceptable risk) for current/active breast cancers and category 3 (risks outweigh benefits) for personal history of breast cancer in the past 5 years with no active disease [13].

Other contraindications include active liver disease and those which are related to the intrauterine application and also include severe deformity of the uterine cavity, acute STIs, unexplained vaginal bleeding, and known or suspected pregnancy (see the "Copper intrauterine device" section).

Summary

LNG-IUD is a very effective and safe method for obese women independent of the degree of obesity. It can be used in obese women with additional risk factors (hypertension, diabetes, history of VTE) and is especially effective and indicated in obese women suffering from HMB.

Contraindications include those for intrauterine contraceptives and current/active breast cancer (category 4) or (category 3) for women with a personal history of breast cancer in the past 5 years with no active disease [12,13].

Etonogestrel-releasing implant
Progestogen implants

Hormonal implants are subdermally inserted contraceptives that provide reliable contraception for 3–5 years [14,33].

The matrices are inert or biologically degradable rods or capsules that release the respective steroid continuously over a lengthy period of time. The hormone implants consist of one or several small flexible rods or a capsule inserted under the skin of the upper arm. Depending on the product, they release the progestin megestrol acetate, norethindrone, norgesterone, or etonogestrel (ENG) for a period of 1–5 years.

Norplant was composed of six rods. Each rod contains 36 mg of levonorgestrel. The total duration of action of these six rods was 5 years. The product has not been marketed since 2002.

Norplant II (Jadelle), Norplant's successor product, is composed of two flexible silicone rods (43 mm Å ∼ 2.5 mm) each containing 75 mg of levonorgestrel and also has a duration of action of 5 years. The same product is commercially available in China under the name Sinoplant.

Implanon is an ENG-releasing hormonal implant. The rod is 4 cm long and 2 mm in diameter and is composed of 40% ethylene-vinyl acetate and 60% (68 mg) ENG (3-keto-desogestrel). The duration of action after subdermal implantation is 3 years [34].

Efficacy

The PI is 0.05–0.38, which is similar to that of other long-acting methods of contraception.

The concentration falls over time at a rate that depends on body weight, due the increased volume of distribution, effects on plasma protein binding, and altered clearance in individuals with obesity. In this context, it is important to note that that the absolute ENG serum level required for the suppression of ovulation has not been defined and may be lower than the usually suggested 90 pg/mL [35].

These pharmacological findings have led to some controversy about the real-world efficacy of the implant in obese patients especially because clinical experience with

Implanon in women weighing more than 80 kg is limited, and plasma levels of ENG are lower in obese women and come close to the concentration, which is necessary to effectively prevent ovulation [35].

Two more recent analyses have included larger proportions of women with raised BMI and have looked at extended implant use.

A secondary analysis of 1168 implant users in the US Contraceptive CHOICE project showed that pregnancy rates in users of the ENG implant are extremely low and similar in women who were overweight and had obesity compared with normal-weight women over 4 years of implant use. In the CHOICE study the efficacy of this implant was not reduced in obese women [36].

In a study of prolonged ENG implant use (up to 5 years), 237 women continued ENG implant use beyond 3 years, of whom 25% were overweight and 46% had obesity (88). No pregnancies occurred during the period of prolonged use, leading to an estimated failure rate of 0 (97.5% one-sided CI 0−1.61) for all BMIs. Serum ENG evaluation showed that median levels remain above the suggested ovulation threshold of 90 pg/mL for women in all BMI classes at 4 and 5 years [37].

Nonetheless, there remains some controversy about the practical use of the implant with respect to efficacy.

The Summary of Product Characteristics (SPC) for the ENG progestogen-only implant advises "the clinical experience in heavier women in the third year of use is limited." It therefore states, "it cannot be excluded that the contraceptive effect may be lower than for women of normal weight." It advises that health professionals may therefore consider the earlier replacement of the implant in "heavier" women. The SPC91 does not specify a definition of heavier weight or after what duration of use replacement may need to be considered [38].

The FSRH advises that there is no direct evidence to support a need for earlier implant replacement, and recent data assessing continued use in women with raised BMI beyond 3 years are very reassuring. Therefore the GDG recommends that the ENG implant can be considered to provide very effective contraception for 3 years for women in all weight/BMI categories [39].

Health risks

No major health risks are known. There is no concern regarding bone loss [40].

Based on the limited epidemiological data, there is no evidence that the implant increases the risk of thromboembolic events or myocardial infarction [41].

Studies on hemostatic parameters did not show significant procoagulatory values nor an increase in blood pressure was found [42,43].

Metabolic studies did not show negative effects on carbohydrate or lipid parameters used as markers of risk (fasting glucose, insulin, triglycerides, and HDL) [43].

The implant has minimal clinically nonrelevant impact on fasting glucose and insulin in obese women. In women with diabetes, HbA1c did not change in implant users, neither did the daily insulin requirement [44].

There are not enough long-term follow-up studies as regards the risk of breast cancer.

It has to be taken into account that obese women have a higher risk for breast cancer that theoretically could increase their sensibility to the progestogen action on the breast. There is however until now no consensus about the impact of progestogen-only contraception on breast cancer risk in normal-weight women.

Whether the implant leads to an additional weight increase has not yet been completely clarified. In a comparative 12-month study of three different progestin-only methods, including the ENG implant, a small increase in weight (2.1 kg) was observed with no difference between the three methods as compared to copper IUD [45].

In the National Danish Registry study, five confirmed venous thrombosis events were observed during the use of ENG implants, corresponding to an incidence rate of 1.7 per 10,000 exposure years and a nonsignificant aRR of 1.40 (95% CI 0.58−3.38) compared with nonusers of hormonal contraception [23].

Side effects

The most frequent side effect is unscheduled bleeding. Rarely reported side effects (around 5%) are acne, headache, weight gain, mastalgia, vaginal infections, and bleeding disorders. Interactions with broad-spectrum antibiotics, St. John's wort, a number of antiepileptic agents and mood-altering drugs have been documented [46].

The implant is associated with changes in bleeding pattern and bleeding intensity. Studies show that these changes lead to discontinuation during the first year by up to 20% of user [47]. As the majority of studies about implant tolerability excluded women with BMI beyond 35, little is known about the size of this problem in obese women.

It may however have a negative impact, and further studies are needed [20].

Health benefits

The main noncontraceptive benefit of implant is that it may help alleviate dysmenorrhea and ovulatory pain that are not associated with any identifiable pathological condition.

While there is theoretically no reason why this would not be the case for women who are overweight or with obesity, this has not been specifically studied in women of different weight categories [20].

Practical issues

There are no data to suggest placement or removal of progestogen only implant (IMP) is problematic in women who are overweight or women with obesity. Correct subdermal placement of the implant is important in women of all BMIs. Insertion or removal difficulties should not be presumed in women with raised BMI. Removal of appropriately placed implants (i.e., subdermal placement) should not be affected by BMI, including the case of weight gain after insertion.

Contraindications

There are very few contraindications that the implant shares with other progestogen-only contraceptives such as active or recent breast cancer and chronic liver disease.

Safe prescribing

Special consideration

It should be borne in mind that placement and removal require special training.

Accurate placement is crucial to the product's reliability. There are reports of incorrect insertion of the Implanon rod, possibly making the contraceptive rod impossible to palpate and difficult to find. To make the product easier to use safely and simpler to locate the system was upgraded with Implanon NXT. Efficacy may be hampered by drugs affecting the metabolism of ENG such as antiviral drugs. Contraindications include breast cancer, active liver disease, and benign and malignant liver tumors (except nodular hyperplasia).

Summary

The most frequently used and studied progestogen implant is the ENG-containing implant. It provides highly efficient protection against unwanted pregnancies in obese women. The safety profile is favorable, and at this moment, there is no evidence of major cardiovascular or other health risks, including cancer or osteoporosis in obese patients.

The question regarding earlier replacement of the implant (2 years instead of 3 years) in women over 80 kg weight or at a high BMI is still controversial. But recent studies indicate that the efficacy is maintained over the recommended time of use, independent of BMI.

References

[1] World Health Organization. BMI classification. Global database on body mass index. 2018. <http://apps.who.int/bmi/index.jsps>.

[2] Grimes DA, Shields WC. Family planning for obese women: challenges and opportunities. Contraception 2005;72:1–7.

[3] Wormser D, Kaptoge S, , et al.Emerging Risk Factors Collaboration Separate and combined associations of body-mass index and abdominal adiposity with cardiovascular disease: collaborative analysis of 58 prospective studies. Lancet 2011;377:1085–95.

[4] Sinicrope FA, Dannenberg AJ. Obesity and breast cancer prognosis: weight of the evidence. J Clin Oncol 2011;29:4–7.

[5] Zhang Y, Liu H, Yang S, et al. Overweight, obesity and endometrial cancer risk: results from a systematic review and meta-analysis. Int J Biol Markers 2014;29:21–9.

[6] Holst AG, Jensen G, Prescott E. Risk factors for venous thromboembolism: results from the Copenhagen City Heart Study. Circulation 2010;121:1896–903.

[7] Wattanakit K, Lutsey PL, Bell EJ, et al. Association between cardiovascular disease risk factors and occurrence of venous thromboembolism. A time-dependent analysis. Thromb Haemost 2012;108:508–15.

[8] Edelman AB, Cherala G, Munar MY, et al. Prolonged monitoring of ethinyl estradiol and levonorgestrel levels confirms an altered pharmacokinetic profile in obese oral contraceptives users. Contraception 2013;87:220–7.

[9] Edelman AB, Carlson NE, Cherala G, et al. Impact of obesity on oral contraceptive pharmacokinetics and hypothalamic-pituitary-ovarian activity. Contraception 2009;80:119–24.

[10] Westhoff CL, Torgal AH, Mayeda ER, et al. Pharmacokinetics of a combined oral contraceptive in obese and normal-weight women. Contraception 2010;81:474–82.

[11] Shaw KA, Edelman KE. Obesity and oral contraceptives: a clinician's guide. Best Pract Res Clin Endocrinol Metab 2013;27:55–65.

[12] National Institute for Health and Care Excellence (NICE). Long-acting reversible contraception (Update); 2014. <https://www.nice.org.uk/guidance/cg30>.

[13] Faculty of Sexual and Reproductive Healthcare Clinical Guidance. Intrauterine contraception; 2015. ISSN 1755-103 (updated June 2015).

[14] American College of Obstetricians and Gynecologists. ACOG practice bulletin no. 121: long-acting reversible contraception: implants and intrauterine devices. Obstet Gynecol 2011;118:184–96.

[15] Vu Q, Micks E, McCoy E, et al. Efficacy and safety of long-acting reversible contraception in women with cardiovascular conditions. Am J Cardiol 2016;117:302–4.

[16] Hubacher D. Intrauterine devices & infection: review of the literature World Health Organization Indian J Med Res 2014;140 (Suppl. 1):S53–7.

[17] Ashley SF, Gaudet MM, La Veccia C. Intrauterine devices and endometrial cancer risk: A pooled analysis of the Epidemiology of Endometrial Cancer Consortium. Int J Cancer 2015;136(5): E410–22.

[18] Beatty MN, Blumenthal PD. The levonorgestrel-releasing intrauterine system: Safety, efficacy, and patient acceptability. Ther Clin Risk Manage 2009;5:561–74.

[19] <www.hpra.ie/img/uploaded/swedocuments/Kyleena_HCP_Differentiation%20of%20Kyleena%2C%20Jaydess%20and%20Mirena.Pdf>. Accessed on 05.05.2019.

[20] Merki-Feld GS, Skouby S, Serfaty D, et al. European Society of Contraception statement on contraception in obese women. Eur J Contracept Reprod Health Care 2015;20:19–28.

[21] Saito-Tom LY, Soon RA, Harris SC, et al. Levonorgestrel intrauterine device use in overweight and obese women. Hawaii J Med Public Health 2015;74:369–74.

[22] Gemzell-Danielsson K, Apter D, Hauck B, et al. The effect of age, parity and body mass index on the efficacy, safety, placement and user satisfaction associated with two low-dose levonorgestrel intrauterine contraceptive systems: subgroup analyses of data from a phase III trial. PLoS One 2015;10:e0135309.

[23] Lidegaard O, Nielson L, Skovlund CW, et al. Venous thrombosis in users of non-oral hormonal contraception: follow-up study, Denmark 2001-10. BMJ 2012;344:e2990.

[24] Mantha S, Karp R, Raghavan V, et al. Assessing the risk of venous thromboembolic events in women taking progestin-only contraception: a meta-analysis. BMJ 2012;345:e4944.

[25] Backman T, Rauramo I, Jaakkola K, et al. Use of the levonorgestrel-releasing intrauterine system and breast cancer. Obstet Gynecol 2005;106:813–17.

[26] Dinger J, Bardenheuer K, Minh TD. Levonorgestrel-releasing and copper intrauterine devices and the risk of breast cancer. Contraception 2011;83:211–17.

[27] Soini T, Hurskainen R, Grenman S, Maenpaa J, Paavonen J, Pukkala E. Cancer risk in women using the levonorgestrel-releasing intrauterine system in Finland. Obstet Gynecol 2014;124:292–9.

[28] Soini T, Hurskainen R, Grenman S, et al. Levonorgestrel-releasing intrauterine system and the risk of breast cancer: a nationwide cohort study. Acta Oncol 2016;55:188–92.

[29] Mørch L, Skovlund C, Hannaford P, et al. Contemporary contraception and the risk of breast cancer. N Engl J Med 2017;377:2228–39.

[30] Faculty of Sexual & Reproductive Healthcare (FSRH). Problematic bleeding with hormonal contraception; 2015. <http://www.fsrh.org/standards-and-guidance/documents/ceuguidanceproblematicbleedinghormonalcontraception/>.

[31] Skovlund CW, Mørch LS, Kessing LV, et al. Association of hormonal contraception with depression. JAMA Psychiatry 2016;73:1154–62.

[32] Bitzer J, Rapkin A, Soares CN. Managing the risks of mood symptoms with LNG-IUS: a clinical perspective. Eur J Contracept Reprod Health Care 2018 Oct;23(5):321–5.

[33] Silva Dos Santos PN, Madden T, Omvig K, Peipert JF. Changes in body composition in women using long-acting reversible contraception. Contraception 2017;95(4):382–9.

[34] Darney P, Patel A, Rosen K, Shapiro LS, Kaunitz AM. Safety and efficacy of a single-rod etonogestrel implant (Implanon): results from 11 international clinical trials. Fertil Steril 2009;91:1646–53.

[35] Díaz S, Pavez M, Moo-Young AJ, et al. Clinical trial with 3-keto-desogestrel subdermal implants. Contraception 1991;44:393–408.

[36] Xu H, Wade JA, Peipert JF, et al. Contraceptive failure rates of etonogestrel subdermal implants in overweight and obese women. Obstet Gynecol 2012;120:21–6.

[37] Graesslin O, Korver T. The contraceptive efficacy of Implanon: a review of clinical trials and marketing experience. Eur J Contracept Reprod Health Care 2008;13(Suppl. 1):4–12.

[38] SPC. Nexplanon 68 mg implant for subdermal use—summary of product. <https://www.medicines.org.uk/emc/product/5720/smpc>.

[39] FSRH Guideline. Overweight, Obesity and Contraception. FSRH guideline [April 2019] overweight, obesity and contraception. BMJ Sex Reprod Health 2019;45(Suppl. 2):1–69.

[40] Curtis KM, Martins SL. Progestogen-only contraception and bone mineral density: a systematic review. Contraception 2006;73:470–87.

[41] Chakhtoura Z, Canonico M, Gompel A, et al. Progestogen-only contraceptives and the risk of acute myocardial infarction: a meta-analysis. J Clin Endocrinol Metab 2011;96:1169–74.

[42] Edelman AB, Jensen JT. Obesity and hormonal contraception: safety and efficacy. Semin Reprod Med 2012;30(6):479–85.

[43] Bender NM, et al. Effects of progestin-only long-acting contraception on metabolic markers in obese women. Contraception 2013;88(3):418–25.

[44] Vicente L, et al. Etonogestrel implant in women with diabetes mellitus. Eur J Contracept Reprod Health Care 2008;13(4):387–95.

[45] Vickery Z, Madden T, Zhao Q, Secura GM, Allsworth JE, Peipert JF. Weight changes at 12 months in users of three progestin-only contraceptive methods. Contraception 2013;88(4):503–8.

[46] Croxatto HB, Makarainen L. The pharmacodynamics and efficacy of Implanon®. 1: An overview of the data. Contraception 1998;58(6 Suppl.):91S–7S.

[47] Bitzer J, Tschudin S, Alder J. Acceptability and side-effects of Implanon in Switzerland: a retrospective study by the Implanon Swiss Study Group. Eur J Contracept Reprod Health Care 2004;9(4):278–84.

Section 3

Male and female Infertility

Chapter 8

Obesity and hirsutism

Mostafa Metwally
Consultant in Reproductive Medicine and Surgery, Sheffield Teaching Hospitals, University of Sheffield, Sheffield, United Kingdom

Introduction

Obesity is a growing worldwide epidemic with associated significant adverse reproductive effects. The prevalence of obesity has consistently increased to the point where more than 35% of adults are now considered to be obese. Women are generally more prone to obesity than men possibly as a result of their lower basal metabolic rate. The effect of obesity on reproductive function reflects through complex endocrinological changes resulting from an interaction between the fat compartment and hypothalamic–pituitary–gonadal axis with an ultimate effect on sex steroids [1,2] mediated through the effect of circulating adipokines, including leptin, adiponectin, and resistin. The cutaneous manifestations of obesity are the results of the associated metabolic syndrome resulting from these interactions and include the skin lesions known as acanthosis nigracans and hyperandrogenic manifestations, namely, acne and hirsutism [3].

Obesity and ovarian function

The mutual communication between adipose and other tissue being has been hypothesized since the 1940s [4]. It has long been recognized that gonadal steroids can influence adipose tissue as evidenced by the presence of estrogen, progesterone, and androgen receptors on adipose cells, and this interaction has also long been recognized to be mediated through both genomic and nongenomic mechanisms. Genomic mechanisms include the transcription of certain genes that control leptin and lipoprotein lipase through the activation of cyclic adenosine monophosphate and, hence, activation of protein kinase C, hormone-sensitive lipase, and lipolysis [5]. Over the last few decades, our understanding of this interaction has evolved more to reveal a complex interaction between the adipose compartment and the hypothalamic–pituitary–gonadal axis mediated through the effect of adipokines, insulin, and insulin-like growth factor–binding protein-1 (IGFBP-1).

Obesity and androgen production

The most important endocrine change in obesity is hyperinsulinemia where high insulin concentrations aim to inhibit feeding by inhibiting neuropeptide Y, a potent stimulant of food intake by direct stimulation of hypothalamic receptors [6]. This insulin resistance and hyperinsulinemia lead to an increased production of ovarian androgens through the effect and insulin-like growth factor 1 (IGF-1), which leads to an increase in testosterone production. Testosterone is then converted by the granulosa cells to estradiol. This transition from an androgen-dominant to estrogen-dominant environment is vital to normal ovulation and ovarian function. Obesity also results in the inhibition of the hepatic production of sex hormone–binding globulin (SHBG) and IGFBP-1 and stimulation of ovarian P450c17α activity [7–9].

Obesity can also increase the local activity of 5 ∝ reductase enzyme through the influence of IGF-1. 5 ∝ reductase is responsible for the activation of testosterone at the level of the hair follicle by conversion to dihydrotestosterone (DHT). Two forms of this enzyme exist, type 1 and type 2.

Type 1 is mainly present in the sebaceous gland while type 2 is found mainly in the hair follicle. Relative activity of these isoenzymes can lead to a discrepancy between the severity of hirsutism and acne in women with hyperandrogenism [10].

These processes are manifested more in women with visceral rather than peripheral obesity as the visceral fat compartment is highly metabolically active leading to a higher risk of insulin resistance, hyperandrogenism, type 2 diabetes, and cardiovascular disease.

Hirsutism

Hirsutism is defined as male pattern terminal hair growth in a female as a result of increased androgen production or increased skin sensitivity to circulating androgens. Hypertrichosis on the other hand is a generalized nonsexual (vellus) hair growth. Vellus hair is the fine lightly pigmented hair that covers most areas of the body during the prepubertal years. Excessive growth of vellus hair may be hereditary, a result of various medications or malignancies, and should not be confused with hirsutism. As opposed to villus hair, terminal hair is the androgen-dependent thick pigmented hair normally present on the face, limbs, axilla, and pubic area. Its density and distribution are influenced by genetic and racial factors. Although thought to affect 5%–10% of reproductive age [11], the exact prevalence of hirsutism that is highly variable depending on the ethnic group studied is most prevalent of women of Mediterranean descent [11].

Obesity and polycystic ovarian syndrome

Women suffering from obesity often have associated polycystic ovarian syndrome (PCOS), which is a common endocrine disorder affecting approximately 6%–10% of women of reproductive age [12]. The syndrome is characterized by chronic oligo/anovulation and a variable combination of symptoms, including variable degrees of menstrual disturbances, obesity, and hyperandrogenism. Approximately 50% of patients with PCOS are overweight or obese [13]. Patients with PCOS may have difficulty in dealing with energy balance. They may report strong hunger, greater cravings for sweets, and eating disorders such as bulimia nervosa [14,15]. One study [15] compared meal-related appetite and secretion of the satiety peptide cholecystokinin (CCK) together with glucose regulatory hormones in women with and without PCOS and found that women with PCOS have reduced postprandial CCK secretion and deranged appetite regulation associated with increased concentrations of testosterone. CCK secretion may therefore play a role in the greater frequency of binge eating and overweight in women with PCOS.

Similar to obesity, women with PCOS also suffer from the metabolic syndrome with the most common denominator between PCOS and obesity being insulin resistance and the consequent associated hyperandrogenism. Insulin resistance has been reported to occur in between 30% and 70% of women depending on the characteristics of the studied population and the presence or absence of obesity [16]. Insulin resistance in women with PCOS may be a result of abnormalities in pituitary gonadotropins and excessive stimulation of IGF-1 receptors and an increased conversion of 17-hydroxy (OH)-progesterone into androstenedione through the activity of 17α-hydroxylase enzyme [16].

However, in women with PCOS, other mechanisms may also be involved in the dysregulation of androgen metabolism, including dysregulation of theca cell function and increased adrenal androgen production [17]. Consequently 5%–25% of women with PCOS consequently suffer from hirsutism [18].

From the fact mentioned previously, it is clear that there is often a strong overlap between obesity, PCOS, and hirsutism and often the primary pathology whether PCOS or obesity is unclear and a cause for confusion. Although not conclusive, there is some evidence from early data that PCOS is the primary problem and obesity is its consequence [14], but it may also be that peripubertal obesity and associated hyperandrogenism are the forerunners of adult PCOS [19]. Further studies are needed to answer this difficult question.

However, the question as to which is the primary pathology is of mainly academic interest since both share a common pathology and the underlying principles of the treatment of associated hirsutism remain the same, where weight reduction remains the key factor toward decreasing the impact of metabolic syndrome, decreasing insulin resistance, and improving the manifestations of hyperandrogenism.

The role of adrenal androgens in obese women with hirsutism

The obesity with or without PCOS should not be immediately considered as the primary cause for hirsutism in a woman suffering from this condition since other endocrine disturbances can be associated with both obesity and hirsutism, as a result of increased production of adrenal androgens. Adrenal androgens include androstenedione and dehydroepiandrosterone (DHEA) and are peripherally converted to testosterone, which together with that produced from the ovaries circulates in two forms, an inactive form bound to SHBG and a metabolically active free form. Free testosterone to stimulate hair growth needs to be further metabolized at the level of the hair follicle into a more active form, DHT by the enzyme, 5∝ reductase. Women with Cushing's syndrome would therefore suffer from an increased production of adrenal androgens leading to hirsutism in addition to other symptoms of the syndrome such as weight gain. Obesity may therefore be the underlying primary disorder causing hirsutism or an association due to a common underlying pathology.

Management of hirsutism associated with obesity

As previously discussed, hirsutism can be associated with obesity but this should not preclude the need to conduct a full systematic examination and appropriate investigations to exclude other potential causes. These include increased adrenal or ovarian androgens such as delayed onset congenital adrenal hyperplasia (CAH) and androgen-producing adrenal or ovarian tumors. Iatrogenic hirsutism can be caused by the administration of certain medications such as danazol, androgen therapy, sodium valproate, and anabolic steroids. In order to understand the expected effects of any treatment, it is first essential to have a firm understanding of the normal phases of hair growth.

The hair cycle

Hair growth is a dynamic process and can be divided into three distinct phases. The relative duration of these different stages influences the length and appearance of hair in different parts of the body. The first stage is known as anagen during which active mitotic division occurring in the basal matrix of the hair follicle leads to hair growth. This stage is relatively long in areas such as the scalp where hair appears to be continuously growing. Of particular relevance to hirsutism is the fact that facial hair has a relatively long anagen phase; hence, any therapeutic intervention would require around 6–9 months before becoming apparent [20]. This is then followed by a stage where hair growth stops and the hair follicle prepares to enter the resting phase (catagen stage). Finally, the hair becomes short and loosely attached (telogen stage) and will ultimately become detached in the preparation for the start of a new cycle.

Clinical assessment of hirsutism

Detailed history should include the severity and duration of hirsutism as well as the presence of any other symptoms of virilization. Severe rapidly progressive hirsutism or associated virilization point to the possibility of a more serious underlying abnormality such as an ovarian or adrenal tumor. The presence of history suggestive of other endocrine abnormalities such as Cushing's syndrome or hypothyroidism or the use of medications such as steroids or danazol is also highly relevant. The presence of irregular anovulatory periods and history of infertility may also point to the presence of associated PCOS.

General examination may show other manifestations of androgen excess such as acne or signs of virilization such as clitoromegaly. The presence of velvety, pigmented skin patches (acanthosis nigricans) in the groin, neck, or axillae should be noted as they may indicate associated insulin resistance. The combination of hirsutism, acanthosis nigricans, and insulin resistance is a hereditary condition known as HAIR-AN syndrome, possibly due to an insulin receptor defect and can be associated with severe hirsutism [21].

Evaluation of the severity of hirsutism is commonly performed using the Ferriman–Gallwey scoring system that includes an evaluation of nine androgen-sensitive body areas. Each area is assigned a score from 0 to 4, and then the scores are added. A minimal score of 8 is required for the diagnosis of hirsutism. In order to improve the quality of assessment using this scoring system, women are advised not to use hair removal methods such as laser and electrolysis for 3 months, waxing for 4 weeks, or shaving for 5 days prior to the assessment [22]. The Ferriman–Gallwey, however, does not account for focal hirsutism and furthermore ignores some androgen-sensitive areas such as the buttocks and side burns [20]. Finally, pelvic examination may reveal the presence of a pelvic mass such as an androgen-producing ovarian tumor.

Investigations include measurement of testosterone concentrations, SHBG, and the calculation of the free androgen index (FAI). It is important to note that testosterone concentrations correlate poorly with the severity of hirsutism due to individual variations in hair follicle response. The FAI takes into consideration testosterone concentrations as well as SHBG and therefore gives an account on the concentration of the relevant free active testosterone component. Obese and PCOS patients may have an elevated FAI when the testosterone concentrations are normal due to a decrease in SHBG. High testosterone concentrations (>5 mmol/L) should also trigger immediate investigations for the potential presence of an androgen-producing tumor.

Other androgens that can be measured include DHEA, which, if markedly elevated, may point to an adrenal cause. Baseline 17-OH-progesterone can also be measured in suspected cases of late onset CAH. Results of 17-OH-progesterone may then lead to the need for a short synacthen test to confirm the diagnosis. This is performed by the intramuscular administration of 250 mg of synacthen after baseline measurement of 17-OH-progesterone. 17-OH-progesterone measurements are taken again after an hour, and a significant rise over baseline measurement is diagnostic of CAH. A dexamethasone suppression test or 24-hour urinary free cortisol may also be performed for the suspected cases of Cushing's syndrome. Since obese women have a high risk of insulin resistance, metabolic syndrome (31%–35% of obese women with PCOS) and type 2 diabetes (7.5%–10% of obese women with PCOS) [23], a test for insulin resistance should be performed. Finally, and depending on the results of history, examination, and blood tests, pelvic imaging such as ultrasound, CT, or MRI may be required and may show the presence of polycystic ovaries or an androgen-producing ovarian or adrenal tumor. In cases where imaging is inconclusive, selective venous sampling from the ovarian and adrenal veins may also be performed to determine the exact source of the androgen production.

Treatment

It is important to note that the treatment of hirsutism can prevent or slow further hair growth but will not treat the already existent hair growth, which will need to be physically removed using a variety of methods, including electrolysis, plucking, waxing, shaving, and laser removal. Targeting the hair follicles in the anagen stage can lead to permanent hair removal.

1. *Treatment of obesity:* The key to the treatment of hirsutism associated with obesity is to treat the obesity.

As little as 5% weight loss can improve the ovulatory status and restore menstrual regularity [23]. Lifestyle interventions are therefore recommended as a first-line treatment. A Cochrane review examining the effects of lifestyle changes in women with PCOS showed that even a small amount of weight loss also resulted in a significant improvement in the metabolic profile and can result in an improvement in the FAI [24].

In addition to a suitable diet and lifestyle changes, both insulin-sensitizing agents and weight loss medications can be used to improve on hirsutism associated with obesity. Metformin is a synthetic biguanide that reduces hepatic glucose production, stimulates hepatic insulin-mediated glucose uptake, reduces serum lipid levels, and inhibits gluconeogenesis [25]. Metformin may therefore improve hirsutism through an improvement in insulin resistance. In a small randomized controlled study, it was found that the use of 1500 mg of metformin alone in addition to exercise and lifestyle modifications resulted in a significant improvement in the testosterone concentrations in insulin-resistant women with PCOS [26]. Myoinositol is also considered as a natural insulin sensitizer and has been compared to metformin. However, while myoinositol was demonstrated in studies to have a beneficial effect on insulin sensitivity in obese women with PCOS, metformin had a better effect on the metabolic features with notable weight loss and improvement in menstrual regularity and androgenic symptoms [27].

On the other hand, orlistat is a peripherally acting antiobesity drug that acts through inhibition of gastric and pancreatic lipase and leads to a 30% reduction in fat absorption from the intestinal lumen. The main side effects that may limit patient compliance are gastrointestinal, which are particularly pronounced if a low-fat diet is not adhered to. Regarding the relative efficacies of orlistat or metformin, a randomized controlled study comparing both these drugs in obese anovulatory women showed that both had a similar effect on weight loss, ovulation rates, and androgen concentrations [28].

Bariatric surgery can also be used in women with severe obesity and where lifestyle changes and pharmacological agents alone are not sufficient. In addition to being an effective method for weight loss in the patients with morbid obesity, bariatric surgery has been shown to lead to an improvement in ovulatory and menstrual function and hyperadrogenemia in obese women with PCOS [29,30].

2. *The combined oral contraceptive pill*

The combined oral contraceptive pill (COCP) has traditionally been the first-line medication for women with hirsutism and acts by increasing SHBG, thus decreasing the free effective testosterone fraction. Additional benefits include antagonizing luteinizing hormone−stimulated androgen production by the theca cells, a mild decrease in adrenal androgen production and a mild blockage of the androgen receptors [20]. A preference should be given to pills with an estrogen-dominant effect, such as those containing desogestrel, gestodene, or norgestimate while avoiding preparations containing first- and second-generation progestins that have a stronger androgenic effect such as norethindrone and levonorgestrel [31]. Levonorgestrel may also oppose the estrogen-driven increase in SHBG.

Dianette is an OCP containing cyproterone acetate (2 mg), a progestogen having additional antiandrogenic effects through gonadotropin inhibition and increased hepatic clearance of androgens. Cyproterone acetate can be given separately in higher doses (50−100 mg/day), but without the use of a concomitant effective contraception, there is risk of feminization of a male fetus should pregnancy occur. One particular regimen involves a combination of cyproterone acetate and ethinyl estradiol in a reverse sequential regimen. This involves the administration of ethinyl estradiol 25−50 mg/day from days 5 to 25 and cyproterone acetate from days 5 to 15. After improvement the dose of cyproterone acetate can be decreased to 5 mg/day [20].

Yasmin is yet another OCP that contains the progestogen drospirenone that inhibits ovarian androgen production and blocks androgen receptors and compared to desogestrel-containing OCPs has a better antiandrogenic effect [10,32]. However, a study comparing desogestrel, cyproterone acetate, and drospirenone found that although in the short term (6 months) all three were very similar, in the long term (12 months), cyproterone acetate was the most effective [33].

The use of the COCP has to be considered carefully in obese women, where they need to be counseled regarding the increased risk of venous thromboembolism. In women with a BMI of ≥ 35 kg/m^2, the use of the COCP is not recommended as the risks outweigh any potential benefits [34].

3. *Androgen antagonists*: The use of any of the following medications, usually reserved as second-line agents, should be combined with the use of an effective contraceptive due to the risk of feminization of a male fetus should a pregnancy occur during treatment.
   a. *Spironolactone*: This is a commonly used agent that acts by blocking androgen receptors and by inhibiting 5 $\propto$ reductase. Although relatively safe, side effects include diuresis and postural hypotension in early stages as well as menstrual irregularities and rarely hyperkalemia. When combined with the COCP, spironolactone has been shown to be more effective than metformin alone in improving hirsutism [35].

b. *Finasteride*: This is an inhibitor of 5 ∝ reductase used at a dose of 5 mg/day. Finasteride can also be used topically to improve the efficacy of intense pulsed light with radio-frequency hair removal [36]. Finasteride is teratogenic and hence, the use of an effective contraceptive is essential.
   c. *Flutamide*: A potent androgen receptor antagonist that needs to be used with care and only in a specialist setting due to the potential of hepatotoxicity.
   d. *Eflornithine*: A topical antiprotozoal drug that inhibits local hair growth through the inhibition of ornithine decarboxylase enzyme. It is mainly used to enhance the effect of laser treatment for hair removal [37,38]. It can also result in the obstruction of the sebaceous glands and, hence, worsening of acne. For those practicing in the United Kingdom, it is important to note that the Scottish Medicines Consortium has restricted the use of eflornithine to women in whom alternative therapies are not possible [39,40].

Conclusion

Obesity is a worldwide epidemic that can result in a number of significant reproductive effects, including increased androgen production and consequently hirsutism. There is also a strong overlap between obesity and PCOS, which is another common finding of women of reproductive age and which again can result in hyperandrogenemia and hirsutism. Furthermore, obesity may be the manifestation of a larger endocrine abnormality that can be associated with androgenic manifestations such as adrenal conditions. Evaluation of obese women with hirsutism therefore requires a careful clinical evaluation that recognizes that hirsutism is merely an endocrine manifestation that may underline a larger problem. Treatments should therefore focus on the primary pathology, and weight loss is the key behind successful treatment in this group of women. Further interventions will depend largely on the severity of the condition and the wish to conceive and will vary from the simple use of the COCP to the addition of an antiandrogen medication.

References

[1] Metwally M, Ledger WL, Li TC. Reproductive endocrinology and clinical aspects of obesity in women. Ann NY Acad Sci 2008;1127:140–6.
[2] Mitchell S, Shaw D. The worldwide epidemic of female obesity. Best Pract Res Clin Obstet Gynaecol 2015;29:289–99.
[3] Uzuncakmak TK, Akdeniz N, Karadag AS. Cutaneous manifestations of obesity and the metabolic syndrome. Clin Dermatol 2018;36:81–8.
[4] Miner JL. The adipocyte as an endocrine cell. J Anim Sci 2004;82:935–41.
[5] Mayes JS, Watson GH. Direct effects of sex steroid hormones on adipose tissues and obesity. Obes Rev 2004;5:197–216.
[6] Schwartz MW, Seeley RJ. Seminars in medicine of the Beth Israel Deaconess Medical Center. Neuroendocrine responses to starvation and weight loss. N Engl J Med 1997;336:1802–11.
[7] Douglas CC, Gower BA, Darnell BE, Ovalle F, Oster RA, Azziz R. Role of diet in the treatment of polycystic ovary syndrome. Fertil Steril 2006;85:679–88.
[8] Broughton DE, Moley KH. Obesity and female infertility: potential mediators of obesity's impact. Fertil Steril 2017;107:840–7.
[9] Cooper LA, Page ST, Amory JK, Anawalt BD, Matsumoto AM. The association of obesity with sex hormone-binding globulin is stronger than the association with ageing—implications for the interpretation of total testosterone measurements. Clin Endocrinol (Oxf) 2015;83:828–33.
[10] Archer JS, Chang RJ. Hirsutism and acne in polycystic ovary syndrome. Best Pract Res Clin Obstet Gynaecol 2004;18:737–54.
[11] Mihailidis J, Dermesropian R, Taxel P, Luthra P, Grant-Kels JM. Endocrine evaluation of hirsutism. Int J Womens Dermatol 2017;3:S6–10.
[12] Cooney LG, Dokras A. Beyond fertility: polycystic ovary syndrome and long-term health. Fertil Steril 2018;110:794–809.
[13] Faloia E, Canibus P, Gatti C, Frezza F, Santangelo M, Garrapa GG, et al. Body composition, fat distribution and metabolic characteristics in lean and obese women with polycystic ovary syndrome. J Endocrinol Invest 2004;27:424–9.
[14] Linne Y. Effects of obesity on women's reproduction and complications during pregnancy. Obes Rev 2004;5:137–43.
[15] Hirschberg AL, Naessen S, Stridsberg M, Bystrom B, Holtet J. Impaired cholecystokinin secretion and disturbed appetite regulation in women with polycystic ovary syndrome. Gynecol Endocrinol 2004;19:79–87.
[16] Lewandowski KC, Skowronska-Jozwiak E, Lukasiak K, Galuszko K, Dukowicz A, Cedro M, et al. How much insulin resistance in polycystic ovary syndrome? Comparison of HOMA-IR and insulin resistance (Belfiore) index models. Arch Med Sci 2019;15:613–18.
[17] Crespo RP, Bachega T, Mendonca BB, Gomes LG. An update of genetic basis of PCOS pathogenesis. Arch Endocrinol Metab 2018;62:352–61.
[18] Glintborg D. Endocrine and metabolic characteristics in polycystic ovary syndrome. Dan Med J 2016;63.
[19] McCartney CR, Prendergast KA, Chhabra S, Eagleson CA, Yoo R, Chang RJ, et al. The association of obesity and hyperandrogenemia during the pubertal transition in girls: obesity as a potential factor in the genesis of postpubertal hyperandrogenism. J Clin Endocrinol Metab 2006;91.
[20] Martin KA, Chang RJ, Ehrmann DA, Ibanez L, Lobo RA, Rosenfield RL, et al. Evaluation and treatment of hirsutism in premenopausal women: an endocrine society clinical practice guideline. J Clin Endocrinol Metab 2008;93:1105–20.
[21] Somani N, Harrison S, Bergfeld WF. The clinical evaluation of hirsutism. Dermatol Ther 2008;21:376–91.
[22] Lizneva D, Gavrilova-Jordan L, Walker W, Azziz R. Androgen excess: investigations and management. Best Pract Res Clin Obstet Gynaecol 2016;37:98–118.
[23] Goodman NF, Cobin RH, Futterweit W, Glueck JS, Legro RS, Carmina E, et al. American Association of Clinical Endocrinologists,

American College of Endocrinology, and Androgen Excess and PCOS Society disease state clinical review: guide to the best practices in the evaluation and treatment of polycystic ovary syndrome—Part 2. Endocr Pract 2015;21:1415—26.

[24] Lim SS, Hutchison SK, Van Ryswyk E, Norman RJ, Teede HJ, Moran LJ. Lifestyle changes in women with polycystic ovary syndrome. Cochrane Database Syst Rev 2019;3:CD007506.

[25] Sivalingam VN, Myers J, Nicholas S, Balen AH, Crosbie EJ. Metformin in reproductive health, pregnancy and gynaecological cancer: established and emerging indications. Hum Reprod Update 2014;20:853—68.

[26] Fux Otta C, Wior M, Iraci GS, Kaplan R, Torres D, Gaido MI, et al. Clinical, metabolic, and endocrine parameters in response to metformin and lifestyle intervention in women with polycystic ovary syndrome: a randomized, double-blind, and placebo control trial. Gynecol Endocrinol 2010;26:173—8.

[27] Tagliaferri V, Romualdi D, Immediata V, De Cicco S, Di Florio C, Lanzone A, et al. Metformin vs myoinositol: which is better in obese polycystic ovary syndrome patients? A randomized controlled crossover study. Clin Endocrinol (Oxf) 2017;86:725—30.

[28] Metwally M, Amer S, Li TC, Ledger WL. An RCT of metformin versus orlistat for the management of obese anovulatory women. Hum Reprod 2009;24:966—75.

[29] Li YJ, Han Y, He B. Effects of bariatric surgery on obese polycystic ovary syndrome: a systematic review and meta-analysis. Surg Obes Relat Dis 2019;15.

[30] Skubleny D, Switzer NJ, Gill RS, Dykstra M, Shi X, Sagle MA, et al. The impact of bariatric surgery on polycystic ovary syndrome: a systematic review and meta-analysis. Obes Surg 2016;26:169—76.

[31] Schmidt TH, Shinkai K. Evidence-based approach to cutaneous hyperandrogenism in women. J Am Acad Dermatol 2015;73:672—90.

[32] Kriplani A, Periyasamy AJ, Agarwal N, Kulshrestha V, Kumar A, Ammini AC. Effect of oral contraceptive containing ethinyl estradiol combined with drospirenone vs. desogestrel on clinical and biochemical parameters in patients with polycystic ovary syndrome. Contraception 2010;82:139—46.

[33] Bhattacharya SM, Jha A. Comparative study of the therapeutic effects of oral contraceptive pills containing desogestrel, cyproterone acetate, and drospirenone in patients with polycystic ovary syndrome. Fertil Steril 2012;98:1053—9.

[34] FSRH. Overweight, obesity and contraception. Faculty of Sexual and Reproductive Healthcare; 2019.

[35] Alpanes M, Alvarez-Blasco F, Fernandez-Duran E, Luque-Ramirez M, Escobar-Morreale HF. Combined oral contraceptives plus spironolactone compared with metformin in women with polycystic ovary syndrome: a one-year randomized clinical trial. Eur J Endocrinol 2017;177:399—408.

[36] Farshi S, Mansouri P, Rafie F. A randomized double blind, vehicle controlled bilateral comparison study of the efficacy and safety of finasteride 0.5% solution in combination with intense pulsed light in the treatment of facial hirsutism. J Cosmet Laser Ther 2012;14:193—9.

[37] Hamzavi I, Tan E, Shapiro J, Lui H. A randomized bilateral vehicle-controlled study of eflornithine cream combined with laser treatment versus laser treatment alone for facial hirsutism in women. J Am Acad Dermatol 2007;57:54—9.

[38] Smith SR, Piacquadio DJ, Beger B, Littler C. Eflornithine cream combined with laser therapy in the management of unwanted facial hair growth in women: a randomized trial. Dermatol Surg 2006;32:1237—43.

[39] BNF. indicationsAndDosesAvailable from: https://bnf.nice.org.uk/drug/triptorelin.html.

[40] BNF. Eflornithine. 2019.

Chapter 9

Obesity and female infertility

Suresh Kini[1], Mythili Ramalingam[1] and Tahir A. Mahmood[2]

[1]*Assisted Conception Unit, Department of Obstetrics and Gynaecology, Ninewells Hospital, Dundee, United Kingdom,* [2]*Department of Obstetrics and Gynaecology, Victoria Hospital, Kirkcaldy, United Kingdom*

Introduction

Overweight is defined by the World Health Organization (WHO) [1] as a body mass index (BMI) ≥ 25 kg/m^2 and obesity as ≥ 30 kg/m^2. Obesity exerts a negative influence on female fertility. Obese women are more likely to have ovulatory dysfunction due to dysregulation of the hypothalamic–pituitary–ovarian (HPO) axis. Obese women have reduced fecundity even when eumenorrheic and demonstrate poorer outcomes with the use of assisted conception.

Obesity brings out many problems such as social, psychological, demographic, and health problems. It is related to increased health risks such as diabetes mellitus, hypertension, coronary heart disease, and osteoarthritis and is linked to various malignancies, particularly endometrium, breast, and colon cancers. Obesity also plays a significant role in reproductive disorders in women. It is associated with anovulation, menstrual disorders, infertility, difficulties in assisted reproduction, miscarriage, and adverse pregnancy outcomes. In obese women, gonadotropin secretion is affected because of the increased peripheral aromatization of androgens to estrogens. The insulin resistance and hyperinsulinemia in obese women lead to hyperandrogenemia. The sex hormone–binding globulin, growth hormone, and insulin-like growth factor–binding proteins are decreased, and leptin levels are increased. Thus the neuroregulation of the HPO axis deteriorates. These alterations may explain impaired ovulatory function and so reproductive health. Because of lower implantation and pregnancy rates, higher miscarriage rates, and increased maternal and fetal complications during pregnancy, obese women have a lower chance to give birth to a healthy newborn [2].

Epidemiology

Obesity has become a global epidemic, affecting more than 650 million adults worldwide. The prevalence of obesity has increased in developed countries because of a change in lifestyle, including reduced physical activity, changes in nutrition style, and an increased calorie intake. However, some other factors such as endocrine disorders, hormonal disorders, psychological disorders, and use of some drugs such as steroids and antidepressants may lead to obesity. Rates of obesity in the United States are significantly higher than in other developed nations, with more than one-third of adult Americans affected. The number of obese Americans has doubled since 1960. The WHO reported that in the United States and most European countries, 60% of women are overweight (≥ 25 kg/m^2), of these, 30% are obese (≥ 30 kg/m^2) and 6% are morbidly obese (≥ 35 kg/m^2) [3].

Transgenerational inheritance

There is a mounting body of evidence suggesting that maternal obesity may confer a risk of metabolic dysfunction through multiple generations. We know that obesity affects intergenerational risk, exposing the offspring to develop noncommunicable disease later in life. Children of obese mothers are more likely to develop obesity, type 2 diabetes, and cardiovascular disease as adults. This may be due to epigenetic modifications in utero. A recent study [4] in a diet-induced obesity (DIO) mouse model showed that metabolic dysfunction mediated through impaired mitochondrial dynamics can be passed through the maternal germ line to second- and third-generation offspring. The transmission appeared to be germ line and through aberrant oocytes [3].

Pathophysiological basis of infertility in obese women

Obesity and reproductive functions

It is difficult to describe the mechanism of how obesity affects the reproductive system because it is complex and multifactorial. Several mechanisms are involved in the relationship between fertility and obesity. The insulin resistance and leptin levels are increased, and hyperandrogenemia occurs in obese women. Similarly, anovulation, changes in adipokine levels and the HPO axis, and steroidogenesis in obese women affects the reproductive system.

Adipose tissue and adipokines

White adipose tissue is an important endocrine organ that regulates energy homeostasis and metabolism by secreting adipokines. Adipokines are signaling molecules (hormones), and abnormalities in adipokines can cause inflammation and abnormal cell signaling and thus can lead to deterioration in cell metabolism and function. Although adipose tissue is necessary for reproductive function and normal development, the excessive adipose tissue causes some reproductive disturbances. Some of these adipokines are leptin, adiponectin, interleukin-6 (IL-6), plasminogen activator inhibitor-1 (PAI-1), tumor necrosis factor-α (TNF-α), resistin, visfatin, chemerin, omentin, and ghrelin (Table 9.1).

Leptin

Leptin is a 16 kDa protein that is secreted almost exclusively by the adipocytes and is produced by the obese (*ob*) gene. In obesity, circulating leptin levels are high due to leptin resistance. Leptin may affect reproductive function at many levels. Physiologically, in women, by central action, leptin seems to be important for the hypothalamic–pituitary function and puberty. At the level of the ovary, leptin was found to modulate basal and follicle-stimulating hormone (FSH) stimulated steroidogenesis in cultured human lutein granulosa cells, with high concentrations suppressing the secretion of estradiol and progesterone. High levels of leptin, representing hyperleptinemia of obesity, may inhibit folliculogenesis. Leptin may have a role in regulation of embryo implantation and endometrial receptivity, and it has been suggested that obesity-related perturbations of the leptin system can possibly interfere with embryo implantation, therefore causing infertility. In obese women with polycystic ovary syndrome (PCOS), abnormalities in the relationship between leptin and luteinizing hormone (LH) secretory characteristics have been found. Leptin levels have been found to be positively correlated with insulin resistance in women with PCOS.

Adiponectin

Adiponectin, a 30 kDa protein, is the most abundant serum adipokine, secreted exclusively by the adipose tissue. In obese women, unlike the other adipose tissue hormones, adiponectin levels increase with weight loss. Adiponectin stimulates glucose uptake in the liver and muscle and decreases hepatic gluconeogenesis. As a result, insulin sensitivity is impaired. Adiponectin also affects lipid synthesis, energy homeostasis, vasodilatation, and atherogenic activity. Thus adiponectin decreases triglyceride accumulation and improves insulin sensitivity. In the absence of adiponectin in obese women, plasma insulin levels increase. Consequently, high levels of insulin lead to hyperandrogenemia.

Interleukin-6

IL-6 is an inflammatory cytokine, and approximately 30% of circulating levels are derived from adipose tissue. Circulating IL-6 levels increase in obesity, and they are associated with increased insulin resistance. It seems that IL-6, in the high levels seen in obese women, may contribute to impaired fertility in women with PCOS.

Plasminogen activator inhibitor type 1

PAI-1 is a regulator of blood fibrinolytic activity and is mainly produced by white adipose tissue and visceral fat.

TABLE 9.1 The effects of the adipokines on reproduction.

Adipokines	Effects on reproduction in obesity	Serum levels in obesity
Leptin	Inhibits insulin-induced ovarian steroidogenesis	Increases
	Inhibits LH-stimulated estradiol production by the granulosa cells	
Adiponectin	Plasma insulin levels increase	Decreases
IL-6	Causes insulin resistance	Increases
PA1-1	Causes Insulin resistance	Increases
TNF-α	Impairs insulin action—hyperinsulinemia	Increases
	Inhibits insulin signaling	
Resistin	Causes insulin resistance	Increases
Visfatin	Increased insulin sensitivity	Increases
Omentin	Increased insulin sensitivity	Decreases
Chemerin	Negatively regulates FSH-induced follicular steroidogenesis	Increases

IL-6, Interleukin-6; *TNF-α*, tumor necrosis factor-α; *LH*, luteinizing hormone; *FSH*, follicle-stimulating hormone; *PA1-1*, plasminogen activator inhibitor type-1.

Circulating PAI-1 levels increase in obesity and correlate with the elements of the metabolic syndrome. PAI-1 has been associated with miscarriage in women with PCOS.

Tumor necrosis factor-α

TNF-α is synthesized in adipose tissue by adipocytes and other cells in the tissue matrix. Blood levels and adipocyte production of TNF-α correlate with BMI and hyperinsulinemia, and TNF-α impairs insulin action by inhibiting insulin signaling. TNF-α may affect several levels of the reproductive axis: inhibition of gonadotrophin secretion, ovulation, steroidogenesis, corpus luteum regression, and endometrial development.

The mechanism of other adipokines on reproductive functions such as resistin and ghrelin has not been fully understood. Resistin is a protein secreted by the adipose tissue. As a result of increased resistin levels in obesity, insulin resistance occurs and this leads to decreased insulin sensitivity. Another adipokine, visfatin, is secreted from several cell types and tissues, including adipose tissue and adipocytes, bone marrow, lymphocytes, muscle, liver, trophoblast, and fetal membranes. The association between visfatin and obesity and insulin action is not fully understood. It has been reported that visfatin shows insulin-mimetic effects, increases glucose uptake in adipocytes and muscle cells, and decreases glucose release from hepatocytes. Chemerin is another adipokine that affects the adipocyte and glucose metabolism. It has been shown that chemerin levels increase during the metabolic syndrome; therefore it is associated with obesity, metabolic syndrome, and type 2 diabetes mellitus. Chemerin also can impair FSH-induced follicular steroidogenesis and thus can play a role in the pathogenesis of PCOS.

Adipose tissue also affects follicular development by the inhibition of gonadotropin secretion through the conversion of androgens to estrogens in the adipose tissue. Therefore almost all of the adipokines seem to have their effects on reproduction by causing insulin resistance [2,5].

The clinical effects of obesity on female infertility

Obesity has a negative effect on reproductive potential, primarily thought to be due to functional alteration of the HPO axis. Obese women often have higher circulating levels of insulin, which is a known stimulus for increased ovarian androgen production. These androgens are aromatized to estrogen at high rates in the periphery owing to excess adipose tissue, leading to negative feedback on the HPO axis and affecting gonadotropin production. This manifests as menstrual abnormalities and ovulatory dysfunction. Hyperinsulinemia is highly implicated in the pathogenesis of the PCOS, characterized by oligomenorrhea and hyperandrogenism. Obesity contributes to insulin resistance and appears to exacerbate the symptoms of PCOS, with obese women often demonstrating a more severe phenotype. Elevated androgen levels in PCOS lead to deposition of visceral fat, leading to insulin resistance and hyperinsulinemia, further stimulating ovarian and adrenal androgen production in a perpetual cycle. The prevalence of PCOS in some obese populations approaches 30%, although a causative role of obesity in the development of PCOS has not been established.

Menstrual irregularity occurs more frequently in women above 175% of ideal body weight compared with women below 150% of ideal body weight (54% vs 19%, respectively). Obese women with a BMI > 27 kg/m^2 have a relative risk (RR) of anovulatory infertility of 3.1 (95% CI, 2.2–4.4) compared with their lean counterparts with a BMI 20.0–24.9 kg/m^2. Obese women have a lower chance of conception within 1 year of stopping contraception compared with normal-weight women (i.e., 66.4% of obese women conceive within 12 months, compared with 81.4% of those of normal weight) [6].

Multiple studies have demonstrated that obese women have increased time to pregnancy. Two studies in large cohorts of Danish women planning pregnancies showed a decline in fecundability ratios with increasing BMI [7,8]. Interestingly, obese women remain subfertile even in the absence of ovulatory dysfunction. Examination of a large American cohort of more than 7000 women by Gesink Law et al. [9] showed reduced fecundity in eumenorrheic obese women, and van der Steeg et al. [10] presented data from a large Dutch cohort of more than 3000 women with normal cycles, in which the probability of spontaneous conception declined linearly with each BMI point >29 kg/m^2. Anovulatory women have a greater waist circumference and more abdominal fat than ovulatory women of similar BMI. In normogonadotropic anovulatory women, increased BMI and abdominal obesity are associated with decreased odd ratios (OR) of ovulation in response to clomiphene citrate [increased BMI: OR 0.92 (0.88–0.96) and increased waist-to-hip ratio: OR 0.60 (0.40–0.89)]. A systematic review of 27 in vitro fertilization (IVF) studies, 23 of which were retrospective, shows that overweight women (BMI, >25 kg/m^2) undergoing IVF have a 10% lower live birth rate than women of normal weight (BMI, <25 kg/m^2) [6].

Effect on the hypothalamic–pituitary–ovarian axis

Obese women have higher circulating levels of leptin than normal-weight control subjects that may lead to chronic downregulation of leptin receptor in the hypothalamus. Women with high serum concentrations of leptin and elevated leptin–BMI ratios have lower rates of pregnancy with IVF. Jain et al. [11] studied eumenorrheic obese women and found that the amplitude of LH pulsatility was significantly decreased, again pointing to a central defect that may be unique to this disease.

Effects on the oocyte

There is abundant literature supporting an effect of obesity on the oocyte. The DIO mice models reveal high rates of meiotic aneuploidy with fragmented disorganized meiotic spindles and chromosomes not properly aligned on the metaphase plate. Obesity also appears to alter mitochondrial function in the oocyte. Mitochondria in DIO mice have disrupted architecture with fewer cristae, more vacuoles, and evidence of swelling. There is also a change in mitochondrial distribution, with clumping throughout the ooplasm compared with uniform perinuclear localization in control subjects. These abnormal mitochondria show evidence of metabolic stress, with lower levels of citrate, a tricarboxylic acid cycle end product. There is also evidence of endoplasmic reticulum (ER) stress in the obese state. There is evidence that women with PCOS also exhibit impaired oocyte competence, with lower rates of conception with ovulation induction and altered follicular fluid biomarkers [3].

One potential mechanism for oocyte organelle damage in obesity is lipotoxicity. The continued dietary excess of fatty acids accumulates in the tissues other than adipocytes (storage compartment) and exerts toxic effects, which is termed lipotoxicity. Obese women have higher levels of circulating free fatty acids, which damage nonadipose cells by increasing reactive oxygen species (ROS) that induce mitochondrial and ER stress leading to apoptosis. The oocytes of obese mice have twofold increased production of ROS and depleted levels of glutathione, an important intracellular defense against ROS damage. Lipotoxicity plays a role in the development of insulin resistance and a heightened inflammatory state in obese women. Obesity is considered to be a chronic low-grade inflammatory state. Obese women have higher circulating levels of C-reactive protein (CRP), a marker of systemic inflammation. Adipose tissue produces many proinflammatory adipokines, including leptin, TNF-α, and IL-6. Obese women have lower circulating levels of adiponectin. The tissues of the reproductive tract are not immune to the inflammatory state. CRP levels are elevated in the follicular fluid of obese women. Inflammatory pathways are critically important in reproductive events such as follicle rupture at the time of ovulation and invasion of the trophoblast into the receptive endometrium. The developing blastocyst produces adiponectin, IL-1, and IL-6. The altered inflammatory milieu in obese women likely exerts an influence on these processes. Higher serum levels of leptin in obese women correlate with higher levels of leptin in the follicular fluid. In vitro studies have shown that leptin affects steroidogenic pathways in granulosa cells, decreasing estrogen and progesterone production in a dose-dependent manner. This effect of obesity at the level of the oocyte could have downstream effects on endometrial receptivity and embryo implantation [3].

Effects on the embryo

The preimplantation embryo is also affected by an obese environment. Obese women are more likely to create poor-quality embryos. Leary et al. [12] noted that embryos from women with BMI ≥ 25 kg/m^2 were less likely to develop after fertilization, and those that did reached the morula stage more quickly. In addition, those that reached the blastocyst stage had fewer cells in the trophectoderm and demonstrated poor glucose uptake and increased levels of triglycerides. Embryos may also be susceptible to lipotoxicity. In women undergoing IVF, elevated levels of a specific omega-3 fatty acid, α-linoleic acid, were associated with decreased pregnancy rates. An increased ratio of linoleic acid, an omega-6 fatty acid, to α-linoleic acid correlated with improved pregnancy rates in the same population. This suggests that the balance of certain free fatty acids is important in mediating lipotoxicity in human reproduction. In addition to acting centrally, elevated leptin levels in obese women may exert a direct negative effect on the developing embryo. Tonically elevated levels of leptin in obesity may decrease the sensitivity of the trophoblast to its effects [3].

Effect on the endometrium

There are conflicting data as to whether obesity has a significant effect on the endometrium. Some studies conclude that obesity does not negatively affect endometrial receptivity [13,14]. Others found BMI to be an independent predictor of clinical pregnancy [15]. Some other studies suggest the importance of decidualization defects. Such defects may contribute to compromised endometrial receptivity and poor implantation. Decidualization and implantation defects may negatively affect the placentation process. Many of the pregnancy complications seen in obese women are linked to placental dysfunction. An Italian study of 700 women undergoing donor oocyte cycles found significantly higher spontaneous abortion rates in obese women: 38.1% compared with 13.3% in normal-weight control women [16]. A large nested case--control study showed an increased risk of recurrent miscarriage in the obese group, with an odds ratio of 3.5 [17]. In women with a history of recurrent pregnancy loss (RPL), obesity is a known risk factor for miscarriage in a subsequent pregnancy [18]. A chromosomal analysis of 117 miscarriage specimens from patients with RPL demonstrated that obese women had a much higher rate of euploid miscarriage, again suggesting a potential independent effect of obesity on the endometrium [19]. Leptin may also modulate endometrial receptivity, as evidenced by upregulation of markers of receptivity with leptin exposure in both epithelial and stromal cells [20]. Chronic dysregulation of leptin pathways in obesity may negatively affect implantation.

Challenges of managing obese women

Overweight and obese subfertile women have a reduced probability of successful fertility treatment.

Examination: The utility of the clinical examination is often limited in the obese woman, which results in a greater reliance on imaging. Obese patients have difficult venous access.

Funding: Around the world, fertility treatment is withheld from women above a certain BMI, with a threshold ranging from 25 to 40 kg/m^2. The proponents of this policy use three different arguments to justify their restrictions: risks for the woman, health and well-being of the future child, and importance of society. The opponents feel that the obese women should be informed about the consequences and encouraged to lose weight. If, however, they are unable to lose weight despite effort, there should not be any argument to withhold their treatment. However, based on available evidence, it may be appropriate to consider morbid obesity as a contraindication for public-funded treatment where the aim is to maximize the value for money [21,22].

Pelvic ultrasound: Obesity can contribute to missed diagnoses, nondiagnostic results, imaging examination cancellation because of weight or girth restrictions and scheduling of inappropriate examinations. Recognition of equipment limitations, imaging artifacts, optimization techniques, and appropriateness of modality choices is critical to providing good patient care to this health-challenged group.

Ovulation induction: The RR of anovulatory infertility is 2.7 (95% CI, 2.0–3.7) in women with BMI ≥ 32 kg/m^2 at age 18, while in ovulatory but subfertile woman, the chance of spontaneous conception decreases by 5% for each unit increase in the BMI. Overweight and obese women respond poorly to clomiphene induction of ovulation and require higher doses of gonadotropins for ovulation induction and superovulation. Obese women undergoing IVF require higher doses of gonadotropins, respond poorly to ovarian stimulation. A BMI above the normal range is an independent negative prognostic factor for multiple outcomes, including cycle cancellation, oocyte and embryo counts, and ongoing clinical pregnancy. These negative outcomes were most profound in women with class-II/III obesity, ovulatory dysfunction, or PCOS. The procedure of oocyte recovery is more challenging in women with high BMI. General anesthesia can be more hazardous while response to conscious sedation may be erratic with a higher risk of hypoxemia [22].

Tubal investigations: Operators have encountered difficulty in completing the HyCoSy procedures in obese participants, when the uterus was acutely retroverted or oblique, when multiple loops of active bowel were present, or the adnexa were located beyond the penetration of the ultrasound signal. Significant technical difficulty and increased radiation exposure have been associated with hysterosalpigography. Laparoscopy is not contraindicated in obese patients. Despite being associated with increased operating times, complication rates in obese patients are comparable to their nonobese counterparts. However, these procedures should be performed by a skilled surgeon in a special hospital setting.

Ovarian reserve: The systematic review and meta-analysis [6] suggest that antimullerian hormone and FSH are significantly lower in obese than in nonobese women and are inversely correlated with BMI. There is also some evidence supporting an association between BMI and inhibin B, although meta-analysis for this marker is limited by the limited number of published studies. Antral follicle count does not appear to differ according to BMI [23].

Treatment options

Because of pregnancy complications related to obesity, obese women wishing to conceive should consider a weight management program that focuses on preconception weight loss (to a BMI < 35 kg/m^2), prevention of excess weight gain in pregnancy, and long-term weight reduction. Weight management in all individuals is best achieved through a lifestyle modification program that combines dietary modification, physical activity, and behavioral interventions, including psychological, behavioral, and stress management strategies.

Weight loss

The body of literature on the effect of weight loss in obese women desiring conception is mixed. With a goal of 10% weight loss, some studies [24] had significantly higher conception rates and live birth rates (LBRs). In a small randomized controlled trial (RCT) of 49 obese women [25] undergoing fertility treatment, those randomized to an intensive 12-week lifestyle intervention had an average 6.6 kg weight loss and a significantly higher LBR than the control group and required fewer treatment cycles. Other studies examined the effect of weight loss in obese women with PCOS, demonstrating improved ovulation and LBRs in the treatment group with lifestyle intervention and weight loss. Many would argue that we should not be delaying infertility treatment for attempts at weight loss. These studies had a clinically realistic design but were unable to answer the mechanistic question as to whether weight loss improves fertility outcomes. We must also keep in mind that weight loss before conception in the obese population can ameliorate risks in pregnancy [3].

Physical activity

Attempts have been made to examine the effect of physical activity in the obese infertile population, independently from weight loss. Current recommendations are to increase physical activity to at least 150 minutes weekly of moderate activity such as walking. In a retrospective cohort of obese infertile women undergoing 216 cycles of IVF/ICSI (intracytoplasmic sperm injection), the outcomes of patients that engaged in regular physical activity were compared with those who were sedentary. There were significantly higher pregnancy rates and LBRs in the active group. Moderate physical activity was associated with a small increase in fecundity across the cohort. Physical activity has been shown to decrease systemic inflammatory mediators that may contribute to the improvement in fertility [3].

Dietary factors

It is highly likely that fertility is not affected solely by excess caloric intake but by the distribution of those calories across food groups. A 500–1000 kcal/day decrease from usual dietary intake should lead to a 1- to 2-lb weight loss per week, with a low-calorie diet of 1000–1200 kcal/day, achieving an average 10% decrease in total body weight over 6 months. There is potential therapeutic benefit of the "Mediterranean" diet, characterized by higher intake of unsaturated fats, lower intake of animal fats, and lower ratios of omega-6 to omega-3 fatty acids. Adherence to a Mediterranean diet for 2 years in patients with metabolic syndrome significantly decreased insulin resistance and serum concentrations of inflammatory markers, including CRP and IL-6. Increasing adherence to a Mediterranean diet correlated with an increased chance of pregnancy. A lower risk of infertility was observed in women in the highest quartile of adherence to the Mediterranean diet. Chavarro et al. [26] have published extensively on the "fertility diet," a pattern of dietary intake that has been associated with lower risk of ovulatory infertility and characterized by less consumption of trans fats and animal protein and more consumption of low-glycemic carbohydrates, high-fat dairy, and multivitamins. Better understanding of the mechanisms underlying obesity's impact on fertility has led to investigation of targeted dietary supplementation. An RCT of 100 patients undergoing ovulation induction that had previously been resistant to clomiphene showed that CoQ-10 supplementation improved ovulation and pregnancy rates. These studies have yet to be undertaken in the obese infertile population [3].

Bariatric surgery

Bariatric surgery in women can restore menstrual regularity, correct ovulation, shorten folliculogenesis with ovulation, reduce serum testosterone levels, diminish percent body fat, and improve both sexual function and chance of pregnancy. Eumenorrheic women with a BMI ≥ 35 kg/m^2 have deficient luteal LH and progesterone. Surgically induced weight loss only partially improves deficient luteal progesterone production with a rise in LH secretion, suggesting the persistence of corpus luteum dysfunction.

Delaying pregnancy until 1–2 years after bariatric surgery has been recommended to avoid fetal exposure to nutritional deficiencies from rapid maternal weight loss, although limited data suggest that pregnancy within the first year after bariatric surgery may not necessarily increase the risk for adverse maternal or perinatal outcomes.

A retrospective cohort study [27] examining pregnancy outcomes after bariatric surgery demonstrated lower risk of gestational diabetes and large-for-gestational-age infants. However, it also showed a concerning increased risk of small-for-gestational-age infants and a trend toward higher risks of stillbirth and neonatal death with no improvement in preterm birth. The only significant change noted after surgery was a shortening of the follicular phase, but the impact of this is unclear in the absence of fertility outcomes. Bariatric surgery does appear to improve the PCOS phenotype. Metabolic parameters, including insulin sensitivity and blood pressure, were also improved. This again demonstrates that obesity has a significant impact on the pathophysiology of PCOS. Clearly, more studies are needed regarding the effect of bariatric surgery on obesity-related infertility [3].

Barriers to weight loss in infertile women

Based on what little evidence is available, overweight infertile women appear most deterred from exercise by the perception that it causes fatigue and it is a hard work. These perceptions, as well as depression, seem to decrease with continuation of an exercise program [28].

Types of exercise: The majority of exercise interventions sought to increase weekly aerobic activity in participants in order to increase caloric expenditure. As compliance is key to success, coached sessions of achievable frequency, for example, weekly, for up to 6 months, should be considered. Motivational interviewing techniques might also be useful. An advantage of dual enrolment may result in better adherence as partners tend to motivate each other.

Types of diet: The dietary interventions used were based on caloric restriction and were usually consistent with weight reduction advice from national guidelines [29]. It is hard to say whether any particular degree of restriction was superior for achievement of weight loss, given the frequent pairing of diet with exercise. The RCT with a vegan diet in one of its study arms had the lowest recruitment and highest discontinuation rate [28].

Types of weight-loss medication: Oral medications used in the studies included orlistat, a lipase inhibitor; sibutramine, a selective serotonin and norepinephrine reuptake inhibitor; and acarbose, an alpha-glucosidase inhibitor shown to induce modest weight loss, though not suitable for weight maintenance. Of these, sibutramine, which has been withdrawn in Europe and the United States but is still available on the Internet, has been shown in a large study to have a risk of cardiovascular defects in unborn infants [30], while the same study showed no risk of birth defects from orlistat use. The safety of acarbose in pregnancy is not established. Orlistat was shown to be superior to a control with respect to achievement of pregnancy and ovulation in a single study [31].

Metformin: The meta-analyses showed that weight-loss interventions have a nonsignificant advantage over metformin with respect to achievement of pregnancy or improvement of ovulation status. There was also no significant difference in menstrual regularity improvement.

Herbal: Many obese women may also self-medicate with herbal supplements, although their safety and effectiveness have not been demonstrated. Ephedra containing supplements have potentially life-threatening cardiovascular side effects and have been banned by the FDA.

Conclusion

Overweight and obese women seeking fertility should be educated on the effects of being overweight or obese on the ability to achieve pregnancy and the benefits of weight reduction, including improvement in pregnancy rates, and a reduced need for ovulation induction and assisted conception. A combination of a reduced calorie diet, which is not overly restrictive and aerobic exercise, intensified gradually, should form the basis of programs designed for such individuals. Until further evidence is available, lifestyle interventions should still be considered the first line therapy, with drug use reserved for monitored trials.

References

[1] World Health Organization. Obesity and overweight fact sheet no 311. 2016: February 2018.
[2] Dağ ZÖ, Dilbaz B. Impact of obesity on infertility in women. J Turk Ger Gynecol Assoc 2015;16(2):111–17.
[3] Broughton DE, Moley KH. Obesity and female infertility: potential mediators of obesity's impact. Fertil Steril 2017;107(4):840–7.
[4] Saben JL, Boudoures AL, Asghar Z, Thompson A, Drury A, Zhang W, et al. Maternal metabolic syndrome programs mitochondrial dysfunction via germline changes across three generations. Cell Rep 2016;16(1):1–8.
[5] Mahmood TA, Arulkumaran S, editors. Obesity: a ticking time bomb for reproductive health. London: Elsevier; 2012. ISBN-13: 978-0-124-16045-3.
[6] Practice Committee of the American Society for Reproductive Medicine. Obesity and reproduction: a committee opinion. Fertil Steril 2015;104(5):1116–26.
[7] Wise LA, Rothman KJ, Mikkelsen EM, Sorensen HT, Riis A, Hatch EE. An internet-based prospective study of body size and time-to-pregnancy. Hum Reprod 2010;25:253–64.
[8] Ramlau-Hansen CH, Thulstrup AM, Nohr EA, Bonde JP, Sorensen TI, Olsen J. Subfecundity in overweight and obese couples. Hum Reprod 2007;22:1634–7.
[9] Gesink Law DC, Maclehose RF, Longnecker MP. Obesity and time to pregnancy. Hum Reprod 2007;22:414–20.
[10] van der Steeg JW, Steures P, Eijkemans MJ, Habbema JD, Hompes PG, Burggraaff JM, et al. Obesity affects spontaneous pregnancy chances in subfertile, ovulatory women. Hum Reprod 2008;23:324–8.
[11] Jain A, Polotsky AJ, Rochester D, Berga SL, Loucks T, Zeitlian G, et al. Pulsatile luteinizing hormone amplitude and progesterone metabolite excretion are reduced in obese women. J Clin Endocrinol Metab 2007;92(7):2468–73.
[12] Leary C, Leese HJ, Sturmey RG. Human embryos from overweight and obese women display phenotypic and metabolic abnormalities. Hum Reprod 2015;30(1):122–32.
[13] Styne-Gross A, Elkind-Hirsch K, Scott Jr RT. Obesity does not impact implantation rates or pregnancy outcome in women attempting conception through oocyte donation. Fertil Steril 2005;83:1629–34.
[14] Wattanakumtornkul S, Damario MA, Stevens Hall SA, Thornhill AR, Tummon IS. Body mass index and uterine receptivity in the oocyte donation model. Fertil Steril 2003;80:336–40.
[15] Dessolle L, Darai E, Cornet D, Rouzier R, Coutant C, Mandelbaum J, et al. Determinants of pregnancy rate in the donor oocyte model: a multivariate analysis of 450 frozen-thawed embryo transfers. Hum Reprod 2009;24:3082–9.
[16] Bellver J, Rossal LP, Bosch E, Zuniga A, Corona JT, Melendez F, et al. Obesity and the risk of spontaneous abortion after oocyte donation. Fertil Steril 2003;79:1136–40.
[17] Lashen H, Fear K, Sturdee DW. Obesity is associated with increased risk of first trimester and recurrent miscarriage: matched case-control study. Hum Reprod 2004;19:1644–6.
[18] Metwally M, Saravelos SH, Ledger WL, Li TC. Body mass index and risk of miscarriage in women with recurrent miscarriage. Fertil Steril 2010;94:290–5.
[19] Boots CE, Bernardi LA, Stephenson MD. Frequency of euploid miscarriage is increased in obese women with recurrent early pregnancy loss. Fertil Steril 2014;102:455–9.
[20] Gonzalez RR, Leavis P. Leptin upregulates beta3-integrin expression and interleukin-1beta, upregulates leptin and leptin receptor expression in human endometrial epithelial cell cultures. Endocrine 2001;16:21–8.
[21] Koning A, Mol BW, Dondorp W. It is not justified to reject fertility treatment based on obesity. Hum Reprod Open 2017;2017(2):1–4.
[22] Pandey S, Pandey S, Maheshwari A, Bhattacharya S. The impact of female obesity on the outcome of fertility treatment. J Hum Reprod Sci 2010;3(2):62–7.

[23] Moslehi N, Shab-Bidar S, Ramezani Tehrani F, Mirmiran P, Azizi F. Is ovarian reserve associated with body mass index and obesity in reproductive aged women? A meta-analysis. Menopause 2018;25(9):1046−55.

[24] Kort JD, Winget C, Kim SH, Lathi RB. A retrospective cohort study to evaluate the impact of meaningful weight loss on fertility outcomes in an overweight population with infertility. Fertil Steril 2014;101(5):1400−3.

[25] Sim KA, Dezarnaulds GM, Denyer GS, Skilton MR, Caterson ID. Weight loss improves reproductive outcomes in obese women undergoing fertility treatment: a randomized controlled trial. Clin Obes 2014;4(2):61−8.

[26] Chavarro JE, Willett WC, Skerrett PJ. McGraw-Hill The fertility diet: groundbreaking research reveals natural ways to boost ovulation & improve your chances of getting *pregnant*. J Clin Invest 2008;118(4):1210.

[27] Johansson K, Cnattingius S, Näslund I, Roos N, Trolle Lagerros Y, Granath F, et al. Outcomes of pregnancy after bariatric surgery. N Engl J Med. 2015;372(9):814−24.

[28] Best D, Avenell A, Bhattacharya S. How effective are weight-loss interventions for improving fertility in women and men who are overweight or obese? A systematic review and meta-analysis of the evidence. Hum Reprod Update 2017;23(6):681−705.

[29] Dietz WH, Baur LA, Hall K, Puhl RM, Taveras EM, Uauy R, et al. Management of obesity: improvement of health-care training and systems for prevention and care. Lancet 2015;385(9986):2521−33.

[30] Källén BA. Antiobesity drugs in early pregnancy and congenital malformations in the offspring. Obes Res Clin Pract 2014;8: e571−6.

[31] Kumar P, Arora S. Orlistat in polycystic ovarian syndrome reduces weight with improvement in lipid profile and pregnancy rates. J Hum Reprod Sci 2014;4:255−61.

Chapter 10

Obesity and recurrent miscarriage

Andrew C. Pearson and Tahir A. Mahmood
Department of Obstetrics and Gynaecology, Victoria Hospital, Kirkcaldy, United Kingdom

Introduction

Recurrent miscarriage (RM) is usually defined as loss of three or more consecutive pregnancies prior to 20 weeks of gestation, though it can be described as two or more consecutive pregnancy losses [1]. It can be described as primary (where no previous pregnancy has gone beyond 20 weeks of gestation) or secondary (where it has). Incidence has been reported to be 0.5%−2.3% [2]. The incidence of RM has found to be rising in some studies [3], possibly related to the rising prevalence of obesity and decision to start childbearing later in life [3]. The risk of ongoing RM, after three miscarriages, occurring by chance alone is 0.34%, taking a loss rate for each clinically recognized pregnancy of 15% [4].

RM is a pregnancy complication with a multifactorial etiology. Genetic abnormalities in the couple, anatomical abnormalities of the uterus and antiphospholipid syndrome hormone disorders, are common causes [1,4]. Miscarriage is a highly distressing event, and RM is a huge blight on the lives of many couples. Approximately in 50% of couples, the underlying cause remains unidentified after investigation [5−7]. The heterogeneity of the condition and the existence of conflicting evidence for the treatment of underlying associated etiologies and, hence, conflicting advice given to couples contribute to the challenge in the management of RM [8].

In addition to direct causes, there are several risk factors and numerous possible etiologies. Increased body mass index (BMI) was the second most significant factor predicting early pregnancy loss, after advanced female age [9].

Obesity is defined by the World Health Organization as a BMI of >30 kg/m^2 [10] and overweight as BMI > 25 kg/m^2. The current obesity pandemic has resulted in 1.9 billion overweight adults and 650 million obese adults worldwide, it having nearly tripled in the last 40 years [11]. 23% of American women of reproductive age are now obese [12]. Reasons for this pandemic are discussed elsewhere in the book.

Possible etiologies of RMs include:
1. genetic
   a. embryonic chromosomal abnormalities
   b. parental balanced reciprocal translocations
   c. sperm DNA fragmentation
   d. maternal age
2. thrombotic
   a. hereditary thrombophilia
   b. antiphospholipid syndrome
   c. alloimmunity
3. uterine factors
   a. congenital uterine abnormalities
   b. acquired uterine anomalies (uterine fibroids)
   c. cervical incompetence
4. endometrial
   a. endometrial receptivity disorders
   b. luteal phase defect
   c. decidualization defects
5. hormonal
   a. hypothyroidism
   b. diabetes mellitus
   c. hyperprolactinemia
   d. PCO
6. metabolic
   a. obesity
   b. metabolic syndrome of obesity
7. environmental
   a. excessive smoking
   b. caffeine consumption
   c. cocaine use
   d. heavy alcohol consumption

Adapted from: Barrenetxea G, Ortuzar M, Barrenetxea J. Endocrinological and environmental causes of recurrent miscarriage: a review. Open Access J Endocrin 2017;1(2):1−6.

Obesity and miscarriage

Obesity is associated with reduced fertility; higher the BMI, greater is the effect on fecundity. Several mechanisms may be involved affecting the development of egg, embryo, and the receptive nature of endometrium. Obesity is associated with low levels of adiponectin but raised levels of leptin in both serum and follicular fluid. Higher levels of leptin can impair ovarian steroidogenesis. Low levels of adiponectin lead to higher levels of serum insulin levels, which is one of the factors to increase levels of circulating androgens [13]. So obesity per se creates quite a hostile biochemical environment for the early stages of developing gamete and pregnancy.

In general, it has been recognized that visceral obesity has adverse effects on women's health and pregnancy. It is generally believed that obesity is associated with increased risk of miscarriage, in obese and overweight women, both in general population and in women undergoing assisted conception, thus postulating a possible link between obesity and increased risk of miscarriage [14,15], but others have not agreed with this observation specifically during in vitro fertilization (IVF) treatment [16]. Bellever et al.'s study [15] is interesting as healthy oocytes from women with normal BMI were used in an egg donation program and women with high BMI had low implantation rates, thus providing an evidence about the role of the endometrial receptivity.

A metaanalysis of 16 studies concluded that women with BMI $\geq$ 25 kg/m^2 had significantly higher odds of miscarriage regardless of the method of conception. Subgroup analysis from a limited number of studies in the same metaanalysis suggested that this group of women may also have significantly higher odds of miscarriage after oocyte donation and ovulation induction, but there was no evidence for increased odds of miscarriage after IVF or IVF−ICSI (intracytoplasmic sperm injection) [13]. Another systematic review of 11 studies [14] again confirmed a lower likelihood of pregnancy and an increased risk of miscarriage after IVF among women with BMI $\geq$ 25 kg/m^2. Obese women also have reduced oocytes retrieved despite requiring higher doses of gonadotrophins [14]. The risk of miscarriage in obese women may be as high as 25%−37% before the first live-born child [17].

Obesity and recurrent miscarriage

With adequate evidence for obesity's role in increasing the risk of miscarriage, it would be reasonable to hypothesize that it is a likely contributory factor in RM. A recent systemic review and metaanalysis [2] has reported a significant association between excess weight and RM, independent of age (though subgroup analysis confirms this association with the obese group only). Obese women with a history of RM have a high risk of future pregnancy losses [2].

The exact mechanism by which obesity increases the risk of miscarriage and RM is still unclear, possible association between obesity and RM includes the effect of obesity on endometrial development and effect on oocyte quality. Interestingly, a study has found that obese women with RM had a much higher chance of their miscarriage being euploid than in nonobese women with RM, this would support a strong role for endometrial involvement as an underlying mechanism [18].

Polycystic ovarian syndrome

Polycystic ovarian syndrome (PCOS) is associated with menstrual irregularity, ovarian dysfunction, and symptoms of hyperandrogenism. It is the most common endocrine disorder in women of reproductive age. Around half of women with PCOS are obese. High serum levels of androgens (testosterone, dehydroepiandrosterone sulfate, androstenedione) are present in women with PCOS. There is peripheral insulin resistance (IR) and compensatory hyperinsulinemia. Raised leptin has been reported in obese women, causing a detrimental effect on ovarian steroidogenesis.

Some studies looking at obesity and RM have excluded women with PCOS, while others have included them. There is a clear association between PCOS and RM [13], the prevalence of PCOS in RM remains highly uncertain. In women with PCOS, multiple metabolic and endocrine changes could be responsible for miscarriage and RM. (1) Elevated concentrations of luteinizing hormone (LH) in women with PCOS have been found to be unlikely responsible for RM [19]. (2) Hyperandrogenemia secondary to low levels of adiponectin and hyperinsulinemia may have detrimental effect on follicular growth and oocyte quality. There is a possible negative association between free androgen index (FAI) and oocyte quality or fertilization in women with PCOS [20]. 11% of women with RM have hyperandrogenemia and in this group of women, there is a significantly increased risk of miscarriage in a subsequent pregnancy [21]. (3) Hyperinsulinemia and IR in PCOS are attributed to obesity as well as IR independent of body weight [22]. The suggested prevalence, of IR in RM, is between 17% and 27% [19]. Hyperinsulinemia is associated with increased levels of plasminogen activator inhibitor−1 (PAI-1) [23] that is strongly associated with an increased risk of miscarriage and RM [24,25], and also hyperinsulinemia by itself is a significant independent risk factor for miscarriage [26] and also believed to play a key role in implantation failure by suppression of circulating glycodelin and insulin-like growth factor−binding protein-1 [27]. (4) Alterations in secretion of other hormones, including ghrelin, leptin, resistin, and adiponectin, in obese women may be responsible for affecting early embryo development and implantation [22].

Ovarian dysfunction

Good fertilization rates depend on embryo development, and implantation requires mature good quality oocytes. Maheshwari et al. [14] showed that there is a 30% increased risk of miscarriage if women are overweight (BMI > 2 kg/m^2) compared to those with normal BMI. A recent metaanalysis comparing outcome of assisted reproduction in obese women with nonobese women found that obese women have a significantly lower number of mature oocytes and oocytes with reduced diameter [2]. Overweight women's embryos also had a lower potential for development and IVF and a lower rate of blastocyst formation [2]. Intrafollicular human chorionic gonadotrophin (hCG) concentration appears to be inversely related to BMI and may be related to concurrent decrease in embryo quality and rates of pregnancy [28]. These effects on the ovarian function and endometrial receptivity could be the part of the reason for increased risk of RM in obese women.

Endometrial changes in obesity

Although consecutive pregnancies with fetal chromosomal aberrations can account for recurrent pregnancy loss (RPL), the frequency of euploid miscarriages increases with each additional loss. Implantation of the embryo and a successful pregnancy require receptive endometrium and obesity may have its effects on the endometrium or its environment causing implantation failure and pregnancy losses. The epidemiological observations suggest that uterine factors are a major cause of higher order miscarriages [29].

Lucas et al. [30] have recently reported that RPL is associated with loss of a conspicuous epigenetic stem cell signature in endometrial stromal cells obtained from mid luteal phase endometrial biopsies. Colony-forming unit (CFU) assays confirmed a reduction in endometrial mesenchymal stem cell (eMSC)–like progenitor cell in RPL. The eMSC loss has been linked to impaired decidualization that denotes the process of intense remodeling that transforms the cycling endometrium upon embryo implantation into the decidua to accommodate the rapidly expanding placenta.

Tewary et al. [31] carried out a double-blind, randomized, placebo-controlled trial to demonstrate the effect of DPP4 inhibitor sitagliptin on eMSC in women with RPL by determining the impact on endometrial decidualization. Women treated with sitagliptin 100 mg daily for 3 consecutive months showed an increase in eMSC and a decrease in the expression of DIO2 that is a gene marker of senescent decidual cells but no effect on uterine DPP4. However, a large randomized trial is warranted to test the safety and effectiveness of sitagliptin in preconception cycles.

The precise impact of obesity on molecular and histopathological aspects of endometrium is not fully understood.

Progesterone induces secretory changes in the lining of the uterus, which is needed for implantation of the embryo. It has been suggested that a causative factor in many cases of miscarriage may be inadequate secretion of progesterone. It is not known whether obesity may specifically interact with this.

Interestingly, however, a recent metaanalysis found that in women having IVF using normal-weight oocyte donors, obesity of the reciprocant undergoing IVF did not influence IVF results [2]. Extrapolating this would support oocyte quality rather than endometrial receptivity being a more important mechanism for increased RM in obese women.

Immunological factors

Immunological factors in RM and other obstetric complications have been of strong recent interest. One is the role of uterine natural killer (uNK) cells [32], which have been investigated as playing a role in later stages on implantation, which may in turn result in miscarriage. Studies of peripheral blood and endometrial uNK cells have suggested that they may play a role, but it is unclear at present. Obesity is associated with chronic inflammation. Women with idiopathic RM and some other obstetric complications are known to have higher levels of inflammatory markers (e.g., IL-6 and CRP). Chronic inflammation, therefore, could be a mechanism contributing to higher risk of RM in obese women [2].

Leptin

Leptin is a hormone, predominantly produced by adipocytes. Its role is to suppor regulation of body weight and food intake [33]. It has various other roles, including a possible role in implantation by stimulating the effect on matrix metalloproteinase expression in the cytotrophoblast [34]. One study found that women with RM who subsequently miscarried had lower levels of leptin than women with RM who subsequently went onto have a live birth [35]. Higher BMI is associated with defective leptin and insulin signaling in the hypothalamus, leading to a condition termed leptin/IR [36]. This could be a mechanism for RM in obese women.

Male obesity and recurrent miscarriage

There are no known studies directly evaluating the impact of male weight on RM [1]. Male obesity is associated with impaired semen parameters and sperm DNA damage [37]. In one study of 520 men a positive correlation between BMI and sperm DNA fragmentation was reported, with a 20% increase in sperm DNA damage in obese men [38]. It is possible, therefore, that male obesity could be a risk for RM, though this is yet to be established.

Management

Gradual weight loss has been shown to improve fertility and the outcomes of fertility treatments [39]. There are currently no good quality studies looking at the effect of weight loss on RM [1]. As there is an association between obesity and RM [2], it would be rational to recommend women with RM aim to attain a normal weight [1], despite a lack of direct evidence for female weight loss programs to improve outcomes in couples with RM [2].

The management of obesity is important because exercise and diet modification are of low cost. Alternatively, bariatric surgery is increasingly offered to obese patients with comorbidities. Weight loss decreases body fat, reduces truncal–abdominal fat, and improves metabolism and hormonal balance. Weight loss in obese women with PCOS, through protein-rich very low–calorie diet, has been shown to significantly reduce serum fasting glucose and insulin, improve insulin sensitivity, and decrease PAI-1 activity [40]. Weight loss also significantly decreases testosterone levels and increases sex hormone–binding globulin levels [41]. Normalizing hyperinsulinemia, improving insulin sensitivity, and reduction in hyperandrogenemia through weight reduction could be important for these women in achieving positive reproductive outcomes by improving ovarian function and endometrial receptivity.

Weight loss can be challenging, particularly in the setting of RM. The woman will be attempting to lose weight in an obesogenic environment (inexpensive calorie-dense food, reduced physical activity, and inexpensive nonphysical entertainment) (obesity pandemic paper), at a time of potential psychological angst. Those attempting weight loss will be surrounded with an excess emphasis on reduction of dietary fats, at the expense of continued excess intake of simple carbohydrates and sugar [42].

There is an association between raised inflammatory markers and RM, and a "Mediterranean" *diet* is associated with lower inflammatory markers than one that is high in red meat, high-fat dairy products, and simple carbohydrates. A case–control study [43] has shown higher dietary inflammation index among women with RM. This evidence would seem not strong enough to offer specific dietary advice for RM, although the contents of a low inflammatory diet may be in line with weight loss and healthy diet advice anyway.

Bariatric surgery is popular among morbidly obese women of childbearing age [44]. There has been an increasing uptake of bariatric surgery in women of childbearing age across Europe, the United States, Asia, and Australia [45]. One study observed that women with PCOS who underwent bariatric surgery had increased rates of ovulation, regular menstrual cycle, and decreased hirsutism and FAI [46]. There are no known studies looking at miscarriage or RM rate after bariatric surgery; however, given what is discussed earlier regarding the role of PCOS and hyperinsulinemia in RM and lack of evidence for dietary interventions, bariatric surgery as part of a wider health strategy in an obese woman could be promising, further studies are warranted.

Insulin-sensitizing agents such as metformin have been used in the treatment of obese PCOS women who have shown improvements in hyperinsulinemia and hyperandrogenism. Metformin use appears to be safe in the first trimester [47].

Metformin has been found to reduce miscarriage rate in women with PCOS [1].

A case–control study [47] suggested that the spontaneous miscarriage was significantly lower in obese women with PCOS who conceived while taking metformin and continued metformin throughout pregnancy, when compared to "controls" who either conceived without metformin or metformin was stopped soon after the confirmation of pregnancy. The same study also suggested that the risk of subsequent miscarriage in women with RM was less in "cases" when compared to "controls." Though this study was based on Pakistani women but has provided some evidence about the use of metformin to minimize risk of miscarriage in these women. A large-scale RCT of weight management is desirable to provide robust evidence for the management of obese women with RM. It would also be important to consider ethnicity as it is found that African and Hispanic American women have higher IR than Whites after adjusting for BMI and women of South Asian and Asian descent have a higher risk of IR compared with those of European origin with a similar BMI.

A small study looked at treating women with RM who had an abnormal glucose tolerance, and the miscarriage rate was reduced significantly in the metformin group [48].

Indirect evidence would, therefore, suggest that metformin in women with PCOS and RM or even in obese women with RM who do not have PCOS may decrease the chances of a concurrent miscarriage in women with RM but there is as yet no evidence strong enough to support this [1].

Immunological agents have been suggested as potentially useful generally in RM, although not specifically in obese women with RM.

A Cochrane review for intravenous immunoglobulin for RM showed no significant effect [32]. It is expensive, not without risk and in limited supply. Use of glucocorticoids has been shown to reduce uNK cells in women with RM [32], it could be a promising area of investigation.

Given obesity is associated with chronic inflammation, it could be hypothesized that immunotherapy could be more effective in obese women with RM than nonobese women. There is, however, no evidence at present to support this.

A recent Cochrane review [49] concluded that for women with unexplained RM, supplementation with *progestogen therapy* may reduce the rate of miscarriage in subsequent pregnancies. *Prism trial* has shown that for women who had three or more miscarriages, there was a 15% increase in births who were given progesterone compared to those who were given a placebo. It is not known whether this is more or less relevant for obese women with RM [50].

Given the association with male obesity with sperm DNA fragmentation, it would also be prudent to recommend normal BMI for the male partner, while acknowledging the lack of direct evidence for this.

Conclusion

There is a positive correlation between obesity and RM based on current evidence. Exact mechanisms are uncertain [2]. Weight loss would seem prudent for obese women contemplating pregnancy after RM. There is limited other specific advice or treatment for women at present based on current evidence.

References

[1] European Society of Human Reproduction and Embryology. Recurrent pregnancy loss. Guideline; 2017.
[2] Cavalcante MB, et al. Obesity and recurrent miscarriage: a systematic review and meta-analysis. J Obstet Gynaecol Res 2019;45(1):30–8.
[3] Rasmark Roepke E, Matthiesen L, Rylance R, Christiansen OB. Is the incidence of recurrent pregnancy loss increasing? A retrospective register-based study in Sweden. Acta Obstet Gynecol Scand 2017;96:1365–72.
[4] Stirrat GM. Recurrent miscarriage I: definition and epidemiology. Lancet 1990;336:673–5.
[5] Yang CJ, Stone P, Stewart AW. The epidemiology of recurrent miscarriage: a descriptive study of 1214 prepregnant women with recurrent miscarriage. Aust N Z J Obstet Gynaecol 2006;46(4):316–22.
[6] Li TC, Makris M, Tomsu M, Tuckerman E, Laird S. Recurrent miscarriage: aetiology, management and prognosis. Hum Reprod Update 2002;8(5):463–81.
[7] Quenby SM, Farquharson RG. Predicting recurring miscarriage: what is important? Obstet Gynecol 1993;82:132–8.
[8] Tang AW, Quenby S. Recent thoughts on management and prevention of recurrent early pregnancy loss. Curr Opin Obstet Gynecol 2010;22(6):446–51.
[9] Metwally M, Saravelos SH, Ledger WL, Li TC. Body mass index and risk of miscarriage in women with recurrent miscarriage. Fertil Steril 2010;94(1):290–5.
[10] World Health Organisation. 2020. Body mass index. <http://www.euro.who.int/en/health-topics/disease-prevention/nutrition/a-healthy-lifestyle/body-mass-index-bmi> [accessed January 2020].
[11] WHO. 2020. 2018 WHO obesity fact sheet. <https://www.who.int/news-room/fact-sheets/detail/obesity-and-overweight> [accessed January 2020].
[12] Vahratian A. Prevalence of overweight and obesity among women of childbearing age: results from the 2002 National Survey of Family Growth. Matern Child Health J 2009;13:268–73.
[13] Metwally M, Ong KJ, Ledger WL, Li TC. Does high body mass index increase the risk of miscarriage after spontaneous and assisted conception? A meta-analysis of the evidence. Fertil Steril 2008;90(3):714–26.
[14] Maheshwari A, Stofberg L, Bhattacharya S. Effect of overweight and obesity on assisted reproductive technology—a systematic review. Hum Reprod Update 2007;13(5):433–44.
[15] Bellever J, Melo MA, Bosch E, et al. Obesity and poor reproductive outcome: potential role of the endometrium. Fertil Steril 2007;88:446–51.
[16] Lashen H, Ledger W, Bernal AL, Barlow D. Extremes of body mass do not adversely affect the outcome of superovulation and in-vitro fertilization. Hum Reprod 1999;14(3):712–15.
[17] Hamilton-Fairley D, Kiddy D, Watson H, Paterson C, Franks S. Association of moderate obesity with a poor pregnancy outcome in women with polycystic ovarian syndrome treated with low dose gonadotrophin. Br J Obstet Gynaecol 1992;99:128–31.
[18] Boots CE, Bernardi LA, Stephenson MD. Frequency of euploid miscarriage is increased in obese women with recurrent early pregnancy loss. Fertil Steril 2014;102:455–9.
[19] Cocksedge KA, Li TC, Saravelos SH, Metwally M. A reappraisal of the role of polycystic ovary syndrome in recurrent miscarriage. Reprod Biomed Online 2008;17(1):151–60.
[20] van Wely M, Bayram N, van der Veen F, Bossuyt PMM. Predicting ongoing pregnancy following ovulation induction with recombinant FSH in women with polycystic ovarian syndrome. Hum Reprod 2005;20:1827–32.
[21] Cocksedge KA, Saravelos SH, Wang Q, Tuckerman E, Laird SM, Li TC. Does free androgen index predict subsequent pregnancy outcome in women with recurrent miscarriage? Hum Reprod 2008;23(4):797–802.
[22] Fedorcsak P, Dale PO, Storeng R, Tanbo T, Åbyholm T. The impact of obesity and insulin resistance on the outcome of IVF or ICSI in women with polycystic ovarian syndrome. Hum Reprod 2001;16:1086–91.
[23] Palomba S, Orio Jr F, Falbo A. Plasminogen activator inhibitor 1 and miscarriage after metformin treatment and laparoscopic ovarian drilling in women with polycystic ovarian syndrome. Fertil Steril 2005;84:761–5.
[24] Glueck CJ, Wang P, Bornovali S. Polycystic ovary syndrome, the G1691A factor V Leiden mutation, and plasminogen activator inhibitor activity: associations with recurrent pregnancy loss. Metab Clin Exp 2003;52:1627–32.
[25] Glueck CJ, Wang P, Fontaine RN. Plasminogen activator inhibitor activity: an independent risk factor for the high miscarriage rate during pregnancy in women with polycystic ovarian syndrome. Metab Clin Exp 1999;48:1589–95.
[26] Glueck CJ, Wang P, Goldenberg N. Pregnancy outcomes among women with polycystic ovarian syndrome treated with metformin. Hum Reprod 2002;17:2858–64.
[27] Jakubowicz DJ, Seppälä M, Jakubowicz S, Rodriguez-Armas O, Rivas-Santiago A, Koistinen H, et al. Insulin reduction with metformin increases luteal phase serum glycodelin and insulin-like growth factor-binding protein 1 concentrations and enhances uterine vascularity and blood flow in the polycystic ovary syndrome. J Clin Endocrinol Metab 2001;86(3):1126–33.
[28] Carrell DT, Jones KP, Peterson CM, Aoki V, Emery BR, Campbell BR. Body mass index is inversely related to intrafollicular HCG concentrations, embryo quality and IVF outcomes. Reprod Biomed Online 2001;3(2):109–11.
[29] Robberrecht C, Pexsters A, Dprest J, et al. Cytogenetic and morphological analysis of early products of conception following

hystero-embryoscopy from couples with recurrent pregnancy loss. Prenat Diagn 2012;32(10):933–42.
[30] Lucas ES, Dyer NP, Murakami K, et al. Loss of endometrial plasticity in recurrent pregnancy loss. Stem Cells 2016;34(2):346–56.
[31] Tewary S, Lucas ES, Fujihara R, et al. Impact of sitagliptin on endometrial mesenchymal stem-like progenitor cells: a randomised, double-blind placebo-controlled feasibility trial. EBioMedicine 2019. Available from: https://doi.org/10.1016/j.ebjom.2019.102597.
[32] RCOG. The role of natural killer cells in human fertility. Scientific impact paper no. 53. RCOG; 2016.
[33] Henson MC, Castracane VD. Leptin in pregnancy. Biol Reprod 2000;63(5):1219–28.
[34] Sagawa N, Yura S, Itoh H, Kakui K, Takemura M, Nuamah MA, et al. Possible role of placental leptin in pregnancy: a review. Endocrine 2002;19(1):65–71.
[35] Laird SM, Quinton ND, Anstie B, Li TC, Blakemore AI. Leptin and leptin-binding activity in women with recurrent miscarriage: correlation with pregnancy outcome. Hum Reprod 2001;16(9):2008–13.
[36] Clarke IJ. Whatever way weight goes, inflammation shows. Endocrinology 2010;151(3):846–8.
[37] Du Plessis SS, Cabler S, McAlister DA, Sabanegh E, Agarwal A. The effect of obesity on sperm disorders and male infertility. Nat Rev Urol 2010;7:153–61.
[38] Chavarro JE, Toth TL, Wright DL, Meeker JD, Hauser R. Body mass index in relation to semen quality, sperm DNA integrity, and serum reproductive hormone levels among men attending an infertility clinic. Fertil Steril 2010;93:2222–31.
[39] Pandey S, Maheshwari A, Bhattacharya S. The impact of female obesity on the outcome of fertility treatment. J Hum Reprod Sci 2010;3:62–7.
[40] Andersen P, Seljeflot I, Abdelnoor M, Arnesen H, Dale PO, Løvik A, et al. Increased insulin sensitivity and fibrinolytic capacity after dietary intervention in obese women with polycystic ovary syndrome. Metabolism 1995;44(5):611–16.
[41] Clark AM, Ledger W, Galletly C, Tomlinson L, Blaney F, Wang X, et al. Weight loss results in significant improvement in pregnancy and ovulation rates in anovulatory obese women. Hum Reprod 1995;10:2705–12.
[42] Meldrum DR, et al. Obesity pandemic: causes, consequences, and solutions – but do we have the will? Fertil Steril 2017;833–9.
[43] Aminianfar A, Vahid F, Shayanfar M, et al. Association between Dietary Inflammatory Index (DII) and risk of breast cancer: a case-control study. Asian Pac J Cancer Prev 2018;19(5):1215–21.
[44] American Society for Metabolic & Bariatric Surgery. Bariatric surgical society takes on new name, new mission and new surgery. Gainesville, FL: American Society for Metabolic & Bariatric Surgery; 2007. <http://asmbs.org/benefits-of-bariatric-surgery/>.
[45] Ibiebele I, et al. Perinatal outcomes following bariatric surgery between a first and second pregnancy: a population data linkage study. BJOG 2020;127:345–54.
[46] Escobar-Morrreale HF, Botella-Carretero JL, Alvarez-Blasco F, et al. The polycystic ovary syndrome associated with morbid obesity may resolve after weight loss induced by bariatric surgery. J Clin Endocrinol Metab 2005;90:6364–9.
[47] Andrade C. Major malformation risk, pregnancy outcomes, and neurodevelopmental outcomes associated with metformin use during pregnancy. J Clin Psychiatry 2016;77:e411–14.
[48] Zolghadri J, Tavana Z, Kazerooni T, Soveid M, Taghieh M. Relationship between abnormal glucose tolerance test and history of previous recurrent miscarriages, and beneficial effect of metformin in these patients: a prospective clinical study. Fertil Steril 2008;90:727–30.
[49] Haas DM, Hathaway TJ, Ramsey PS. Progestogen for preventing miscarriage in women with recurrent miscarriage of unclear etiology. Cochrane Database Syst Rev 2019;(11):CD003511. Available from: https://doi.org/10.1002/14651858.CD003511.pub5.
[50] Coomararsamy A, Devall A, Brosens JJ et al. Micronized vaginal progesterone to prevent miscarriage: a critical evaluation of randomized evidence. Am J Obstet Gynaecol 2020. <https://doi.org/10.1016/j.ajog.2019.12.006>.

Chapter 11

Obesity and assisted reproduction

Mark Hamilton[1] and Abha Maheshwari[2]

[1]*University of Aberdeen, Aberdeen, United Kingdom,* [2]*Reproductive Medicine, NHS Grampian, Aberdeen, United Kingdom*

Introduction

Obesity is a major issue in the Western society and can have profound effects on reproductive health. There is persuasive evidence that the prevalence of obesity has increased over the past 35 years with more than half of women in the United Kingdom who are either overweight or obese. Many of these are in the reproductive age-group and a significant number present with infertility. There is a convincing literature, which suggests that there are genuine issues of concern with respect to adverse clinical outcomes, increased health risks, and expense associated with assisted reproduction treatment (ART) in women who are obese. Obstetrics data suggest that maternal and fetal risks increase in the obese individuals, and there has been debate in recent times as to whether it is appropriate to offer access to ART for this group of patients. This chapter sets out to explore these issues highlighting the need to use limited state resources to maximum effectiveness with the safety of women and children being the prime concern.

Prevalence of obesity in the assisted reproduction sector

The association of obesity with infertility is well described. However, many overweight women conceive without difficulty, though pregnancies in these circumstances may be associated with increased risks to the mother and child.

A normal body mass index (BMI) is considered to lie between 19 and 24.9 kg/m^2. Overweight is defined as BMI $\geq$ 25 kg/m^2. Obesity is subdivided into moderate (BMI 30–34.9 kg/m^2), severe (BMI 35–39.9 kg/m^2), or morbid (BMI $\geq$ 40 kg/m^2) according to the level of BMI.

It is sometimes suggested that obesity is a disease of the modern age. There is certainly some evidence to suggest that proportion of obese individuals within the population is changing. In the United States the prevalence of obesity (BMI > 30 kg/m^2) in young adults (18–29 years) has tripled from 8% in 1971–74 to 24% in 2004–06 (NCHS, 2009). In the United Kingdom, more than half of women are either overweight or obese [1,2].

Data from the United States have suggested that, specific to users of in vitro fertilization (IVF) services, 40% had a BMI > 25 kg/m^2, while over 6% had a BMI $\geq$ 35 kg/m^2; in other words, they were severely obese (Fig. 11.1).

National data specific to those accessing IVF treatment in the United Kingdom are lacking, although a study analyzing economic costs of IVF relevant to BMI in a UK center showed a similar proportion (41.3%) with a BMI $\geq$ 25 kg/m^2 [3].

Evidence of reduced fertility in the obese

The association of obesity with impaired fertility has been described in many reviews. There are several mechanisms whereby the obese may have reduced fertility potential. Psychosocial factors are likely to be important, but pathophysiological mechanisms linked to disturbed ovulation patterns as well as issues of egg, embryo, and endometrial receptivity have also been implicated. A positive correlation between increasing BMI and infertility has been described with a relative risk at a BMI $\geq$ 30 kg/m^2 of 2.7 [4]. The time to conception in the overweight (BMI $\geq$ 25 kg/m^2) has also been observed to be longer [5]. Sexual dysfunction has also been described as occurring with greater frequency in the obese, which could be related to physical or psychological disturbances [6].

Cycle effects

Obesity is linked to disturbances in the hypothalamic–pituitary–ovarian axis. Increased levels of serum and follicular fluid leptin are described with increasing BMI. High levels of leptin impair follicular development and reduce ovarian steroidogenesis through direct effects on theca and granulose cells [6]. There is also an inverse

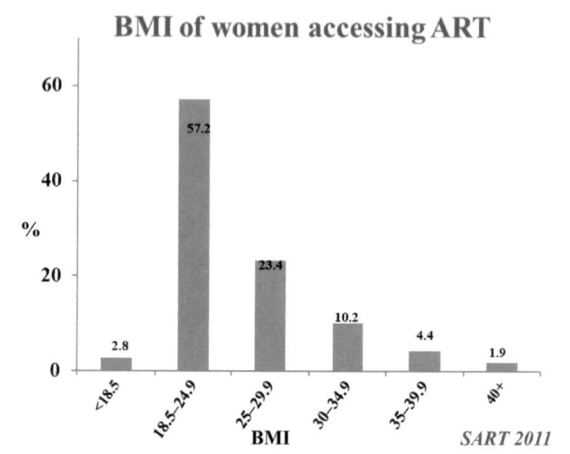

FIGURE 11.1 BMI of women accessing IVF treatment. *BMI*, Body mass index; *IVF*, in vitro fertilization.

relationship of increasing BMI with reduced serum adiponectin levels. The low adiponectin levels are associated with elevated serum insulin levels, which increase circulating androgen levels in part linked to a reduction in the production of sex hormone—binding globulin by the liver. The trend to hyperandrogenism in the obese is also contributed by IGF-1-mediated effects on LH-induced steroidogenesis by theca cells. Enhanced androgen production causes granulosa cell apoptosis with direct consequences for follicle function. The increased availability of androgens for peripheral conversion to estrogens in adipose tissue has pituitary effects with impaired FSH production affecting the ovarian follicular development [7].

The clinical manifestations of the biochemical disturbances described include anovulatory cycles and subfertility. Ovarian dysregulation associated with hyperandrogenism, insulin resistance, menstrual irregularity, and infertility is commonly found in women with polycystic ovarian syndrome, many of whom are obese [2].

Specific issues relating to assisted reproduction treatment

ART nowadays has a therapeutic role in the management of all causes of infertility. The use of IVF may also provide some insight into the pathophysiology underlying impaired reproductive performance in the obese.

Effects on the oocyte

A number of studies have suggested that oocyte yield after stimulation for IVF may be affected in the obese. Quantitative effects have been described where increased doses of gonadotrophins are required to elicit an ovarian response, and the ultimate yield of cumulus—oocyte complexes may be less than in normal weight controls [9,10]. This may be linked to disturbances in leptin production or sensitivity as described earlier. Some studies have suggested that fertilization rates of oocytes retrieved may be impaired in the obese, but this observation has not been consistent. Prospective studies are needed to clarify this issue [6]. The observation of increased risks of miscarriage in the obese after IVF has been attributed by some to qualitative effects on oocytes leading to aberrant embryo development [8].

Effects on embryos

As with oocytes, the literature is not consistent with respect to the effects of obesity on embryonic development. Some studies have suggested that markers of embryo quality differ in the obese. Furthermore, there may be less available surplus embryos for cryostorage potentially having an impact on cumulative pregnancy rates per episode of ovarian stimulation. Some have suggested that these observed effects are unreliable since studies may not have taken into account potential confounders such as age, parity, and duration of infertility [12]. Further work is required to inform this controversial debate.

Effects on the endometrium

There is an increase in miscarriage rate in the obese both in natural conception and that associated with infertility treatment. Specific to IVF, a 50% increased risk of miscarriage in women with a BMI > 30 kg/m^2 has been described [12]. While embryo quality will be an important determinant of implantation potential, studies using an egg donation model [13] suggest that endometrial factors are likely to be involved in this phenomenon as well. The precise mechanism is not understood but ovarian steroid regulation of endometrial development, perturbations in inflammatory and coagulation pathways, perhaps linked to insulin resistance have been suggested to be involved.

Rationale for the use of assisted reproduction

The main cause of infertility in the obese relates to disturbances in ovulation. These, for the most part, can be resolved with a combined approach involving weight reduction strategies together with pharmacologically induced ovulation induction. Refractory dysovulation occurs with greater frequency in the obese, and for them, the use of ART has to be considered. ART, specifically IVF, will address the issues of egg and sperm availability, as well as tubal infertility for the obese, just as it does for the general infertile population. There is no definitive evidence that unexplained infertility occurs with greater frequency in the obese; though given the abovementioned remarks with respect to oocyte, embryo, and endometrial factors, one might have expected this to be the case. The practical issues that arise through the use of these techniques in overweight women need to be considered carefully.

Practical management of obese women undertaking assisted reproduction treatment

Patient selection

The selection of which patients to treat, and in whom treatment should be deferred until weight loss is achieved, should ideally depend on age, tests of ovarian reserve, and the presence of comorbidities [14].

If tests of ovarian reserve, which might include age, serum anti-Müllerian hormone (AMH), and/or antral follicle count, suggest that there is good ovarian reserve with no other comorbidities, then it is appropriate to defer treatment up until the desired BMI is obtained. However, if there is evidence of ovarian aging, there is a limited time for weight loss. In these circumstances, it might be wiser to proceed with the treatment [15]. Despite this, health-care professionals have a duty of care not just to the patient but also to the potential child, and treatment should arguably not be provided if there are significant obstetric and perinatal risks such as in cases of extreme morbid obesity. There are data showing that levels of AMH are reduced and hence the egg number in obese women [16].

Stimulation regimes

As alluded to the abovementioned fact, there are observational data suggesting that the requirement of gonadotrophins is increased by at least 20% if BMI is ≥ 30 kg/m^2. Chong et al. [17] demonstrated that patients who have normal $\pm 10\%$ ideal body weight (IBW) are more likely to respond to lower doses of hMG than patients whose weight is >10% above IBW and, in particular, those who are >25% above their IBW. A high BMI was associated with a higher FSH threshold dose. This observation is supported by findings that the total dose of gonadotrophins needed to induce ovulation is increased in parallel with body weight [18,19]. Why heavier women may need more hormones to induce ovulation or for controlled ovarian hyperstimulation is not clear. It may be related to the greater amount of body surface, inadequate estradiol metabolism, and decreased sex hormone−binding globulin. Also, the intramuscular absorption of the drug may be slower and incomplete in obese patients because of increased subcutaneous fat or fat infiltration of the muscle.

The effect of FSH at the ovarian level is dependent on plasma concentrations of the hormone. This, in turn, is influenced not just by the dose administered but also by endogenous FSH secretion, metabolic clearance rate, and the volume of distribution, which are individual and differ from woman to woman and are influenced by BMI. Elimination of FSH is carried out largely by the kidneys and the liver. The clearance rate is dependent on filtration, secretion, and reabsorption. The extent to which a drug is bound to plasma proteins also determines the fraction of drug extracted by the eliminating organs, which, in turn, is dependent on BMI and weight.

Pooled analysis from the observational studies has shown that the duration of gonadotrophin stimulation was significantly longer (weighted mean difference (WMD) 0.27, 95% CI: 0.26−0.28, $P < .00001$) and the dose was higher (WMD 406.77, 95% CI: 169.26−644.2, $P = .0008$) in women with BMI ≥ 30 kg/m^2 compared to women with normal BMI [17]. However, there is no randomized controlled trial in the literature testing the hypothesis that increasing the dose of gonadotrophins in obese women improves the live birth rates.

There is no evidence to suggest that one regime of pituitary suppression (agonist or antagonist) is better in obese women compared to those with normal BMI.

Monitoring of stimulation

While there are no data in the literature quantifying differences of monitoring in those with higher BMI, it is accepted generally that the performance and interpretation of ultrasound scans can be difficult in the obese. Theoretically, were estradiol to be used in monitoring response to stimulation, the levels might be expected to differ in the obese from those with a normal BMI. However, there is no evidence to suggest that with overweight patients, it is advantageous to use both ultrasound and estradiol in monitoring stimulated cycles [21].

Clinical procedures

Egg collection

Clinical staff will be sensitive to the challenges which the care of women with high BMI undergoing surgical procedures present. While there is no evidence from the literature that there are more problems in caring for those who are obese, this is probably because most units are not treating morbidly obese women. That said, obese women will require higher dose of sedation, due to increased surface area, which potentially may lead to a higher risk of exposure to the side effects of the drugs utilized, but in the absence of any published data in the literature, the perceived increase in risk remains theoretical.

Embryo transfer

For the most part the procedure of embryo transfer (ET) is simple. However, with moves toward ultrasound-guided (USG) ET that may involve the use of abdominal ultrasound, USG-guided ET will be difficult in obese women due to poor views. Whether this would lead to lower pregnancy rate remains unknown as there are no data in the literature to explore either difficulties with the procedure or lower pregnancy rates.

hCG trigger

Theoretically, bioactive levels of hCG used for the ovulatory trigger will be less in obese women. However, as long as more than equivalent of 1000 IU of recombinant hCG is given as the ovulatory trigger, oocyte fertilization rates and luteal function are unlikely to be influenced by differences in bioavailable gonadotrophin. Most ovulatory triggered preparations now contain at least 6500 IU of hCG.

Luteal support

Luteal support for obese women should be the same as that for women with normal BMI. This is because the vaginal pessaries are locally absorbed and bypass first-pass metabolism. There are no data comparing luteal support and outcomes in various BMI groups.

Effect of obesity on the results of assisted reproduction treatment

Pregnancy rate

Pooling of data from observational studies is associated with inherent bias as one cannot adjust for confounding factors such as age and number of embryos transferred. Moreover, the number of cases in these studies, where BMI was in morbid obesity range, was extremely small. However, systematic reviews of observational studies [12,20,22] have repeatedly demonstrated a detrimental impact of obesity on pregnancy rates.

There is a reduction in pregnancy rates [risk ratio (RR) 0.87, 95% CI: 0.80–0.95, $P = .002$] in obese women (BMI $\geq$ 30 kg/m^2) when compared with those that have normal BMI ($<$25 kg/m^2). This reduction in pregnancy rate was also observed in women who were overweight (BMI $\geq$ 25 kg/m^2) compared to those with normal BMI (RR 0.90, 95% CI: 0.85–0.94, $P < .0001$) [17].

However, there was no difference in pregnancy rates when BMI $<$ 30 kg/m^2 was compared with BMI $\geq$ 30 kg/m^2, thereby indicating that there is no further detrimental effect of obesity (BMI $\geq$ 30 kg/m^2) when compared to those who are overweight (BMI $\geq$ 25 kg/m^2) [12].

The largest single series comes from the Society of Assisted Reproduction (SART) in the United States [23]. This analysis showed that failure to achieve a clinical intrauterine gestation was significantly more likely among obese women (1.22; 95% CI: 1.13–1.32). This was based on 31,672 ETs from a single database, and the analysis permitted adjustment for age, parity, number of embryos transferred, and the day of ET. The denominator used in this analysis was per ET rather than per woman.

Miscarriage rate

Of those who conceive after ART, there is a 30% increased risk of miscarriage if women are overweight (BMI $\geq$ 25 kg/m^2) compared to those with normal BMI (1.33; 95% CI: 1.06–1.68). This risk further increases to just over 50% when miscarriage rates are compared in those who are obese to those with a BMI $<$ 30 kg/m^2 [odds ratio (OR) 1.53; 95% CI: 1.27–1.84] [12] (Fig. 11.2).

As discussed earlier, it is uncertain whether the cause of increased miscarriages is linked to oocyte quality or other factors within the endometrium involved in implantation.

Live birth rate

The SART data [23] demonstrated that the OR (95% CI) of failure to achieve live birth was 1.27 (1.10–1.47) in obese women (BMI $\geq$ 30 kg/m^2) as compared to those with normal BMI. However, the results also indicate that there are significant differences in pregnancy and live birth rates after ART when analyzed by race and ethnicity, even within the same BMI categories. Moreover, from the same data, there was no difference identified in live birth rate based on BMI if donor oocytes were used [24]—a finding supported by others [25]. In contrast, some other studies exploring the use of donor oocytes in those with high BMI have suggested a lower chance of conception [22].

A previous systematic review also showed a 9% reduction in live birth rate in overweight women (BMI 25–29.9 kg/m^2) when compared with those with a normal BMI (OR, 95% CI: 0.91, 0.85–0.98). The reduction in live birth rate in women who are obese (BMI $\geq$ 30 kg/m^2) was 20% compared to those with a normal BMI (RR 0.80, 95% CI: 0.71–0.90). These reductions in live birth rates were statistically significant ($P < .0002$) [17]. This

Review: Effect of obesity on ART
Compaison: 05 Miscarriage rates, per pregnancy achieved
Outcome: 02 BMI>30

Study or subcategory	BMI>30 n/N	BMI<30 n/N	OR(fixed) 95% CI	Weight (%)	OR(fixed) 95% CI
Krizanovska (2002)	1/3	7/40		0.39	2.36 [0.19, 29.75]
Nichols et al. (2003)	1/12	11/189		0.72	1.47 [0.17, 12.45]
van Swieten (2005)	2/11	6/54		0.99	1.78 [0.37, 10.25]
Wang et al. (2002)	74/268	382/2011		38.64	1.63 [1.22, 2.17]
Winter et al. (2002)	26/153	153/944		21.05	1.06 [0.67, 1.67]
Fedorcsak et al. (2004)	55/138	377/1325		25.42	1.67 [1.16, 2.39]
Dechaud et al. (2006)	1/8	11/125		0.69	1.48 [0.17, 13.16]
Dokras et al. (2006)	26/154	49/461		12.12	1.71 [1.02, 2.86]
Total (95% CI)	747	5149		100.00	1.53 [1.27, 1.84]

Total events: 186 (BMI>3), 996 (BMI<30)
Test for heterogeneity: Chi^2 = 3.21, df = 7 (P =.86), I^2 = 0%
Test for overall effect: Z = 4.50 (P<.00001)

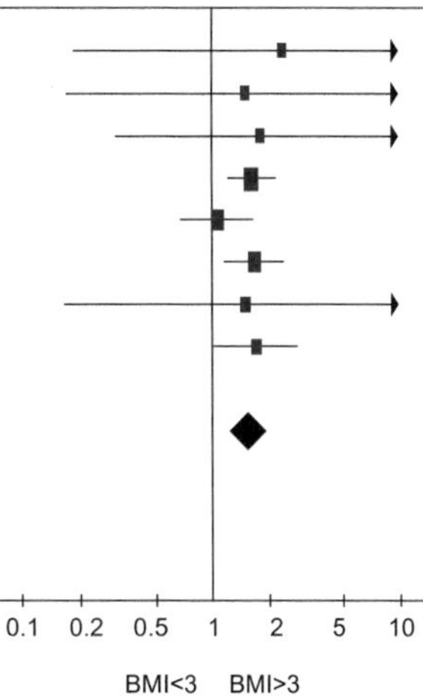

0.1 0.2 0.5 1 2 5 10
BMI<3   BMI>3

FIGURE 11.2 Miscarriage per pregnancy achieved after ART in obese women. *ART*, Assisted reproduction treatment.

was confirmed again in a recent updated systematic review and metaanalysis of 21 studies. A decreased probability of live birth following IVF was observed in obese (BMI ≥ 30 kg/m^2) women when compared with normal weight (BMI 18.5−24.9 kg/m^2) women: RR (95% CI) 0.85 (0.82−0.87). Analyses done by the subgroups demonstrated that prognosis was poorer when obesity was associated with polycystic ovary syndrome, while the oocyte origin (donor or nondonor) did not modify the overall interpretation [22] (Fig. 11.3).

Safety issues for mothers and offspring

Many significant health issues and chronic medical conditions for women are associated with obesity. The risk of diabetes increases with the degree and duration of being overweight. Coronary artery disease risks increase with obesity, and weight reduction is an important adjunct to the management of those at risk. There is evidence that gynecological, particularly endometrial, and breast cancer risks increase in obese women, and increased weight may influence the outcome of disease.

Many adverse maternal, fetal, and neonatal outcomes are known to be associated with obesity. Management of the infertile thus poses complex questions linked to the welfare of potential mothers and their offspring. Many pregnancy-associated complications occur with greater frequency in the obese, for example, pregnancy-induced hypertension and gestational diabetes [27]. Need for intervention carries with the specifics of the difficulty of surgery in those who are morbidly obese together with the potential for complications such as infection, venous thromboembolism, and anesthetic hazards. Maternal mortality, while a rare occurrence, has associations with obesity and in a recent report highlighted the fact that many maternal deaths occurred in women with preexisting medical conditions, including obesity (BMI ≥ 30 kg/m^2), which seriously affected the outcome of their pregnancies [27].

Fetal risks in pregnancy are a concern with observed increased occurrence of fetal abnormality, macrosomia, low birth weight, neonatal mortality, and stillbirth. An influential report suggested that obesity is the principal modifiable risk factor for stillbirth in the developed world, greater than increased maternal age and smoking. Recent evidence suggests that maternal BMI was a significant risk factor for preterm delivery, even in pregnancies as a result of frozen ETs, following freeze all cycles [28].

Beyond these short-term outcomes the long-term health of individuals born to obese mothers is a public health issue of concern. Children of the obese will grow up with greater risks of coronary heart disease, hypertension, glucose intolerance, and diabetes as well as themselves being obese, thereby perpetuating the problem for the subsequent generation.

The management of the obese infertile raises economic issues of note given increased costs not only associated with treatment but also those associated with the

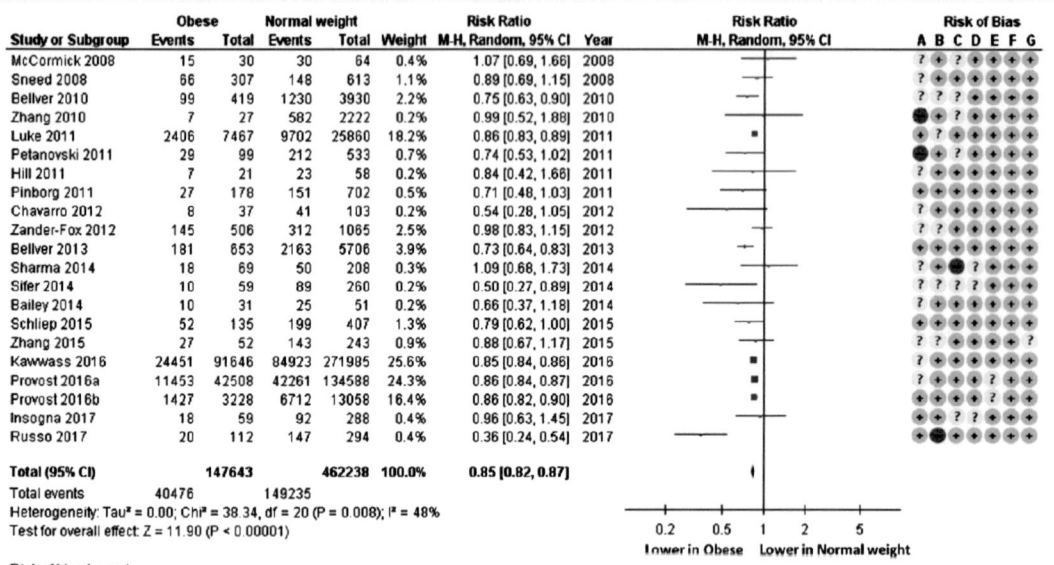

Study or Subgroup	Obese Events	Total	Normal weight Events	Total	Weight	Risk Ratio M-H, Random, 95% CI	Year
McCormick 2008	15	30	30	64	0.4%	1.07 [0.69, 1.66]	2008
Sneed 2008	66	307	148	613	1.1%	0.89 [0.69, 1.15]	2008
Bellver 2010	99	419	1230	3930	2.2%	0.75 [0.63, 0.90]	2010
Zhang 2010	7	27	582	2222	0.2%	0.99 [0.52, 1.88]	2010
Luke 2011	2406	7467	9702	25860	18.2%	0.86 [0.83, 0.89]	2011
Petanovski 2011	29	99	212	533	0.7%	0.74 [0.53, 1.02]	2011
Hill 2011	7	21	23	58	0.2%	0.84 [0.42, 1.68]	2011
Pinborg 2011	27	178	151	702	0.5%	0.71 [0.48, 1.03]	2011
Chavarro 2012	8	37	41	103	0.2%	0.54 [0.28, 1.05]	2012
Zander-Fox 2012	145	506	312	1065	2.5%	0.98 [0.83, 1.15]	2012
Bellver 2013	181	653	2163	5706	3.9%	0.73 [0.64, 0.83]	2013
Sharma 2014	18	69	50	208	0.3%	1.09 [0.68, 1.73]	2014
Sifer 2014	10	59	89	260	0.2%	0.50 [0.27, 0.89]	2014
Bailey 2014	10	31	25	51	0.2%	0.66 [0.37, 1.18]	2014
Schliep 2015	52	135	199	407	1.3%	0.79 [0.62, 1.00]	2015
Zhang 2015	27	52	143	243	0.9%	0.88 [0.67, 1.17]	2015
Kawwass 2016	24451	91646	84923	271985	25.6%	0.85 [0.84, 0.86]	2016
Provost 2016a	11453	42508	42261	134588	24.3%	0.86 [0.84, 0.87]	2016
Provost 2016b	1427	3228	6712	13058	16.4%	0.86 [0.82, 0.90]	2016
Insogna 2017	18	59	92	288	0.4%	0.96 [0.63, 1.45]	2017
Russo 2017	20	112	147	294	0.4%	0.36 [0.24, 0.54]	2017
Total (95% CI)		**147643**		**462238**	**100.0%**	**0.85 [0.82, 0.87]**	
Total events	40476		149235				

Heterogeneity: Tau² = 0.00; Chi² = 38.34, df = 20 (P = 0.008); I² = 48%
Test for overall effect: Z = 11.90 (P < 0.00001)

Risk of bias legend
(A) Confounding
(B) Selection of participants
(C) Classification of intervention
(D) Deviations from intervention
(E) Missing data
(F) Measurement of outcome
(G) Selection of reported results

FIGURE 11.3 Live birth rate in obese women compared to those with normal BMI. *BMI*, Body mass index. *From Sermondade N, Huberlant S, Bourhis-Lefebvre V, Arbo E, Gallot V, Colombani M, et al. Female obesity is negatively associated with live birth rate following IVF: a systematic review and meta-analysis. Hum Reprod Update 2019;25(4):439—51.*

management of complicated pregnancies, particularly the need for increased surveillance, higher rates of operative delivery, and the management of women with gestational diabetes and hypertension.

Ethical issues relevant to access to services

Debate within the last few years has taken place as to whether these morbidities and adverse outcomes, together with higher costs, should play a part in whether the obese should be permitted the same access to infertility services as those who are not overweight [2,3,29,30].

It could be argued with the prevalence of obesity being at the level it is that in fact the boundaries of what can be considered normal in the population have changed. Adverse outcomes however would suggest this is not the case. It has been argued that a restrictive policy would lead to stigmatization of the obese, but genuine health hazards are being increasingly identified, which carry significant implications for the individuals concerned. It has even been suggested that the autonomy of the individual to determine their own health would be being infringed by policies to deny access to care. On the other hand, the identification of long-term health risks could be considered as an opportunity for the empowerment of the individual to make lifestyle adjustments that may have real health benefits for themselves. Bearing in mind the issues described earlier, it is clear that patients have responsibilities beyond themselves, and health-care professionals similarly have responsibilities to offspring and to society at large. Scarce resources, particularly at the present time, should be used to maximum effect. Interventions to assist individuals to achieve and sustain weight loss are not always effective [31]. However, it would be anomalous for the reproductive health sector not to share with other areas of medical practice the public health responsibility for health promotion messages relevant to weight. Losing weight may of course delay the initiation of treatment and this is important particularly in those who seek assistance in later reproductive years [14]. However, in the younger patient the amount of weight loss to make a difference may not be substantial and the time taken to achieve a target may not adversely affect the chance of treatment being successful. That said, there is no randomized trial evidence at present that weight loss programs prior to IVF treatment have an appreciable effect on outcomes or pregnancy-associated complications.

Conclusion

There is irrefutable evidence that fertility potential is adversely affected in the obese. The proportion of patients accessing infertility services who are obese is increasing. Natural fecundity, responses to treatment, and pregnancy outcomes are suboptimal in this group of patients. The mechanisms whereby these effects are manifested are not fully understood, but it is likely that the causes are multifactorial, including endocrine, inflammatory pathways, as well as effects on oocyte quality. Interventions to address subfertility while offering increased potential for conception raise important questions relevant to the safety of mothers and offspring. While adverse outcomes are increased in this group of patients, the absolute risk to the individual of complications remains relatively small. Most conceptions will result in healthy live-born, but offspring will have increased lifetime health risks. Ethical issues in this sphere of reproductive medicine challenge principles of helping the individual while taking account of consequences for others, not least the potential child but, bearing in mind the costs of treatment, pregnancy care and beyond, the views of society at large.

References

[1] Ogden CL, Carroll MD, Curtin LR, McDowell MA, Tabak CJ, Flegal KM. Prevalence of overweight and obesity in the United States, 1999-2004. J Am Med Assoc 2006;295:1549—55.

[2] Pandey S, Pandey S, Maheshwari A, Bhattacharya S. The impact of female obesity on the outcome of fertility treatment. J Hum Reprod Sci 2010;3(2):62—7.

[3] Maheshwari A, Scotland G, Bell J, McTavish A, Hamilton M, Bhattacharya S. The direct health services costs of providing assisted reproduction services in overweight or obese women: a retrospective cross-sectional analysis. Hum Reprod 2009;24:633—9.

[4] Rich-Edwards JW, Goldman MB, Willett WC, Hunter DJ, Stampfer MJ, Colditz GA, et al. Adolescent body mass index and infertility caused by ovulatory disorder. Am J Obstet Gynecol 1994;171:171—7.

[5] Hassan MA, Killick SR. Negative lifestyle is associated with a significant reduction in fecundity. Fertil Steril 2004;81:384.

[6] Shah M. Obesity and sexuality in women. Obstet Gynecol Clin N Am 2009;36:347—60.

[7] Brewer CJ, Balen AH. The adverse effects of obesity on conception and implantation. Reproduction 2010;140(3):347—64.

[8] Metwally M, Ong KJ, Ledger WL, Li TC. Does high body mass index increase the risk of miscarriage after spontaneous and assisted conception? A meta-analysis of the evidence. Fertil Steril 2008;90(3):714—26.

[9] Esinler I, Bozdag G, Yarali H. Impact of isolated obesity on ICSI outcome. Reprod Biomed Online 2008;17:583—7.

[10] Dokras A, Baredziak L, Blaine J, Syrop C, VanVoorhis BJ, Sparks A. Obstetric outcomes after in vitro fertilization in obese and morbidly obese women. Obstet Gynecol 2006;108:61—9.

[11] Minge CE, Bennett BD, Norman RJ, Robker RL. Peroxisome proliferator-activated receptor-gamma agonist rosiglitazone reverses the adverse effects of diet-induced obesity on oocyte quality. Endocrinology 2008;149:2646—56.

[12] Maheshwari A, Stofberg L, Bhattacharya S. Effect of overweight and obesity on assisted reproductive technology a systematic review. Hum Reprod Update 2007;1:433—44.

[13] Bellver J, Melo MA, Bosch E, Serra V, Remohi J, Pellicer A. Obesity and poor reproductive outcome: the potential role of the endometrium. Fertil Steril 2007;88:446—51.

[14] Sneed ML, Uhler ML, Grotjan HE, Rapisarda JJ, Lederer KJ, Beltsos AN. Body mass index: impact on IVF success appears age-related. Hum Reprod 2008;23:1835—9.

[15] Goldman RH, Farland LV, Thomas AM, Zera CA, Ginsburg ES. The combined impact of maternal age and body mass index on cumulative live birth following in vitro fertilization. Am J Obstet Gynecol 2019;221:617.e1—617.e13.

[16] Vitek W, Sun F, Baker VL, Styer AK, Christianson MS, Stern JE, et al. Lower anti-Mullerian hormone is associated with lower oocyte yield but not live-birth rate among women with obesity. Am J Obstet Gynecol 2019;. Available from: https://doi.org/10.1016/j.ajog.2019.09.046 S0002-9378(19)31212-31218.

[17] Chong AP, Rafael RW, Forte CC. Influence of weight in the induction of ovulation with human menopausal gonadotropin and human chorionic gonadotropin. Fertil Steril 1986;46:599—603.

[18] Hamilton-Fairly D, Kiddy D, Watson H, Paterson C, Franks S. Association of moderate obesity with a poor pregnancy outcome in women with polycystic ovary syndrome treated with low dose gonadotrophin. Br J Obstet Gynaecol 1992;99:128—33.

[19] Balen A, Platteau P, Nyboe Andersen A, Devroey P, Sørensen P, Helmgaard L, et al. The influence of body weight on response to ovulation induction with gonadotrophins in 335 women with World Health Organization Group II anovulatory infertility. Br J Obstet Gynaecol 2006;113:1195—202.

[20] Rittenberg V, Seshadri S, Sunkara SK, Sobaleva S, Oteng-Ntim E, El-Toukhy T. Effect of body mass index on IVF treatment outcome: an updated systematic review and meta-analysis. Reprod Biomed Online 2011;23(4):421—39.

[21] Kwan I, Bhattacharya S, McNeil A. Monitoring of stimulated cycles in assisted reproduction (IVF and ICSI). Cochrane Database Syst Rev 2008;(2):CD005289. Available from: https://doi.org/10.1002/14651858.CD005289.pub2.

[22] Sermondade N, Huberlant S, Bourhis-Lefebvre V, Arbo E, Gallot V, Colombani M, et al. Female obesity is negatively associated with live birth rate following IVF: a systematic review and meta-analysis. Hum Reprod Update 2019;25(4):439—51.

[23] Luke B, Brown MB, Stern JE, Missmer SA, Fujimoto VY, Leach R. Racial and ethnic disparities in assisted reproductive technology pregnancy and live birth rates within body mass index categories. Fertil Steril 2011;95:1661—6.

[24] Luke B, Brown MB, Stern JE, Missmer SA, Fujimoto VY, Leach RA, et al. Female obesity adversely affects assisted reproductive technology (ART) pregnancy and live birth rates. Hum Reprod 2011;26:245—52.

[25] Styne-Gross A, Elkind-Hirsch K, Scott RT. Obesity does not impact implantation rates or pregnancy outcome in women attempting conception through oocyte donation. Fertil Steril 2005;83:1629—34.

[26] Bellver J, Ayllón Y, Ferrando M, Melo M, Goyri E, Pellicer A, et al. Female obesity impairs in vitro fertilization outcome without affecting embryo quality. Fertil Steril 2010;93(2):447—54.

[27] Centre for Maternal and Child Enquiries and the Royal College of Obstetricians and Gynaecologists. Management of women with obesity in pregnancy. London: RCOG; 2010.

[28] Ozgur K, Bulut H, Berkkanoglu M, Humaidan P, Coetzee K. Increased body mass index associated with increased preterm delivery in frozen embryo transfers. J Obstet Gynaecol 2019;39(3):377–83.

[29] Koning AMH, Kuchenbecker WKH, Groen H, Hoek A, Land JA, Khan KS, et al. Economic consequences of overweight and obesity in infertility: a framework for evaluating the costs and outcomes of fertility care. Hum Reprod Update 2010;16:246–54.

[30] ESHRE Task Force on Ethics and Law, including, Dondorp W, de Wert G, Pennings G, Shenfield F, Devroey P, et al. Lifestyle-related factors and access to medically assisted reproduction. Hum Reprod 2010;25:578–83.

[31] National Institute for Clinical Excellence. Dietary interventions and physical activity interventions for weight management before, during and after pregnancy. In: NICE public health guidance 27. 2010.

Further reading

Andersen AN, Balen A, Platteau P, Devroey P, Helmgaard L, Arce JC, et al. Predicting the FSH threshold dose in women with WHO Group II anovulatory infertility failing to ovulate or conceive on clomiphene citrate. Hum Reprod 2008;23:1424–30.

Bellver J, Rossal LP, Bosch E, Zuniga A, Corona JT, Melendez F, et al. Obesity and the risk of spontaneous abortion after oocyte donation. Fertil Steril 2003;79:1136–40.

Centre for Maternal and Child Enquiries (CMACE). Saving mothers' lives: reviewing maternal deaths to make motherhood safer: 2006–08 the eighth report of the confidential enquiries into maternal deaths in the United Kingdom. BJOG 2011;118(Suppl. 1):1–203.

Chapter 12

Obesity and sexual dysfunction in men

Darius A. Paduch[1,2,3] and Laurent Vaucher[2,3]

[1]Consulting Research Services, Inc, Red Bank, NJ, United States, [2]Department of Urology, The Smith Institute for Urology, Northwell Health, New Hyde Park, NY, United States, [3]Clinique de Genolier, Genolier, Switzerland

Prevalence of obesity has been rising over the last several decades in the industrialized countries, and according to a study, 68% of US adults are overweight and 35% are obese [1]. WHO data indicate that worldwide 41 million children and 1.9 billion adults aging 18 years or older are affected by obesity. Obesity affects every aspect of daily life, reproduction, and health in general.

Obesity has been linked to increased risk of diabetes mellitus (DM); hypertension (HTN); and risk of cardiovascular events, such as stroke, cancers, and osteoarthritis. As normal sexual function is a part of normal reproductive system, health consequences of obesity on low testosterone and erectile dysfunction (ED) are becoming of significant clinical importance.

Diagnosis of obesity is based on body mass index (BMI). BMI is calculated by dividing weight by square of height. BMI is very useful and has been extensively used in epidemiological studies but should not be used as a sole measure of obesity as it does not measure per se the percentage of total body fat. Using WHO criteria for adults older than 20, normal BMI is between 18.5 and 25; subjects are overweight if their BMI is between 25 and 30, they are obese if their BMI is 30–40, and morbidly obese if BMI > 40.

Physiology of sexual function

The normal erectile function depends on intact neurovascular function of the penis—arterial blood flow and venous closing mechanism, normal sensory innervation of the penis, normal central nervous system processing and integration of sexual cues in brain and spinal cord, and intact autonomous nervous system responses (Fig. 12.1).

Sexual cues (tactile, auditory, olfactory, visual, and recollection of past experience) are integrated into the hypothalamic and cortical centers and descending signals are sent through autonomic nervous and spinal cord to thoracic and lumbosacral regions of sexual response

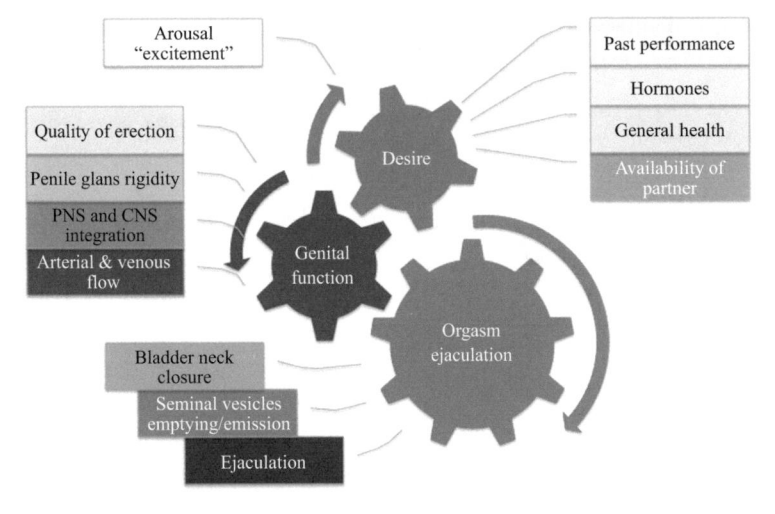

FIGURE 12.1 Elements of normal sexual response.

(Fig. 12.2). Somatosensory input from penis is carried through dorsal nerve of the penis through pudendal nerve and ends in the sacral region S2–S4. Posterocentral gyrus in cortex represents primary penile sensory activation. Scrotum and pubic area are innervated by the branches of ilioinguinal (L1) and femoral cutaneous nerves (L2–L3) (Figs. 12.3 and 12.4). DM can impair normal conduction in pudendal and sensory nerves and has been postulated as one of the reasons for diabetic ED [2].

The descending signals reach penile erectile tissues through cavernous nerves carrying parasympathetic and sympathetic fibers (Fig. 12.4). Parasympathetic stimulation results in relaxation of smooth muscles within penis, opening of bilateral cavernosal arteries, and sudden increase in blood flow through penis, which results in tumescence (penile erection).

At the same time, subtunical veins that normally drain blood from penis get closed (Fig. 12.5).

Any pathology that impedes the muscle relaxation with an increase in penile blood flow will result in ED.

Over the last two decades, most of the researches have focused on ED; however, it is well known that the presence of rigid erection is not sufficient for satisfactory reproductive and sexual performance, thus both evaluation and therapeutic interventions need to embrace different aspects of sexuality.

Sexual Dysfunction

Sexual dysfunction can be divided into ED (problems sustaining erections adequate for penetration), disorder of sex

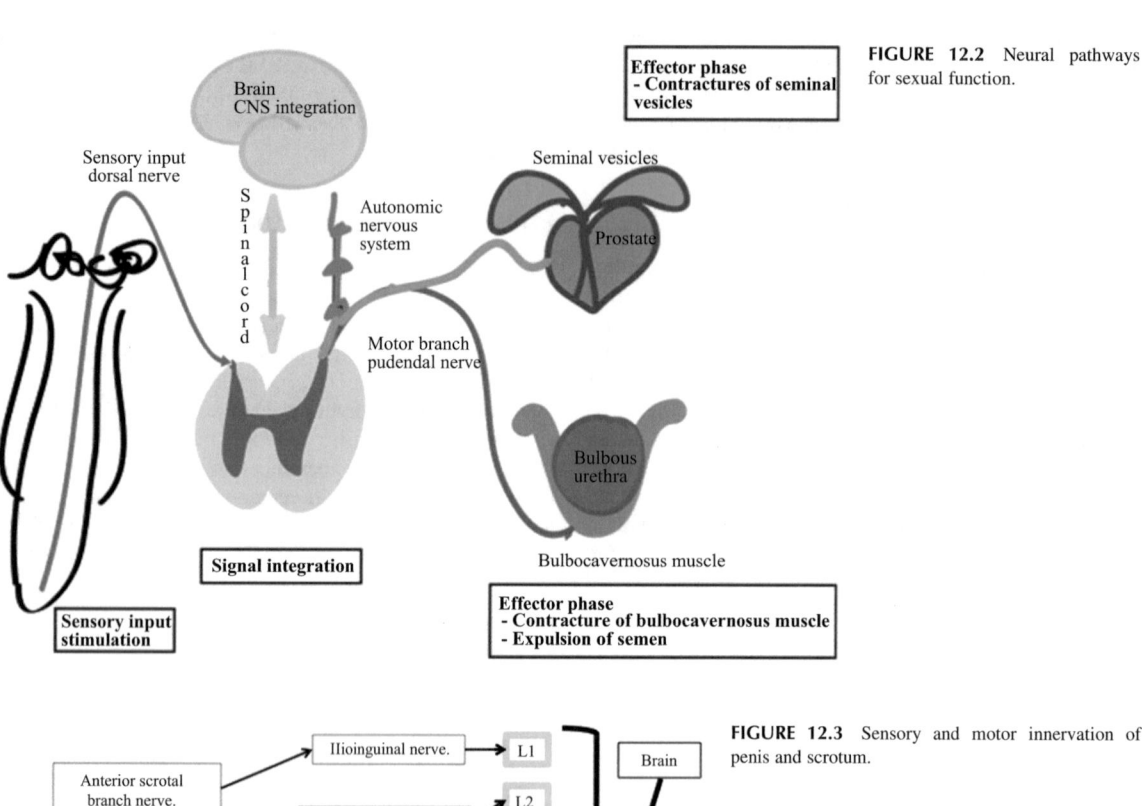

FIGURE 12.2 Neural pathways for sexual function.

FIGURE 12.3 Sensory and motor innervation of penis and scrotum.

FIGURE 12.4 Sympathetic and parasympathetic innervation of penis.

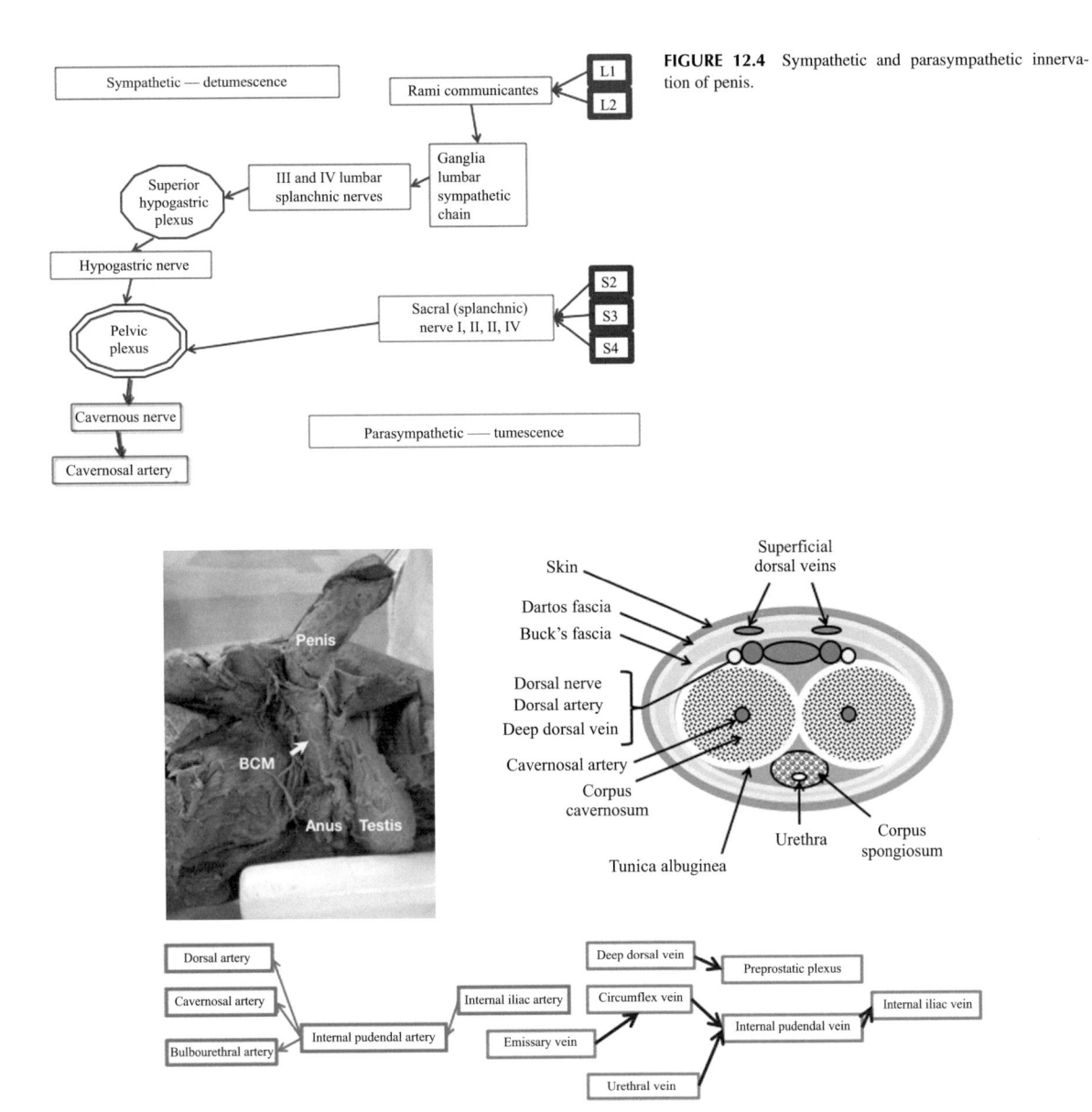

FIGURE 12.5 Vascular anatomy of penis.

drive (libido), disorder of arousal (excitement), and disorders of ejaculation and orgasm. The ejaculatory dysfunction can present itself as premature ejaculation (intravaginal latency time is less than 2 minutes), delayed ejaculation (subjective prolonged time to ejaculate), and anejaculation (lack of ejaculation). Lack of ejaculation can be the result of retrograde ejaculation (semen goes back into bladder) or lack of emission. Orgasm is a subjective sensation of enhanced pleasure followed by postcoital refractory period. Orgasm is typically associated with ejaculation in normal subjects. Orgasmic dysfunction can range from lack of orgasm through decreased sensation of orgasm (Fig. 12.6). Patients and health professionals often use the words "orgasm" and "ejaculation" as synonyms even though biologically the two phenomena are regulated by two different neurobiological pathways.

Sex Drive

Sex drive is a complex neuropsychological phenomenon affected by hormones, energy status, health, social norms,

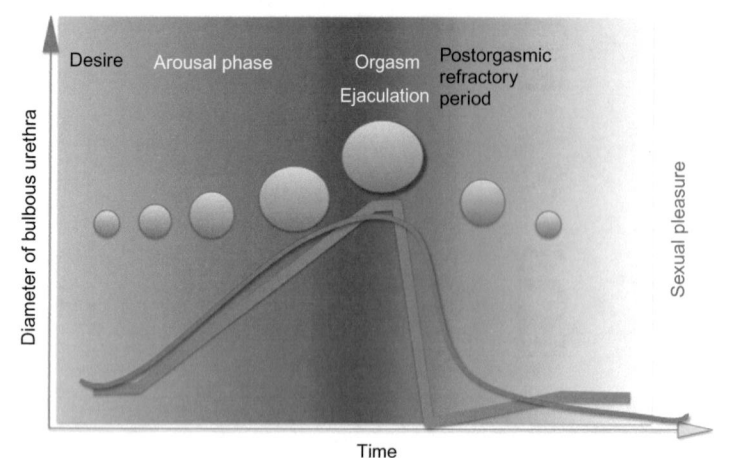

FIGURE 12.6 Phases of normal sexual response in man.

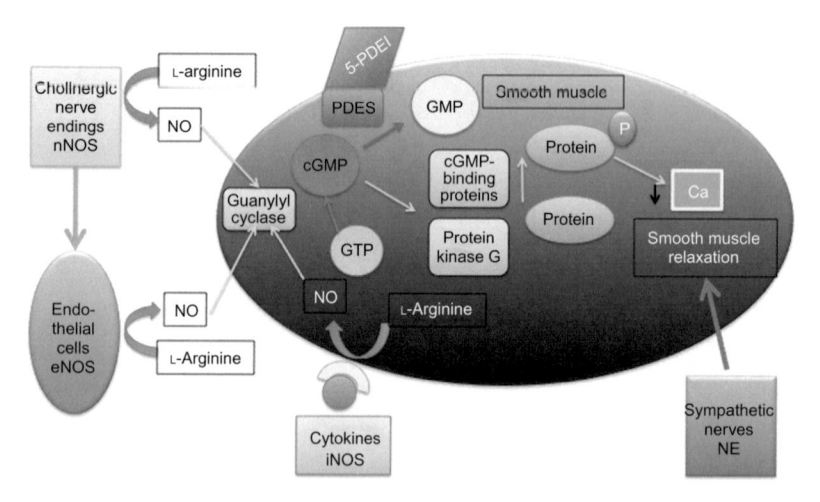

FIGURE 12.7 Molecular mechanism of erection.

and emotional well-being. Body image and recollection of positive sexual experience play important roles in sex drive and normal sexual response in both men and women. Obesity with change in body image, decrease in functional length of penis because of pubic fat pad, and limitation in physical capabilities may further erode one's confidence in sexual performance and have detrimental effect on sex drive leading to withdrawal from interpersonal relationships. Pain from osteoarthritis in knees as a consequence of obesity further limits the physical stamina leading further complicating sexual activities. Pain medications are known to lower testosterone level. Hence, consequences of obesity on body mechanics and function often result in vicious circle with weight gain—depression—functional restrains—leading to sedentary lifestyle—and further increase in weight.

Erection (tumescence) starts with stimulation parasympathetic nerves with release of acetylcholine and activation of nitric oxide synthase within nerve endings (nNOS) and endothelial nitric oxide synthase (eNOS). NOS converts L-arginine into nitric oxide—a potent vasodilator that is transported by diffusion to smooth muscles. NO activates guanylate (guanylyl) cyclase converting GTP to active cGMP. cGMP activates phosphorylation of target proteins resulting in decrease in intracellular calcium, smooth muscle relaxation, and erection (Fig. 12.7). Sympathetic nervous system and local factors such as endothelin-1 (ET-1) oppose smooth muscle relaxation and result in detumescence.

Obesity related sexual dysfunction

Obesity is known to affect hormonal levels, specifically increase in circulating estradiol (E2) level and decrease in total testosterone (TT) level. In addition, obesity-related medical comorbidities such as HTN, dyslipemia, and DM are linked to ED.

Exact mechanism of obesity-related sexual dysfunction is an area of intense research and multifactorial model with local impairment of penile tissue relaxation with global endocrinopathy and metabolic changes in nerve signaling seem to best describe clinically observed sexual dysfunction in obese men [3,4].

In rodents that are fed high-fat diet (HFD) versus standard diet, there was a clinically and statistically significant drop in cavernous strip relaxation after induction with acetylcholine (Ach) [5]. This and other studies showed a decreased concentration of cGMP in penile tissue from obese mice as well as decrease in endothelial and acetylcholine and nonadrenergic–noncholinergic nitrergic signaling thus impairing the relaxation of penis, which is necessary for normal tumescence [6]. The tissues from obese penis are much more sensitive to adrenergic stimulation and ET-1 [7]. Thus obesity decreases penile ability to relax smooth muscles and increases sensitivity to stress-related signaling impairing the balance between tumescence and detumescence, which is necessary for normal erection.

Hyperglycemia decreases NO production by eNOS by O-linked glycosylation of eNOS at the Akt target S1177 [8]. Diabetes is clearly associated with diminished endothelial production of NO [9,10].

Sexual response in men is initiated by the desire to have sexual activity. Desire is a complex neurobiological prenomen that is first experienced with initiation of puberty, thus sex steroids have always been considered a critical component of normal sexual desire. Although testosterone level and desire do not have linear correlation, multiple studies in diverse ethnical groups showed a similar association between low testosterone and low sexual desire [11–13].

It is well established that in obese men the sexual desire is negatively correlated with total and free testosterone [14]. Gastric bypass surgery improved testosterone levels and improved sexual desire.

The negative effects of obesity on sexual desire are most likely related to low testosterone and elevated estradiol, but fear of failure, changed body image, and functional limitations may be significant contributing factors.

Testosterone is produced in testes under control of luteinizing hormone (LH) and local paracrine control. Testosterone is metabolized to estradiol (E2) by aromatase CYP19 expressed in testes and abdominal fat (Fig. 12.8). Elevated levels of E2 have been found in most obese men. E2 suppresses not only central release of LH but it also has negative effects on Leydig cell function and normal T production thus explaining high prevalence of low testosterone in men with hyperestrogenism [15,16]. Elevated estradiol level in men is also associated with decrease in sperm production.

Normal sexual function depends on adequate balance between testosterone that in men should be high and estradiol that needs to be low. Treatment with aromatase inhibitors may help increasing testosterone level in some obese men with low T and elevated E2 (see next).

During sexual activity—progressive and linear increase in level of excitement—arousal is necessary to achieve orgasm and sustain erection. Little is known about effect of obesity on arousal in men, but obesity severely impairs arousal in females [17,18]. As arousal is central neurobiological process, it is very plausible that obese men suffer from similar issues with arousal. Here again, low testosterone, hormonal abnormalities, physical

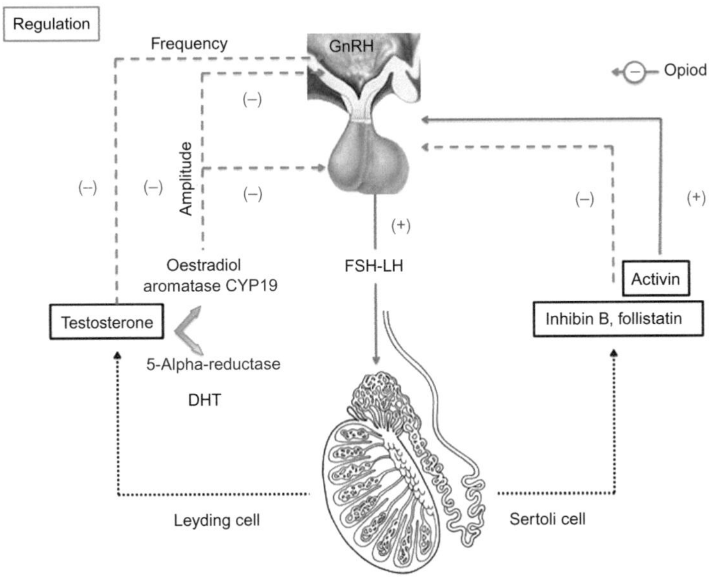

FIGURE 12.8 Continuous tonic inhibition of peripheral steroids.

limitations, and body image may interfere with normal processing of sexual cues necessary to achieve normal progression of arousal.

Sexual dysfunction and obesity-related comorbidities

It is often difficult to dissect pure effects of obesity from associated obesity-related medical comorbidities such as HTN, peripheral vascular disease, hypercholesterolemia, and diabetes.

Based on our clinical practice, we use the following functional theorem of obesity-related sexual dysfunction: (1) genital effects (relative shortening of the penis), (2) body habitus limitations (decreased accessibility to vagina due to abdominal obesity), (3) positional restrictions (due to excessive weight on partner), (4) poor muscular stamina (strength and time), (5) poor cardiovascular and pulmonary fitness (shortness of breath, poor exercise tolerance), (6) joint dysfunction due to obesity (physical limitations and pain), (7) obesity-related neuroendocrine effects (hypogonadism and hyperestrogenism), (8) obesity-related chronic inflammation cascade, (9) poor body self-image, (10) perceived perception of one's attractiveness in eyes of partner, (11) partner's concern for possible negative events during sexual activity, (12) partner's actual perception of obese male attractiveness, (13) anxiety-related to past failures in sexual domain, (14) depression either primary or reactive, and (15) metabolic consequences of obesity on overall health (HTN, DM, cardiovascular disease, etc.).

We find it useful in practice to go over different, often interconnected, dimensions how obesity relates to sexual desire and performance. Most of the published researches focused on metabolic effects of obesity but discussing with the couple's physical aspects of sexual performance and understanding specific wants and needs within couple dynamics are critically important aspects of care of patients.

Hypercholesterolemia—defined as elevated total cholesterol and low-density lipoprotein (LDL)—has been linked to increased risk of CAD in Framingham Heart Study and many others [19]. Massachusetts Male Aging Study (MMAS) showed that ED is inversely related to baseline high-density lipoprotein and prevalence of ED doubled over 9 years in men who were obese at baseline as compared to men who were not obese at baseline or follow-up [20,21]. Obesity increases risk of vasculogenic ED as assessed by penile Doppler ultrasonography [22]. But from published literature, it appears that it is a long-term obesity with obesity-related comorbidities, which correlates with ED rather than elevated BMI by itself. In Rancho Bernardo Study, which followed men over 25 years the age, hypercholesterolemia and obesity were independent predictors of severity of ED in logit model, but obesity was not an independent predictor of the presence of ED by itself.

Small nonrandomized clinical trial showed that reduction in total cholesterol to below 200 mg/dL and LDL to less than 120 mg/dL had positive effect on ED after 3.7 months of treatment [23]. However, randomized double-blinded trial (STED TRAIL) of simvastatin failed to show improvement in erectile function by simvastatin as compared to placebo. It is possible that statins have role in men with ED and hypercholesterolemia who fail to respond to sildenafil [24,25]. However, a recent study of tadalafil 20 mg three times a week versus 10 mg atorvastatin showed that tadalafil was better in restoring ED than atorvastatin, whereas atorvastatin showed some improvement in erectile function especially in men with hypercholesterolemia [26,27]. French Pharmacovigilance System Database study showed that statins may actually induce or worsen ED [28]. So far, no conclusive evidence supports the routine use of statins in patients with ED without a clear indication for the use of statins because of cardiovascular protective effects [29].

Rather conflicting results of statin treatments on improving ED in men with hypercholesterolemia may be due to short follow-up and severity of peripheral vascular disease. Reduction in prevalence of cardiovascular events with statins has been showed after 5 years of therapy by most studies (CAPS, CARE, LIPID), and there is some evidence that with longer treatment the positive effects of statin treatment on ED may be noticeable [30,31]. Another complicating factor may be lowering of TT in some men taking statins [32–34]. Thus testosterone level should be monitored in men on statins therapy. Randomized clinical trial of simvastatin 80 mg or placebo showed decrease in bioavailable testosterone by 10%. Bioavailable testosterone best correlates with biological activity of testosterone; hence, one may assume that at least in some men the drop in bioavailable testosterone may be clinically significant. In summary, it is clear that dyslipidemia is a risk factor for ED but not clear if obesity by itself without its sequel has a similar detrimental effect on ED. Often ignored but important for patient is muscle ache associated with statin therapy. Obese men have physical limitations already which can be further exacerbated by muscle aches and pains. In general, men withdraw from sex if they are in pain or discomfort. The consensus statement from AHA on statin safety indicates that 10% of men discontinue statin therapy because of muscle ache without laboratory evidence of elevation in creatinine kinase. Most of the patients will not restart statins despite normal laboratory evaluation—thus managing expectations and side effects of statins in men early on may help with compliance. Clearly further studies are needed.

Hypertension

Approximately 50%–70% of men with HTN report varied degree of ED [35,36]. The underlying mechanism of ED in HTN seems to be HTN-induced peripheral vascular disease, and severity of ED in HTN men is correlated with the duration of HTN [37]. HTN impairs neurogenic-induced smooth muscle relaxation and reduction in superoxide dismutase in animal models [38].

Treatment of HTN with some of medications may further worsen ED and contribute to poor compliance with antihypertensive medications [39,40]. Nonselective beta-blockers, hydrochlorothiazide, spironolactone, and angiotensin II antagonists, are known to result in ED in significant number of patients thus angiotensin converting enzyme inhibitors, whereas selective beta-blockers are a better choice for men with HTN and preexisting ED [41,42]. In men who require multidrug therapy for HTN, adding 5-phosphodiesterase inhibitor improves compliance with antihypertensive therapy and results in better blood pressure control [43].

As HTN and hypercholesterolemia often occur together, most of the obese men are both antihypertensive and cholesterol-lowering agents.

Large randomized (2153 men) clinical trial of effect of rosuvastatin and candesartan plus hydrochlorothiazide versus placebo showed no difference in erectile function after 5.8 years of follow-up as measured by IIEF-EF score. Hence, clinicians have to stress on the importance of controlling cholesterol and blood pressure as a way to prevent cardiovascular events, but there is no clear evidence that the use statins or blood pressure medications can improve ED per se.

Peripheral vascular disease

Classic epidemiological study by Blumentals showed that men with ED have increased risk of peripheral vascular disease by 75% [44]. Even after adjusting for the presence of other risk factors for stroke, men with ED have a higher risk of stroke and lower risk of stroke-free survival over 5-year follow-up study in Taiwan [45]. Often ED is the first sign of peripheral vascular disease prior to claudication, thus men with ED should be screened for PVD [46]. It is believed that endothelial dysfunction combined with hypercholesterolemia is responsible for PVD and vasculogenic ED. Large epidemiological studies showed that PVD is prevalent in obesity [47]. In our practice, men with obesity and vascular ED are referred for cardiovascular evaluation and risk assessment.

Coronary artery disease

Intracavernosal (IC) arteries in penis are less than 1 mm in diameter, thus it is no surprise that often ED is first manifestation and predates coronary artery disease [48,49]. It has been shown that the presence of ED is an additional cardiovascular risk, which should be considered in stratification assessment and decision-making for need for further invasive coronary evaluation [50]. Men with ED have a higher volume of coronary calcifications as compared to men without ED [51]. The presence of ED has been demonstrated to be an independent predictor and risk factor for cardiovascular events, cardiovascular-related morality, and all-cause mortality, thus obese patients with ED should be evaluated by cardiologists to determine if they need stress test or further testing to assess their cardiovascular risks. It is our practice that in obese men with multiple risk factors for cardiovascular disease who have vascular ED, we do not initiate treatment for ED unless a patient is seen by a cardiologist, as for many men, erectile function seems to be more important than general health. The role of urologists, general practitioners, and endocrinologists in the evaluation of cardiovascular risk in men with ED is evolving, but there is no question of significant opportunity to improve one's general health.

Diabetes mellitus

Effects of DM on ED is multifactorial but neurogenic dysfunction as well as vasculopathy with impairments of NO signaling decrease in vasodilation are best understood at this point.

DM can affect sensory signaling and result in autonomic nervous system dysfunction. In obese diabetic subjects the peripheral neuropathy in lower extremities assessed by increased vibratory threshold is strongly associated with ED [52]. This should not be a surprise that tactile stimulation is one of the most important sexual cues. Micro- and macrovascular disease has been linked to DM using standardized instruments to measure sexual function [53]. Similar changes in microvascular environment with decrease in endothelial function within penis were found in animal models of DM [6].

Evaluation—General considerations

Evaluation of sexual dysfunction in obese males should include detailed medical, social, family, and sexual history combined with physical examination and laboratory testing.

Sexual history should focus how changes in BMI affect sexual drive and performance, both sexual activities with partner and alone, penetrative and other forms of sexual activities and mechanics should also be taken into consideration. Body habitus disproportion between partners has to be assessed in culturally sensitive way. Cognitive–affective aspects of BMI and body image, self-acceptance and attractiveness, have been most exclusively studied in females, but it is established that dissatisfaction with body image is similar in men (59.5%) and among women (55.2%).

One has to consider cultural and ethnical differences in BMI and their effects on health. The concept of BMI-related metabolic transition toward HTN, hypercholesterolemia, and diabetes is affected by ethnicity. Ethnic groups with lower BMI have a higher rate of metabolic transition (Asians) compared to ethnic group with historically high BMI (Whites). This is important when advising patients about their overall risks of developing obesity-related health consequences. Practitioners should avoid projecting their own stereotypes of what is desirable and acceptable when it comes to body size and build.

Based on the high prevalence of low testosterone and sexual dysfunction in obese men, it is prudent to measure testosterone level in most obese men, especially those with loss of morning erections, decreased sex drive, fatigue, and ED. Morning total sex hormone globulin levels, and free testosterone should be obtained using the most reliable and sensitive method. In our practice, we use liquid chromatography and mass spectrometry (LC–MS) to measure TT. LC–MS has been recommended as the preferred method to measure testosterone in hypogondal men because of its lower coefficient of variance as compared to other methods [54].

FDA uses cutoff point of 300 ng/dL to establish hypogonadism in pharmacological studies. In Europe, 10.2 nmol/L is often used. One needs to understand that testosterone levels change over age, and 300 ng/dL cutoff point is probably most appropriate in older men >65 as most of the studies on testosterone level were done in older men. Free testosterone and bioavailable testosterone are useful in men with normal or normal low TT who present with symptoms. In men who have low TT, FSH, LH, PRL, thyroid profile, estradiol, and cortisol as well as baseline PSA should be obtained. CBC and liver function tests should be obtained at baseline.

In clearly hypogondal men who have elevated PRL or unexplained low or low normal LH and FSH, one may consider CT or MRI of brain with attention to pituitary and hypothalamus to exclude pituitary or hypothalamic mass.

However, obesity is associated with hypogonadotropic hypogonadism because of elevated estradiol, thus decision to obtain additional imaging studies has to be considered on individual basis.

In our practice, we also obtain ultrasensitive CRP and HgA1c at baseline in obese men who present with sexual dysfunction. In men with type I DM in the Diabetes Control and Complications Trial the risk of ED was correlated with HgA1c and men who had tight control of DM with insulin had a significantly lower rate of ED [55]. Severity of ED increases dramatically in men whose HgA1c is above 8% and with DM type II of 6 or more years [56]. Considering that a significant number of men with early diabetes are not aware of their abnormal sugars, it is critical to establish diagnosis of DM early to prevent microvascular sequels that with time may be difficult to reverse or control.

Erectile and endothelial dysfunction may have similar pathways through impaired nitric oxide activity; in obese men the endothelial dysfunction may be further impaired through increase in interleukins (IL-6, IL-8, IL-18) and CRP. Obese men with ED but not obese men without ED had a significant increase in CRP and inflammatory procytokines [57].

Elevated CRP is associated with increased risk of cardiovascular events, and in many men the presence of such objectively measured risk may aid in behavioral modifications to lose weight and to exercise.

This study also points to the fact that it is not obesity itself that results in the ED but associated vasculopathy and proinflammatory response. It is not known at this point why some obese men do not suffer from ED. Possible explanation may be limitation in BMI as marker of obesity, ethnical differences in effect of BMI on metabolic profile, and further studies with better assessment of lean body mass and percent of body fat may help us to understand the link between obesity and ED better.

We strongly advocate active screening for sexual dysfunction among obese and overweight men as it has been shown that correction of ED with medical therapy may have positive impact on glycemic control through better adherence to medical therapy and diet.

Evaluation of sexual function

Depending on practice, one can consider using screening questionnaires to assess ED. Questionnaires that assess broad aspects of sexuality specifically sex drive, orgasmic, and ejaculatory dysfunction in addition to ED may be better suited for obese men considering that they often suffer from complex sexual dysfunction when ED is only one of the domains. In our practice, we use MSHQ during initial visit and follow-up, but each physician has to choose the instrument which is most appropriate to his or her patient population and cultural and social norms.

Although BMI is most commonly used measure of obesity, we use lean body mass and percentage body fat as well as waist circumference to better stratify metabolic risks of higher BMI. Ethnical differences in BMI norms may be considered. BMI is very useful in large epidemiological studies but may be less useful in managing individual patients.

Typically in obese patient with ED by history and verified by questionnaires other than hormonal evaluation, no further workup is needed especially if testosterone was normal. These men may be started on trial of one of 5-phosphodiesterase inhibitors—see later—combined with diet, exercise, and management of existing comorbidities. Smoking cessation is critical.

However, in men with failed response to oral therapy for ED and in younger men with risks for peripheral vascular disease or neuropathy, referral for penile Doppler ultrasound and neurosensory testing may be prudent. Often patients especially younger ones want to know why they suffer from ED.

Penile Doppler ultrasound is a minimally invasive and very well-tolerated procedure. After checking blood pressure explaining risks such as need for redosing IC injection, bruise, prolonged erection, and embarrassment, patient is brought to sexual medicine laboratory when IC injection is administered in the penis to induce erection. We typically use from 5 to 10 U of Trimix and take continuous measurement of blood flow through cavernosal arteries. Patient is allowed to self-stimulate and watch adult audiovisual aids as needed. In men with severe vasculopathy the dose of medication is escalated up to 60 U; however, risk of priapism increase with higher doses. It is critical to achieve rigid erection in the sexual medicine laboratory to exclude venous leak, which is often overdiagnosed because of inadequate dosing and stimulation. We measure peak systolic velocity (PSV), end diastolic velocity (EDV), and resistive index. At the same time the length of time it takes to ejaculate or orgasm (in men with anejaculation) can be measured and the amount of force to achieve tactile stimulation to sustain erection is recorded. In men with decreased sensation to the penis, penile biothesiometry is performed and vibratory thresholds measured. Typically, once the subject achieves nonbending erection, the study is completed and the patient is observed at 15 minute intervals to assure detumescence. No patient is allowed to leave office unless his penis is flaccid. In the case of prolonged erection the 500–1000 μg of Neo-Synephrine is injected into the penis with blood pressure monitoring.

In men who present with delayed ejaculation or anejaculation the study is continued till patient achieves orgasm or at least 30 minutes passes to exclude lack of normal arousal as a reason for anejaculation.

In diabetic men who complain of anejaculation, postorgasmic urine analysis is performed to determine if subjects suffer from retrograde ejaculation.

Based on penile Doppler ultrasound, diagnosis of arteriogenic ED (PSV < 35 cm/s) or venous leak (RI < 0.75) can be established. At the same time, optimal dose of IC pharmacotherapy can be established. Often it is very relieving to the patient to show him that he can achieve erection even with IC therapy.

Men with venous leak will require typically much higher doses of IC therapy, and they are not likely to respond to oral medication initially, thus PDUS helps with directing therapy in individual patients. In men with venous leak the penile rings may help; however, they are difficult to apply in men with severe obesity and panus.

In men with neurogenic ED or ED related to low testosterone, one should consider oral 5-PDE as initial therapy. Similarly in men with mild arteriogenic ED, 5-PDE should be initial choice.

Multidisciplinary approach to treatment

Decrease in BMI, management of comorbidities, and improvement in hormonal profile through hormone replacement therapy, with medical and behavioral therapy to improve sexual function and reinforce positive behavior, should be the goals of therapy. Multidisciplinary approach, including internist, dietician, bariatric surgeon, sexual medicine expert, endocrinologist, and mental health practitioner, may be necessary for complicated patients with multiple comorbidities.

Weight management

Two approaches can be employed depending on the extent of patients' obesity, willingness to lose weight, and acceptance of medical therapy. Some practitioners will defer medical therapy to improve erectile function and focus on weight loss initially. There is no question that decrease in BMI improves erectile function, body image, and sexual desire, thus this approach can be tried in selective, motivated patients with mild-to-moderate ED; however, most of the men who present to sexual medicine specialists will present with reactive depression, poor self-esteem, marriage issues, and withdrawal from sexual activities. Thus in our practice, we strongly advocate the combination of diet and exercise with medical therapy to help with erectile function and replace testosterone in men with hypogonadism.

As sexuality plays an important role in male self-esteem, by improving sexual function we often achieve improvement in mood and dedication to weight loss [58].

High-protein, carbohydrate-reduced, low-fat diet and low-calorie diet (1000 kcal/day) over 52 weeks have similar degree in improving sexual function, sexual desire, and urinary symptoms in obese men. High-protein diet may be easier to tolerate by younger men when combined with exercise program then restrictive calorie intake diet; however, the choice of diet should be established based on the basic metabolic rate, level of daily activities, and patients' own goals in consultation and under the supervision of dietician. Both diets help to reduce systemic inflammation [59]. Mediterranean diet helps to reduce ED [58].

Exercise has been shown to reduce elevated blood pressure and improve ED and should be combined with diet and medical therapy [60].

In severely obese patients, bariatric surgery improves erectile function, increases testosterone levels, and decreases prolactin and estradiol level, but bariatric surgery does not reverse obese-related impairment in spermatogenesis. Bariatric surgery in obese patients with severe sexual dysfunction and other comorbidities may be considered especially in patients who have contraindications or fail to respond to first- and second-line therapy for sexual dysfunction.

Pharmacological therapy

The mainstay of therapy of ED remains 5-PDEs.

5-PDE blocks the inactivation of cGMP in smooth muscles and thus improves smooth muscle relaxation.

Following 5-PDE inhibitors are approved in the United States: sildenafil (Viagra), vardenafil (Levitra and Staxyn), tadalafil (Cialis), and avanafil (Stendra). After patent for sildenafil expired in 2019, generic forms of it are widely available in the United States, but bioequivalence studies are lacking so far.

The efficacy of these agents has been established in large multinational trials in men with varied degrees of ED and broad spectrum of etiologies.

The drugs differ in their time and duration of onset, side effect profile, and effects of food on bioavailability. This is especially an important issue in men with diabetes who may suffer from gastroparesis and men with frequent snacking as in obesity. Bioavailability of sildenafil and vardenafil is significantly decreased by food so they should be taken on empty stomach. But bioavailability of tadalafil and avanafil is not affected by food so can be taken on empty stomach as well as with food. All information about pharmacological agents described next have been extracted from the latest official FDA labels for each of medications.

Sildenafil (Viagra) was the first 5-PDE approved for ED. Viagra comes in 25, 50, and 100 mg tablets. Sildenafil should be used prior to sexual activity "on-demand." Sildenafil is rapidly absorbed with median absorption of 60 minutes when taken in fasting state. High-fat food delays peak level of sildenafil by 60 minutes and mean reduction in Cmax of 29%. Although sildenafil's effect may last up to 4 hours, most of effectiveness is observed within 2 hours after oral intake. Sildenafil has been studied in men with ED and chronic stable angina limited by exercise, there was no difference in exercise-induced angina episodes between men receiving sildenafil to men on placebo. Sildenafil has dose-related impairment of color discrimination (blue/green) without change in visual acuity or intraocular pressure. 82% of men taking 100 mg of sildenafil, 74% 50 mg of sildenafil, and 63% of 25 mg reported improvement in sexual function.

Sildenafil and all 5-PDEs are contraindicated in patients taking nitrates either regularly or intermittently. Sildenafil has systemic vasodilator properties and can cause transient decrease in supine blood pressure of 8.4/5.5 in healthy volunteers. The decrease in blood pressure may be more pronounced with aortic stenosis and severely impaired autonomic control of blood pressure. Men on antihypertensive therapy may be at increased risk of hypotensive episodes; thus in these groups the therapy should be initiated at lower doses and increased slowly. Sildenafil is contraindicated in men with retinitis pigmentosa.

Important issues to discuss with all 5-PDEs are risk of developing prolonged erection ($>$4 hours) and priapism. Sudden loss of vision in one or both eyes may be a sign of nonarteritic anterior ischemic optic neuropathy (NAION). Most of the cases on NAION occurred in men with diabetes, HTN, coronary artery disease, hyperlipidemia, and over 50 years. Similarly, sudden decrease or loss of hearing that may be accompanied by tinnitus and dizziness has been reported as well. However, at this point, it has not been determined that the vision and hearing loss are related to the use of 5-PDE or to other factors. Sildenafil is associated with headache 16%, flushing 10%, dyspepsia 7%, and nasal congestion 4%. At 100 mg dose, 17% of patients reported dyspepsia and 11% abnormal vision.

It is critical to consider that 5-PDE should not be used in men with angina, recent acute MI, and limited cardiac reserve as normal cardiovascular function is necessary to have sexual activities.

Diabetic patients have somehow reduced response to sildenafil, as 63% of men with type II DM and 67% of men with type I DM reported improvement in erections in pulled data from 11 trials. 86% of men with hyperlipidemia reported improvement and 69% of men with HTN.

Vardenafil (Levitra) in oral form has been approved in 2003 by FDA. Levitra comes in 2.5, 5, 10, and 20 mg tablets. A 10 mg tablet taken 60 minutes prior to sexual actives is a recommended starting dose. Vardenafil as all other 5-PDEs is contraindicated in men taking nitrates, with unstable angina and hypotension.

Levitra is contraindicated in patients with congenital QT syndrome or taking class IA (quinidine, procainamide, disopyramide) or III antiarrhythmic (amiodarone, sotalol, ibutilide, dofetilide). Levitra should not be used in men with unstable angina, hypotension, uncontrolled HTN ($>$170/110 mmHg), recent history of stroke, life-threatening arrhythmias, recent myocardial infarction ($<$6 months), and severe cardiac failure. Levitra shares the same warnings about prolonged erections and NAION as sildenafil. Headache in 15% of subjects, flushing in 11%, rhinitis in 9%, and dyspepsia in 4% are the most common side effects of Levitra. After oral intake of Levitra the peak concentration is noticed at median of 60 minutes in fasted state. The high-fat meals reduced in Cmax between 18% and 50%, which is of significance in obese patients who often have HFDs. Levitra shows dose-related response that in general, population of men with ED is 65% for 5 mg, 75% for 10 mg, and 80% for 20 mg compared to placebo response of 52%. However, in men with ED and DM the response was 51% at 5 mg, 64% at 10 mg, and 65% at 20 mg, thus showing a decreased response rate in patients with DM and ED by almost 15% in higher dose as compared to normal men.

Staxyn is orally disintegrable table of 10 mg of vardenafil. It is not interchangeable with 10 mg of Levitra as it achieves higher systemic exposure compared to 10 mg Levitra. In clinical studies the side effects profile was similar to Levitra, but the rate of per patient rate of achieving erection sufficient for penetration was 47% for 10 mg and 22% for placebo.

Tadalafil (Cialis) has been approved by FDA in 2003 for use in men with ED and was also approved in 2011 for treatment of men with benign prostatic hyperplasia.

Tadalafil can be used on demand at dose of 5, 10, or 20 mg or as daily 5 mg medication.

Because of its selectivity for 5-PDE, tadalafil has much less flushing and nasal congestion 2% for 5 mg and 3% for 20 mg than sildenafil or vardenafil. Headache can occur in 6% of men on daily dose of 5 mg, and 15% of men taking highest 20 mg dose. Dyspepsia is seen in 5% of men taking daily Cialis and 10% of men with 20 mg on-demand Cialis. Back pain can be observed in 6% of men taking 20 mg dose and 3% men (3%—placebo) on daily 5 mg tadalafil. 77% of men on 20 mg tadalafil versus 43% on placebo experienced improvement in ability to penetrate. 64% versus 23% on placebo reported ability to sustain erection. 57% of men with DM-related ED taking 10 mg Cialis versus 30% men taking placebo reported ability to insert. Daily 5 mg tadalafil was also successful in improving ED as 67% of men on daily tadalafil versus 37% men on placebo reported ability to insert. Similar effectiveness was seen in men with DM.

Obesity increases the frequency of *lower urinary tract symptoms and ED*. Tadalafil is the only medication in this group which is approved for men with BPH with or without ED; thus it represents good option in obese men who present with both. Bioavailability of tadalafil is not affected by food; hence, it may be a better option in general for men with obesity. However, no head-to-head study compared differences in effectiveness between three available 5-PDEs in men with delayed stomach emptying or morbid obesity.

Daily tadalafil has been now shown to improve peak systolic velocity in diabetic patients with vascular disease as compared to on-demand tadalafil [61]. Considering that daily treatment with tadalafil has recently been shown to improve endothelial function, which is impaired in men with obesity-related diseases, it seems prudent to initiate therapy with daily 5 mg tadalafil or fixed dose of 20 mg of tadalafil three times a week [62].

Avanafil (Stendra) was approved in 2012 in the United States and is considered to have unique properties of quick on quick off action. Avanafil may be taken 15 minutes before sexual activities with or without food. Avanafil comes in 50, 100, and 200 mg tablets. Recommended initial dose is 100 mg once daily as needed.

Patients with left ventricular outflow obstruction (aortic stenosis, idiopathic hypertrophic subaortic stenosis) and autonomic blood pressure dysregulation may be at risk for vasodilating effect of avanafil.

In healthy volunteers, avanafil 200 mg resulted in modest decrease in blood pressure—8 mmHg systolic and 3.3 mmHg diastolic.

All 5-PDE inhibitors are contraindicated in patients taking GC stimulators such as riociguat. CYP3A4 inhibitors taken together with avanafil may increase its level and result in significant drop in blood pressure. If therapeutically indicated, one should start at lowest dose, 50 mg. Avanafil should not be used in patients taking ritonavir because of significant increase in Cmax of avanafil. Alpha-blocker therapy together with avanafil needs to be monitored initially as combination therapy increases risks of hypotensive episodes. Headache occurred in 10.5% of patients taking 200 mg of Stendra, flushing in 4.0%, nasal congestion in 2%, and back pain in 1.1%. Studies in animals showed decreased fertility and abnormal sperm motility and morphology in animals exposed to avanafil. However, study of 181 healthy volunteers taking 100 mg of avanafil showed no effect on semen parameters. Avanafil is quickly absorbed with median Tmax of 30–45 minutes. Most of avanafil is eliminated within 8 hours from intake. Avanafil has similar effectiveness in improving erectile function as other 5-PDEs. Overall, vaginal penetration was reported by 77.3% patients on 200 mg of avanafil compared to 53.8% on placebo. Successful intercourse was reported by 57% of men on 200 mg of avanafil as compared to 27% of men taking placebo. However, for men with DM, 63% were able to penetrate vaginally (42% placebo) and 40% had successful intercourse (20.5% placebo).

All 5-PDEs share similar contraindications described earlier.

Penile intracavernosal injection therapy

Subjects who fail oral therapy may benefit from *penile IC injection therapy* [63]. Self-administered penile injection therapy is well tolerated and may be more successful in diabetic men with severe ED as compared to oral agents [64]. IC therapy can be started with premixed Caverject (prostaglandin E1). PGE1 relaxes smooth muscles in penis through EP receptor [65]. *Caverject* comes in prefilled vials of 5, 10, 20, and 40 µg to be used with self-injector or single-use syringe, which can deliver from 6 to 20 µg of medication. Side effects of Caverject are penile pain which can occur in 37% of patients, penile fibrosis in 3%, prolonged erection in 4%, and injection site hematoma in 3%. Priapism has been reported in 0.4% subjects according to FDA label. Over 72% of men with diabetes responded to Caverject. Many practices, including ours, use multidrug combination of vasoactive agents for IC injections, which may decrease pain and improve efficacy as compared to prostaglandin E1 [66]. Most commonly used preparations are *Trimix* (papaverine 30 mg, phentolamine 1 mg, and prostaglandin E1 10 µg/1 mL) and *super-Trimix* (papaverine 30 mg, phentolamine 2 mg, and prostaglandin E1 20 µg/1 mL). Typically, we use 0.05–0.1 mL of Trimix to start. It is critical to remember that all forms of IC therapy have to be initiated and dose titrated in physician office prior to prescribing an adequate dose.

Penile prosthesis

Penile prosthesis can be considered in men who failed injection therapy; however, it is more important for obese men that they should first consider weight loss and aggressive treatment of hypercholesterolemia and DM prior to penile prosthesis as once placed, it has to remain in the penis to prevent scarring and dramatic penis shortening. Obesity is associated with decreased satisfaction with penile prosthesis [67]. Poorly controlled diabetes is associated with increased risk of penile prosthesis infections [68]. Long-term follow-up studies showed that at 5 years, 1:10 penile prosthesis has to be replaced [69].

Testosterone replacement therapy

It is well established that obese men suffer from hypogonadism but it is less clear if testosterone replacement therapy (TRT) should be considered in all obese men [70]. A significant amount of data exists which supports combining weight loss strategy with the use of TRT [71]. TRT may decrease BMI and improve metabolic markers of cardiovascular risks [72,73].

TRT may be administered using injectable forms of therapy—such as testosterone enanthate or cypionate, testosterone pellets, or topical forms of replacement. Injectable testosterone is typically administered as intramuscular injection at 2 weeks intervals at 200–300 mg of depot-testosterone. Topical preparations can be applied to underarm (Axrion), shoulders (Androgel and Testim), and inner tight (Fortesta).

Goal of therapy is to achieve normal testosterone levels. Based on recent large data analysis of patients enrolled in MrOS study in Sweden, the cardioprotective effect of testosterone was observed if levels were above 550 ng/dL; thus we typically set goal of therapy in upper normal range [54]. We use exclusively topical preparations because of significantly less risk of polycythemia as compared to injectable agents.

TRT may have positive metabolic effect in obese men but also improve their sexual drive, ejaculatory volume, and orgasmic function and improve patients' quality of life. Further studies are needed to prove the positive long-term benefits of TRT in obese men. In men who failed to improve TT despite use of adequate dose of TRT, aromatase inhibitors such as anastrozole 1 mg daily may be used for set amount of time (6–12 months) to overcome excessive conversion of testosterone into E2 seen in many obese men [74,75].

Prevention

A longitudinal study of close to 4 million patients in the United Kingdom showed that obesity is diagnosed at young people, 20–29 years, and ED diagnosis increases dramatically in men, 40–59 years. Thus there seems to be a window of opportunity early on to manage obesity in young men. Perhaps, disseminating information that obesity today may lead to loss of erectile function in future may be an impetus for younger men to lose weight.

In summary, sexual dysfunction is common in obese men. The risk of ED is related to obesity-related comorbidities such as hypogonadism, peripheral vascular disease, hypercholesterolemia, and diabetes rather than simple increase in BMI. Reduction in BMI combined with successful treatment of sexual dysfunction improves the quality of life and has a positive motivator effect on compliance with control of comorbidities. Further placebo control studies are needed to establish a role of multimodal therapy in obese men with sexual dysfunction.

References

[1] Flegal KM, et al. Prevalence and trends in obesity among US adults, 1999–2008. JAMA 2010;303(3):235–41.

[2] Daniels JS. Abnormal nerve conduction in impotent patients with diabetes mellitus. Diabetes Care 1989;12(7):449–54.

[3] Villalba N, et al. Differential structural and functional changes in penile and coronary arteries from obese Zucker rats. Am J Physiol Heart Circ Physiol 2009;297(2):H696–707.

[4] Corona G, et al. Is obesity a further cardiovascular risk factor in patients with erectile dysfunction? J Sex Med 2010;7(7):2538–46.

[5] Toque HA, et al. High-fat diet associated with obesity induces impairment of mouse corpus cavernosum responses. BJU Int 2011;107(10):1628–34.

[6] Albersen M, et al. Functional, metabolic, and morphologic characteristics of a novel rat model of type 2 diabetes-associated erectile dysfunction. Urology 2011;78(2):476.e1–8.

[7] Contreras C, et al. Insulin resistance in penile arteries from a rat model of metabolic syndrome. Br J Pharmacol 2010;161(2):350–64.

[8] Du XL, et al. Hyperglycemia inhibits endothelial nitric oxide synthase activity by posttranslational modification at the Akt site. J Clin Invest 2001;108(9):1341–8.

[9] De Angelis L, et al. Erectile and endothelial dysfunction in Type II diabetes: a possible link. Diabetologia 2001;44(9):1155–60.

[10] Akingba AG, Burnett AL. Endothelial nitric oxide synthase protein expression, localization, and activity in the penis of the alloxan-induced diabetic rat. Mol Urol 2001;5(4):189–97.

[11] Hofstra J, et al. High prevalence of hypogonadotropic hypogonadism in men referred for obesity treatment. Neth J Med 2008;66(3):103–9.

[12] Kalucy RS, Crisp AH. Some psychological and social implications of massive obesity. A study of some psychosocial accompaniments of major fat loss occurring without dietary restriction in massively obese patients. J Psychosom Res 1974;18(6):465–73.

[13] Chao JK, et al. A survey of obesity and erectile dysfunction of men conscripted into the military in Taiwan. J Sex Med 2011;8(4):1156–63.

[14] Hammoud A, et al. Effect of Roux-en-Y gastric bypass surgery on the sex steroids and quality of life in obese men. J Clin Endocrinol Metab 2009;94(4):1329–32.

[15] Cohen PG. Obesity in men: the hypogonadal-estrogen receptor relationship and its effect on glucose homeostasis. Med Hypotheses 2008;70(2):358–60.

[16] Wake DJ, et al. Intra-adipose sex steroid metabolism and body fat distribution in idiopathic human obesity. Clin Endocrinol (Oxf) 2007;66(3):440–6.

[17] Ostbye T, et al. Sexual functioning in obese adults enrolling in a weight loss study. J Sex Marital Ther 2011;37(3):224–35.

[18] Bond DS, et al. Prevalence and degree of sexual dysfunction in a sample of women seeking bariatric surgery. Surg Obes Relat Dis 2009;5(6):698–704.

[19] Kannel WB, Castelli WP, Gordon T. Cholesterol in the prediction of atherosclerotic disease. New perspectives based on the Framingham study. Ann Intern Med 1979;90(1):85–91.

[20] Feldman HA, et al. Impotence and its medical and psychosocial correlates: results of the Massachusetts Male Aging Study. J Urol 1994;151(1):54–61.

[21] Derby CA, et al. Modifiable risk factors and erectile dysfunction: can lifestyle changes modify risk? Urology 2000;56(2):302–6.

[22] Kim SC, Kim SW, Chung YJ. Men's health in South Korea. Asian J Androl 2011;13(4):519–25.

[23] Saltzman EA, Guay AT, Jacobson J. Improvement in erectile function in men with organic erectile dysfunction by correction of elevated cholesterol levels: a clinical observation. J Urol 2004;172(1):255–8.

[24] Herrmann HC, et al. Can atorvastatin improve the response to sildenafil in men with erectile dysfunction not initially responsive to sildenafil? Hypothesis and pilot trial results. J Sex Med 2006;3(2):303–8.

[25] Filippi S, et al. Testosterone partially ameliorates metabolic profile and erectile responsiveness to PDE5 inhibitors in an animal model of male metabolic syndrome. J Sex Med 2009;6(12):3274–88.

[26] Gokce MI, et al. Effect of atorvastatin on erectile functions in comparison with regular tadalafil use. A prospective single-blind study. Int Urol Nephrol 2012;44.

[27] Mastalir ET, Carvalhal GF, Portal VL. The effect of simvastatin in penile erection: a randomized, double-blind, placebo-controlled clinical trial (Simvastatin treatment for erectile dysfunction-STED TRIAL). Int J Impot Res 2011;23(6):242–8.

[28] Abdel Aziz MT, et al. Effects of losartan, HO-1 inducers or HO-1 inhibitors on erectile signaling in diabetic rats. J Sex Med 2009;6(12):3254–64.

[29] La Vignera S, et al. Statins and erectile dysfunction: a critical summary of current evidence. J Androl 2011;33.

[30] Long-Term Intervention with Pravastatin in Ischaemic Disease (LIPID) Study Group. Prevention of cardiovascular events and death with pravastatin in patients with coronary heart disease and a broad range of initial cholesterol levels. N Engl J Med 1998;339(19):1349–57.

[31] Lewis SJ, et al. Effect of pravastatin on cardiovascular events in older patients with myocardial infarction and cholesterol levels in the average range. Results of the Cholesterol and Recurrent Events (CARE) trial. Ann Intern Med 1998;129(9):681–9.

[32] Stanworth RD, et al. Statin therapy is associated with lower total but not bioavailable or free testosterone in men with type 2 diabetes. Diabetes Care 2009;32(4):541–6.

[33] Dobs AS, et al. Effects of high-dose simvastatin on adrenal and gonadal steroidogenesis in men with hypercholesterolemia. Metabolism 2000;49(9):1234–8.

[34] Corona G, et al. The effect of statin therapy on testosterone levels in subjects consulting for erectile dysfunction. J Sex Med 2010;7(4 Pt 1):1547–56.

[35] Kloner R. Erectile dysfunction and hypertension. Int J Impot Res 2007;19(3):296–302.

[36] Mittawae B, et al. Incidence of erectile dysfunction in 800 hypertensive patients: a multicenter Egyptian national study. Urology 2006;67(3):575–8.

[37] Prisant LM, Loebl Jr. DH, Waller JL. Arterial elasticity and erectile dysfunction in hypertensive men. J Clin Hypertens (Greenwich) 2006;8(11):768–74.

[38] Ushiyama M, et al. Erectile dysfunction in hypertensive rats results from impairment of the relaxation evoked by neurogenic carbon monoxide and nitric oxide. Hypertens Res 2004;27(4):253–61.

[39] Bener A, et al. Prevalence of erectile dysfunction among hypertensive and nonhypertensive Qatari men. Medicina (Kaunas) 2007;43(11):870–8.

[40] Karavitakis M, et al. Evaluation of sexual function in hypertensive men receiving treatment: a review of current guidelines recommendation. J Sex Med 2011;8(9):2405–14.

[41] Engbaek M, et al. The effect of low-dose spironolactone on resistant hypertension. J Am Soc Hypertens 2010;4(6):290–4.

[42] Shiri R, et al. Cardiovascular drug use and the incidence of erectile dysfunction. Int J Impot Res 2007;19(2):208–12.

[43] Scranton RE, et al. Effect of treating erectile dysfunction on management of systolic hypertension. Am J Cardiol 2007;100(3):459–63.

[44] Blumentals WA, et al. Is erectile dysfunction predictive of peripheral vascular disease? Aging Male 2003;6(4):217–21.

[45] Chung SD, et al. Increased risk of stroke among men with erectile dysfunction: a nationwide population-based study. J Sex Med 2011;8(1):240–6.

[46] Polonsky TS, et al. The association between erectile dysfunction and peripheral arterial disease as determined by screening ankle-brachial index testing. Atherosclerosis 2009;207(2):440–4.

[47] Ylitalo KR, Sowers M, Heeringa S. Peripheral vascular disease and peripheral neuropathy in individuals with cardiometabolic clustering and obesity: National Health and Nutrition Examination Survey 2001–2004. Diabetes Care 2011;34(7):1642–7.

[48] Fukuhara S, et al. Vardenafil and resveratrol synergistically enhance the nitric oxide/cyclic guanosine monophosphate pathway in corpus cavernosal smooth muscle cells and its therapeutic potential for erectile dysfunction in the streptozotocin-induced diabetic rat: preliminary findings. J Sex Med 2011;8(4):1061–71.

[49] Jackson G, et al. Erectile dysfunction and coronary artery disease prediction: evidence-based guidance and consensus. Int J Clin Pract 2010;64(7):848–57.

[50] Awad H, et al. Erectile function in men with diabetes type 2: correlation with glycemic control. Int J Impot Res 2010;22(1):36–9.

[51] Mayo Clinic womens Healthsource. Too much, too little sleep associated with adult weight gain. Mayo clin womens healthsource 2008;12(10):3. Available from: https://www.ncbi.nih.gov/pubmed/18772835.

[52] Amano T, et al. The usefulness of vibration perception threshold as a significant indicator for erectile dysfunction in patients with diabetes mellitus at a primary diabetes mellitus clinic. Urol Int 2011;87(3):336–40.

[53] Fukui M, et al. Five-item version of the international index of erectile function correlated with albuminuria and subclinical atherosclerosis in men with type 2 diabetes. J Atheroscler Thromb 2011;18(11):991–7.

[54] Ohlsson C, et al. High serum testosterone is associated with reduced risk of cardiovascular events in elderly men. The MrOS (Osteoporotic Fractures in Men) study in Sweden. J Am Coll Cardiol 2011;58(16):1674–81.

[55] Wessells H, et al. Effect of intensive glycemic therapy on erectile function in men with type 1 diabetes. J Urol 2011;185(5):1828–34.

[56] Rhoden EL, et al. Glycosylated haemoglobin levels and the severity of erectile function in diabetic men. BJU Int 2005;95(4):615–17.

[57] Giugliano F, et al. Erectile dysfunction associates with endothelial dysfunction and raised proinflammatory cytokine levels in obese men. J Endocrinol Invest 2004;27(7):665–9.

[58] Giugliano F, et al. Adherence to Mediterranean diet and erectile dysfunction in men with type 2 diabetes. J Sex Med 2010;7(5):1911–17.

[59] Khoo J, et al. Comparing effects of a low-energy diet and a high-protein low-fat diet on sexual and endothelial function, urinary tract symptoms, and inflammation in obese diabetic men. J Sex Med 2011;8(10):2868–75.

[60] Lamina S, Okoye CG, Dagogo TT. Managing erectile dysfunction in hypertension: the effects of a continuous training programme on biomarker of inflammation. BJU Int 2009;103(9):1218–21.

[61] La Vignera S, et al. Tadalafil and modifications in peak systolic velocity (Doppler spectrum dynamic analysis) in the cavernosal arteries of patients with type 2 diabetes after continuous tadalafil treatment. Minerva Endocrinol 2006;31(4):251–61.

[62] Aversa A, et al. Relationship between chronic tadalafil administration and improvement of endothelial function in men with erectile dysfunction: a pilot study. Int J Impot Res 2007;19(2):200–7.

[63] Hatzimouratidis K, et al. Guidelines on male sexual dysfunction: erectile dysfunction and premature ejaculation. Eur Urol 2010;57(5):804–14.

[64] Perimenis P, et al. Switching from long-term treatment with self-injections to oral sildenafil in diabetic patients with severe erectile dysfunction. Eur Urol 2002;41(4):387–91.

[65] Angulo J, et al. Regulation of human penile smooth muscle tone by prostanoid receptors. Br J Pharmacol 2002;136(1):23–30.

[66] Montorsi F, et al. Clinical reliability of multi-drug intracavernous vasoactive pharmacotherapy for diabetic impotence. Acta Diabetol 1994;31(1):1–5.

[67] Akin-Olugbade O, et al. Determinants of patient satisfaction following penile prosthesis surgery. J Sex Med 2006;3(4):743–8.

[68] Selph JP, Carson 3rd CC. Penile prosthesis infection: approaches to prevention and treatment. Urol Clin North Am 2011;38(2):227–35.

[69] Chung E, et al. Penile prosthesis implantation for the treatment for male erectile dysfunction: clinical outcomes and lessons learnt after 955 procedures. World J Urol 2012;.

[70] Drewa T, Olszewska-Slonina D, Chlosta P. Testosterone replacement therapy in obese males. Acta Pol Pharm 2011;68(5):623–7.

[71] Jones TH. Effects of testosterone on type 2 diabetes and components of the metabolic syndrome. J Diabetes 2010;2(3):146–56.

[72] Corona G, et al. Testosterone and metabolic syndrome: a meta-analysis study. J Sex Med 2011;8(1):272–83.

[73] Jones TH, Saad F. The effects of testosterone on risk factors for, and the mediators of, the atherosclerotic process. Atherosclerosis 2009;207(2):318–27.

[74] Cohen PG. The hypogonadal-obesity cycle: role of aromatase in modulating the testosterone-estradiol shunt—a major factor in the genesis of morbid obesity. Med Hypotheses 1999;52(1):49–51.

[75] Simpson ER, Mendelson CR. Effect of aging and obesity on aromatase activity of human adipose cells. Am J Clin Nutr 1987;45(1 Suppl.):290–5.

Chapter 13

Male obesity—impact on semen quality

Vanessa Kay[1] and Sarah Martins da Silva[2]

[1]Assisted Conception Unit, Ninewells Hospital, Dundee, United Kingdom, [2]Reproductive and Developmental Biology, School of Medicine, Ninewells Hospital and Medical School, University of Dundee, Dundee, United Kingdom

Introduction

Male infertility is the commonest cause of infertility, underlying or contributing to at least 40% of all cases of infertility [1]. The treatment of male infertility almost exclusively comprises intracytoplasmic sperm injection (ICSI) that now represents nearly half of all in vitro fertilization (IVF) cycles in the United Kingdom [1]. While ICSI is a highly successful treatment, it bypasses the underlying reason(s) for male infertility, and in many cases the etiology is not identified. However, it should be remembered that ICSI treatment has medical risks and is expensive, and long-term safety data is relatively scarce [2,3]. Therefore it is important to identify modifiable risk factors for male fertility, to enable both preventative and curative treatments for this condition.

The incidence of obesity is rapidly rising in almost every region of the world. The World Health Organization (WHO) defines obesity as a body mass index (BMI) over 30 kg/m^2, while overweight is defined as a BMI over 25 kg/m^2. In Europe the International Obesity Task Force has indicated that obesity rates in adult men range from 10% to 27% [4]. Notably, Scotland has one of the highest rates of obesity among developed countries. Recent data has shown that Scottish men are significantly more likely to be overweight or obese (68%) in comparison to women (61%), with female prevalence remaining static since 2008 [5].

The adverse influence of obesity on various aspects of female reproduction and fertility has been acknowledged for some time [6], and management guidelines are now available [7]. More recently, data regarding male obesity and infertility has been accumulating [8]. There are now several population-based studies showing that overweight and obese men have up to 50% higher rate of subfertility when compared with normal-weight men [9–12]. This risk is particularly high if the female partner is also overweight or obese. One could argue that the association between male obesity and infertility could be due to confounding factors such as male age, diet, smoking and alcohol use, and female partner obesity. However, once these factors have been accounted for, couples were still 10% more likely to be infertile for every three-point increase in the man's BMI [13].

Raised male BMI has also been shown to adversely affect infertility treatment outcomes. A recent systematic review and metaanalysis of 14,372 treatment cycles showed that elevated BMI in the male partner was associated with a significant reduction in both clinical pregnancy rate and live birth rate following IVF or ICSI treatment [14]. However, this finding conflicts with an earlier metaanalysis, including 5262 men, which concluded that raised male BMI did not affect outcomes of assisted reproduction [15].

Impact on semen quality

Semen quality is accepted as a surrogate marker for male fertility, with established reference ranges [16]. Although the majority of obese men have normal semen parameters, a correlation between poor semen quality and obesity has been recognized for some time [17], with recent large-scale data (4440 men) confirming statistically significant relationships between obesity and semen analysis characteristics [12].

Sperm concentration and count

In a cross-sectional study of 1558 military recruits undergoing compulsory physical examination, men with normal BMI were compared with those with BMI over 25 kg/m^2. The sperm concentration and total sperm count per ejaculate were reduced by 22% and 24%, respectively, in men with a BMI of over 25 kg/m^2, with no change in semen volume. In addition, an associated decrease in testosterone, follicle-stimulating hormone (FSH), inhibin B, and sex hormone—binding globulin (SHBG) was observed in the overweight group [18]. A WHO surveillance study on male partners of pregnant women confirmed that obese men had significantly lower total sperm count than nonobese men (mean 231×10^6 vs 324×10^6, respectively), although other sperm parameters were not shown to be affected [19]. This finding was supported in a review by Mah and Wittert [20] who concluded that obese men have reduced sperm concentration and total sperm count, compared to lean men, but with motility and morphology unaffected.

Clearly, a simplistic link between fertility and BMI underestimates the complexities of male reproduction. Reported effects of BMI on sperm count are also variable, with analysis of a database of 2139 men finding only a slightly lower total sperm count among overweight men and a nonsignificant change in obese men (mean total sperm count in BMI 20–25: 231×10^6, BMI 25–30: 216×10^6, and BMI > 30: 265×10^6) [21]. A further study of men attending an infertility clinic showed that despite major changes in reproductive hormones, only extreme obesity (BMI > 35 kg/m^2) correlated to lower total sperm counts [22]. Another study [23] showed that this effect was only observed in men with BMI > 30 kg/m^2. Conversely, Qin et al. found a correlation between BMI and sperm count, but this was only significant in men with low BMI [24]. In another study of 349 men, no correlation was found between BMI and total sperm count, although few men with extreme BMIs were included in this study [25]. Magnusdottir et al. observed an association between BMI and sperm parameters but only in subfertile men and not in fertile men [26].

Sperm motility

An observational study of 520 men showed a significantly reduced number of normal motile sperm (normal BMI 18.6×10^6, overweight 3.6×10^6, and obese 0.7×10^6) and an increase in sperm DNA fragmentation in overweight and obese men [27]. The findings of this study have been confirmed in a retrospective study of 526 infertile men in which the incidence of oligozoospermia and low progressive motile sperm concentration was higher in overweight and obese men compared with those of normal weight [28] and later confirmed by Sekhavat and Moein who studied 852 healthy men [29]. Multivariate analysis also confirms a negative association between BMI and motility, progressive motility, and neutral alpha-glucosidase levels (considered an epididymal functional marker) [30]. Other large population-based studies on humans have shown no such association [18–21]; however, convincing evidence linking male obesity and sperm motility has been provided from animal studies, using a model of obesity created by feeding rodents a high-fat diet (HFD) for a number of weeks. First, a study demonstrated that the percentage of motile sperm was significantly reduced from 44% to 36% in mice fed on an HFD [31]. A similar finding was then shown in rats, where those on an HFD had a decrease in the percentage of sperm with progressive movement [32].

Sperm morphology

The WHO criteria for male fertility assessment recommend $\geq 4\%$ normal morphology ($\geq$ 5th centile) in observed sperm [33]. Morphology evaluation is somewhat controversial, not least because it is influenced by the subjectiveness of the observer and lacks objective measurement. Perhaps, unsurprisingly it is this area of diagnostic semen analysis where studies have found quite conflicting effects of obesity. While some studies suggest adverse effects of elevated BMI on sperm morphology [12,28–30,34,35], others show no such association [29,36–38]. The percentage of normal sperm morphology was adversely affected by either high or low BMIs in a study of 1558 military conscripts, although this did not reach statistical significance [18]. In a smaller study comparing obese fertile and obese infertile men, a significant positive correlation between BMI and abnormal sperm morphology was observed, with mean sperm abnormal forms being 21.4% in obese fertile men compared to 35.5% in obese infertile men [34]. A more recent study has contradicted this finding, with no relationship between self-reported BMI and sperm morphology, despite showing a significant inverse correlation between BMI and sperm count and motility [29].

Combined semen parameters

A few studies have looked at combined measures of semen parameters. In these studies, elevated BMI was shown to be related to poorer sperm quality, for example, an association with low progressively motile sperm count [28] and normal motile sperm count [18].

Sperm DNA damage

It is possible that the traditional semen parameters described previously are not the most appropriate characteristics to assess the effects of obesity on sperm function. Recently, there has been some enthusiasm for the assessment of sperm DNA damage, using a variety of different assays. The integrity of sperm DNA is associated with fertility [39,40]. However, fundamental questions remain regarding the nature of sperm DNA damage and the lack of standardization of clinical assays to assess this [41].

Kort et al. demonstrated a significant negative relationship between BMI and total number of motile sperm cells. In this study, obese men had a significantly higher percentage of sperm with DNA damage, assessed by sperm chromatin structure assay (SCSA), when compared with normal-weight men (DNA fragmentation index 19.9%, 25.8%, and 27.0% in normal, overweight and obese men, respectively) [27]. This was confirmed by Chavarro et al. who identified significantly more sperm with high DNA damage in obese men compared with men of normal weight (adjusted mean difference from normal BMI of cells with high DNA damage 4, 5, 7 in normal, overweight, and obese men, respectively), although there was no difference with three other standard measures of sperm DNA integrity [22]. Of note, these two studies measured different aspects of DNA integrity, and therefore data should be interpreted with caution (SCSA measures susceptibility of sperm chromatin to DNA denaturation, and the comet assay measures the extent of sperm DNA fragmentation in individual sperm). A study by Tunc et al. attempted to correlate BMI with seminal reactive oxygen species (ROS) production (nitroblue tetrazolium assay), sperm DNA damage (TUNEL), markers of semen inflammation (CD45, seminal plasma polymorphonuclear elastase and neopterin concentration), and routine sperm parameters, as well as reproductive hormones [35]. The study confirmed an increase in oxidative stress with increase in BMI, primarily attributed to increase in seminal macrophage activation. However, the magnitude was small and deemed to be of only minor clinical significance as there was no associated decline in sperm DNA integrity or sperm motility. Increased BMI was, however, linked to decreased sperm concentration, as well as lower serum testosterone and increased serum estrogen. Mice fed on HFD for 9 weeks show elevated intracellular ROS (692 ± 83 vs 409 ± 22; $P < .01$), as well as increase in sperm DNA damage ($1.64\% \pm 0.6\%$ vs $0.17\% \pm 0.06\%$; $P < .01$). HFD mice also had a significantly lower percentage of noncapacitated sperm that resulted in lower fertilization rates (25.9% vs 43.9%; $P < .01$) [31]. However, data is conflicting. A further study examining basic semen parameters, chromatin integrity, and chromatin condensation concluded no proof of impact of BMI [42]. Conversely, a 3-year multicenter study assessing the impact of BMI on sperm DNA integrity in 330 men observed an increased rate of sperm DNA damage in obese men (odds ratio 2.5; 95% CI 1.2–5.1) [43].

It is important to note that most obese men have been shown to have normal semen quality and fertility. In a small study of 87 men, no association between BMI and semen parameters was identified, although correlations with several endocrine markers were observed [44]. A further study observed no correlation between self-reported BMI and sperm concentration, motility or morphology, or pregnancy rates with IVF [45]. There are various possible reasons to explain these discrepancies, including the bias from confounding lifestyle factors, insufficient sample size, the presence of genetic causes of male infertility, appropriateness of the sperm parameter assessed, and whether the effect is only observed at extremes of BMI. For example, in a study on 483 infertile men, only the most obese men (BMI > 35 kg/m^2) had a lower sperm count when compared with normal-weight men [22]. A systematic review with metaanalysis of 6800 men concluded that there was no strong evidence for a relationship between BMI and semen parameters; however, it was accepted that their data analysis was restricted to only five studies that had comparable outcome measures, with many studies excluded from the metaanalysis even though their evidence may have been useful [46]. Another limitation recognized in this metaanalysis was that BMI is not the most accurate measure of an individual's body fat, but rather waist circumference and/or bioimpedance. The relationship between sperm quality and obesity is still not clearly understood and is complex with various factors involved. It is interesting to explore evidence from animal studies, which avoids the issues of correlational data and confounding factors present in many human studies. A metaanalysis of 52 animal studies showed that HFDs reduced testicular and sex accessory gland size, semen quality, mating activity and fertilization success [47]. This suggests that the negative effects on human male infertility with obesity are likely to be a true effect.

Etiological theories

It has not yet been clearly established how excess weight relates to the biological changes that underlie male infertility, although there are several theories worth exploring.

Endocrine theory

The endocrine abnormalities associated with obesity in women are well known, with an increase androgen metabolism and elevated estrogen levels [48,49]. Similarly, obese men are known to have relative hyperestrogenic hypogonadotropic hypogonadism. Aggerholm et al. [21] showed that serum testosterone concentrations were 25%–32% lower in obese men than in normal-weight men, whereas the estrogen concentration was 6% higher in obese men. An association between male obesity and decreased free and total testosterone and SHBG have all been well documented, with possible mechanisms, including (1) decreased luteinizing hormone (LH), (2) central inhibitory effects of estrogen, and (3) effects of leptin and other peptides both centrally and on Leydig cells (for review see Mah and Wittert [20]). Another hormone that may be of relevance is inhibin that is secreted from the Sertoli cells and suppresses FSH production by the pituitary gland. Severely obese men have reduced inhibin B levels that may be important due to effects on spermatogenesis [50].

The reason why obesity causes hypoandrogenism is thought to be multifactorial. Circulating levels of estrogens are increased due to aromatization of testicular and adrenal androgens in adipose tissue. Indeed, when the aromatase inhibitor letrozole was administered to obese men, testosterone levels increased and serum estradiol levels decreased [51]. High estrogen levels cause inappropriate suppression of the hypothalamic—pituitary—gonadal axis, resulting in decreased testosterone production. It is also possible that the elevated estrogen levels have a direct adverse influence on spermatogenesis, although the exact nature of this is as yet undetermined.

Another factor to consider is the direct secretion of hormones from adipose cells. Leptin is such a hormone, with plasma concentrations rising in parallel with fat reserves. It is primarily responsible for satiety, but it has also been shown to influence male reproduction, both locally and the hypothalamic—pituitary—gonadal axis. Leptin is present in semen [52], and leptin receptors have been demonstrated on sperm plasma membranes [53], suggesting a direct endocrine effect on sperm. But what is really interesting is that human spermatozoa have also been shown to produce leptin [54], with the suggestion that leptin can act as an autoregulator in spermatozoa by managing the energy status of the cell. Spermatozoa also have insulin receptors and secrete insulin suggesting that insulin may act as a metabolic regulator in the spermatozoa. While these results require confirmation, they suggest that the effect of any changes in insulin and leptin in overweight men may act directly on the function of the spermatozoa following ejaculation, and that investigating semen analysis may be insufficient to detect a subtle effect on fertility. At a hypothalamic—pituitary—gonadal level, high levels of leptin are known to decrease basal and LH-stimulated androgen level, with obese men demonstrating a 30%—40% reduced testosterone level that is inversely correlated with leptin levels [52]. More recent studies have confirmed this link between leptin to obesity and male infertility, with obese oligozoospermic men having higher plasma leptin levels than obese fertile controls [55].

Insulin resistance is known to be associated with obesity and is negatively correlated with testosterone levels [56]. Interestingly, in a large metaanalysis of 742 publications, men with type 2 diabetes had a lower level of testosterone compared to controls [57]. However, conventional semen parameters (concentration, motility, and morphology) in men with type 2 diabetes do not differ significantly from those unaffected, although these may have a significantly higher level of DNA fragmentation [58]. This DNA damage impairs sperm function and adversely affects male fertility [59]. Insulin has also been demonstrated to influence the level of SHBG, with several studies showing that SHBG and total testosterone are inversely correlated with BMI and insulin levels [56,58,60,61].

Another mechanism for these endocrine changes may relate to sleep apnea that is more common in those who are obese. Apneic episodes may lead to regularly interrupted sleep that can have a big impact on quality of life and increases the risk of developing certain conditions. Indeed, sleep is increasingly recognized to influence a growing array of physiological processes. Testosterone secretion follows a diurnal pattern, with rise in testosterone coinciding with REM sleep rather than changes in melatonin [62]. Decrease in nocturnal rise of testosterone associated with sleep apnea results in lower morning testosterone levels, which can be reversed by weight loss [63]. However, whether this modest decrease in testosterone level is responsible for the suppression of spermatogenesis remains debatable, given that the endocrine control of spermatogenesis is usually maintained even in morbidly obese men. Conversely, it has been suggested that defective spermatogenesis causes obesity, rather than obesity causing impaired testicular function. This is supported by the observation that men administered therapies to reduce testosterone, for example, treatment for prostatic cancer, tend to become obese [64,65].

Genetic theory

Could a genetic abnormality independently cause both obesity and male infertility? There are a number of rare genetic conditions in which these conditions occur together, such as Laurence—Moon and Prader—Willi syndromes. More specifically, the gene encoding for leptin, which is synthesized in adipose tissue, has been shown to be associated with obesity in both humans and mice, with leptin administration in leptin-deficient mice improving male reproductive function [66]. There has been a growing understanding about the role epigenetics may play in paternal obesity, with evidence of reprograming of spermatogonial stem cells in obese men, resulting in changes in the metabolic and reproductive phenotypes of children [67]. This has led to concerns regarding potential long-term metabolic problems in future generations [68].

Sexual dysfunction theory

Another hypothesis to explain why obese men are less fertile relates to reduced coital frequency. Certainly, it has been demonstrated that overweight and obese men have fewer sexual partners than normal-weight men [69]. However, when men with reduced coital frequency are excluded from analysis, overweight and obese men still have an increased incidence of subfertility [11]. This indicates that reduced coital frequency is not necessarily the cause of infertility in obese men.

Obese men have been observed to have a higher incidence of erectile dysfunction (penile rigidity grade 1.32 vs

1.62 in nonobese men) [70]. However, when cardiovascular risk factors (such as heart disease and diabetes) were excluded, there was no difference in erectile dysfunction between obese and nonobese men, suggesting that vascular impairment rather than obesity is responsible.

Testicular hyperthermia theory

Excess fat in the inner thighs and suprapubic region may result in testicular hyperthermia (over 35°C). Certainly, it is known that a rise in testicular temperature of a few degrees is sufficient to hinder sperm production in rats [71], and that occupational heat exposure has been shown to reduce sperm output and quality in men [72]. Moreover, testicular hyperthermia related to tight-fitting underwear has been shown to adversely affect semen parameters with possible impact on fertility [73]. There are limited studies specifically examining testicular hyperthermia and obesity, although hyperthermia is increasingly recognized to detrimentally affect spermatogenesis [74]. In men presenting with excess suprapubic fat due to scrotal lipomatosis, scrotal lipectomy can result in improved semen analysis [75].

Reactive oxygen species theory

Obesity results in an increase in oxidative stress, resulting from an imbalance between free radicals, ROS, and antioxidants. This may be a major contributor to many of the underlying comorbidities of obesity, including male infertility [76]. There are two main mechanisms by which ROS affect sperm function: DNA damage resulting in defective paternal DNA being passed onto children and sperm membrane damage resulting in decreased motility and ability to fuse with the oocyte [77]. The recent Cochrane systematic review "Antioxidants for male subfertility" suggested that infertile men may benefit from taking an oral antioxidant to increase the chance of conceiving. However, whether this is of benefit to obese men has not been specifically evaluated [78].

Treatment

Less is known about the treatment of infertility in obese men than in obese women. In obese men, loss of weight in a 4-month program has been shown to improve hormonal profile, with an increase in SHBG and testosterone and a decrease in insulin and leptin [79], as well as free testosterone [80]. There is also evidence that reduction in caloric intake corrects erectile dysfunction in men with metabolic syndrome especially when combined with physical exercise [81,82].

Despite the lack of evidence, if the changes by which obesity causes infertility are reversible, then weight loss is likely to be an effective intervention. However, it is first important to identify that a man is overweight. Unfortunately, individuals increasingly fail to recognize that they are obese, and this is more so in men. A British survey showed that 53% of the population was overweight or obese, yet only 67% obese men recognized themselves to be overweight or obese [83]. Therefore it is important that height and weight are measured and BMI calculated, for all men attending for advice on infertility. Overweight and obese men should be informed regarding the association between increased weight and infertility and advised to lose weight, aiming for a BMI of below 25 kg/m^2 before starting fertility treatment.

The primary proven method to achieve weight loss is by reducing calorie intake and increasing physical activity. Weight loss can be improved by providing dietary advice, psychological support, exercise classes, and weight-reducing agents [84]. However, sustained weight loss in patients is very difficult to achieve. As such, considerable support and active monitoring are needed for men (and women) attempting to achieve a significant reduction in BMI. In morbid obesity, bariatric surgery, such as gastric banding or bypass, may be considered. In a systematic review, all current bariatric operations lead to major weight loss, lasting for 10 years [85]. Although surgery has been shown to improve hormonal profile [86] and improve quality of sexual life [87], initial data indicates that it may reduce sperm numbers. For example, in a small study of six severely obese previously fertile men, all men had developed azoospermia when followed-up at a mean of 17 months following Roux-en-Y gastric bypass surgery [88]. In contrast, a prospective study on 31 morbidly obese men showed improved semen parameters 6 months following laparoscopic Roux-en-Y-gastric bypass surgery, compared to those who had not undertaken bariatric surgery [89]. Long-term data is required to explore these associations further.

In those couples that require assisted conception treatment, it could be argued that ICSI should be offered when there is male obesity, in order to overcome the negative effects demonstrated on sperm function. However, it has been shown that raised male BMI is associated with decreased success of both IVF and ICSI treatment [15]. The authors suggested this may be related to ICSI treatment not overcoming key obesity-related impairment of sperm function, in particular, increased sperm DNA damage, chromatin decondensation, and impaired spermatogenesis.

More specific treatments to target known abnormalities that cause infertility in obese men may also be considered. In men with hypogonadotropic hypogonadism, treatments with aromatase inhibitors and testosterone have both been shown to improve testosterone to estradiol ratio and increase semen parameters [90]. Studies are required to evaluate whether these treatments could improve fertility in obese men. Another interesting area of research is in leptin replacement therapy. In leptin-deficient mice the administration of leptin has been shown to restore fertility [66]. As yet this has not been evaluated as a fertility treatment in men.

In an obese woman, there are good reasons to deny access to infertility treatment until weight loss has been achieved, namely, the increased risk of maternal and perinatal complications associated with maternal obesity [91]. It has been recommended by Balen et al. that women with polycystic ovarian syndrome should be denied fertility treatment until their BMI is less than 35 kg/m^2 [92]. Obviously, obesity in men does not carry such risks, and it is perhaps not surprising that male BMI is not usually used to restrict access to fertility treatment. However, there are growing concerns regarding potential epigenetic programing affecting future generations [67]. It is disappointing that therapeutic options are lacking, despite the known detrimental effect of male obesity on fertility. This may be related to the perception that male obesity has only a modest effect on fertility, compared with female obesity. However, as the incidence of male obesity increases, even a modest effect on fertility will become more clinically relevant.

Conclusion

With the ever-increasing incidence of obesity, it is important to be aware of the adverse impact of obesity on male fertility. Observational studies have shown associations between male obesity and a variety of sperm parameters, including concentration, motility, abnormal morphology, and DNA damage. However, the association between obesity and sperm parameters is not conclusive, with some showing this association only in a proportion of men with severe obesity. Endocrine abnormalities (including increased plasma levels of estrogen, leptin, insulin resistance, and reduced androgens and inhibin B levels) are likely to be important in the etiology of sperm dysfunction in obese men. However, other factors may also contribute, including genetic abnormalities, sexual dysfunction, testicular hyperthermia, and oxidative stress. The primary management must be to achieve weight reduction, using a reduced calorie diet in combination with an exercise program. Such regimes are difficult for patients to follow, and considerable support is required. In extreme cases, bariatric surgery can be considered, although, at present, there is no long-term data on semen analysis or fertility outcomes following surgery. More specific treatments to correct endocrine abnormalities associated with obesity are being evaluated, but disappointingly, no effective treatment has yet been proven.

References

[1] HFEA. Fertility treatment 2014–2016: trends and figures. 2017.

[2] Hart R, Norman RJ. The longer-term health outcomes for children born as a result of IVF treatment: Part I—General health outcomes. Hum Reprod Update 2013;19:232–43.

[3] Hart R, Norman RJ. The longer-term health outcomes for children born as a result of IVF treatment. Part II—Mental health and development outcomes. Hum Reprod Update 2013;19:244–50.

[4] Swinburn B, Gill T, Kumanyika S. Obesity prevention: a proposed framework for translating evidence into action. Obes Rev 2005;6:23–33.

[5] Tod E, Bromley C, Millard AD, Boyd A, Mackie P, McCartney G. Obesity in Scotland: a persistent inequality. Int J Equity Health 2017;16:135.

[6] Best D, Bhattacharya S. Obesity and fertility. Horm Mol Biol Clin Investig 2015;24:5–10.

[7] Balen AH, Anderson RA, Policy & Practice Committee of the BFS. Impact of obesity on female reproductive health: British Fertility Society, Policy and Practice Guidelines. Hum Fertil 2007;10:195–206.

[8] Kahn BE, Brannigan RE. Obesity and male infertility. Curr Opin Urol 2017;27:441–5.

[9] Nguyen RH, Wilcox AJ, Skjaerven R, Baird DD. Men's body mass index and infertility. Hum Reprod 2007;22:2488–93.

[10] Ramlau-Hansen CH, Thulstrup AM, Nohr EA, Bonde JP, Sorensen TI, Olsen J. Subfecundity in overweight and obese couples. Hum Reprod 2007;22:1634–7.

[11] Sallmen M, Sandler DP, Hoppin JA, Blair A, Baird DD. Reduced fertility among overweight and obese men. Epidemiology 2006;17:520–3.

[12] Bieniek JM, Kashanian JA, Deibert CM, Grober ED, Lo KC, Brannigan RE, et al. Influence of increasing body mass index on semen and reproductive hormonal parameters in a multi-institutional cohort of subfertile men. Fertil Steril 2016;106:1070–5.

[13] Glenn T, Harris AL, Lindheim SR. Impact of obesity on male and female reproductive outcomes. Curr Opin Obstet Gynecol 2019;31.

[14] Mushtaq R, Pundir J, Achilli C, Naji O, Khalaf Y, El-Toukhy T. Effect of male body mass index on assisted reproduction treatment outcome: an updated systematic review and meta-analysis. Reprod Biomed Online 2018;36:459–71.

[15] Le W, Su SH, Shi LH, Zhang JF, Wu DL. Effect of male body mass index on clinical outcomes following assisted reproductive technology: a meta-analysis. Andrologia 2016;48:406–24.

[16] Cooper TG, Noonan E, von Eckardstein S, Auger J, Baker HW, Behre HM, et al. World Health Organization reference values for human semen characteristics. Hum Reprod Update 2010;16:231–45.

[17] Sermondade N, Faure C, Fezeu L, Shayeb AG, Bonde JP, Jensen TK, et al. BMI in relation to sperm count: an updated systematic review and collaborative meta-analysis. Hum Reprod Update 2013;19:221–31.

[18] Jensen TK, Andersson AM, Jorgensen N, Andersen AG, Carlsen E, Petersen JH, et al. Body mass index in relation to semen quality and reproductive hormones among 1,558 Danish men. Fertil Steril 2004;82:863–70.

[19] Stewart TM, Liu DY, Garrett C, Jorgensen N, Brown EH, Baker HW. Associations between andrological measures, hormones and

semen quality in fertile Australian men: inverse relationship between obesity and sperm output. Hum Reprod 2009;24:1561–8.

[20] Mah PM, Wittert GA. Obesity and testicular function. Mol Cell Endocrinol 2010;316:180–6.

[21] Aggerholm AS, Thulstrup AM, Toft G, Ramlau-Hansen CH, Bonde JP. Is overweight a risk factor for reduced semen quality and altered serum sex hormone profile? Fertil Steril 2008;90:619–26.

[22] Chavarro JE, Toth TL, Wright DL, Meeker JD, Hauser R. Body mass index in relation to semen quality, sperm DNA integrity, and serum reproductive hormone levels among men attending an infertility clinic. Fertil Steril 2010;93:2222–31.

[23] Koloszar S, Fejes I, Zavaczki Z, Daru J, Szollosi J, Pal A. Effect of body weight on sperm concentration in normozoospermic males. Arch Androl 2005;51:299–304.

[24] Qin DD, Yuan W, Zhou WJ, Cui YQ, Wu JQ, Gao ES. Do reproductive hormones explain the association between body mass index and semen quality? Asian J Androl 2007;9:827–34.

[25] Nicopoulou SC, Alexiou M, Michalakis K, Ilias I, Venaki E, Koukkou E, et al. Body mass index vis-à-vis total sperm count in attendees of a single andrology clinic. Fertil Steril 2009;92:1016–17.

[26] Magnusdottir EV, Thorsteinsson T, Thorsteinsdottir S, Heimisdottir M, Olafsdottir K. Persistent organochlorines, sedentary occupation, obesity and human male subfertility. Hum Reprod 2005;20:208–15.

[27] Kort HI, Massey JB, Elsner CW, Mitchell-Leef D, Shapiro DB, Witt MA, et al. Impact of body mass index values on sperm quantity and quality. J Androl 2006;27:450–2.

[28] Hammoud AO, Wilde N, Gibson M, Parks A, Carrell DT, Meikle AW. Male obesity and alteration in sperm parameters. Fertil Steril 2008;90:2222–5.

[29] Sekhavat L, Moein MR. The effect of male body mass index on sperm parameters. Aging Male 2010;13:155–8.

[30] Martini AC, Tissera A, Estofan D, Molina RI, Mangeaud A, de Cuneo MF, et al. Overweight and seminal quality: a study of 794 patients. Fertil Steril 2010;94:1739–43.

[31] Bakos HW, Mitchell M, Setchell BP, Lane M. The effect of paternal diet-induced obesity on sperm function and fertilization in a mouse model. Int J Androl 2011;34:402–10.

[32] Fernandez CD, Bellentani FF, Fernandes GS, Perobelli JE, Favareto AP, Nascimento AF, et al. Diet-induced obesity in rats leads to a decrease in sperm motility. Reprod Biol Endocrinol 2011;9:32.

[33] World Health Organization. Laboratory manual for examination and processing of human semen. 5th ed. World Health Organization; 2010.

[34] Hofny ER, Ali ME, Abdel-Hafez HZ, Kamal Eel D, Mohamed EE, Abd El-Azeem HG, et al. Semen parameters and hormonal profile in obese fertile and infertile males. Fertil Steril 2010;94:581–4.

[35] Tunc O, Bakos HW, Tremellen K. Impact of body mass index on seminal oxidative stress. Andrologia 2011;43:121–8.

[36] Rufus O, James O, Michael A. Male obesity and semen quality: any association? Int J Reprod Biomed (Yazd) 2018;16:285–90.

[37] Alshahrani S, Ahmed AF, Gabr AH, Abalhassan M, Ahmad G. The impact of body mass index on semen parameters in infertile men. Andrologia 2016;48:1125–9.

[38] Eisenberg ML, Kim S, Chen Z, Sundaram R, Schisterman EF, Louis GM. The relationship between male BMI and waist circumference on semen quality: data from the LIFE study. Hum Reprod 2015;30:493–4.

[39] Spano M, Bonde JP, Hjollund HI, Kolstad HA, Cordelli E, Leter G. Sperm chromatin damage impairs human fertility. The Danish First Pregnancy Planner Study Team. Fertil Steril 2000;73:43–50.

[40] Sakkas D, Alvarez JG. Sperm DNA fragmentation: mechanisms of origin, impact on reproductive outcome, and analysis. Fertil Steril 2010;93:1027–36.

[41] Cissen M, Wely MV, Scholten I, Mansell S, Bruin JP, Mol BW, et al. Measuring sperm DNA fragmentation and clinical outcomes of medically assisted reproduction: a systematic review and meta-analysis. PLoS One 2016;11:e0165125.

[42] Rybar R, Kopecka V, Prinosilova P, Markova P, Rubes J. Male obesity and age in relationship to semen parameters and sperm chromatin integrity. Andrologia 2011;43:286–91.

[43] Dupont C, Faure C, Sermondade N, Boubaya M, Eustache F, Clement P, et al. Obesity leads to higher risk of sperm DNA damage in infertile patients. Asian J Androl 2013;15:622–5.

[44] Pauli EM, Legro RS, Demers LM, Kunselman AR, Dodson WC, Lee PA. Diminished paternity and gonadal function with increasing obesity in men. Fertil Steril 2008;90:346–51.

[45] Relwani R, Berger D, Santoro N, Hickmon C, Nihsen M, Zapantis A, et al. Semen parameters are unrelated to BMI but vary with SSRI use and prior urological surgery. Reprod Sci 2011;18:391–7.

[46] MacDonald AA, Herbison GP, Showell M, Farquhar CM. The impact of body mass index on semen parameters and reproductive hormones in human males: a systematic review with meta-analysis. Hum Reprod Update 2010;16:293–311.

[47] Crean AJ, Senior AM. High-fat diets reduce male reproductive success in animal models: a systematic review and meta-analysis. Obes Rev 2019;20:921–33.

[48] Broughton DE, Moley KH. Obesity and female infertility: potential mediators of obesity's impact. Fertil Steril 2017;107:840–7.

[49] Klenov VE, Jungheim ES. Obesity and reproductive function: a review of the evidence. Curr Opin Obstet Gynecol 2014;26:455–60.

[50] Globerman H, Shen-Orr Z, Karnieli E, Aloni Y, Charuzi I. Inhibin B in men with severe obesity and after weight reduction following gastroplasty. Endocr Res 2005;31:17–26.

[51] de Boer H, Verschoor L, Ruinemans-Koerts J, Jansen M. Letrozole normalizes serum testosterone in severely obese men with hypogonadotropic hypogonadism. Diabetes Obes Metab 2005;7:211–15.

[52] Isidori AM, Caprio M, Strollo F, Moretti C, Frajese G, Isidori A, et al. Leptin and androgens in male obesity: evidence for leptin contribution to reduced androgen levels. J Clin Endocrinol Metab 1999;84:3673–80.

[53] Jope T, Lammert A, Kratzsch J, Paasch U, Glander HJ. Leptin and leptin receptor in human seminal plasma and in human spermatozoa. Int J Androl 2003;26:335–41.

[54] Aquila S, Gentile M, Middea E, Catalano S, Morelli C, Pezzi V, et al. Leptin secretion by human ejaculated spermatozoa. J Clin Endocrinol Metab 2005;90:4753–61.

[55] Malik IA, Durairajanayagam D, Singh HJ. Leptin and its actions on reproduction in males. Asian J Androl 2019;21:296–9.

[56] Tsai EC, Matsumoto AM, Fujimoto WY, Boyko EJ. Association of bioavailable, free, and total testosterone with insulin resistance: influence of sex hormone-binding globulin and body fat. Diabetes Care 2004;27:861–8.

[57] Corona G, Monami M, Rastrelli G, Aversa A, Sforza A, Lenzi A, et al. Type 2 diabetes mellitus and testosterone: a meta-analysis study. Int J Androl 2011;34:528–40.

[58] Agbaje IM, Rogers DA, McVicar CM, McClure N, Atkinson AB, Mallidis C, et al. Insulin dependant diabetes mellitus: implications for male reproductive function. Hum Reprod 2007;22:1871–7.

[59] Roessner C, Paasch U, Kratzsch J, Glander HJ, Grunewald S. Sperm apoptosis signalling in diabetic men. Reprod Biomed Online 2012;25:292–9.

[60] Vermeulen A, Kaufman JM, Giagulli VA. Influence of some biological indexes on sex hormone-binding globulin and androgen levels in aging or obese males. J Clin Endocrinol Metab 1996;81:1821–6.

[61] Osuna JA, Gomez-Perez R, Arata-Bellabarba G, Villaroel V. Relationship between BMI, total testosterone, sex hormone-binding-globulin, leptin, insulin and insulin resistance in obese men. Arch Androl 2006;52:355–61.

[62] Luboshitzky R, Zabari Z, Shen Orr Z, Herer P, Lavie P. Disruption of the nocturnal testosterone rhythm by sleep fragmentation in normal men. J Clin Endocrinol Metab 2001;86:1134–9.

[63] Semple PA, Graham A, Malcolm Y, Beastall GH, Watson WS. Hypoxia, depression of testosterone, and impotence in Pickwickian syndrome reversed by weight reduction. Br Med J (Clin Res Ed) 1984;289:801–2.

[64] Smith MR. Changes in fat and lean body mass during androgen-deprivation therapy for prostate cancer. Urology 2004;63:742–5.

[65] Chen Z, Maricic M, Nguyen P, Ahmann FR, Bruhn R, Dalkin BL. Low bone density and high percentage of body fat among men who were treated with androgen deprivation therapy for prostate carcinoma. Cancer 2002;95:2136–44.

[66] Hoffmann A, Manjowk GM, Wagner IV, Kloting N, Ebert T, Jessnitzer B, et al. Leptin within the subphysiological to physiological range dose dependently improves male reproductive function in an obesity mouse model. Endocrinology 2016;157:2461–8.

[67] Craig JR, Jenkins TG, Carrell DT, Hotaling JM. Obesity, male infertility, and the sperm epigenome. Fertil Steril 2017;107:848–59.

[68] Hur SSJ, Cropley JE, Suter CM. Paternal epigenetic programming: evolving metabolic disease risk. J Mol Endocrinol 2017;58:R159–68.

[69] Nagelkerke NJ, Bernsen RM, Sgaier SK, Jha P. Body mass index, sexual behaviour, and sexually transmitted infections: an analysis using the NHANES 1999-2000 data. BMC Public Health 2006;6:199.

[70] Chung WS, Sohn JH, Park YY. Is obesity an underlying factor in erectile dysfunction? Eur Urol 1999;36:68–70.

[71] Chowdhury AK, Steinberger E. Early changes in the germinal epithelium of rat testes following exposure to heat. J Reprod Fertil 1970;22:205–12.

[72] Thonneau P, Bujan L, Multigner L, Mieusset R. Occupational heat exposure and male fertility: a review. Hum Reprod 1998;13:2122–5.

[73] Tiemessen CH, Evers JL, Bots RS. Tight-fitting underwear and sperm quality. Lancet 1996;347:1844–5.

[74] Kastelic JP, Wilde RE, Bielli A, Genovese P, Rizzoto G, Thundathil J. Hyperthermia is more important than hypoxia as a cause of disrupted spermatogenesis and abnormal sperm. Theriogenology 2019;131:177–81.

[75] Shafik A, Olfat S. Lipectomy in the treatment of scrotal lipomatosis. Br J Urol 1981;53:55–61.

[76] Vincent HK, Innes KE, Vincent KR. Oxidative stress and potential interventions to reduce oxidative stress in overweight and obesity. Diabetes Obes Metab 2007;9:813–39.

[77] Tremellen K. Oxidative stress and male infertility—a clinical perspective. Hum Reprod Update 2008;14:243–58.

[78] Smits RM, Mackenzie-Proctor R, Yazdani A, Stankiewicz MT, Jordan V, Showell MG. Antioxidants for male subfertility. Cochrane Database Syst Rev 2019;3:CD007411.

[79] Kaukua J, Pekkarinen T, Sane T, Mustajoki P. Sex hormones and sexual function in obese men losing weight. Obes Res 2003;11:689–94.

[80] Niskanen L, Laaksonen DE, Punnonen K, Mustajoki P, Kaukua J, Rissanen A. Changes in sex hormone-binding globulin and testosterone during weight loss and weight maintenance in abdominally obese men with the metabolic syndrome. Diabetes Obes Metab 2004;6:208–15.

[81] Duca Y, Calogero AE, Cannarella R, Giacone F, Mongioi LM, Condorelli RA, et al. Erectile dysfunction, physical activity and physical exercise: recommendations for clinical practice. Andrologia 2019;51:e13264.

[82] Esposito K, Giugliano D. Review: lifestyle modifications and pharmacotherapy for cardiovascular risk factors are associated with improvements in erectile dysfunction. Evid Based Nurs 2012;15:71–2.

[83] Johnson F, Cooke L, Croker H, Wardle J. Changing perceptions of weight in Great Britain: comparison of two population surveys. BMJ 2008;337:a494.

[84] Best D, Avenell A, Bhattacharya S. How effective are weight-loss interventions for improving fertility in women and men who are overweight or obese? A systematic review and meta-analysis of the evidence. Hum Reprod Update 2017;23:681–705.

[85] Laffin M, Karmali S. An update on bariatric surgery. Curr Obes Rep 2014;3:316–20.

[86] Mancini MC. Bariatric surgery—an update for the endocrinologist. Arq Bras Endocrinol Metabol 2014;58:875–88.

[87] Hammoud A, Gibson M, Hunt SC, Adams TD, Carrell DT, Kolotkin RL, et al. Effect of Roux-en-Y gastric bypass surgery on the sex steroids and quality of life in obese men. J Clin Endocrinol Metab 2009;94:1329–32.

[88] di Frega AS, Dale B, Di Matteo L, Wilding M. Secondary male factor infertility after Roux-en-Y gastric bypass for morbid obesity: case report. Hum Reprod 2005;20:997–8.

[89] Samavat J, Cantini G, Lotti F, Di Franco A, Tamburrino L, Degl'Innocenti S, et al. Massive weight loss obtained by bariatric surgery affects semen quality in morbid male obesity: a preliminary prospective double-armed study. Obes Surg 2018;28:69–76.

[90] Schlegel PN. Aromatase inhibitors for male infertility. Fertil Steril 2012;98:1359–62.

[91] Catalano PM, Shankar K. Obesity and pregnancy: mechanisms of short term and long term adverse consequences for mother and child. BMJ 2017;356:j1.

[92] Balen AH, Dresner M, Scott EM, Drife JO. Should obese women with polycystic ovary syndrome receive treatment for infertility? BMJ 2006;332:434–5.

Chapter 14

Evidence-based assisted reproduction in obese women

Brenda F Narice and Mostafa Metwally

[1]*Academic Unit of Reproductive and Developmental Medicine, The University of Sheffield and Sheffield Teaching Hospitals, The Jessop Wing, Sheffield, United Kingdom*

Introduction

As the prevalence of obesity continues to rise worldwide [1], so does the number of childbearing-age women who become obese [2]. In 2010, 26.1% of the female adult population in England was estimated to be obese, which represented over a 10% increase from 1993 [3]. Unrestrictive access to high-caloric food as well as an increased sedentary lifestyle have been proposed as contributory factors to the obesity epidemic [4,5].

Even though the complex biochemical and physiological association between excessive adipose tissue and reproductive function remains to be fully elucidated [6,7], a series of neuroendocrine pathways have been proposed [8–10]. Adipokines such as leptin and adiponectin and gut and pancreatic hormones, including ghrelin, insulin, and glucagon-like peptide-1 (GLP-1), have all been implicated in the regulatory cross-talks between caloric intake and reproductive function [5]. Hormonal secretion is tightly regulated by nutrition [11]. When energy is available, leptin as well as insulin and GLP-1 act centrally on the hypothalamus and induce satiety by altering the secretion of neuropeptide Y [12] and stimulate the hypothalamic–pituitary axis by increasing the circulation of gonadotrophin-releasing hormone (GnRh) [9,13]. Peripherally, these hormones act synergistically on the ovarian follicles stimulating theca proliferation and androgen production [14].

On the contrary, in cases of fasting or caloric restriction, leptin and insulin secretion decrease thus inhibiting GnRh pulsatility, and the circulating levels of ghrelin and adiponectin increase encouraging food intake [15–17]. Inhibited reproductive function seems to be an adaptive response to starvation as energy is mainly preserved for other essential functions rather than reproduction [13,18,19].

In obesity the excess of energy is thought to lead to a state of hyperleptinemia [20] and leptin resistance [21,22] with suppressed satiety response and anovulation and hyperandrogenemia due to the persistent release of GnRH, theca cell proliferation, and granulosa cell apoptosis [5,23,24]. Excess insulin also decreases steroid hormone–binding globulin and contributes to increased free androgens [25]. In addition, in morbidly obese patients with hypogonadotropic hypogonadal phenotype, a series of mutations in the leptin and its related receptors genes [26,27] have been identified suggesting an alternative pathophysiological mechanism in these individuals [28,29].

The detrimental effect of obesity on the female reproduction function is well recognized and can affect both natural and assisted conceptions with a growing body of evidence suggesting that adverse reproductive performance may stem from defective folliculogenesis, poorer oocyte/embryo quality, altered fertilization and implantation, as well as impaired endometrial receptivity [23,30–42]. It is estimated that the probability of spontaneous conception linearly decreases by 5% per unit of BMI above 29 kg/m^2 [43]. Women undergoing various assisted reproductive technologies (ART) such as in vitro fertilization/intracytoplasmic sperm injection (IVF/ICSI), frozen-thawed embryo transfer (FET), and intrauterine insemination (IUI) [5,44–46] have been found to have poorer reproductive outcomes such as decreased pregnancy rates, poorer ovarian stimulation, and higher miscarriage rates [47,48].

Impaired ovarian folliculogenesis

Embryo development

The abnormal development of ovarian follicles has been proposed as a contributory mechanism to explain the lower conception rate and increased risk of miscarriage observed in overweight and obese women [49–51]. Studies in diet-induced animals have reported a significantly higher number of apoptotic follicles and smaller oocytes with delayed maturation in obese female mice compared to the nonobese control group [52]. A higher rate of meiotic aneuploidy, disorganized spindles, and metaphase II chromosome misalignment have

Obesity and Gynecology. DOI: https://doi.org/10.1016/B978-0-12-817919-2.00014-0
© 2020 Elsevier Inc. All rights reserved.

been reported both in animal models and obese women with failed fertilized oocytes [53,54]. However, follicular defects in obesity do not seem to be confined to the nuclei as mounting evidence suggests that other cellular organelles may also play a role. In diet-induced mice, mitochondria have been shown to adopt an uneven intracellular distribution with a tendency to aggregate in perinuclear clusters. Other structural and physiological changes include altered cristae with increased swelling and vacuoles, and lower citrate levels with higher mitochondrial stress [53,55].

The lipotoxicity theory has also been proposed as a pathophysiological mechanism to explain impaired follicular development in obese women [49]. In normal conditions, fat that is stored as droplets of triglycerides in cells is presumed to be biologically inert [56]. However, when the storage capacity of adipocytes is surpassed, fatty acids accumulate in other tissues and disrupt the cellular metabolism by inducing the production of reactive oxygen species and apoptosis [56–58]. An increased concentration of fatty acids in the follicular free fluid has been associated with poorer cumulus oocyte complex morphology which may have an impact on fertility [58]. Further research is required before a firm conclusion between impaired fatty acid oxidation and altered folliculogenesis can be drawn in obese women.

Because the early development of the embryo appears to be signposted by oocyte quality, embryos in overweight and obese women are less likely to develop after fertilization [40,48,59]. The fewer embryos that do develop appear to reach the morula stage faster with significantly fewer cells in the trophectoderm. Blastocysts in obese women have also been found to display an abnormal metabolic fingerprint with reduced glucose consumption and altered amino acids metabolism [59,60]. Early embryonic exposure to obesity is thought to lead to long-term metabolic implications in the offspring, including hypertension, cardiovascular disease, type 2 diabetes, and chronic kidney disease [60–63].

Altered endometrial receptivity

In addition to the potentially suboptimal oocyte and embryo development proposed in obese women, researchers have hypothesized that women with raised BMI may also display altered endometrial receptivity [64–67]. Several studies have shown that implantation rates following oocyte donor IVF cycles are considerably lower in overweight and obese women compared to those of normal weight despite no differences in oocytes or embryo quality [66,68,69]. However, the oocyte donation model is not free from controversies as the requirements for oocyte donation may constitute a crucial confounding factor within obese women, which may explain their lower implantation rates [37,64,70].

Observational evidence has been complemented with further research in animal models and in vitro human primary cultures that have suggested the lower implantation rates and higher miscarriages seen in obesity could be secondary to defects in endometrial decidualization [71,72]. The analysis of endometrial biopsies in women with recurrent miscarriages did not reveal any significant differences in morphology, leukocyte population, and/or steroid receptor based on BMI [73]. However, it did show a significant increase in the endometrial expression of transthyretin, beta-globulin, and haptoglobin in overweight and obese women, which may suggest an ongoing inflammatory reaction in the endometrium as a potential cause for miscarriage in women with raised BMI [74].

Obesity and in vitro fertilization

In 2011 Rittenberg et al. [75] published one of the largest systematic reviews to date exploring the impact of obesity on IVF/ICSI clinical outcomes. The metaanalysis, which included 47,967 cycles from 33 primary studies, concluded that overweight and obese women had significantly lower clinical pregnancy (RR = 0.90, $P < .0001$) and live birth rates (RR = 0.84, $P = .0002$) when compared to those with a BMI < 25 kg/m^2. One of the limitations of this review, however, was the lack of adjustment for potential confounding factors such as age, smoking, and polycystic ovary syndrome (PCOS) [5]. The prevalence of overweight and obesity in PCOS women has been reported to be as high as 50% [76,77]. Recent studies have shown that PCOS can significantly affect follicular development and pregnancy rate in women undergoing IVF/ICSI [78,79] and increase the risk of complications such as ovarian hyperstimulation syndrome [80,81].

Some of these shortcomings were subsequently addressed by Moragianni et al. [82] in their extensive retrospective cohort study which included data from over 4600 patients undergoing their first cycle of IVF/ICSI. When outcomes were adjusted by BMI, age, ovarian reserve, and quality and number of embryos transferred, women with BMI > 30 kg/m^2 were found to have up to 68% lower odds of having a live birth than their lower BMI counterparts after their first IVF cycle. A subsequent systematic review by Jungheim et al. [83] did not find any statistical association between obesity and adverse IVF outcomes in oocyte donor recipients. However, the authors recommended to exercise caution when interpreting these results as many of the primary studies included in the review had not formally assessed live birth rate. More recently, a systematic review and metaanalysis by Sermondade et al. [84], which included 21 studies with 682,532 cycles, reported a significantly lower birth rate in women with a BMI > 30 kg/m² when compared with normal weight controls after IVF (RR 0.85, 95% CI: 0.84−0.87).

- *Controlled ovarian stimulation and final follicular maturation*

  In women with a raised BMI undergoing IVF/ICSI, the duration of gonadotrophin stimulation has been shown to be significantly longer when compared to normal weight controls [25,69,75,85−88]. Obese women have also been found to be suboptimal responders to the final follicular maturation trigger requiring larger doses of human chorionic gonadotrophin hormone (βhCG), recombinant LH (rLH), or GnRh agonist [89−92].

- *Oocyte retrieval and fertilization rates*

  The effect of obesity on oocyte retrieval still remains unclear as the number of oocytes to be retrieved is subjected to follicular maturation and a successful response to stimulation [93,94]. While some studies have reported a higher rate of IVF cancellation in women with raised BMI [87,95−98], others have not found any significant differences in oocyte retrieval between BMI groups [75,86,99,100]. Equally conflicting has been the observational data regarding fertilization, and embryo transfer and implantation rates based on BMI. Some authors have reported a lower fertilization rate [94,97] and higher implantation failure in women with raised BMI [46], whereas other researchers have found no statistically significant differences based on body adipose tissue [40,69,83,87,101,102]. The lack of standardization in obesity classification, and the diverse inclusion criteria and IVF protocols used across the studies are partly responsible for the large heterogeneity seen that renders some of the data noncomparable [64].

Obesity and frozen-thawed embryo transfer

Similar to fresh embryo transfers, increasing BMI appears to negatively impact on the outcomes of frozen−thawed embryo transfer (FET), even though not all studies have reached the same conclusion [100,103]. When results from a large retrospective study, including data from over 22,043 first FET cycles, were adjusted for potential confounders such as maternal age, infertility duration and cause, number of embryos transferred, and quality and stage of the embryo at the time of transfer, obese women (defined as BMI ≥ 27.5 kg/m²) had significantly lower live birth (OR = 0.70, 95% CI: 0.62−0.80) and implantation rates (OR = 0.80, 95% CI: 0.73−0.87) and a higher risk of pregnancy loss (first trimester OR = 1.46, 95% CI: 1.15−1.87; second trimester OR = 2.76, 95% CI: 1.67−4.58) than those in the reference group [31]. Raised BMI was also shown to be associated with lower rates of embryo implantation and live birth in obese women with PCOS undergoing FET when compared to normal weight PCOS controls [104].

In fresh embryo transfer the supraphysiologic hormonal level resulting from controlled ovarian stimulation and the trauma following oocyte retrieval may adversely affect embryo implantation and placentation [105,106]. Conversely, FET has been reported to have better IVF outcomes as the uterine milieu becomes more receptive for a successful embryo transfer, implantation, and clinical pregnancy [107−109]. Despite higher implantation and clinical pregnancy rates in FET, the transfer of frozen embryos has also been associated with higher risk of pregnancy-induced hypertension, preeclampsia, and abnormal placentation compared to fresh embryo transfer IVF [105,110]. The benefits of transferring frozen versus fresh embryos may, therefore, not be applicable to all individuals (which would support a freeze-all policy) but rather to a certain cohort of patients [31,105,111]. It has been hypothesized that obese women may be one of the cohorts that could benefit from FET. In a recent retrospective study by Monica et al. [112] with data from 527 blastocyst transfers, IVF outcomes such as implantation and clinical pregnancy were shown to be significantly higher in obese women who underwent FET compared to those who had fresh embryo transfers even after embryo-grade adjustment.

Obesity and intrauterine insemination

Evidence on how a raised BMI affects the success rate of IUI remains contradictory [113]. Several studies have reported that the clinical outcomes of IUI, either with homologous or frozen donor semen, do not seem to be significantly influenced by maternal BMI [114−118]. Conversely, other authors have shown that pregnancy rates seemed to progressively reduce with increasing BMI [119−121]. However, some studies have shown that when results are adjusted for confounding factors such as age and ovarian reserve, BMI does not seem to significantly influence success rate [114,122].

Conclusion

Excessive adipose tissue is thought to have adverse effects on reproductive function. As the number of fertile women with raised BMI continues to grow around the world, so does their need to access ART. Observational data has shown that overweight and obese women have significantly lower conception rates following ART than their normal weight counterparts. Even though the exact mechanisms through which obesity affects fertility are yet to be elucidated, a growing body of evidence suggests that impaired folliculogenesis and poorer embryo quality as well as altered endometrial receptivity may play a role.

A series of systematic reviews and metaanalysis have consistently reported worse clinical outcomes for IVF/ICSI in obese women but have failed to agree on how obesity affects each step of the IVF/ICSI protocol. Raised BMI has also been shown to be associated with lower FET success rate. Finally, regarding the effects of obesity on IUI, the evidence is less robust and requires further investigation.

Given the large heterogeneity among studies and the paucity of data from certain interventions, there is currently insufficient evidence to fully understand how obesity impacts on ART. Future primary research should include standardized definitions of BMI and ART protocols, and more homogenous inclusion criteria to enable further systematic data analysis and a better understanding of how obesity affects assisted reproduction.

References

[1] Mitchell NS, et al. Obesity: overview of an epidemic [in English] Psychiatr Clin North Am 2011;34(4):717 32.

[2] Norman RJ, et al. Improving reproductive performance in overweight/obese women with effective weight management [in English] Hum Reprod Update 2004;10(3):267—80.

[3] NHS. Statistics on obesity, physical activity and diet: England, 2012. NHS Information Centre for Health and Social Care; 2012, Available from: <https://digital.nhs.uk/data-and-information/publications/statistical/statistics-on-obesity-physical-activity-and-diet/statistics-on-obesity-physical-activity-and-diet-england-2012>.

[4] Agha M, Agha R. The rising prevalence of obesity: part A: Impact on public health. IJS Oncol 2017;2(7):e17.

[5] Khairy M, Rajkhowa M. Effect of obesity on assisted reproductive treatment outcomes and its management: a literature review. Obstetrician Gynaecol 2017;19(1):47—54.

[6] Garcia-Galiano D, et al. Role of the adipocyte-derived hormone leptin in reproductive control [in English] Hormone Mol Biol Clin Invest 2014;19(3):141—9.

[7] Coelho M, et al. Biochemistry of adipose tissue: an endocrine organ [in English] Arch Med Sci 2013;9(2):191—200.

[8] Brannian JD, Hansen KA. Leptin and ovarian folliculogenesis: implications for ovulation induction and ART outcomes [in English] Semin Reprod Med 2002;20(2):103—12.

[9] Moschos S, et al. Leptin and reproduction: a review [in English] Fertil Steril 2002;77(3):433—44.

[10] Goldsammler M, et al. Role of hormonal and inflammatory alterations in obesity-related reproductive dysfunction at the level of the hypothalamic-pituitary-ovarian axis [in English] Reprod Biol Endocrinol 2018;16(1):45.

[11] Tong Q, Xu Y. Central leptin regulation of obesity and fertility. Curr Obes Rep 2012;1(4):236—44.

[12] Ozcan L. A new player in hunger games. Sci Transl Med 2019;11 (499):eaay3569.

[13] Comninos AN, et al. The relationship between gut and adipose hormones, and reproduction. Hum Reprod Update 2013;20(2):153—74.

[14] Duleba AJ, et al. Effects of insulin and insulin-like growth factors on proliferation of rat ovarian theca-interstitial cells [in English] Biol Reprod 1997;56(4):891—7.

[15] Makris MC, et al. Ghrelin and obesity: identifying gaps and dispelling myths. A reappraisal [in English] In Vivo 2017;31(6):1047—50.

[16] Adamska-Patruno E, et al. The relationship between the leptin/ghrelin ratio and meals with various macronutrient contents in men with different nutritional status: a randomized crossover study. Nutr J 2018;17(1):118.

[17] Tsutsumi R, Webster NJG. GnRH pulsatility, the pituitary response and reproductive dysfunction [in English] Endocr J 2009;56 (6):729—37.

[18] Evans JJ, Anderson GM. Balancing ovulation and anovulation: integration of the reproductive and energy balance axes by neuropeptides. Hum Reprod Update 2012;18(3):313—32.

[19] Fontana R, Torre SD. The deep correlation between energy metabolism and reproduction: a view on the effects of nutrition for women fertility [in English] Nutrients 2016;8(2):87.

[20] Knight ZA, et al. Hyperleptinemia is required for the development of leptin resistance [in English] PLoS One 2010;5(6):e11376.

[21] Myers Jr. MG, et al. Obesity and leptin resistance: distinguishing cause from effect [in English] Trends Endocrinol Metab 2010;21 (11):643—51.

[22] Enriori PJ, et al. Leptin resistance and obesity [in English] Obes (Silver Spring) 2006;14(Suppl. 5):254s—8s.

[23] Metwally M, et al. The impact of obesity on female reproductive function. Obes Rev 2007;8(6):515—23.

[24] Billig H, et al. Gonadal cell apoptosis: hormone-regulated cell demise [in English] Hum Reprod Update 1996;2(2):103—17.

[25] Purcell SH, Moley KH. The impact of obesity on egg quality [in English] J Assist Reprod Genet 2011;28(6):517—24.

[26] Dubern B, Clement K. Leptin and leptin receptor-related monogenic obesity [in English] Biochimie 2012;94(10):2111—15.

[27] Wasim M, et al. Role of leptin deficiency, inefficiency, and leptin receptors in obesity [in English] Biochem Genet 2016;54(5):565—72.

[28] Hannema SE, et al. Novel leptin receptor mutations identified in two girls with severe obesity are associated with increased bone mineral density. Hormone Res Paediatr 2016;85(6):412—20.

[29] Paz-Filho G, et al. Congenital leptin deficiency: diagnosis and effects of leptin replacement therapy. Arq Bras Endocrinol Metabol 2010;54:690—7.

[30] Silvestris E, et al. Obesity as disruptor of the female fertility [in English] Reprod Biol Endocrinol 2018;16(1):22 pp..

[31] Zhang J, et al. Effect of body mass index on pregnancy outcomes in a freeze-all policy: an analysis of 22,043 first autologous frozen-thawed embryo transfer cycles in China. BMC Med 2019;17(1):114.

[32] Lashen H, et al. Obesity is associated with increased risk of first trimester and recurrent miscarriage: matched case-control study [in English] Hum Reprod 2004;19(7):1644—6.

[33] Linne Y. Effects of obesity on women's reproduction and complications during pregnancy [in English] Obes Rev 2004;5 (3):137—43.

[34] Yogev Y, Catalano PM. Pregnancy and obesity [in English] Obstet Gynecol Clin North Am 2009;36(2):285—300 viii.

[35] Boots C, Stephenson MD. Does obesity increase the risk of miscarriage in spontaneous conception: a systematic review [in English] Semin Reprod Med 2011;29(6):507—13.

[36] Cavalcante MB, et al. Obesity and recurrent miscarriage: a systematic review and meta-analysis. J Obstet Gynaecol Res 2019;45 (1):30—8.

[37] Bellver J, et al. Obesity and poor reproductive outcome: the potential role of the endometrium [in English] Fertil Steril 2007;88(2):446–51.

[38] Rittenberg V, et al. Influence of BMI on risk of miscarriage after single blastocyst transfer [in English] Hum Reprod 2011;26(10):2642–50.

[39] Metwally M, et al. Does high body mass index increase the risk of miscarriage after spontaneous and assisted conception? A meta-analysis of the evidence [in English] Fertil Steril 2008;90(3):714–26.

[40] Metwally M, et al. Effect of increased body mass index on oocyte and embryo quality in IVF patients [in English] Reprod Biomed Online 2007;15(5):532–8.

[41] Fedorcsak P, et al. Obesity is a risk factor for early pregnancy loss after IVF or ICSI [in English] Acta Obstet Gynecol Scand 2000;79(1):43–8.

[42] Jungheim ES, et al. Weighing the impact of obesity on female reproductive function and fertility [in English] Nutr Rev 2013;71(Suppl. 1):S3–8.

[43] van der Steeg JW, et al. Obesity affects spontaneous pregnancy chances in subfertile, ovulatory women [in English] Hum Reprod 2008;23(2):324–8.

[44] Maheshwari A, et al. Effect of overweight and obesity on assisted reproductive technology—a systematic review [in English] Hum Reprod Update 2007;13(5):433–44.

[45] Mahutte N, et al. Obesity and reproduction. J Obstet Gynaecol Can 2018;40(7):950–66.

[46] Provost MP, et al. Pregnancy outcomes decline with increasing recipient body mass index: an analysis of 22,317 fresh donor/recipient cycles from the 2008-2010 Society for Assisted Reproductive Technology Clinic Outcome Reporting System registry [in English] Fertil Steril 2016;105(2):364–8.

[47] Rich-Edwards JW, et al. Adolescent body mass index and infertility caused by ovulatory disorder [in English] Am J Obstet Gynecol 1994;171(1):171–7.

[48] Depalo R, et al. Oocyte morphological abnormalities in overweight women undergoing in vitro fertilization cycles [in English] Gynecol Endocrinol 2011;27(11):880–4.

[49] Broughton DE, Moley KH. Obesity and female infertility: potential mediators of obesity's impact. Fertil Steril 2017;107(4):840–7.

[50] Wittemer C, et al. Does body mass index of infertile women have an impact on IVF procedure and outcome? [in English] J Assist Reprod Genet 2000;17(10):547–52.

[51] Carrell DT, et al. Body mass index is inversely related to intrafollicular HCG concentrations, embryo quality and IVF outcome [in English] Reprod Biomed Online 2001;3(2):109–11.

[52] Jungheim ES, et al. Diet-induced obesity model: abnormal oocytes and persistent growth abnormalities in the offspring [in English] Endocrinology 2010;151(8):4039–46.

[53] Luzzo KM, et al. High fat diet induced developmental defects in the mouse: oocyte meiotic aneuploidy and fetal growth retardation/brain defects [in English] PLoS One 2012;7(11):e49217.

[54] Machtinger R, et al. The association between severe obesity and characteristics of failed fertilized oocytes [in English] Hum Reprod 2012;27(11):3198–207.

[55] Igosheva N, et al. Maternal diet-induced obesity alters mitochondrial activity and redox status in mouse oocytes and zygotes [in English] PLoS One 2010;5(4):e10074.

[56] Sorensen TI, et al. Obesity as a clinical and public health problem: is there a need for a new definition based on lipotoxicity effects? [in English] Biochim Biophys Acta 2010;1801(3):400–4.

[57] Broughton DE, Jungheim ES. A focused look at obesity and the preimplantation trophoblast [in English] Semin Reprod Med 2016;34(1):5–10.

[58] Jungheim ES, et al. Associations between free fatty acids, cumulus oocyte complex morphology and ovarian function during in vitro fertilization [in English] Fertil Steril 2011;95(6):1970–4.

[59] Leary C, et al. Human embryos from overweight and obese women display phenotypic and metabolic abnormalities [in English] Hum Reprod 2015;30(1):122–32.

[60] Comstock IA, et al. Increased body mass index negatively impacts blastocyst formation rate in normal responders undergoing in vitro fertilization [in English] J Assist Reprod Genet 2015;32(9):1299–304.

[61] Catalano PM, Ehrenberg HM. The short- and long-term implications of maternal obesity on the mother and her offspring [in English] BJOG 2006;113(10):1126–33.

[62] Glastras SJ, et al. Maternal obesity increases the risk of metabolic disease and impacts renal health in offspring [in English] Biosci Rep 2018;38(2). p. BSR20180050.

[63] Li M, et al. Maternal obesity and developmental programming of metabolic disorders in offspring: evidence from animal models [in English] Exp Diabetes Res 2011;2011:592408.

[64] Brewer CJ, Balen AH. The adverse effects of obesity on conception and implantation [in English] Reproduction 2010;140(3):347–64.

[65] Styne-Gross A, et al. Obesity does not impact implantation rates or pregnancy outcome in women attempting conception through oocyte donation [in English] Fertil Steril 2005;83(6):1629–34.

[66] Dessolle L, et al. Determinants of pregnancy rate in the donor oocyte model: a multivariate analysis of 450 frozen-thawed embryo transfers [in English] Hum Reprod 2009;24(12):3082–9.

[67] Robker RL. Evidence that obesity alters the quality of oocytes and embryos [in English] Pathophysiology 2008;15(2):115–21.

[68] Loveland JB, et al. Increased body mass index has a deleterious effect on in vitro fertilization outcome [in English] J Assist Reprod Genet 2001;18(7):382–6.

[69] Bellver J, et al. Female obesity impairs in vitro fertilization outcome without affecting embryo quality. Fertil Steril 2010;93(2):447–54.

[70] Howards PP, Cooney MA. Disentangling causal paths between obesity and in vitro fertilization outcomes: an intractable problem? [in English] Fertil Steril 2008;89(6):1604–5.

[71] Rhee JS, et al. Diet-induced obesity impairs endometrial stromal cell decidualization: a potential role for impaired autophagy [in English] Hum Reprod 2016;31(6):1315–26.

[72] Schulte MM, et al. Obese women experience impaired *in vitro* decidualization due to decreased autophagy. Fertil Steril 2016;106(3):e213–14.

[73] Metwally M, et al. Impact of high body mass index on endometrial morphology and function in the peri-implantation period in women with recurrent miscarriage [in English] Reprod Biomed Online 2007;14(3):328–34.

[74] Metwally M, et al. A proteomic analysis of the endometrium in obese and overweight women with recurrent miscarriage: preliminary evidence for an endometrial defect [in English] Reprod Biol Endocrinol 2014;12:75.

[75] Rittenberg V, et al. Effect of body mass index on IVF treatment outcome: an updated systematic review and meta-analysis [in English] Reprod Biomed Online 2011;23(4):421–39.

[76] Gambineri A, et al. Obesity and the polycystic ovary syndrome [in English] Int J Obes Relat Metab Disord 2002;26(7):883–96.

[77] Sam S. Obesity and polycystic ovary syndrome [in English] Obes Manag 2007;3(2):69–73.

[78] Marquard KL, et al. Polycystic ovary syndrome and maternal obesity affect oocyte size in in vitro fertilization/intracytoplasmic sperm injection cycles [in English] Fertil Steril 2011;95(6):2146–9 2149.e1.

[79] Cui N, et al. Impact of body mass index on outcomes of in vitro fertilization/intracytoplasmic sperm injection among polycystic ovarian syndrome patients [in English] Cell Physiol Biochem 2016;39(5):1723–34.

[80] Tso LO, et al. Metformin treatment before and during IVF or ICSI in women with polycystic ovary syndrome. Cochrane Database Syst Rev 2014;(11).

[81] Tanbo T, et al. Ovulation induction in polycystic ovary syndrome. Acta Obstetricia et Gynecologica Scandinavica 2018;97(10):1162–7.

[82] Moragianni VA, et al. The effect of body mass index on the outcomes of first assisted reproductive technology cycles [in English] Fertil Steril 2012;98(1):102–8.

[83] Jungheim ES, et al. IVF outcomes in obese donor oocyte recipients: a systematic review and meta-analysis [in English] Hum Reprod (Oxford, Engl) 2013;28(10):2720–7.

[84] Sermondade N, et al. Female obesity is negatively associated with live birth rate following IVF: a systematic review and meta-analysis [in English] Hum Reprod Update 2019;25(4):439–51.

[85] Weng SF, et al. Prediction of premature all-cause mortality: a prospective general population cohort study comparing machine-learning and standard epidemiological approaches [in English] PLoS One 2019;14(3) pp. e0214365-e0214365.

[86] Ozekinci M, et al. Does obesity have detrimental effects on IVF treatment outcomes? [in English] BMC Women's Health 2015;15:61.

[87] Esinler I, et al. Impact of isolated obesity on ICSI outcome. Reprod BioMed Online 2008;17(4):583–7.

[88] Dechaud H, et al. Obesity does not adversely affect results in patients who are undergoing in vitro fertilization and embryo transfer [in English] Eur J Obstet Gynecol Reprod Biol 2006;127(1):88–93.

[89] Lu X, et al. Dual trigger for final oocyte maturation improves the oocyte retrieval rate of suboptimal responders to gonadotropin-releasing hormone agonist. Fertil Steril 2016;106(6):1356–62.

[90] Irani M, et al. Dose of human chorionic gonadotropin to trigger final oocyte maturation. Fertil Steril 2016;106(3):e262–3.

[91] Detti L, et al. Serum human chorionic gonadotropin level after ovulation triggering is influenced by the patient's body mass index and the number of larger follicles [in English] Fertil Steril 2007;88(1):152–5.

[92] Elkind-Hirsch KE, et al. Serum human chorionic gonadotropin levels are correlated with body mass index rather than route of administration in women undergoing in vitro fertilization—embryo transfer using human menopausal gonadotropin and intracytoplasmic sperm injection [in English] Fertil Steril 2001;75(4):700–4.

[93] Bhandari HM, et al. An overview of assisted reproductive technology procedures. Obstetrician Gynaecol 2018;20(3):167–76.

[94] Vural F, et al. The role of overweight and obesity in in vitro fertilization outcomes of poor ovarian responders [in English] BioMed Res Int 2015;2015:781543.

[95] Matalliotakis I, et al. Impact of body mass index on IVF and ICSI outcome: a retrospective study [in English] Reprod Biomed Online 2008;16(6):778–83.

[96] Dokras A, et al. Obstetric outcomes after in vitro fertilization in obese and morbidly obese women [in English] Obstet Gynecol 2006;108(1):61–9.

[97] van Swieten EC, et al. Obesity and Clomiphene Challenge Test as predictors of outcome of in vitro fertilization and intracytoplasmic sperm injection [in English] Gynecol Obstet Invest 2005;59(4):220–4.

[98] Pinborg A, et al. Influence of female bodyweight on IVF outcome: a longitudinal multicentre cohort study of 487 infertile couples [in English] Reprod Biomed Online 2011;23(4):490–9.

[99] Christensen MW, et al. Effect of female body mass index on oocyte quantity in fertility treatments (IVF): treatment cycle number is a possible effect modifier. A Register-Based Cohort Study [in English] PLoS One 2016;11(9):e0163393.

[100] Banker M, et al. Effect of body mass index on the outcome of in-vitro fertilization/intracytoplasmic sperm injection in women [in English] J Hum Reprod Sci 2017;10(1):37–43.

[101] Espinós JJ, et al. Weight decrease improves live birth rates in obese women undergoing IVF: a pilot study. Reprod BioMed Online 2017;35(4):417–24.

[102] Farhi J, et al. High-quality embryos retain their implantation capability in overweight women [in English] Reprod Biomed Online 2010;21(5):706–11.

[103] Bishop LA, et al. BMI does not affect live birth outcomes in euploid frozen embryo transfers (FET). Fertil Steril 2017;108(3):e253.

[104] Chen R, et al. Pregnancy outcomes of PCOS overweight/obese patients after controlled ovarian stimulation with the GnRH antagonist protocol and frozen embryo transfer [in English] Reprod Biol Endocrinol 2018;16(1):36.

[105] Roque M, et al. Obstetric outcomes after fresh versus frozen-thawed embryo transfers: a systematic review and meta-analysis [in English] JBRA Assist Reprod 2018;22(3):253–60.

[106] Weinerman R, Mainigi M. Why we should transfer frozen instead of fresh embryos: the translational rationale [in English] Fertil Steril 2014;102(1):10–18.

[107] Maheshwari A, et al. Obstetric and perinatal outcomes in singleton pregnancies resulting from the transfer of frozen thawed versus fresh embryos generated through in vitro fertilization treatment: a systematic review and meta-analysis [in English] Fertil Steril 2012;98(2):368–377.e1-9.

[108] Chen ZJ, et al. Fresh versus frozen embryos for infertility in the polycystic ovary syndrome [in English] N Engl J Med 2016;375(6):523–33.

[109] Wang A, et al. Freeze-only versus fresh embryo transfer in a multicenter matched cohort study: contribution of progesterone and maternal age to success rates [in English] Fertil Steril 2017;108(2):254–261.e4.

[110] Maheshwari A, et al. Is frozen embryo transfer better for mothers and babies? Can cumulative meta-analysis provide a definitive answer? [in English] Hum Reprod Update 2018;24(1):35–58.

[111] Wong KM, et al. Fresh versus frozen embryo transfers in assisted reproduction [in English] Cochrane Database Syst Rev 2017;3: Cd011184.

[112] Monica S, et al. Clinical pregnancy outcomes of fresh versus frozen blastocyst transfers by body mass index. Fertil Steril 2019;111(4): e54–5.

[113] Pandey S, et al. The impact of female obesity on the outcome of fertility treatment [in English] J Hum Reprod Sci 2010;3(2):62–7.

[114] Souter I, et al. Women, weight, and fertility: the effect of body mass index on the outcome of superovulation/intrauterine insemination cycles [in English] Fertil Steril 2011;95(3):1042–7.

[115] Isa AM, et al. Age, body mass index, and number of previous trials: are they prognosticators of intra-uterine-insemination for infertility treatment? [in English] Int J Fertil Steril 2014;8 (3):255–60.

[116] Dodson WC, et al. Association of obesity with treatment outcomes in ovulatory infertile women undergoing superovulation and intrauterine insemination [in English] Fertil Steril 2006;86 (3):642–6.

[117] Thijssen A, et al. Predictive factors influencing pregnancy rates after intrauterine insemination with frozen donor semen: a prospective cohort study [in English] Reprod Biomed Online 2017;34(6):590–7.

[118] Thijssen A, et al. Predictive value of different covariates influencing pregnancy rate following intrauterine insemination with homologous semen: a prospective cohort study [in English] Reprod Biomed Online 2017;34(5):463–72.

[119] Koloszar S, et al. Effect of female body weight on efficiency of donor AI [in English] Arch Androl 2002;48(5):323–7.

[120] Zaadstra BM, et al. Fat and female fecundity: prospective study of effect of body fat distribution on conception rates [in English] BMJ 1993;306(6876):484–7.

[121] Yavuz A, et al. Predictive factors influencing pregnancy rates after intrauterine insemination [in English] Iran J Reprod Med 2013;11(3):227–34.

[122] Huyghe S, et al. Influence of BMI and smoking on IUI outcome with partner and donor sperm [in English] Facts Views Vis Obgyn 2017;9(2):93–100.

Chapter 15

Obesity, bariatric surgery, and male reproductive function

Man-wa Lui[1], Jyothis George[2,3] and Richard A. Anderson[2]

[1]Department of Obstetrics and Gynaecology, Queen Mary Hospital, The University of Hong Kong, Hong Kong, P.R. China, [2]MRC Centre for Reproductive Health, The Queen's Medical Research Institute, University of Edinburgh, Edinburgh, United Kingdom, [3]Boehringer Ingelheim, Frankfurt, Germany

Introduction

Obesity can be viewed in a reductionist manner as the biological consequence that naturally occurs when energy consumption in an individual is persistently in excess of energy expenditure but is, in fact, a complex pathophysiological state arising from an interplay of genetic, economic, environmental, and lifestyle factors [1]. The age of onset and the progression of weight gain differ between individuals with sex, ethnicity, and baseline body mass index (BMI) being predictors of progressive weight gain [2,3]. Spontaneous regression of weight gain is uncommon [2,3]. Interventions ranging from lifestyle and behavioral education to surgical methods are, therefore, required to affect weight loss in obese individuals. The combination of dietary modification and exercise regime affects weight loss and improves cardio-metabolic risk markers such as blood pressure, glucose, and triglycerides, but the magnitude of weight loss observed in clinical trials is modest—with a mean weight loss of 1.5 kg (95% confidence interval (CI) −2.3 to −0.7)—reported following intensive exercise [4]. Surgical interventions, on the other hand, have much more pronounced effects on weight loss than conventional approaches [5–7]. Therefore when weight loss in excess of 10% baseline weight is a therapeutic goal, surgical approaches are the only reliable therapeutic options available.

In obese men, there is evidence to suggest that all the three main testicular cell populations—Leydig, Sertoli, and germ—appear to be functionally impaired resulting in diminished serum testosterone and serum inhibin B, and in decreased sperm count and quality. In men, weight loss seems to have an impact on most of these markers of gonadal function as well as on sexual function, although the impact on testosterone secretion has been studied the most. In this chapter, we describe bariatric surgical techniques, the effects of obesity on male reproductive function, and current evidence for the value of surgical approaches to improve male fertility.

Bariatric surgical techniques

Bariatric surgical procedures aim to decrease the intake of food or to induce partial malabsorption of ingested food. Ideally introduced along with an effective patient education and preparation program involving dietary and lifestyle modification, these interventions seek to modify the intake of food—promoting slow consumption of smaller quantities of food [8]. Several surgical techniques have been described, with gastric bypass, gastric banding, biliopancreatic diversion, and sleeve gastrectomy in current use [5], of which sleeve gastrectomy is currently the most common procedure performed worldwide [9,10]. Nowadays, >95% of bariatric surgeries were performed laparoscopically due to the reduced perioperative morbidity and faster recovery [10].

Sleeve gastrectomy

Sleeve gastrectomy irreversibly divides the stomach vertically, affecting an approximately 25% reduction in size. Gastric function and digestion are unaltered as the pyloric valve remains intact. Apart from its restructure nature, the resection of the greater curvature of the stomach reduces the level of ghrelin [11] and thus the feeling of hunger. Although being one of the most recently developed bariatric surgeries, it has gained dominance over the last decade, overtaking other procedures [10]. It leads to a significant weight loss, from BMI 43.8 ± 8 to 30.7 ± 3.9 kg/m^2 over

Obesity and Gynecology. DOI: https://doi.org/10.1016/B978-0-12-817919-2.00015-2
© 2020 Elsevier Inc. All rights reserved.

1 year [12], and has comparative midterm result to the Roux-en-Y gastric bypass [13]. Over time, the stomach may become distended, negating some of the restrictive effects. Other complications include gastroesophageal reflux, anastomotic leakage, and stenosis. However, one of the major drawbacks of this procedure is the lack of long-term data [14].

Gastric bypass

Surgical gastric bypass procedures aim to be both restrictive and malabsorptive and involve the creation of a small gastric pouch (using surgical staples) and a gastroenterostomy stoma that bypasses the duodenum. There are several variations of the technique, a detailed discussion of which is beyond the scope of this chapter. The Roux-en-Y technique, designed to limit biliary reflux that may arise from loop gastroenterostomy, is commonly employed. Moreover, a prosthetic band is sometimes placed at the junction of the gastric pouch and the small intestine to stabilize the stoma [8]. Acute complications include anastomotic leakage and the blockage of efferent limb, while in the long term the stoma may get narrower causing persistent vomiting. Moreover, consumption of refined sugar after gastric bypass can evoke rapid heart rate, nausea, faintness, and diarrhea—a condition termed dumping syndrome [5]. It is technically feasible to reverse a gastric bypass. The Roux-en-Y technique was once the "gold standard" of bariatric surgery but now overtaken by sleeve gastrectomy due to its complexity and related morbidities.

Gastric banding

Gastric banding is a purely restrictive procedure where a ring (band) is placed around the fundus of the stomach to limit food intake. Unlike earlier versions, gastric bands in current use are adjustable, with provision for the addition or removal of saline through a subcutaneous access port to manipulate the size of the stoma. While the surgery, often carried out laparoscopically, is relatively easy to perform, the long-term failure and complication rates are as high as 20%−30% [8,15].

Biliopancreatic diversion

Biliopancreatic diversion is mainly a malabsorptive procedure and involves the removal of part of the stomach and the bypass of part of the small intestine. The resultant gastric pouch is much larger than that created in the typical Roux-en-Y technique allowing the patient to consume larger meals. Despite leading to the highest weight loss, it carries higher perioperative morbidity and mortality [5,9].

Endoluminal procedures

Different endoscopic devices have been investigated, including duodenal−jejunal bypass liner and intragastric balloon [9,16]. The initial results seem promising; however, long-term data are still lacking.

Pathophysiology in obesity

Hypothalamic−pituitary−gonadal axis

Endocrine regulation of the human hypothalamic−pituitary−gonadal axis has been studied extensively in the last three decades since the Nobel Prize−winning experiments involving the extraction of gonadotropin-releasing hormone (GnRH) from porcine hypothalami [17,18]. It is now well established that two hormones secreted by the pituitary, luteinizing hormone (LH) and follicular-stimulating hormone (FSH), regulate the functions of Leydig and germ cells, respectively [19]. Secretion of these pituitary hormones, in turn, is regulated by the modulation of frequency and/or amplitude of pulsatile hypothalamic secretion of GnRH [20,21]. A range of environmental, metabolic, and endocrine factors regulates hypothalamic−pituitary function. In particular, testosterone [22] and estradiol [23] (formed by the aromatization of testosterone in adipose tissue) exert negative feedback inhibition on the reproductive endocrine axis at the level of the hypothalamus and the pituitary. Inhibin B, secreted by the Sertoli cells of the testes, also plays an inhibitory role in gonadotropin secretion [24].

Leydig cell function

A reduction in the concentration of circulating serum testosterone in obese men is now well established [25−30]. Serum concentrations of LH are not elevated in obese men [31], unlike in primary gonadal failure, suggesting that hypothalamic and/or pituitary dysfunction plays a part in the causation of hypogonadism in obese individuals. In addition, in hypogonadal obese men, the testosterone response to exogenous human chorionic gonadotrophin is normal [25−28,31,32] suggesting that Leydig cells per se are not affected by obesity, and thus the primary defect is likely to be at the level of the hypothalamus/pituitary. LH pulse frequency in severely obese men is preserved, while LH pulse amplitude is decreased [31−33]. This would suggest a decrease in pituitary sensitivity and/or a decrease in GnRH pulse amplitude as the central pathophysiological feature characterizing hypogonadism in obesity. In this context, it has to be noted that obese men have higher concentrations of serum estradiol [34] reflecting increased aromatase activity in the larger fat mass, which also exerts inhibitory effects on the hypothalamic−pituitary unit.

Obstructive sleep apnea (OSA) frequently coexists with obesity and has also been implicated in the causation of hypogonadism in obese men. Studies have observed the presence of severe sleep apnea in one-third of men with BMI > 40 kg/m^2, with some degree of disordered sleeping observed in around 98% of the patients studied [35]. In the presence of OSA, nocturnal secretion of LH and testosterone is impaired, being negatively correlated with respiratory distress index [36]. A host of other factors such as systemic inflammation, central leptin resistance, hyperglycemia, and insulin resistance has also been described to play at least a part in mediating central hypogonadism in obesity, and it is likely that the hypothalamic kisspeptin system plays a central role in relaying many of these signals to the hypothalamic GnRH neuronal population [37]. While a detailed review of the mechanisms underpinning hypogonadotropic hypogonadism in obese men is beyond the scope of this chapter, Fig. 15.1 shows the current understanding of the neuroendocrine pathological processes involved.

In a recent metaanalysis, 87% of obesity-associated secondary hypogonadism resolved after bariatric surgery (95% CI 76%−95%) [29]. Weight loss was a predictive factor of recovery of secondary hypogonadism in the prospective European Male Ageing Study [OR 2.24 (1.04; 4.85), P = .042] [30]. Quality of sexual life also showed improvement after gastric bypass surgery [38].

Significant increases in total testosterone have also been observed in clinical studies involving the use of a very low−calorie diet [39,40]. In a randomized controlled

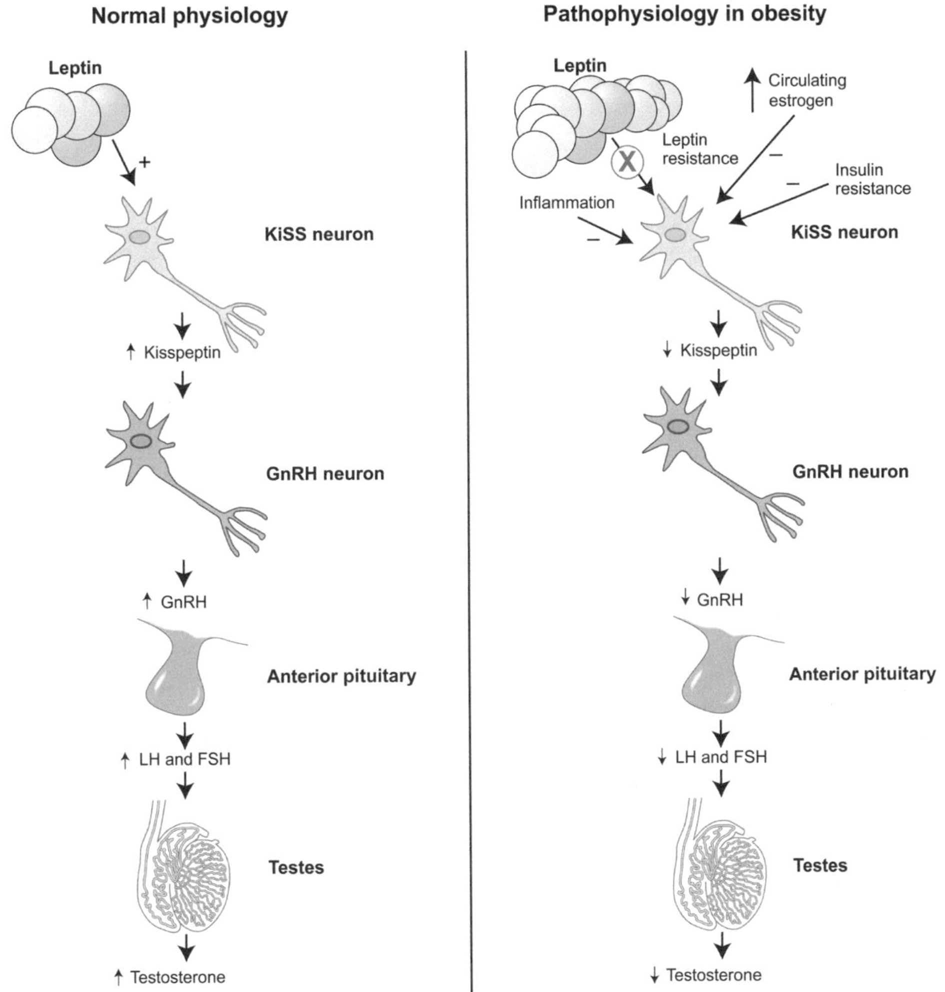

FIGURE 15.1 Physiological regulation of testosterone secretion and putative mechanisms by which pathological processes associated with obesity lead to impaired testosterone secretion. *Reproduced with kind permission from George JT, Millar RP, Anderson RA. Hypothesis: kisspeptin mediates male hypogonadism in obesity and type 2 diabetes. Neuroendocrinology 2010;91(4):302−7.*

trial of a 4-month weight loss program, including 10 weeks on a very low—energy diet (VLED), it was found that the mean weight loss in the intervention group at 8-month follow-up was 17 kg. Increases in sex hormone—binding globulin (SHBG), testosterone, and high-density lipoprotein cholesterol were observed at the end of the follow-up and was found to be associated with decreases in insulin and leptin concentrations. During rapid weight loss observed during the VLED regime, there was further transient improvement in metabolic and reproductive endocrine markers [39]. In another observational study of 58 abdominally obese men (age 46.3 ± 7.5 years; BMI 36.1 ± 3.8 kg/m^2; waist girth 121 ± 10 cm) with metabolic syndrome, 9-week VLED therapy achieved a weight loss of 14.3 ± 9.1 kg after a 12-month maintenance period [40]. Both SHBG and free testosterone increased significantly with the weight loss observed in this study [40]. The effect may be more prominent in hypogonadal men, as suggested by a prospective observation study showing an increase of testosterone level from 6.97 (6.38—11.06) to 13.21 nmol/L (11.44—19.76) ($P = .001$) [41].

It is therefore clear that obesity is associated with decreased testosterone concentrations, with negative effects on overall health that translate into increased mortality [42,43] and that this appears to be potentially reversible.

Obesity and spermatogenesis

A reduction in sperm concentration and total sperm count in overweight men was first reported in a cohort of men drafted for military service in Denmark [44]. This study categorized men into three groups based on BMI: <20, 20—25, and >25 kg/m^2. Men with a BMI < 20 kg/m^2 had a 28.1% (95% CI 8.3—47.9) reduction in sperm concentration and a 36.4% reduction in total sperm count (95% CI 14.6—58.3) compared to those with a BMI of 20—25 kg/m^2. Men with BMI > 25 kg/m^2 also had a reduction in sperm concentration and total sperm count of 21.6% (95% CI 4.0—39.4) and 23.9% (95% CI 4.7—43.2), respectively. Percentages of normal spermatozoa were reduced, although not significantly, among men with high or low BMI. Semen volume and percentage of motile spermatozoa were not affected [44]. It has to be noted, however, that the median BMI in this study was 22.4 kg/m^2 and that the sample was drawn from 18-year-old, presumably fit, men. The extent to which the observations in this population are generalizable to wider population is open to debate, but it would seem likely that greater rather than lesser effects would be apparent in the general population.

Several other studies have reported alterations in sperm count and/or quality as well as quantity. In a study of 81 men, markers of obesity were shown to be inversely associated with sperm count and total motile sperm cell number [45]. This study showed associations between waist and hip circumferences versus sperm count, total motile sperm cell number, and rapid progressive motile sperm count; between weight versus total sperm count and total motile sperm cell number; and between waist circumference and waist/hip ratio versus semen volume [45].

A larger study of 520 male partners who were overweight and obese, of couples presenting for infertility, also showed a dramatic decrease in total motile sperm count in comparison to men with normal BMI [46]. Men in this study were grouped based on BMI values (normal: 20—24 kg/m^2, overweight: 25—30 kg/m^2, obese: >30 kg/m^2). The number of normal motile sperm cells in each of these groups was as follows: normal, 18.6×10^6; overweight, 3.6×10^6; and obese, 0.7×10^6. Linear regression revealed a significant negative relationship between BMI and the total number of normal motile sperm cells. Men with BMI greater than 25 kg/m^2 also had fewer chromatin-intact normal motile sperm cells per ejaculate [46]; increased sperm DNA damage has been associated with reduced chance of successful pregnancy in assisted conception and with increased risk of miscarriage [47], although this is debated [48].

Poor semen quality was three times more likely to be observed in obese men than in men with normal weight in a small study of 72 men with infertility [49]. A significant negative correlation between semen quality parameters and BMI among men with normal semen quality was also observed in that study [49]. Consistent with these observations, a 3 kg/m^2 increase in male BMI was noted to be associated with infertility (OR 1.12; 95% CI 1.01—1.25; $n = 1329$) in men enrolled in the Agricultural Health Study in the United States [50]. The BMI effect was stronger when the data were limited to couples with the most complete data of the highest quality. Moreover, the association between BMI and infertility was similar for older and younger men [50].

While not all studies have demonstrated this association between obesity and qualitative or quantitative changes in spermatogenesis [51—53], a dose-dependent effect of weight on sperm count was demonstrated in a recent metaanalysis, with odds ratio of azoospermia/oligozoospermia of 1.28 (1.06—1.55) and 2.04 (1.59—2.62) for obese and morbid obese men, respectively. The mechanistic pathways affecting these changes in semen parameters are also unclear, as the decrease in circulating serum testosterone in obese men has not been shown to be correlated with marked decrease in intratesticular testosterone. Adverse biochemical effects of obesity on oocyte function are now being elucidated [54], with a reduced oocyte mitochondrial membrane potential in overfed animals; it is possible or perhaps even likely that similar effects are also of relevance to male gametogenesis.

Following a lifestyle modification program that achieved median weight loss of 15% (range 3.5%−25.4%) decline from baseline, the degree of weight loss was positively correlated with an increase in total sperm count and semen volume [55]. Baseline BMI was inversely associated with sperm concentration, total sperm count, sperm morphology and motile sperm, as well as total testosterone and inhibin B and positively associated with serum estradiol. Weight loss, achieved with a 14-week residential weight loss program, was associated with an increase in total sperm count, semen volume, total testosterone, and SHBG. The group with the largest weight loss had a statistically significant increase in total sperm count [193 million (95% CI 45; $n = 341$)] and normal sperm morphology [55].

Despite the well-established favorable effect on obesity-associated secondary hypogonadism, bariatric surgery failed to improve sperm quality [56]. However, only three studies were included in this metaanalysis, and more evidence is required.

Obesity and Sertoli cell function

Inhibin B is a testicular peptide secreted almost exclusively by Sertoli cells. It negatively regulates FSH secretion [57], acting as a readout of the spermatogenic activity of seminiferous tubules [58]. Serum concentrations of inhibin B are correlated with total sperm count, testicular volume, and testicular biopsy score [59]. In that study the receiver operating characteristic analysis showed a diagnostic accuracy of 95% compared to a value of 80% for FSH, suggesting that inhibin B is a useful endocrine marker of spermatogenesis in subfertile men [59].

In keeping with the studies described earlier linking obesity with reduced spermatogenesis, inhibin B concentrations have also been demonstrated to be lower in overweight and obese men [44−46,49−53,60,61] and to increase with weight loss [55]. However, mechanistic studies to explain these findings are yet to be reported.

Obesity and erectile dysfunction

Erectile dysfunction is associated with infertility [62]. Nearly four-fifths of all men reporting erectile dysfunction are found to be overweight or obese. It is unclear whether obesity per se or other commonly occurring confounding variables (such as smoking, dyslipidemia, atherosclerosis, hypertension, and type 2 diabetes) are the prime movers in the pathogenesis of erectile dysfunction in obese men.

In a randomized trial of obese men who achieved 10% weight loss through modification of diet and exercise regimes, one-third of men demonstrated an improvement in sexual function [63]. BMI and physical activity were independently associated with improvements in International Index of Erectile Function scores [63]. However, these findings have not been replicated in other weight loss studies, where despite improvements in testosterone secretion, scores of sexual function remained unchanged [64]. Weight loss after gastric bypass surgery has also been demonstrated to be associated with an improvement in the quality of sexual life in all aspects, as assessed using the Impact of Weight on the Quality of Life-Lite questionnaire [38]. Improvement in erectile function after bariatric surgery has also been demonstrated in a recent metaanalysis [56].

Transgenerational epigenetic effects

In addition to all the negative consequences of obesity on fertility a transgenerational "programing" phenomenon has been observed in both cohort and experimental studies. Paternal BMI has been suggested to correlate with birthweight and risks of insulin resistance and type 2 diabetes [65−68]. This effect can even be observed in the grandchildren, indicating a true epigenetic mechanism. Possible programing through disturbance of glucose−insulin metabolism of an obese father has also been shown in animal studies [69].

Practical considerations

The majority of evidence, although not all, supports a link between obesity and decreased activity of the reproductive axis in men, with adverse effects on endocrine, spermatogenic, and erectile function. Most studies assessing male fertility have used spermatogenic surrogates rather than fertility itself. While there are robust data linking weight loss with increases in serum testosterone, the data to recommend aggressive weight loss, with or without bariatric surgery, to men with a subnormal sperm count seeking fertility remain relatively weak. However, it has to be noted that modest weight loss, as low as 5% of baseline body weight, has been shown to reduce or eliminate disorders associated with obesity [64], and it can be argued that health-care professionals should seize all opportunities to encourage weight loss. There is also a paucity of randomized clinical trials comparing bariatric interventions with diet and exercise-based regimes. Factors, including treatment allocation bias, motivation to achieve weight loss, and compliance with dietary regimes, are potential confounders in current studies of bariatric surgery.

While this chapter has focused on obesity and subfertility in individuals with no known genetic or other causes underlying their obesity, it has to be emphasized that there are several endocrine conditions (e.g., Klinefelter and Prader–Willi syndromes) that may present with obesity and hypogonadism to the clinician.

In conclusion, clinicians while taking care of obese men with hypogonadism and/or infertility should recognize that it is likely that the obesity contributes to the reproductive dysfunction. While weight loss can restore normal testicular endocrine function, the value of weight reduction to improve male fertility remains very uncertain and thus individualization of advice remains the appropriate clinical standpoint.

References

[1] Wilding JPH. Pathophysiology and aetiology of obesity. Medicine 2011;39(1):6–10.

[2] McTigue KM, Garrett JM, Popkin BM. The natural history of the development of obesity in a cohort of young U.S. adults between 1981 and 1998. Ann Intern Med. 2002;136(12):857–64.

[3] Kahn HS, Cheng YJ. Longitudinal changes in BMI and in an index estimating excess lipids among white and black adults in the United States. Int J Obes 2007;32:136.

[4] Shaw KA, Gennat HC, O'Rourke P, Del Mar C. Exercise for overweight or obesity. Cochrane Database Syst Rev 2006;18(4).

[5] Colquitt JL, Pickett K, Loveman E, Frampton GK. Surgery for weight loss in adults. Cochrane Database Syst Rev 2014;8(8).

[6] Gortmaker SL, Swinburn BA, Levy D, Carter R, Mabry PL, Finegood DT, et al. Changing the future of obesity: science, policy, and action. Lancet 2011;378(9793):838–47.

[7] Gloy VL, Briel M, Bhatt DL, Kashyap SR, Schauer PR, Mingrone G, et al. Bariatric surgery versus non-surgical treatment for obesity: a systematic review and meta-analysis of randomised controlled trials. BMJ: Br Med J 2013;347:f5934.

[8] Picot J, Jones J, Colquitt JL, Gospodarevskaya E, Loveman E, Baxter L, et al. The clinical effectiveness and cost-effectiveness of bariatric (weight loss) surgery for obesity: a systematic review and economic evaluation. Health Technol Assess 2009;13 (41):1–190 215–357, iii–iv.

[9] Nguyen NT, Varela JE. Bariatric surgery for obesity and metabolic disorders: state of the art. Nat Rev Gastroenterol Hepatol 2016;14:160.

[10] Khorgami Z, Shoar S, Andalib A, Aminian A, Brethauer SA, Schauer PR. Trends in utilization of bariatric surgery, 2010-2014: sleeve gastrectomy dominates. Surg Obes Relat Dis 2017;13 (5):774–8.

[11] Anderson B, Switzer NJ, Almamar A, Shi X, Birch DW, Karmali S. The impact of laparoscopic sleeve gastrectomy on plasma ghrelin levels: a systematic review. Obes Surg 2013;23(9):1476–80.

[12] Emile SH, Elfeki H, Elalfy K, Abdallah E. Laparoscopic sleeve gastrectomy then and now: an updated systematic review of the progress and short-term outcomes over the last 5 years. Surg Laparosc Endosc Percutan Tech 2017;27(5):307–17.

[13] Shoar S, Saber AA. Long-term and midterm outcomes of laparoscopic sleeve gastrectomy versus Roux-en-Y gastric bypass: a systematic review and meta-analysis of comparative studies. Surg Obes Relat Dis 2017;13(2):170–80.

[14] Juodeikis Z, Brimas G. Long-term results after sleeve gastrectomy: a systematic review. Surg Obes Relat Dis 2017;13(4):693–9.

[15] Suter M, Calmes JM, Paroz A, Giusti V. A 10-year experience with laparoscopic gastric banding for morbid obesity: high long-term complication and failure rates. Obes Surg 2006;16 (7):829–35.

[16] Betzel B, Homan J, Aarts EO, Janssen IMC, de Boer H, Wahab PJ, et al. Weight reduction and improvement in diabetes by the duodenal-jejunal bypass liner: a 198 patient cohort study. Surg Endosc 2017;31(7):2881–91.

[17] Baba Y, Matsuo H, Schally AV. Structure of the porcine LH- and FSH-releasing hormone. II. Confirmation of the proposed structure by conventional sequential analyses. Biochem Biophys Res Commun 1971;44(2):459–63.

[18] Schally AV, Arimura A, Kastin AJ, Matsuo H, Baba Y, Redding TW, et al. Gonadotropin-releasing hormone: one polypeptide regulates secretion of luteinizing and follicle-stimulating hormones. Science 1971;173(4001):1036–8.

[19] Millar RP, Lu ZL, Pawson AJ, Flanagan CA, Morgan K, Maudsley SR. Gonadotropin-releasing hormone receptors. Endocr Rev 2004;25(2):235–75.

[20] McCartney CR, Gingrich MB, Hu Y, Evans WS, Marshall JC. Hypothalamic regulation of cyclic ovulation: evidence that the increase in gonadotropin-releasing hormone pulse frequency during the follicular phase reflects the gradual loss of the restraining effects of progesterone. J Clin Endocrinol Metab 2002;87 (5):2194–200.

[21] Haisenleder DJ, Dalkin AC, Ortolano GA, Marshall JC, Shupnik MA. A pulsatile gonadotropin-releasing hormone stimulus is required to increase transcription of the gonadotropin subunit genes: evidence for differential regulation of transcription by pulse frequency in vivo. Endocrinology 1991;128(1):509–17.

[22] Pitteloud N, Dwyer AA, DeCruz S, Lee H, Boepple PA, Crowley Jr. WF, et al. Inhibition of luteinizing hormone secretion by testosterone in men requires aromatization for its pituitary but not its hypothalamic effects: evidence from the tandem study of normal and gonadotropin-releasing hormone-deficient men. J Clin Endocrinol Metab 2008;93(3):784–91.

[23] Veldhuis JD, Dufau ML. Estradiol modulates the pulsatile secretion of biologically active luteinizing hormone in man. J Clin Invest 1987;80(3):631–8.

[24] de Kretser DM, Buzzard JJ, Okuma Y, O'Connor AE, Hayashi T, Lin SY, et al. The role of activin, follistatin and inhibin in testicular physiology. Mol Cell Endocrinol 2004;225(1–2):57–64.

[25] Glass AR, Swerdloff RS, Bray GA, Dahms WT, Atkinson RL. Low serum testosterone and sex-hormone-binding-globulin in massively obese men. J Clin Endocrinol Metab 1977;45(6):1211–19.

[26] Strain GW, Zumoff B, Kream J, Strain JJ, Deucher R, Rosenfeld RS, et al. Mild Hypogonadotropic hypogonadism in obese men. Metabolism 1982;31(9):871–5.

[27] Zumoff B, Strain GW, Miller LK, Rosner W, Senie R, Seres DS, et al. Plasma free and non-sex-hormone-binding-globulin-bound testosterone are decreased in obese men in proportion to their degree of obesity. J Clin Endocrinol Metab 1990;71(4):929–31.

[28] Tsai EC, Matsumoto AM, Fujimoto WY, Boyko EJ. Association of bioavailable, free, and total testosterone with insulin resistance: influence of sex hormone-binding globulin and body fat. Diabetes Care 2004;27(4):861–8.

[29] Escobar-Morreale HF, Santacruz E, Luque-Ramirez M, Botella Carretero JI. Prevalence of 'obesity-associated gonadal dysfunction'

in severely obese men and women and its resolution after bariatric surgery: a systematic review and meta-analysis. Hum Reprod Update 2017;23(4):390–408.

[30] Rastrelli G, Carter EL, Ahern T, Finn JD, Antonio L, O'Neill TW, et al. Development of and recovery from secondary hypogonadism in aging men: prospective results from the EMAS. J Clin Endocrinol Metab 2015;100(8):3172–82.

[31] Giagulli VA, Kaufman JM, Vermeulen A. Pathogenesis of the decreased androgen levels in obese men. J Clin Endocrinol Metab 1994;79(4):997–1000.

[32] Amatruda JM, Hochstein M, Hsu TH, Lockwood DH. Hypothalamic and pituitary dysfunction in obese males. Int J Obes 1982;6(2):183–9.

[33] Vermeulen A, Kaufman JM, Deslypere JP, Thomas G. Attenuated luteinizing hormone (LH) pulse amplitude but normal LH pulse frequency, and its relation to plasma androgens in hypogonadism of obese men. J Clin Endocrinol Metab 1993;76(5):1140–6.

[34] Schneider G, Kirschner MA, Berkowitz R, Ertel NH. Increased estrogen production in obese men. J Clin Endocrinol Metab 1979;48(4):633–8.

[35] Valencia-Flores M, Orea A, Castano VA, Resendiz M, Rosales M, Rebollar V, et al. Prevalence of sleep apnea and electrocardiographic disturbances in morbidly obese patients. Obes Res 2000;8(3):262–9.

[36] Luboshitzky R, Aviv A, Hefetz A, Herer P, Shen-Orr Z, Lavie L, et al. Decreased pituitary-gonadal secretion in men with obstructive sleep apnea. J Clin Endocrinol Metab 2002;87(7):3394–8.

[37] George JT, Millar RP, Anderson RA. Hypothesis: kisspeptin mediates male hypogonadism in obesity and type 2 diabetes. Neuroendocrinology 2010;91(4):302–7.

[38] Hammoud A, Gibson M, Hunt SC, Adams TD, Carrell DT, Kolotkin RL, et al. Effect of Roux-en-Y gastric bypass surgery on the sex steroids and quality of life in obese men. J Clin Endocrinol Metab 2009;94(4):1329–32.

[39] Kaukua J, Pekkarinen T, Sane T, Mustajoki P. Sex hormones and sexual function in obese men losing weight. Obes Res 2003;11(6):689–94.

[40] Niskanen L, Laaksonen DE, Punnonen K, Mustajoki P, Kaukua J, Rissanen A. Changes in sex hormone-binding globulin and testosterone during weight loss and weight maintenance in abdominally obese men with the metabolic syndrome. Diabetes Obes Metab 2004;6(3):208–15.

[41] Schulte DM, Hahn M, Oberhäuser F, Malchau G, Schubert M, Heppner C, et al. Caloric restriction increases serum testosterone concentrations in obese male subjects by two distinct mechanisms. Horm Metab Res 2014;46(04):283–6.

[42] Khaw KT, Dowsett M, Folkerd E, Bingham S, Wareham N, Luben R, et al. Endogenous testosterone and mortality due to all causes, cardiovascular disease, and cancer in men: European prospective investigation into cancer in Norfolk (EPIC-Norfolk) Prospective Population Study. Circulation 2007;116(23):2694–701.

[43] Araujo AB, Kupelian V, Page ST, Handelsman DJ, Bremner WJ, McKinlay JB. Sex steroids and all-cause and cause-specific mortality in men. Arch Intern Med 2007;167(12):1252–60.

[44] Jensen TK, Andersson AM, Jorgensen N, Andersen AG, Carlsen E, Petersen JH, et al. Body mass index in relation to semen quality and reproductive hormones among 1,558 Danish men. Fertil Steril 2004;82(4):863–70.

[45] Fejes I, Koloszar S, Szollosi J, Zavaczki Z, Pal A. Is semen quality affected by male body fat distribution? Andrologia 2005;37(5):155–9.

[46] Kort HI, Massey JB, Elsner CW, Mitchell-Leef D, Shapiro DB, Witt MA, et al. Impact of body mass index values on sperm quantity and quality. J Androl 2006;27(3):450–2.

[47] Simon L, Zini A, Dyachenko A, Ciampi A, Carrell DT. A systematic review and meta-analysis to determine the effect of sperm DNA damage on in vitro fertilization and intracytoplasmic sperm injection outcome. Asian J Androl 2017;19(1):80–90.

[48] Cissen M, Wely MV, Scholten I, Mansell S, Bruin JP, Mol BW, et al. Measuring sperm DNA fragmentation and clinical outcomes of medically assisted reproduction: a systematic review and meta-analysis. PLoS One 2016;11(11):e0165125.

[49] Magnusdottir EV, Thorsteinsson T, Thorsteinsdottir S, Heimisdottir M, Olafsdottir K. Persistent organochlorines, sedentary occupation, obesity and human male subfertility. Hum Reprod 2005;20(1):208–15.

[50] Sallmen M, Sandler DP, Hoppin JA, Blair A, Baird DD. Reduced fertility among overweight and obese men. Epidemiology 2006;17(5):520–3.

[51] Hammoud AO, Gibson M, Peterson CM, Hamilton BD, Carrell DT. Obesity and male reproductive potential. J Androl 2006;27(5):619–26.

[52] Teerds KJ, de Rooij DG, Keijer J. Functional relationship between obesity and male reproduction: from humans to animal models. Hum Reprod Update 2011;17(5):667–83.

[53] Aggerholm AS, Thulstrup AM, Toft G, Ramlau-Hansen CH, Bonde JP. Is overweight a risk factor for reduced semen quality and altered serum sex hormone profile? Fertil Steril 2008;90(3):619–26.

[54] Wu LL, Dunning KR, Yang X, Russell DL, Lane M, Norman RJ, et al. High-fat diet causes lipotoxicity responses in cumulus-oocyte complexes and decreased fertilization rates. Endocrinology 2010;151(11):5438–45.

[55] Hakonsen LB, Thulstrup AM, Aggerholm AS, Olsen J, Bonde JP, Andersen CY, et al. Does weight loss improve semen quality and reproductive hormones? Results from a cohort of severely obese men. Reprod Health 2011;8:24.

[56] Lee Y, Dang JT, Switzer N, Yu J, Tian C, Birch DW, et al. Impact of bariatric surgery on male sex hormones and sperm quality: a systematic review and meta-analysis. Obes Surg 2019;29(1):334–46.

[57] Anawalt BD, Bebb RA, Matsumoto AM, Groome NP, Illingworth PJ, McNeilly AS, et al. Serum inhibin B levels reflect Sertoli cell function in normal men and men with testicular dysfunction. J Clin Endocrinol Metab 1996;81(9):3341–5.

[58] Anderson RA, Wallace EM, Groome NP, Bellis AJ, Wu FC. Physiological relationships between inhibin B, follicle stimulating hormone secretion and spermatogenesis in normal men and response to gonadotrophin suppression by exogenous testosterone. Hum Reprod 1997;12(4):746–51.

[59] Pierik FH, Vreeburg JT, Stijnen T, De Jong FH, Weber RF. Serum inhibin B as a marker of spermatogenesis. J Clin Endocrinol Metab 1998;83(9):3110–14.

[60] Zini A, Boman JM, Belzile E, Ciampi A. Sperm DNA damage is associated with an increased risk of pregnancy loss after IVF and ICSI: systematic review and meta-analysis. Hum Reprod 2008;23(12):2663–8.

[61] Aitken RJ, De Iuliis GN, McLachlan RI. Biological and clinical significance of DNA damage in the male germ line. Int J Androl 2009;32(1):46–56.

[62] O'Brien JH, Lazarou S, Deane L, Jarvi K, Zini A. Erectile dysfunction and andropause symptoms in infertile men. J Urol 2005;174(5):1932–4.

[63] Esposito K, Giugliano F, Di Palo C, Giugliano G, Marfella R, D'Andrea F, et al. Effect of lifestyle changes on erectile dysfunction in obese men: a randomized controlled trial. JAMA 2004;291 (24):2978–84.

[64] Blackburn G. Effect of degree of weight loss on health benefits. Obes Res 1995;3(Suppl. 2):211s–6s.

[65] Chen Y-P, Xiao X-M, Li J, Reichetzeder C, Wang Z-N, Hocher B. Paternal body mass index (BMI) is associated with offspring intrauterine growth in a gender dependent manner. PLoS One 2012;7(5):e36329.

[66] Bygren LO, Kaati G, Edvinsson S. Longevity determined by paternal ancestors' nutrition during their slow growth period. Acta Biotheor 2001;49(1):53–9.

[67] Kaati G, Bygren LO, Edvinsson S. Cardiovascular and diabetes mortality determined by nutrition during parents' and grandparents' slow growth period. Eur J Hum Genet 2002;10(11):682–8.

[68] Kaati G, Bygren LO, Pembrey M, Sjöström M. Transgenerational response to nutrition, early life circumstances and longevity. Eur J Hum Genet 2007;15:784.

[69] Ng S-F, Lin RCY, Laybutt DR, Barres R, Owens JA, Morris MJ. Chronic high-fat diet in fathers programs β-cell dysfunction in female rat offspring. Nature 2010;467:963.

Chapter 16

Medical interventions to improve outcomes in infertile obese women planning for pregnancy

Vikram Talaulikar

Reproductive Medicine Unit, University College London Hospital, London, United Kingdom

Introduction

The World Health Organization defines being "overweight" as body mass index (BMI) ≥ 25 kg/m^2 and "obesity" as BMI ≥ 30 kg/m^2 [1]. The prevalence of obesity is increasing worldwide with more than 600 million obese adults, including 15% of women, in 2014 [1,2].

Subfertility is defined as "the failure to achieve a clinical pregnancy after 12 months or more of regular unprotected sexual intercourse." Subfertility affects about one in seven couples in the United Kingdom. An increasing numbers of women with high BMI are being referred for the evaluation and treatment of subfertility across the country [3]. Identifying and developing effective long-term reproductive health strategies for overweight and obese women is of paramount importance. This chapter focuses on medical interventions to improve outcomes in subfertile obese women who are planning for pregnancy.

Impact of high body mass index on fertility and pregnancy

High BMI is associated with delayed conception [4]. The mechanisms underlying this are thought to be multifactorial: anovulation [5,6], impaired oocyte development [6,7], menstrual irregularity [5], and unfavorable effects on the endometrium [8]. Women who are overweight or obese need higher doses and longer duration of ovarian stimulation during ovulation induction (OI) treatment [9,10]. Lower implantation and lower success rates of treatment overall have also been reported during in vitro fertilization (IVF) treatment in women with high BMI [11]. Polycystic ovary syndrome (PCOS) is the most common endocrine disorder in women of reproductive age-group (affecting approximately 6%–10% of population), which has diverse reproductive and metabolic manifestations [12]. One of the features of PCOS is hyperinsulinemia and insulin resistance that is exacerbated by obesity. Insulin stimulates ovarian androgen production and inhibits hepatic synthesis of sex hormone–binding globulin (SHBG) [12,13].

A study compared 187 obese and overweight women with PCOS who were immediately treated with clomiphene to induce ovulation with 142 women with PCOS who began a weight-loss program consisting of lower caloric intake, exercise, and weight-loss medication before starting clomiphene treatment. Women who were treated with clomiphene alone had an ovulation rate of 44.7% and a live birth rate of 10.2%. The women who received clomiphene after the weight-loss program had a 62% ovulation rate and a 25% live birth rate [14].

Obesity during pregnancy increases the risk of various complications such as:

1. pregnancy loss—risk of miscarriage, stillbirth, and recurrent miscarriage;
2. gestational diabetes;
3. preeclampsia;
4. cardiac dysfunction;
5. sleep apnea; and
6. difficulties with delivery, shoulder dystocia, need for cesarean section, and operative complications such as wound infections.

Lifestyle interventions to improve outcomes in infertile obese women planning for pregnancy

Weight loss is recommended for women with high BMI before attempting natural conception or fertility treatments to improve fertility outcomes, assist with fertility treatment funding [15], and to reduce the risks of obstetric complications.

Reduced calorie intake and increased physical activity are the two essential pillars of any weight-loss program. The benefits of lifestyle intervention for weight loss and improvement in metabolic and reproductive manifestations of PCOS may be observed with as little as 5% of total body weight loss [12]. Most guidelines recommend a target of 5%–10% body weight loss in overweight/obese PCOS women with long-term goals of 10%–20% weight loss and waist circumference <80–88 cm tailored to the ethnicity [16,17].

A recent systematic review of lifestyle interventions in women with PCOS included six randomized controlled trials (RCTs) comparing lifestyle treatment to minimal or no treatment. Lifestyle interventions consisted of exercise alone or a combination of dietary, exercise, and behavioral interventions. Women who received active treatment experienced reduced total testosterone level, hirsutism, weight, waste circumference, central adiposity, and fasting insulin levels. No significant changes were noted in BMI, bioavailable testosterone, SHBG, glucose, or lipids [18].

Another RCT compared three interventions: oral contraceptive pill (OCP), lifestyle modification with weight-loss agents (sibutramine or orlistat), and combined lifestyle modification and found that the interventions, which included lifestyle modification, yielded significant weight loss and higher ovulation rates [19].

Maintaining long-term weight loss can prove challenging and attention needs to be given to other areas of lifestyle, such as alcohol intake, smoking cessation, and stress-reduction techniques. Strategies to promote sustained weight loss include self-monitoring techniques such as food diaries, pedometers, time management advice, relapse prevention techniques, engagement of social support, and goal setting [16,17]. There is some evidence that intensive weight loss immediately prior to IVF is associated with adverse outcomes, including increased cycle cancellation and decreased rates of fertilization, implantation, ongoing pregnancy, and live births in women with PCOS [19]. Therefore weight-loss strategies should be encouraged well in advance of pregnancy planning by the individual woman. Many women with PCOS and high BMI may present with associated psychological disorders such as anxiety or depression [20]. These women should be offered counseling or referral to a psychologist to help them along their weight-loss journey.

Dietary interventions

Dietary interventions in overweight or obese women should take into account the degree of obesity, dietary preferences, and food availability. If an eating disorder is suspected, referrals to the dietitian and clinical psychologist should be considered. Strategies such as face-to-face education sessions and practical advice on approaches to healthy eating tailored to the patient should be incorporated [17].

Based on the available evidence, it is not possible to conclude that a particular degree of caloric restriction is superior for the achievement of weight loss. Most studies include diet and exercise together as weight-loss interventions making independent assessment of each component challenging. Thomson et al. found that addition of resistance training did not result in a significantly different weight loss in their three-armed RCT of diet alone, diet with aerobic exercise, or diet with aerobic and resistance exercise [21]. It is currently recommended that women with BMI > 25 should aim for weight loss via caloric restriction through balanced dietary approaches irrespective of diet composition [17]. In the general adult population, a target energy deficit of 2500 kJ daily is recommended for weight loss [22].

Diet

General recommendations—There is no single rule or pattern that applies to everyone and an individualized approach works best. The aim should be to lose weight at a safe and sustainable rate of 0.5–1 kg a week, and for most women, the initial advice should be to reduce their energy intake by 600 cal a day. The best way to achieve this goal is to swap unhealthy and high-energy food choices (fast food, processed food and sugary drinks, and alcohol) for healthier choices.

Very low–calorie diets—A very low–calorie diet involves consumption of less than 800 cal a day. These diets can lead to rapid weight loss, but they may not be suitable or desirable for everyone. They are not routinely recommended for weight loss. They are usually recommended in situations of obesity-related complications that would benefit from rapid weight loss. Such diets should not be followed for longer than 12 weeks at a time, and they should only be recommended under the supervision of a suitably qualified health-care professional.

Role of exercise

Evidence shows that exercise benefits overweight women with PCOS even in the absence of significant weight loss [12,16,17]. Thrice-weekly moderate exercise for at least 30 minutes has been demonstrated to reduce BMI, waist circumference, and insulin resistance in young PCOS women [23]. Another study of a 12-week regimen of aerobic exercise for 3 hours weekly in overweight PCOS women resulted in reduced insulin resistance, serum triglyceride levels, and total and abdominal fat mass despite no weight changes [24].

Yet another study compared exercise versus dietary interventions in 40 obese PCOS women with exercise producing greater menstrual regularity and ovulation compared to diet. Data analysis suggests that exercise yielded greater reductions in insulin resistance leading to ovulation, compared with dietary intervention, which led to a greater reductions in body weight, BMI, and waist circumference [25].

Although individual woman's circumstances differ, a general plan to lose weight before pregnancy should include the following:

1. Preconception appointment for pregnancy planning.
2. Healthy diet with planned calorie restriction.
3. Formal dietitian's advice where necessary.
4. Psychological support.
5. At least 150 minutes a week of moderate aerobic activity or 75 minutes a week of vigorous aerobic activity, or a combination of moderate and vigorous activity—preferably spread throughout the week.
6. It is also recommended that adults should do strength exercises and balance training (such as gym workout) 2 days a week and break up their sedentary time. It is important that every individual creates a plan suited to their own personal needs and circumstances, with achievable and motivating goals.

A recent metaanalysis found that the weight-loss interventions, particularly diet and exercise, improved pregnancy rates and ovulatory status [2]. There were improved spontaneous pregnancy rates but not IVF pregnancies. Miscarriage rates remained unchanged by weight-loss interventions. The metaanalyses also showed that weight-loss lifestyle interventions had a nonsignificant advantage over weight-loss medications such as metformin with respect to achievement of pregnancy or improvement of ovulation rates. There was also no significant difference in menstrual cycles' improvement with metformin. In light of these findings, and the gastrointestinal side effects common with metformin, it was recommended that lifestyle interventions should remain the first-line therapy for improvement in ovulation and menstruation.

Weight-loss medications and fertility outcomes

Pharmacological agents are mainly indicated when patients fail to lose significant weight despite lifestyle changes and a low-calorie diet. These agents have been shown to induce modest weight loss but are not suitable for long-term weight maintenance [26]. Several studies have assessed the use of oral medications for weight loss for improving fertility. These have mainly included metformin—an insulin sensitizer, orlistat (a lipase inhibitor), sibutramine (a selective serotonin and norepinephrine reuptake inhibitor), and liraglutide [a glucagon-like peptide-1 (GLP-1) receptor agonist] [2,26]. Of these, *sibutramine* has been withdrawn in Europe and the United States but is still available on the Internet. When prescribing the appropriate weight-losing drug, it is paramount to consider the safety of these drugs should a woman conceive while taking them. Sibutramine has been shown in a large study to have a risk of cardiovascular defects in unborn infants [27]. The same study showed no risk of birth defects from orlistat use. The safety of acarbose in pregnancy is not established. The use of weight-loss medications is contraindicated during pregnancy. Out of all the drugs mentioned previously, pharmacokinetics of the orlistat places it in a favorable position due to its low absorption and first-pass metabolism resulting in a bioavailability of less than 1%.

Overall, the effects of weight-loss medications on weight and on the obesity-related characteristics of PCOS remain unclear, because of limited number of studies with small sample sizes and a short duration of follow-up. Until further evidence is available, lifestyle interventions should still be considered the first-line therapy, with drug use largely reserved for monitored trials.

Metformin

Metformin is an oral antihyperglycemic drug that acts as an insulin sensitizer in the treatment of diabetes mellitus type 2. It is one of the most common medications used to counter the metabolic effects of PCOS. Metformin is a synthetically derived biguanide that decreases hepatic glucose production and intestinal absorption of glucose, while increasing the peripheral uptake and utilization of glucose [28]. It also stimulates fat oxidation and reduces fat synthesis and storage. Metformin decreases "insulin resistance" and has also been reported to act as an aromatase inhibitor. The liver is the primary site of action of metformin. Metformin is not metabolized and is excreted unchanged in the urine and bile. Not only does metformin inhibit the production of hepatic glucose and thereby decrease insulin secretion, but it also enhances insulin sensitivity at the cellular level. These effects result in lowering of blood concentrations of glucose without the associated risk of either hypoglycemia or weight gain. Metformin inhibits the activity of the respiratory electron transport chain in mitochondria leading to reduced oxidative phosphorylation and ATP production and activates a cytoplasmic protein kinase called 5′AMP-activated protein kinase (AMPK) [29,30]. AMPK is present in many tissues and this explains the multiorgan actions of metformin in the human body. Metformin appears to promote weight loss and offers protection from the macrovascular complications of diabetes independently of its hypoglycemic actions [31]. These effects may be achieved through reduced atherogenesis, less oxidative stress, and redistribution of visceral fat [31].

Metformin is administered orally in doses of 1500–2000 mg in divided daily doses. The most common side effects of metformin are gastrointestinal in nature (9%) such as nausea, vomiting, and diarrhea. Such side effects decrease with time and can be lessened by dose reduction and taking the metformin with food. The most serious side effect, lactic acidosis, is rarely seen in patients with normal renal and hepatic function, occurring in just 3/100,000 patient-years of use [32]. Other less common adverse effects include taste changes, elevated liver enzymes, and skin erythema or urticaria. Metformin has altered pharmacokinetics during pregnancy, and it undergoes significant placental transfer. Considering physiological changes of pregnancy and pharmacokinetics of metformin, patients with inadequate glycemic control might require higher doses of metformin during pregnancy [33]. However, the impact of doses exceeding 2500 mg/day during pregnancy on maternal, fetal, and neonatal safety has not been determined. Recent data have shown that the use of metformin during the first 12 weeks of gestation or more reduced the development of gestational diabetes and did not influence the health of babies. No obstetric complications or congenital anomalies were described [34]. Clinical data suggest that metformin use during the first trimester of pregnancy is not teratogenic [35].

A metaanalysis consisting of 608 PCOS women found that combined lifestyle intervention and metformin were associated with reduced BMI and adipose tissue as well as increased menstrual cycles compared to lifestyle intervention alone [36]. However, there were no differences in other parameters such as insulin resistance, lipids, and blood pressure, and weight and BMI were similar in comparison with metformin monotherapy and lifestyle intervention. Metformin is generally recommended as a second-line therapy after lifestyle interventions to suppress metabolic and reproductive issues arising from PCOS and insulin resistance. It is also recommended periconceptionally to reduce the risk of ovarian hyperstimulation syndrome with IVF [37]. Another systematic review and metaanalysis compared the effect of lifestyle modification + metformin versus lifestyle modification and/or placebo versus metformin alone with lifestyle modification and/or placebo in PCOS on anthropometric, metabolic, reproductive, and psychological outcomes [36]. Of 2372 identified studies, 12 RCTs were included for analysis comprising 608 women with PCOS. Lifestyle + metformin were associated with lower BMI [mean difference (MD) -0.73 kg/m^2, 95% confidence intervals (CIs) -1.14, -0.32, $P = .0005$], and subcutaneous adipose tissue (MD -92.49 cm^2, 95% CIs -164.14, -20.84, $P = .01$) and increased number of menstrual cycles (MD 1.06, 95% CIs 0.30, 1.82, $P = .006$) after 6 months compared with lifestyle ± placebo. There were no differences in other anthropometric, metabolic (surrogate markers of insulin resistance, fasting and area under the curve glucose, lipids and blood pressure), reproductive (clinical and biochemical hyperandrogenism), and psychological (quality of life) outcomes after 6 months between lifestyle + metformin compared with lifestyle ± placebo. With metformin alone compared with lifestyle ± placebo, weight and BMI were similar after 6 months, but testosterone was lower with metformin. The authors concluded that lifestyle + metformin is associated with lower BMI and subcutaneous adipose tissue and improved menstruation in women with PCOS compared with lifestyle with/without placebo over 6 months. Metformin alone compared with lifestyle showed similar BMI at 6 months. Although the study was limited by small sample size, the results suggest that the combination of lifestyle with metformin has a role to play in weight management in women with PCOS [36]. In 2006 Tang et al. evaluated the combined effects of lifestyle modification and metformin in 143 obese anovulatory women (BMI > 30 kg/m^2) with PCOS in a prospective randomized, double-blind, placebo-controlled multicenter study [38]. All the patients had an individualized assessment by a research dietitian in order to set a realistic goal that could be sustained for a long period of time with an average reduction of energy intake of 500 kcal/day. As a result, both the metformin-treated and placebo groups managed to lose weight, but the amount of weight reduction did not differ between the two groups. An increase in menstrual cyclicity was observed in those who lost weight but again did not differ between the two arms of the study. Interestingly, it has been observed that nonobese women with PCOS respond better to metformin than obese women [39].

A Cochrane review in 2017 assessed the effectiveness of insulin-sensitizing drugs in improving reproductive outcomes and metabolic parameters for women with PCOS [40]. The review assessed the interventions metformin, clomiphene citrate, metformin + clomiphene citrate, D-chiro-inositol, rosiglitazone, and pioglitazone in comparison to each other, placebo or no treatment. Of the 48 studies (4451 women), 42 of which investigated metformin (4024 women). Evidence quality ranged from very low to moderate. The findings suggested that metformin alone may be beneficial over placebo for live birth, although the evidence quality was low. When metformin was compared with clomiphene citrate, data for live birth rates were inconclusive. Results differed by BMI, emphasizing the importance of stratifying results by BMI.

The clinical pregnancy and ovulation rates suggested that clomiphene citrate remained preferable to metformin alone for OI in obese women with PCOS. However, an improved clinical pregnancy and ovulation rate with a combination of metformin and clomiphene citrate versus clomiphene citrate alone suggested that combined therapy may be useful.

Based on the current best evidence, metformin is mainly used in women with PCOS as a second-line option for OI (either for clomiphene resistance or combined with clomiphene), especially in those with a BMI of 35 and over. Metformin is also used along with weight management strategies as a first-line agent in women with high BMI and wishing to pursue fertility treatment in near future. Metformin should be considered in women with PCOS who are already undergoing lifestyle treatment and do not show improvement in impaired glucose tolerance and those with established abnormal glucose tolerance.

Sibutramine

Sibutramine is no longer recommended in clinical practice because of the risk of serious cardiovascular problems in some patients who take it [27]. Sibutramine was used for the management of obesity, including weight the loss and maintenance of weight loss, and was used in conjunction with a reduced-calorie diet. Sibutramine blocks the reuptake of the neurotransmitters dopamine, norepinephrine, and serotonin. Women taking sibutramine may achieve a 5%–10% reduction from their baseline weight. In addition, sibutramine-assisted weight loss has been accompanied by improvement in serum lipid profile.

Orlistat

Orlistat inhibits pancreatic lipase, resulting in a 30% reduction in the absorption of ingested fat leading to weight loss [41]. Orlistat is usually recommended in conjunction with a mildly hypocaloric diet for the treatment of obese women with a BMI $\geq$ 30 kg/m^2, or overweight patients (BMI $\geq$ 28 kg/m^2) with associated risk factors. Orlistat is recommended once the woman has made a significant effort to lose weight through diet, exercise, or changing her lifestyle.

Treatment with orlistat must be combined with a balanced low-fat diet and other weight-loss strategies, such as exercise. It is important that the diet is nutritionally balanced over three main meals. If a meal is missed or contains no fat, the dose of orlistat should be omitted. The patient should be on a nutritionally balanced, mildly hypocaloric diet that contains approximately 30% of calories from fat. It is recommended that the diet should be rich in fruit and vegetables. A single orlistat capsule (120 mg) should be taken with water immediately before, during, or up to 1 hour after each main meal (up to a maximum of three capsules a day). Treatment with orlistat should continue beyond 3 months if there is loss of 5% of body weight. Women with type 2 diabetes may take longer to lose weight using orlistat, so the target weight loss after 3 months should, therefore, be slightly lower. If weight loss is demonstrated, orlistat should be continued for 12 months or more. The gastrointestinal effects of orlistat result in an increase in fecal fat as early as 24–48 hours after dosing. Upon discontinuation of therapy, fecal fat content usually returns to pretreatment levels within 48–72 hours. Common side effects of orlistat include steatorrhea, diarrhea, flatulence, abdominal discomfort, headaches, and upper respiratory tract infections. Women taking the combined OCP should use an additional method of contraception, if they experience severe diarrhea while taking orlistat. Treatment with orlistat should be discontinued after 12 weeks if patients have been unable to lose at least 5% of the body weight as measured at the start of therapy. The effect of orlistat in patients with hepatic and/or renal impairment, children, and elderly patients has not been studied. Contraindications to the use of orlistat include hypersensitivity to the active drug substance or to any of the excipients, chronic malabsorption syndrome and cholestasis. Orlistat is not recommended for pregnant or breastfeeding women.

Orlistat was shown to be superior to a control with respect to achievement of pregnancy and ovulation in a single study [41]. In a randomized study that compared orlistat with placebo in patients with PCOS ($n = 100$), orlistat reduced weight by 6.4%, reduced testosterone levels, lowered serum levels of low-density lipoprotein C and triglycerides, and increased serum levels of high-density lipoprotein C [42]. In a more recent large randomized study ($n = 149$), orlistat combined with lifestyle modification (caloric restriction and increased physical activity) for 16 weeks induced similar reductions in weight as compared to orlistat and lifestyle modification combined with OCPs (ethinyl estradiol 20 μg/1 mg norethindrone acetate) and greater than the pill alone (6.2%, 6.4%, and 1.0%, respectively) [19]. Rates of ovulation, conception, clinical pregnancy, and live birth did not differ between the two groups but fecundity per patient who ovulated was higher in the group that received only orlistat and lifestyle advice than in the group that received only the pill [19]. Orlistat appears to be equally effective with metformin in reducing weight, IR, and testosterone levels but has not been compared with other antiobesity agents.

Liraglutide

Liraglutide (GLP-1 receptor agonist) action is mediated via a specific interaction with GLP-1 receptors, leading to an increase in cyclic adenosine monophosphate [43]. Liraglutide stimulates insulin secretion in a glucose-dependent manner. Simultaneously, liraglutide lowers inappropriately high glucagon secretion, also in a glucose-dependent manner. Thus when blood glucose is high, insulin secretion is stimulated and glucagon secretion is inhibited. Conversely, during hypoglycemia, liraglutide diminishes insulin secretion and does not impair glucagon secretion. Its mechanism of blood glucose lowering also involves a minor delay in gastric emptying. Liraglutide reduces body weight and body fat mass through mechanisms involving reduced hunger and lowered energy intake as GLP-1 is a physiological regulator of appetite and food intake. It induces weight loss by suppressing appetite and by increasing postprandial satiety and fullness [43]. Liraglutide has been utilized in the management of patients with BMI > 30 kg/m^2 or BMI 27−30 kg/m^2 and obesity-related comorbidities. To improve the gastrointestinal tolerability, the recommended starting dose is 0.6 mg daily. After at least 1 week the dose should be increased to 1.2 mg. Some patients are expected to benefit from an increase in dose from 1.2 to 1.8 mg and based on clinical response; after at least 1 week, the dose can be increased to 1.8 mg. If liraglutide is added to a sulfonylurea or insulin, a reduction in the dose of sulfonylurea or insulin should be considered to reduce the risk of hypoglycemia. Common side effects of liraglutide include nausea, vomiting, stomach upset, decreased appetite, diarrhea, and constipation. In a metaanalysis of randomized controlled studies that evaluated the effects of liraglutide in patients with PCOS, a reduction in weight and in circulating androgens was observed but markers of insulin resistance did not change [44,45]. Regarding the safety of liraglutide, it should be mentioned that this agent should not be used during pregnancy. If pregnancy occurred during treatment, therapy should be discontinued.

In a systematic review and network metaanalysis by Khera et al., direct and indirect evidence from 28 RCTs in 29,018 overweight and obese patients was combined to compare the association of weight-loss drugs with relative weight loss and adverse events [46]. The review found that with at least 1 year of treatment, naltrexone−bupropion, phentermine−topiramate, and liraglutide were all associated with higher odds of achieving weight loss compared with placebo, with moderate confidence in estimates. Liraglutide was associated with higher odds of weight loss of at least 5% and weight loss of at least 10% compared with orlistat and naltrexone−bupropion, with low-to-moderate confidence in estimates, but was associated with higher odds of discontinuation due to adverse events [46]. In conclusion, metformin and orlistat appear to be the two medications that seem to have a role in the management of overweight and obese women who wish to conceive. Most studies with these medications have been conducted in women with PCOS and high BMI. More studies are needed to clarify the role of liraglutide and other medications such as acarbose, phentermine−topiramate, or naltrexone/bupropion in this population.

Barriers to weight loss

Few studies have attempted to explore the barriers that overweight women face when attempting weigh loss. Based on limited evidence, overweight subfertile women appear most deterred from exercise by the perception that it causes tiredness and is hard work. Such perceptions seem to decrease with continuation of an exercise program. Compliance with weight-loss strategies is the key to success, and coached sessions of achievable frequency or motivational techniques should be considered. Effective weight management programs should include behavioral changes to increase the person's physical activity levels [17]. These could include setting goals, stimulus control, and relapse prevention. Unduly restrictive and nutritionally unbalanced diets are ineffective in the long term. A multidisciplinary, holistic approach to weight loss, including primary care physician, gynecologist, endocrinologist, exercise physiologist, dietitian, and psychologist is recommended in women with PCOS who have established metabolic complications.

Conclusion

In overweight and obese subfertile women, weight loss is associated with improved chances of becoming pregnant naturally or through fertility treatment. Weight loss also improves ovulation frequency and aids menstrual regularity.

PCOS is the most common endocrine disorder in women of reproductive age. Obesity is frequently present in these patients and plays a key role in the pathogenesis of both the endocrine and metabolic abnormalities of the syndrome. Recent studies show that weight-loss interventions, particularly those with reducing diets and exercise, are more likely than controls to result in pregnancy. Many women are able to conceive without further assistance through weight-loss alone. Weight loss interventions appear to have an advantage over medications such as metformin with respect to the achievement of pregnancy or improvement of ovulation status. Overweight and obese women seeking fertility should be educated on the effects of high BMI on fertility, pregnancy, and reproductive health. Lifestyle interventions remain the first-line therapy for improvement in ovulation and menstruation. A combination of a reduced-calorie diet, which is not overly restrictive, and aerobic exercise, intensified gradually, should be recommended.

The effects of antiobesity agents on weight and obesity-related characteristics of the PCOS remain unclear. Metformin, in addition to lifestyle changes, appears to be useful in women with PCOS and high BMI who are attempting weight loss prior to achieving pregnancy. Several studies have shown that orlistat induces weight loss, improves insulin resistance, and reduces androgen levels in PCOS. There are limited data, which suggest that liraglutide results in weight loss in women with PCOS. More studies are needed to clarify the role of antiobesity agents as weight-loss intervention prior to fertility treatment.

References

[1] World Health Organization. Obesity and overweight fact sheet no 311. 2016.

[2] Best D, Avenell A, Bhattacharya S. How effective are weight-loss interventions for improving fertility in women and men who are overweight or obese? A systematic review and meta-analysis of the evidence. Hum Reprod Update 2017;23:681–705.

[3] Vahratian A, Smith YR. Should access to fertility-related services be conditional on body mass index? Hum Reprod 2009;7:1532–7.

[4] Gesink Law DC, Maclehose RF, Longnecker MP. Obesity and time to pregnancy. Hum Reprod 2007;22:414–20.

[5] Zain M, Norman R. Impact of obesity on female fertility and fertility treatment. Womens Health 2008;2:183–94.

[6] Klenov V, Jungheim E. Obesity and reproductive function: a review of the evidence. Curr Opin Obstet Gynecol 2014;26:455–60.

[7] Carrell DT, Jones KP, Peterson CM, et al. Body mass index is inversely related to intrafollicular hCG concentrations, embryo quality and IVF outcome. Reprod Biomed Online 2001;2:109–11.

[8] Bellver J, Martinez-Conejero JA, Labarta E, et al. Endometrial gene expression in the window of implantation is altered in obese women especially in association with polycystic ovary syndrome. Fertil Steril 2011;95:2335–41.

[9] Balen A, Dresner M, Scott E, et al. Should obese women with polycystic ovary syndrome receive treatment for infertility? BMJ 2006;332:434–5.

[10] Balen A, Platteau P, Andersen A, et al. The influence of body weight on response to ovulation induction with gonadotrophins in 335 women with World Health Organization group II anovulatory infertility. BJOG 2006;10:1195–202.

[11] Bellver J, Ayllon Y, Ferrando M, et al. Female obesity impairs in vitro fertilization outcome without affecting embryo quality. Fertil Steril 2010;93:447–54.

[12] PCOS Australian Alliance. Evidence-based guideline for the assessment and management of polycystic ovary syndrome. Melbourne: Jean Hailes Foundation for Women's Health on behalf of the PCOS Australian Alliance; 2011.

[13] DeUgarte CM, Bartolucci AA, Azziz R. Prevalence of insulin resistance in the polycystic ovary syndrome using the homeostasis model assessment. Fertil Steril 2005;83:1454–60.

[14] Legro RS, Dodson WC, Kunselman AR, et al. Benefit of delayed fertility therapy with preconception weight loss over immediate therapy in obese women with PCOS. J Clin Endocrinol Metab 2016;101:2658–66.

[15] Farquhar C, Gillett W. Prioritising for fertility treatments? Should a high BMI exclude treatment? BJOG 2006;10:1107–9.

[16] Wild RA, Carmina E, Diamanti-Kandarakis E, et al. Assessment of cardiovascular risk and prevention of cardiovascular disease in women with the polycystic ovary syndrome: a consensus statement by the Androgen Excess and Polycystic Ovary Syndrome (AE-PCOS) Society. J Clin Endocrinol Metab 2010;95:2038–49.

[17] De Sousa SMC, Norman RJ. Metabolic syndrome, diet and exercise. Best Pract Res Clin Obstet Gynaecol 2016;37:140–51.

[18] Moran LJ, Hutchison SK, Norman RJ, et al. Lifestyle changes in women with polycystic ovary syndrome. Cochrane Database Syst Rev 2011; http://dx.doi.org/10.1002/14651858.CD007506.pub3.

[19] Legro RS, Dodson WC, Kris-Etherton PM, et al. Randomized controlled trial of preconception interventions in infertile women with polycystic ovary syndrome. J Clin Endocrinol Metab 2015;100:4048–58.

[20] Hardy T, De Sousa S, Norman RJ. Polycystic ovary syndrome: prognosis and risk of comorbidity. In: Diamanti-Kandarakis E, Nader S, Panidis D, editors. Novel insights into the pathophysiology and treatment of PCOS. London: Future Medicine Ltd; 2013. p. 122–34.

[21] Thomson RL, Buckley JD, Noakes M, et al. The effect of a hypocaloric diet with and without exercise training on body composition, cardiometabolic risk profile, and reproductive function in overweight and obese women with polycystic ovary syndrome. J Clin Endocrinol Metab 2008;9:3373–80.

[22] National Health and Medical Research Council. Clinical practice guidelines for the management of overweight and obesity in adults, adolescents and children in Australia. Melbourne: National Health and Medical Research Council; 2013.

[23] Vigorito C, Giallauria F, Palomba S, et al. Beneficial effects of a three-month structured exercise training program on cardiopulmonary functional capacity in young women with polycystic ovary syndrome. J Clin Endocrinol Metab 2007;92:1379–84.

[24] Hutchison SK, Stepto NK, Harrison CL, et al. Effects of exercise on insulin resistance and body composition in overweight and obese women with and without polycystic ovary syndrome. J Clin Endocrinol Metab 2011;96:E48–56.

[25] Palomba S, Giallauria F, Falbo A, et al. Structured exercise training programme versus hypocaloric hyperproteic diet in obese polycystic ovary syndrome patients with anovulatory infertility: a 24-week pilot study. Hum Reprod 2008;23:642–50.

[26] Hauner H, Petzinna D, Sommerauer B, et al. Effect of acarbose on weight maintenance after dietary weight loss in obese subjects. Diabetes Obes Metab 2001;6:423–7.

[27] Källén BA. Antiobesity drugs in early pregnancy and congenital malformations in the offspring. Obes Res Clin Pract 2014;8:571–6.

[28] Klepser TB, Kelly MW. Metformin hydrochloride: an antihyperglycemic agent. Am J Health Syst Pharm 1997;54:893–903.

[29] El-Mir MY, Nogueira V, Fontaine E, et al. Dimethylbiguanide inhibits cell respiration via an indirect effect targeted on the respiratory chain complex I. J Biol Chem 2000;275:223–8.

[30] Owen MR, Doran E, Halestrap AP. Evidence that metformin exerts its anti-diabetic effects through inhibition of complex 1 of the mitochondrial respiratory chain. Biochem J 2000;348:607–14.

[31] Sivalingam VN, Myers J, Nicholas S, et al. Metformin in reproductive health, pregnancy and gynecological cancer: established and emerging indications. Hum Reprod Update 2014;20:853—68.

[32] Howlett HC, Bailey CJ. A risk-benefit assessment of metformin in type 2 diabetes mellitus. Drug Saf 1999;20:489—503.

[33] Ryu RJ, Hays KE, Hebert MF. Gestational diabetes mellitus management with oral hypoglycemic agents. Semin Perinatol 2014;38:508—15.

[34] Glueck CJ, Goldenberg N, Pranikoff J, et al. Height, weight, and motor-social development during the first 18 months of life in 126 infants born to 109 mothers with polycystic ovary syndrome who conceived on and continued metformin through pregnancy. Hum Reprod 2004;19:1323—30.

[35] Bertoldo MJ, Faure M, Dupont J, et al. Impact of metformin on reproductive tissues: an overview from gametogenesis to gestation. Ann Transl Med 2014;2:55.

[36] Naderpoor N, Shorakae S, de Courten B, et al. Metformin and lifestyle modification in polycystic ovary syndrome: systematic review and meta-analysis. Hum Reprod Update 2015;21:560—74.

[37] Tosca L, Uzbekova S, Chabrolle C, et al. Possible role of 5′AMP-activated protein kinase in the metformin-mediated arrest of bovine oocytes at the germinal vesicle stage during in vitro maturation. Biol Reprod 2007;77:452—65.

[38] Tang T, Glanville J, Hayden CJ, et al. Combined life-style modification and metformin in obese patients with polycystic ovary syndrome (PCOS). A randomised, placebo-controlled, double-blind multi-centre study. Hum Reprod 2006;21:80—9.

[39] Maciel GAR, Junior JMS, Motta ELA, et al. Nonobese women with polycystic ovary syndrome respond better than obese women to treatment with metformin. Fertil Steril 2004;81:355—60.

[40] Morley LC, Tang T, Yasmin E, et al. Insulin-sensitising drugs (metformin, rosiglitazone, pioglitazone, D-chiro-inositol) for women with polycystic ovary syndrome, oligo amenorrhoea and subfertility. Cochrane Database Syst Rev 2017;11:CD003053.

[41] Kumar P, Arora S. Orlistat in polycystic ovarian syndrome reduces weight with improvement in lipid profile and pregnancy rates. J Hum Reprod Sci 2014;4:255—61.

[42] Moini A, Kanani M, Kashani L, et al. Effect of orlistat on weight loss, hormonal and metabolic profiles in women with polycystic ovarian syndrome: a randomized double-blind placebo-controlled trial. Endocrine 2015;49:286—9.

[43] van Can J, Sloth B, Jensen CB, et al. Effects of the once-daily GLP-1 analog liraglutide on gastric emptying, glycemic parameters, appetite and energy metabolism in obese, non-diabetic adults. Int J Obes (Lond) 2014;38:784—93.

[44] Niafar M, Pourafkari L, Porhomayon J, et al. A systematic review of GLP-1 agonists on the metabolic syndrome in women with polycystic ovaries. Arch Gynecol Obstet 2016;293:509—15.

[45] Chatzis P, Tziomalos K, Pratilas GC, et al. The role of antiobesity agents in the management of polycystic ovary syndrome. Folia Med (Plovdiv) 2018;60:512—20.

[46] Khera R, Murad MH, Chandar AK, et al. Association of pharmacological treatments for obesity with weight loss and adverse events: a systematic review and meta-analysis. JAMA 2016;315:2424—34.

Chapter 17

Surgical interventions to improve fertility potential in obese men and women

Joseph Chervenak[1] and Frank A. Chervenak[2]

[1]Obstetrics and Gynecology, New York Presbyterian/Weill Cornell, New York, NY, United States, [2]Obstetrics and Gynecology, Lenox Hill Hospital, Zucker School of Medicine at Hofstra/Northwell, New York, NY, United States

Introduction

Obesity is a worsening worldwide epidemic now affecting as many as 2 billion people worldwide [1]. Women of reproductive age are not immune to this burden—in the United States, over one-third of adult women are obese, and the majority of women who are pregnant have a BMI above 25 [2]. Obesity and associated conditions such as diabetes have an adverse impact on pregnancy from conception to after delivery. It has been associated with decreased fertility, increased time to conceive, and increased morbidity if women do become pregnant [3]. Weight loss can be achieved through conservative, medical, and surgical management. Bariatric procedures have the potential to improve fertility through effective and durable management of obesity.

Obesity and fertility

An essential component of assessment and optimization of natural fertility is consideration of the female partner's weight. Despite this, only one-third of obese patients receive advice from health-care providers regarding weight reduction [4]. Being overweight in early adulthood increases the risk of menstrual irregularities, ovulatory dysfunction, and consequently subfertility. Studies have demonstrated that 12 months of unprotected intercourse is less likely to result in conception in the obese as compared to normal-weight female [5]. Prospective studies have demonstrated that high levels of central and overall adiposity are associated with decreased fecundability, even when adjusting for confounders. Further, vigorous physical activity is found to increase fecundity rates only in the overweight but not normal-weight population [6].

Obesity's established negative impact on reproductive potential is multifactorial, and increased adiposity can influence almost every stage of fertilization from ovulation to successful implantation and development of the embryo. It is thought that one of the primary mediators of obesity's effect is functional alteration in the hypothalamic–pituitary–ovarian axis. Generally, high levels of insulin in obese patients stimulate the ovaries to produce androgens that in turn are aromatized to estrogen in adipose tissue. Increased serum levels of estrogen inhibit gonadotropin release in the pituitary and produce the clinical presentation of menstrual irregularities and ovulatory dysfunction. This process appears to be exacerbated in the setting of polycystic ovarian syndrome (PCOS) of which hyperinsulinemia has been implicated in the development of this disease [7].

Clinically, obesity has been linked to menstrual cycle abnormalities, ovulatory dysfunction, and decreased ovarian responsiveness in patients undergoing IVF. Menstrual irregularity is more than twice as common in the population that is above 175% of ideal body weight compared to those less than 150% of this figure. Cross-sectional studies have shown those who are obese in childhood are more likely than normal-weight children to remain childless later in life [4]. Even obese patients who have regular cycles have decreased fecundity. A Dutch study examining over 3000 women with regular menstrual cycles found that every point increase in BMI above 29 was associated with a 4% lower spontaneous pregnancy rate [8]. Obese patients undergoing IVF require higher doses of medication for ovarian stimulation. Further, systematic review of studies involving IVF of women with BMI > 25 has a 10% lower success rate than those with less than this figure [9]. The deleterious effects of obesity appear to extend to embryo implantation. Obesity has been linked to pathologic changes in oocyte morphology and overall quality in certain studies. Generally, increased BMI > 25 is associated with smaller sized oocytes that are less likely to complete development after fertilization [10].

Obesity and Gynecology. DOI: https://doi.org/10.1016/B978-0-12-817919-2.00017-6
© 2020 Elsevier Inc. All rights reserved.

Nonsurgical management of obesity

The focus of this chapter is surgical interventions that improve fertile potential; however, it is important to note that lifestyle modification and medical therapies represent an important adjunct to any considered bariatric procedure and should be pursued before considering surgery. Weight management can be achieved in many individuals through changes to diet and physical activity. Critically, in those patients undergoing surgery, success will in large part be determined by a patient's ability to adhere to a lifestyle regimen compatible with weight loss. It is important to note that at present, more evidence links weight reduction achieved without surgical intervention to improved spontaneous pregnancy rate and IVF outcomes than weight loss achieved with surgery. As will be addressed, the link between major bariatric procedures and improved fertility-related outcomes is challenging to establish.

Bariatric surgery as a weight loss measure

While dietary and lifestyle interventions remain a critical first step in weight optimization, bariatric surgery represents the most successful treatment that results in sustained long-term weight loss [11]. The percentage of body weight loss at 2 years after bariatric surgery can approach 60%, with continued weight loss benefit demonstrated at intervals up to 20 years after the procedure [12]. As of 2011 over 300,000 of these procedures are performed annually, with over 100,000 occurring in the United States alone. In Asia the absolute growth rate of procedures performed has exceeded 400% in recent years. Surgical complications such as bowel obstruction or hernia are rare. Indications for this procedure are well established by the American Bariatric Society (ABS) and suggested that candidates should have at least a BMI > 40 without serious comorbidities (e.g., diabetes, hypertension) or at least a BMI > 35 in the presence of one serious comorbidity. Candidates with BMI < 35 are considered if they have uncontrolled type 2 diabetes or metabolic syndrome. Absolute contraindications include serious depression or psychosis, eating disorder, alcohol abuse, heart disease, and coagulopathy [13]. It is estimated that only 1% of the population that meets the criteria for these procedures undergoes surgery [14].

Patients should be carefully screened to optimize the success of a considered procedure. Psychological assessment and appropriate setting of expectations before surgery is essential. Sustainable weight loss after a bariatric procedure requires significant lifestyle changes. Preoperative assessments include reviewing a patient's previous attempts at weight loss, eating and dietary styles, physical activity, and history of substance abuse. Patients' expectations should be in line with the reality of post-–bariatric surgery experience, particularly those motivated by fertility. A review of the patients' medical comorbidities and consequent suitability for anesthesia are the key determinants of candidacy for surgery.

Types of bariatric surgery

Bariatric procedures are characterized as restrictive, malabsorptive, or a combination of the two. Restrictive surgeries such as the sleeve gastrectomy aim to create a smaller gastric pouch that reaches capacity soon after food consumption to induce satiety. Malabsorptive procedures such as the Roux-en-Y gastric bypass limit energy intake by effectively shortening the length of the gastrointestinal tract. Since 2013 the sleeve gastrectomy has become the most commonly performed bariatric procedure in the United States [14]. Sleeve gastrectomy has been described as a less complex procedure than bypass with less short-term morbidity. Bypass, with its impact on absorption, can produce a higher degree of weight loss over a shorter time frame. The type of procedure an obese patient undergoes has special relevance in those who desire subsequent pregnancy. Malabsorptive procedures and consequent nutritional deficiencies are thought to drive the increase in intrauterine growth restriction and small for gestational age (SGA) infants seen in initial observational studies of pregnancy in bariatric surgery patients.

The sleeve gastrectomy is a partial gastrectomy in which the majority of the greater curvature of the stomach is removed, thus producing a more "tubular" stomach. This procedure was first offered to patients with so-called super-severe obesity or those with a BMI > 60 as the first-step procedure [15]. It is viewed as a less technically demanding procedure than alternatives and was initially a way to bridge these patients toward more efficacious alternatives. A sea change in the utilization of bariatric surgery has seen the sleeve gastrectomy rise to dominance in recent years. Over a 4-year period between 2008 and 2012, this procedure went from less than 1% of all bariatric procedures performed in the United States to over a third [16]. Much of the increase has been attributed to the relatively decreased 30-day morbidity of this procedure over the primary alternative, the Roux-en-Y bypass. A large study reviewing outcomes of laparoscopic gastrectomy versus bypass revealed that sleeve gastrectomy was associated with shorter operating time, a lower rate of blood loss, and a lower rate of reoperation among other morbidities [17]. With the increase in prevalence of this procedure, evidence has accumulated supporting its use as a primary approach. One of the main limitations of sleeve gastrectomy as a restrictive procedure was consideration of its relatively muted impact on alleviating the burden of metabolic diseases such as diabetes. Recent evidence has accumulated that contradicts this point of view. Several studies have noted an improvement in type 2 diabetes mellitus before weight loss is achieved in patients undergoing sleeve gastrectomies. The etiology for this effect is thought to be due to several factors, such as increased gastric emptying time and decreased secretion of gherlin—a neuropeptide secreted in the stomach that suppresses the sensation of satiety [18].

The second most commonly practiced type of bariatric surgery is the Roux-en-Y gastric bypass, the aforementioned mixed restrictive/malabsorptive procedure. This procedure is characterized by the creation of a small gastric pouch divided from the distal stomach and anastomosed to a "Roux" limb of small bowel. The proximal small intestine is then divided and a limb is created that drains secretions from the gastric remnant, liver, and pancreas. The net effect is to have two stomach chambers that anastomose to the small intestine at different points. The stomach remnant drains proximally, and the newly created smaller functional stomach drains distally through the Roux limb. The decreased size of the stomach acts to restrict caloric intake, and the bypass of a segment of small bowel has a malabsorptive effect. The expected 2-year weight loss after a Roux-en-Y procedure is about 70% compared to 60% for sleeve gastrectomy [19].

Other bariatric procedures which may be considered include the much less commonly performed biliopancreatic diversion with duodenal switch, the intragastric balloon, and vagal blockade. The biliopancreatic diversion involves manipulation of the pylorus, duodenum, and ileum and is considered a technically difficult operation with a significant complication rate. The intragastric balloon has promise as a bridge to another procedure and has been approved for patients at a lower BMI threshold than traditional bariatric surgery (BMI 30–34.9). It involves the placement of a soft saline-filled balloon into the stomach that promotes the sensation of satiety and gradually degrades after about 6 months. Vagal blockade involves the placement of an electric pulse generator that is designed to lead to the decreased sensation of hunger [20].

The impact of bariatric surgery on fertility

Given the efficacy of bariatric surgery on inducing and sustaining weight loss, it stands to reason that the various procedures would increase fertility potential by alleviating the burden of obesity. While existing evidence supports that the nonsurgical treatment of obesity improves fertility, the impact of bariatric surgery on relevant outcomes remains a target of investigation [21]. Initial data on the effects of weight loss surgery on fertility have been encouraging, however limited. Various studies support the conclusion that surgical treatment of obesity tends to reverse the altered reproductive hormone profile seen in this population [22]. For example, postprocedural examinations of relevant markers such as estradiol, sex hormone−binding globulin (SHBG), follicle-stimulating hormone, and luteinizing-hormone levels found relative normalization of these values as compared to the nonobese population. Further, bariatric surgery has been associated with profound changes extending to the hypothalamic−pituitary−adrenal axis. Influence on the secretion and activity of cortisol and thyroid hormones represent a potential mediator of surgery on fertility potential.

Studies examining clinical outcomes have linked weight loss surgery to restoration of ovulation in certain patients. In a survey of 195 patients with a history of bariatric surgery, 71% of women who were previously anovulatory regained normal menses after the procedure [23]. A smaller study involving 24 women with PCOS demonstrated the spontaneous return of ovulation of 5 women after surgical weight loss [24]. Generally, rates of ovulation correlated with success at sustaining postoperative weight loss. Taking it one step further, various small reports have attempted to examine the pregnancy potential of a post−surgical weight loss population. Small studies have described an increased rate of spontaneous conception in the morbidly obese who have achieved sufficient weight loss. A prospective study evaluating 52 patients desiring fertility roughly split between surgical and nonsurgical management of morbid obesity showed a higher pregnancy rate at 2 years in the surgical population. Interpretation in this study as in others is limited by small sample size and assessment of fertile potential prior to intervention [25]. While this supports a trend toward improvement in fertility status, affirming this relationship is challenged by a lack of large well-designed trials. Limited evidence precludes the designation of infertility as an indication for bariatric surgery at present.

Organizations such as the American Congress of Obstetricians and Gynecologists (ACOG) have highlighted the potential for bariatric surgery to improve fertility outcomes through restoration of ovulation and reversing pathologic changes in PCOS, however that it should "not be considered a treatment for infertility" [26]. The American Society for Reproductive Medicine (ASRM) in a committee opinion suggests that bariatric surgery appears safe in a population looking forward to becoming pregnant and has potential to lead to improvement in markers of reproductive health; however, it expresses concern that the benefits may be offset in women late in their reproductive years [4]. The recommended delay after any procedure poses a challenge in women whose window for successful pregnancy may be limited. The American Society for Metabolic and Bariatric Surgery (ASMBS) developed a position statement that was endorsed by ACOG and cited the paucity of data in expanding existing indications for bariatric surgery for fertility-based reasons [27].

Bariatric surgery and polycystic ovarian syndrome

PCOS has been cited as one of the most common causes of infertility. PCOS is generally defined as having at least two of the following: polycystic ovaries on ultrasound, clinical, or biochemical features of hyperandrogenism and/or menstrual dysfunction. PCOS has a well-established association with obesity and insulin resistance. Bariatric surgery represents a mechanism of not only increasing the fertile potential of those with the condition but also alleviating the burden of the disease. Similar to other studies looking at bariatric surgery and reproductive-related outcomes, quantitative data relating to PCOS are small and heterogenous. A prospective study of 14 patients with PCOS undergoing bariatric surgery showed postoperative improvement in clinical symptoms of the disorder, including amenorrhea and hirsutism. This improvement was in line with a decrease in levels of testosterone, fasting glucose, cholesterol, insulin, and triglyceride levels within a year after the bariatric procedure [28]. A large metaanalysis involving over 1000 patients undergoing bariatric surgery with a large proportion of women who met the criteria for PCOS suggested a dramatic decrease in 1-year postoperative rates of hirsutism and anovulation [29]. Improved fertility within the context of PCOS has been described in individual cases. However, data remain insufficient to recommend a bariatric procedure for treatment of either PCOS or its associated infertility beyond established indications.

The potential of bariatric surgery for a negative impact on fertility

Adding to the uncertainty regarding the link between surgery and fertility, there are conflicting reports, which have suggested that bariatric procedures may actually carry a negative impact on fertility in some patients. Notably, a review of pregnancy outcomes after bariatric surgery found that patients receiving the procedure were significantly more likely to require subsequent infertility treatments [30]. Nutritional deficiencies after malabsorptive procedures have been proposed as a potential mechanism for subfertility; however, this has not been proven.

Obesity in general is associated with decreased levels of anti-Müllerian hormone (AMH), an oft-cited proxy of ovarian reserve. This level has been shown to increase after significant weight loss. Contrary to expectations, bariatric surgery is actually noted to cause a significant decrease in AMH levels in women under 35. Importantly, this effect was not seen in those above 35. This change has been attributed to stress involved with operation and decreased absorption of precursors relevant in AMH production [31].

Pregnancy after bariatric surgery

Given the uncertainty of bariatric surgery at improving fertility potential, it is critical to counsel patients regarding the implication of having a procedure on subsequent pregnancies. One of the most important considerations is the interval after surgery for which attempted conception is recommended. Various organizations, including ACOG and the ASMBS, generally recommend delaying pregnancy for 12–24 months. This delay is advised as it allows women to take advantage of a critical window for weight loss and to avoid a high-risk period for nutritional deficiencies [26]. The largest cohort study to date has found a temporal relationship between surgery, pregnancy, and complications. Post–bariatric surgery patients who became pregnant within 2 years had higher rates of prematurity, neonatal intensive care unit admission, and SGA infants [32].

When considering the impact of bariatric surgery on pregnancy in general, it is important to consider that obesity itself is a risk factor for several adverse outcomes. Bariatric surgery may have the ability to alleviate some of this risk, however not reduce to levels seen in the nonobese population. A case in point is gestational diabetes, where a metaanalysis of 2.8 million pregnant women matched for presurgery BMI found that having a procedure was associated with an 80% decreased risk in developing the condition. Rates still remain elevated relative to the overall population [33]. Intuitively, bariatric surgery is significantly associated with decreased rates of large for gestational age infants and increased rates of SGA infants. Studies examining the impact of procedure on miscarriage, rates of preeclampsia, rates of preterm delivery, and congenital anomalies have been equivocal. Whether bariatric surgery can modify risk for miscarriage remains unclear. Rates of cesarean section have been shown to be higher in observational studies of the bariatric surgery population; however, this may be a consequence of these patients representing an older, obese cohort [34].

Assisted reproduction after bariatric surgery

Data are particularly limited in those who undergo IVF treatment after having bariatric surgery. Small case reports have demonstrated successful IVF cycles in these patients [35]. Of particular concern is that while post-bariatric surgery patients are not necessarily at increased risk for ovarian hyperstimulation syndrome (OHSS)—a complication of IVF and ovarian stimulation—they are more susceptible to the complications of this condition. OHSS can produce ascites and increased intraabdominal pressure, which can in turn increase the risk of known bariatric surgery complications such as intestinal obstruction and internal hernia. Another issue of concern in patients requiring assisted reproduction after bariatric surgery is adherence to the recommended postprocedural delay before conception. Age is a critical factor in portending successful IVF outcome and delay represents an important consideration.

Obesity in the male

As with women, the prevalence of obesity in men of reproductive age is increasing. Obesity in the male partner similarly has been associated with decreased reproductive function. Increased adiposity can lead to alterations in several sperm parameters, including decreased sperm count, concentration, and motility. The reproductive sex hormone profile can be impacted as well, as obesity has been linked to suppression of SHBG and the increased aromatization of androgens to estrogen [27]. Whether correcting obesity in the male partner of a couple attempting to conceive improves fertile potential remains an open question. Obese men achieving weight loss through medical and surgical means have been shown to increase quality of life, decrease rates of erectile dysfunction, and possibly improve derangements in reproductive hormone profile [36]. There is no specific data that has been able to characterize improvements in a couple's fertile potential through surgical weight loss in the male.

Conclusion

Bariatric surgery has the potential to induce and sustain significant weight loss in couples looking forward to becoming pregnant. Weight loss after surgery has been shown to improve markers for reproductive health and small studies have demonstrated improved fertility. At present, there is a lack of high-level clinical evidence to consider bariatric surgery primarily for fertility-based indications.

References

[1] Charalampakis V, Tahrani AA, Helmy A, Gupta JK, Singhal R. Polycystic ovary syndrome and endometrial hyperplasia: an overview of the role of bariatric surgery in female fertility. Eur J Obstet Gynecol Reprod Biol 2016;207:220–6.

[2] Johansson K, Cnattingius S, Näslund I, Roos N, Trolle Lagerros Y, Granath F, et al. Outcomes of pregnancy after bariatric surgery. N Engl J Med 2015;372(9):814–24.

[3] Poston L, Caleyachetty R, Cnattingius S, Corvalán C, Uauy R, Herring S, et al. Preconceptional and maternal obesity: epidemiology and health consequences. Lancet Diabetes Endocrinol 2016;4(12):1025–36.

[4] Practice Committee of the American Society for Reproductive Medicine. Obesity and reproduction: a committee opinion. Fertil Steril 2015;104(5):1116–26.

[5] Lake JK, Power C, Cole TJ. Women's reproductive health: the role of body mass index in early and adult life. Int J Obes 1997;21(6):432.

[6] McKinnon CJ, Hatch EE, Rothman KJ, Mikkelsen EM, Wesselink AK, Hahn KA, et al. Body mass index, physical activity and fecundability in a North American preconception cohort study. Fertil Steril 2016;106(2):451–9.

[7] Broughton DE, Moley KH. Obesity and female infertility: potential mediators of obesity's impact. Fertil Steril 2017;107(4):840–7.

[8] van der Steeg JW, Steures P, Eijkemans MJ, Habbema JDF, Hompes PG, Burggraaff JM, et al. Obesity affects spontaneous pregnancy chances in subfertile, ovulatory women. Hum Reprod 2007;23(2):324–8.

[9] Rittenberg V, Seshadri S, Sunkara SK, Sobaleva S, Oteng-Ntim E, El-Toukhy T. Effect of body mass index on IVF treatment outcome: an updated systematic review and meta-analysis. Reprod Biomed Online 2011;23(4):421–39.

[10] Pasquali R, Patton L, Gambineri A. Obesity and infertility. Curr Opin Endocrinol Diabetes Obes 2007;14(6):482–7.

[11] Maciejewski ML, Arterburn DE, Van Scoyoc L, Smith VA, Yancy WS, Weidenbacher HJ, et al. Bariatric surgery and long-term durability of weight loss. JAMA Surg 2016;151(11):1046–55.

[12] Sjöström L, Peltonen M, Jacobson P, Sjöström CD, Karason K, Wedel H, et al. Bariatric surgery and long-term cardiovascular events. JAMA 2012;307(1):56–65.

[13] Burguera B, Agusti A, Arner P, Baltasar A, Barbe F, Barcelo A, et al. Critical assessment of the current guidelines for the management and treatment of morbidly obese patients. J Endocrinol Investig 2007;30(10):844–52.

[14] English WJ, DeMaria EJ, Brethauer SA, Mattar SG, Rosenthal RJ, Morton JM. American Society for Metabolic and Bariatric Surgery estimation of metabolic and bariatric procedures performed in the United States in 2016. Surg Obes Relat Dis 2018;14(3):259–63.

[15] Almogy G, Crookes PF, Anthone GJ. Longitudinal gastrectomy as a treatment for the high-risk super-obese patient. Obes Surg 2004;14(4):492–7.

[16] Khorgami Z, Shoar S, Andalib A, Aminian A, Brethauer SA, Schauer PR. Trends in utilization of bariatric surgery, 2010-2014: sleeve gastrectomy dominates. Surg Obes Relat Dis 2017;13(5):774–8.

[17] Young MT, Gebhart A, Phelan MJ, Nguyen NT. Use and outcomes of laparoscopic sleeve gastrectomy vs laparoscopic gastric bypass: analysis of the American College of Surgeons NSQIP. J Am Coll Surg 2015;220(5):880–5.

[18] Deitel M, Gagner M, Erickson AL, Crosby RD. Third International Summit: current status of sleeve gastrectomy. Surg Obes Relat Dis 2011;7(6):749–59.

[19] Buchwald H, Oien DM. Metabolic/bariatric surgery worldwide 2011. Obes Surg 2013;23(4):427–36.

[20] Kral JG, Paez W, Wolfe BM. Vagal nerve function in obesity: therapeutic implications. World J Surg 2009;33(10):1995.

[21] Slopien R, Horst N, Jaremek JD, Chinniah D, Spaczynski R. The impact of surgical treatment of obesity on the female fertility. Gynecol Endocrinol 2019;35(2):100–2.

[22] Merhi ZO. Impact of bariatric surgery on female reproduction. Fertil Steril 2009;92(5):1501–8.

[23] Teitelman M, Grotegut CA, Williams NN, Lewis JD. The impact of bariatric surgery on menstrual patterns. Obes Surg 2006;16(11):1457–63.

[24] Eid GM, Cottam DR, Velcu LM, Mattar SG, Korytkowski MT, Gosman G, et al. Effective treatment of polycystic ovarian syndrome with Roux-en-Y gastric bypass. Surg Obes Relat Dis 2005;1(2):77–80.

[25] Marceau P, Kaufman D, Biron S, Hould FS, Lebel S, Marceau S, et al. Outcome of pregnancies after biliopancreatic diversion. Obes Surg 2004;14(3):318–24.

[26] American College of Obstetricians and Gynecologists. ACOG practice bulletin no. 105: bariatric surgery and pregnancy. Obstet Gynecol 2009;113(6):1405.

[27] Kominiarek MA, Jungheim ES, Hoeger KM, Rogers AM, Kahan S, Kim JJ. American Society for Metabolic and Bariatric Surgery position statement on the impact of obesity and obesity treatment on fertility and fertility therapy endorsed by the American College of Obstetricians and Gynecologists and the Obesity Society. Surg Obes Relat Dis 2017;13(5):750–7.

[28] Eid GM, McCloskey C, Titchner R, Korytkowski M, Gross D, Grabowski C, et al. Changes in hormones and biomarkers in polycystic ovarian syndrome treated with gastric bypass. Surg Obes Relat Dis 2014;10(5):787–91.

[29] Skubleny D, Switzer NJ, Gill RS, Dykstra M, Shi X, Sagle MA, et al. The impact of bariatric surgery on polycystic ovary syndrome: a systematic review and meta-analysis. Obes Surg 2016;26(1):169–76.

[30] Sheiner E, Menes TS, Silverberg D, Abramowicz JS, Levy I, Katz M, et al. Pregnancy outcome of patients with gestational diabetes mellitus following bariatric surgery. Am J Obstet Gynecol 2006;194(2):431–5.

[31] Merhi ZO, Minkoff H, Feldman J, Macura J, Rodriguez C, Seifer DB. Relationship of bariatric surgery to Müllerian-inhibiting substance levels. Fertil Steril 2008;90(1):221–4.

[32] Parent B, Martopullo I, Weiss NS, Khandelwal S, Fay EE, Rowhani-Rahbar A. Bariatric surgery in women of childbearing age, timing between an operation and birth, and associated perinatal complications. JAMA Surg 2017;152(2):128–35.

[33] Kwong W, Tomlinson G, Feig DS. Maternal and neonatal outcomes after bariatric surgery; a systematic review and meta-analysis: do the benefits outweigh the risks? Am J Obstet Gynecol 2018;218(6):573–80.

[34] Kjaer MM, Nilas L. Pregnancy after bariatric surgery—a review of benefits and risks. Acta Obstet Gynecol Scand 2013;92(3):264–71.

[35] Doblado MA, Lewkowksi BM, Odem RR, Jungheim ES. In vitro fertilization after bariatric surgery. Fertil Steril 2010;94(7):2812–14.

[36] Corona G, Rastrelli G, Monami M, Saad F, Luconi M, Lucchese M, et al. Body weight loss reverts obesity-associated hypogonadotropic hypogonadism: a systematic review and meta-analysis. Eur J Endocrinol 2013;168(6):829–43.

Section 4

General Gynaecology

Chapter 18

Obesity and gynecology ultrasound

Kiran Vanza[1,2], Mathew Leonardi[1,2] and George Condous[1,2]

[1]Acute Gynaecology, Early Pregnancy and Advanced Endoscopy Surgery Unit, Nepean Hospital, Kingswood, New South Wales, Australia,
[2]Sydney Medical School Nepean, The University of Sydney, Sydney, Australia

Key points

- Obesity is a growing epidemic worldwide.
- Clinical examination is often futile, placing greater emphasis on imaging modalities.
- Pelvic ultrasound is a safe and easily accessible imaging tool; however, the imaging modality is most impeded by adiposity.
- Transvaginal or transrectal ultrasound approach together with the optimization of ultrasound settings may be able to improve image quality.
- Obesity is a risk factor for endometrial hyperplasia and cancer, ovarian cancer, chronic anovulation, infertility, and early pregnancy complications. The ultrasound features of these will be discussed.

Introduction

Obesity (defined as body mass index ≥ 30 kg/m^2) is a growing epidemic worldwide, with the number of obese women having doubled over the past four decades [1,2]. Obesity is a risk factor for gynecological pathology, including endometrial hyperplasia, endometrial cancer, and ovarian cancer, and is linked to an increased incidence of polycystic ovarian syndrome (PCOS) and infertility [1,2]. The utility of clinical examination in the assessment of an obese patient is often limited with nonspecific findings. Thus a greater reliance is placed on imaging to assist in achieving an accurate diagnosis. Despite this, obesity is linked to missed diagnoses, suboptimal imaging, with nondiagnostic results, canceled investigations due to attaining maximum weight or girth parameters, and the scheduling of inappropriate imaging studies [1,2].

Pelvic ultrasound remains the first-line imaging modality of choice to investigate for the aforementioned conditions and other gynecological symptoms in women, particularly of reproductive age [1]. However, ultrasound is the imaging modality most impeded by adiposity, and despite best efforts, imaging remains more challenging to acquire and interpret in obese women [2]. There is a paucity of evidence to determine the best ultrasound protocols for ultrasound imaging in obese women.

The purpose of this chapter is to provide an overview of ultrasound as an imaging modality and techniques that can be employed to assist in improving image quality in obese women. In addition, an approach to performing a pelvic ultrasound in benign and malignant conditions more prevalent in obese women will be discussed.

Pelvic ultrasound

Ultrasound remains the first-line imaging modality of choice in gynecology. Ultrasound is easily accessible, portable, and safe lacking ionizing radiation and does not have weight or girth limitations applicable to magnetic resonance imaging (MRI) or computed tomography scanners [3]. However, ultrasound is the imaging modality most affected by excess adipose tissue due to the depth of insonation required and the attenuation of the ultrasound beam [4,5]. Greater depth of insonation is inversely proportional to the image quality, especially at higher frequencies, as the sound beam is absorbed and dissipated by adipose tissue [3,6,7]. A lower frequency setting can be applied; however, this degrades image quality [5,8]. There is a lack of evidence-based literature pertaining to a suggested approach to gynecology ultrasound in the obese population. Basic maneuvers with the ultrasound transducers, optimizing patient positioning and specific transducer settings, may assist in improving image quality. The transabdominal, transvaginal, and transrectal approaches will each be discussed in turn highlighting specific techniques that can be employed to optimize ultrasound images.

Transabdominal approach

Of all the approaches to a pelvic ultrasound in an obese woman, the transabdominal route is most vulnerable due to the attenuation of the ultrasound beam. Modica et al. defined sound attenuation in fat (in decibels) by the product of the attenuation coefficient (in decibels per centimeter at 1 MHz), the transducer frequency (in MHz), and the thickness of fat (in centimeters). Thus the greater the thickness of fat the greater the sound attenuation, with the same being applied to higher ultrasound frequencies.

The transabdominal approach is useful in providing an overview of the pelvis and the relationship between pelvic organs to one another [1]. This route can be particularly beneficial where there is an axial orientation to the uterus, where using a transvaginal probe may provide suboptimal images (Fig. 18.1). However, in order to optimize ultrasound images, the patient requires a full bladder and the often large and heavy panniculus must be navigated to minimize the distance between the ultrasound transducer and the organs of interest [6]. The pannus is often the thickest between the umbilicus and the pubic symphysis, coinciding with the region of interest in a pelvic ultrasound [1].

Sonographic windows can be created around the panniculus through various patient positions. First, the pannus can be lifted up and the transducer placed below it allowing scanning over the pubic bone and minimizing the thickness of adipose tissue the ultrasound beam must traverse (Figs. 18.2 and 18.3).

Second, the transducer can be placed over the umbilicus or above the pannus and firm downward pressure applied to reduce the distance between the transducer and the organ of interest [1,6]. Finally, as a semirecumbent position may be difficult to achieve for a long duration due to aortocaval compression, an oblique or lateral decubitus position can be employed to shift the weight of the pannus onto the examination table (Fig. 18.4). The woman lies prone with the lower leg extended and the upper flexed at the knee. This position creates a sonographic window along the lateral flank and reduces the depth of adipose tissue [1,3,6].

Transvaginal approach

There is no protocol available on how to perform a transvaginal ultrasound in an obese patient. As Glanc et al. state, the transvaginal approach is considered a "rescue" for an inadequate transabdominal study for the assessment of pelvic organs in obese women, allowing for superior images. Fig. 18.5 demonstrates the difference between image quality in transabdominal and transvaginal ultrasounds. This is due to the transducer traversing a shorter distance to image the organs of interest and not having to circumnavigate excess adipose tissue or other obstructions,

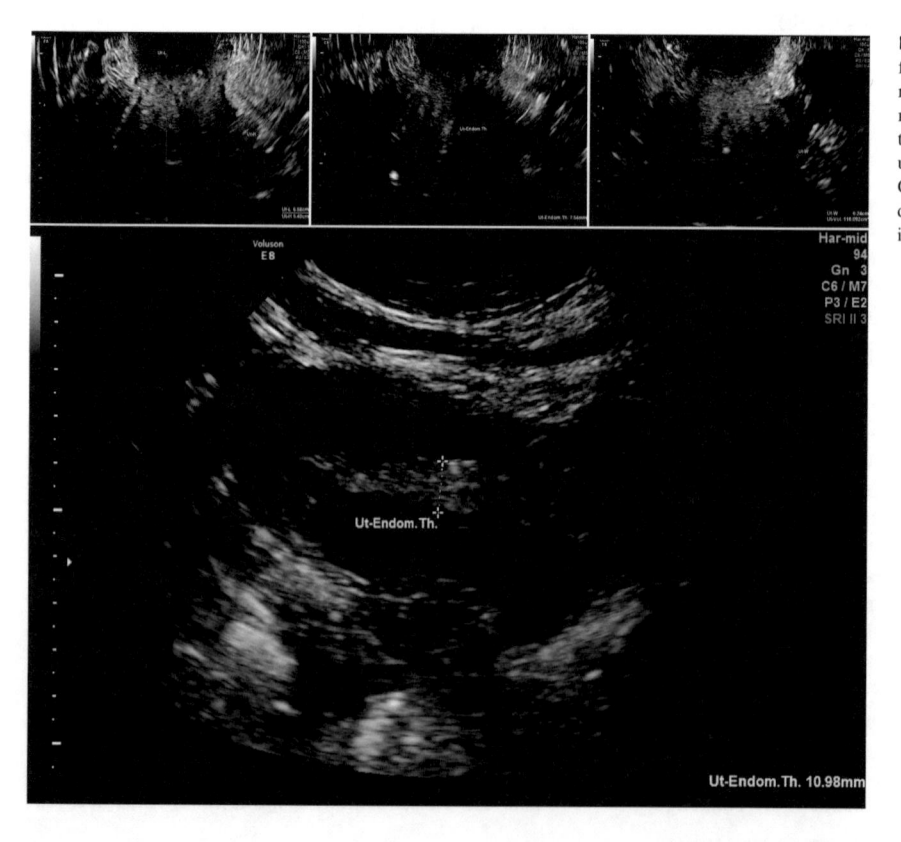

FIGURE 18.1 A 32-year-old female patient presenting for review for possible incomplete miscarriage. Axial uterus, difficult to see endometrium on transvaginal ultrasound (top three images). Global view attained on transabdominal ultrasound (bottom image).

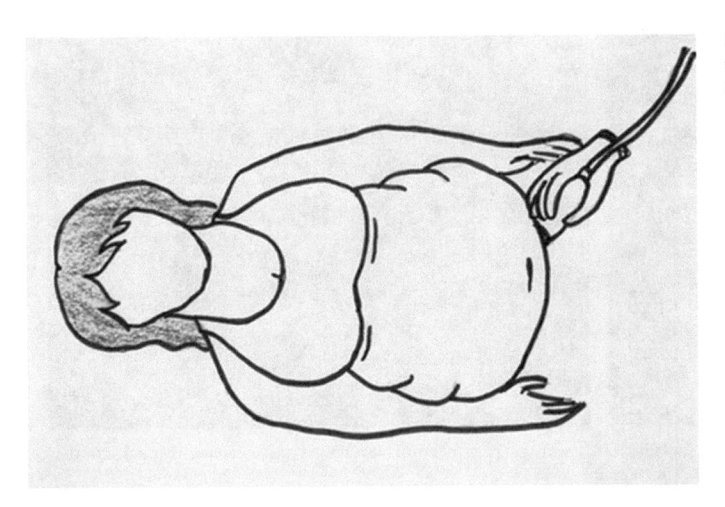

FIGURE 18.2 Transabdominal transducer placed below the pannus and scanning over the pubic bone to minimize thickness of adipose tissue.

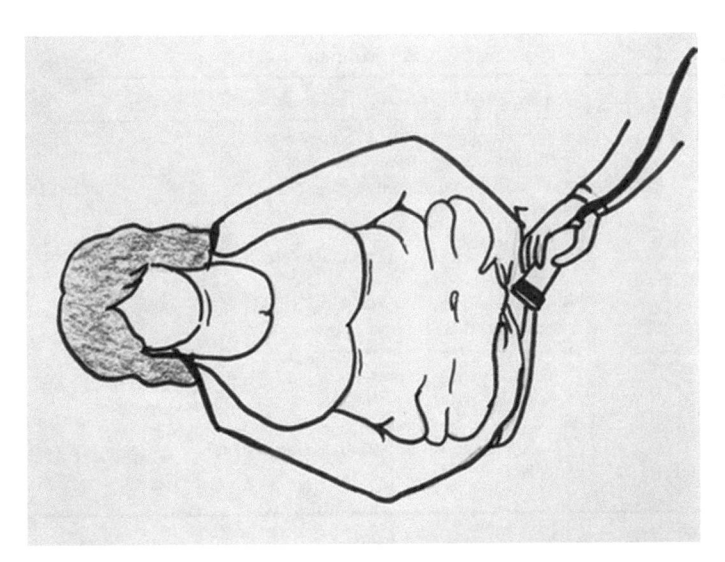

FIGURE 18.3 If the pannus is particularly large or heavy, the patient's assistance can be enlisted to help lift the pannus and minimize strain on the ultrasound operator.

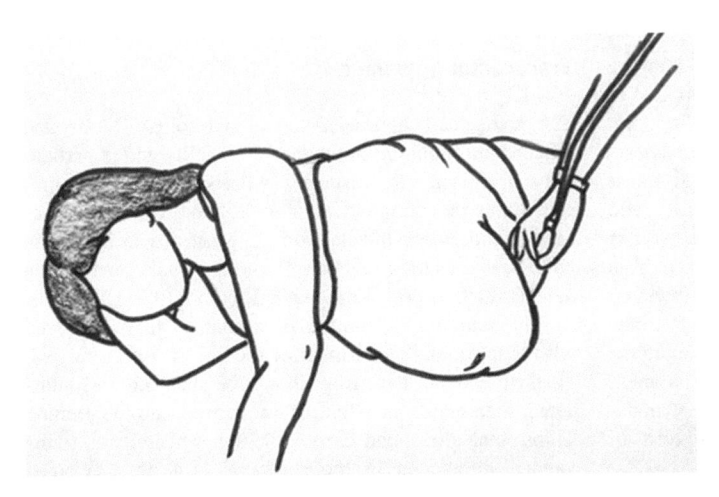

FIGURE 18.4 Lateral decubitus position enables the weight of the pannus to be shifted onto the examination table and creates a sonographic window along the lateral flank.

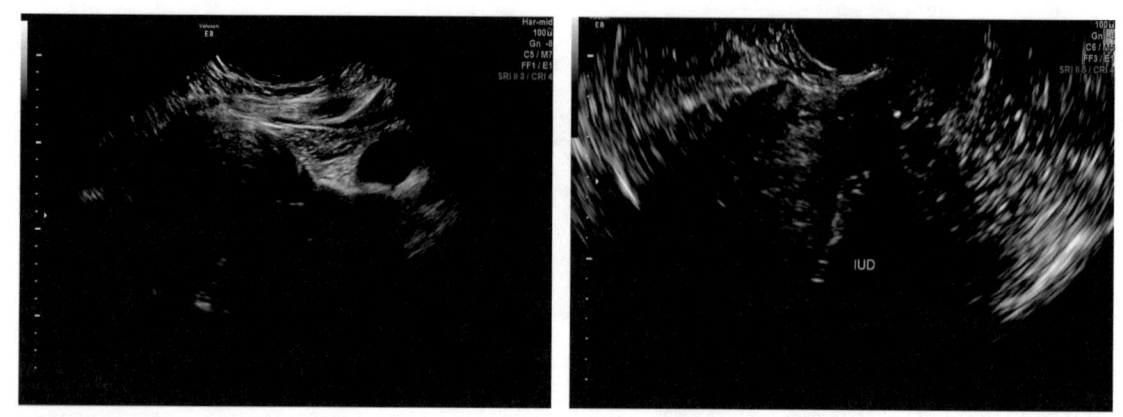

FIGURE 18.5 A 49-year-old female patient presented for the assessment of placement of levonorgestrel-containing intrauterine device. Transabdominal (left) and transvaginal (right) ultrasound view of intrauterine device.

TABLE 18.1 Comparison of transabdominal and transvaginal ultrasound in obese women.

	Transabdominal ultrasound	Transvaginal ultrasound
Ultrasound settings	• Ovaries visualized at 3–5 MHz • Transducer depth 10–15 cm • Full bladder required	• Pelvic organs visualized at 5–7 MHz • Transducer depth 1–8 cm • Bladder filling not required
Sensitivity Specificity	• 91% • 83%	• 96% • 89%
Advantages	• Provides an overview of pelvic structures and their relation to one another • Useful to image an axial uterus	• Provides superior images due to proximity of probe to organs of interest, bypassing adipose tissue • Allows dynamic assessment of pelvic organs (e.g., ovarian mobility, sliding sign) • Provides information regarding sites of tenderness, where pathology may be present
Disadvantages	• Ultrasound route most affected by adipose tissue • Navigate around the pannus • Ergonomically more challenging for the ultrasound operator	• View limited to vaginal depth • May be inappropriate in virgo intacta women, cases of Müllerian anomalies (vaginal agenesis) and canalization defects

cm, Centimeter; MHz, megahertz.

such as bowel gas and adhesions, obscuring the imaging path [1]. A higher frequency setting can also be applied to improve image resolution and quality [1,8]. Furthermore, the transvaginal approach allows for the dynamic assessment of pelvic organs difficult to appreciate through transabdominal ultrasound, particularly ovarian mobility and anterior and posterior compartment obliteration, as well as soft markers such as tenderness to the probe [1]. This route is also less strenuous on the sonographer. However, the transvaginal approach is limited if structures fall outside the scanning window, such as superiorly positioned ovaries, and is not appropriate in virgo intacta women. A summary of the advantages and disadvantages of transabdominal and transvaginal ultrasound is displayed in Table 18.1.

Transrectal approach

The transrectal approach to a gynecology ultrasound remains an important noninvasive modality and is particularly useful in circumstances where transvaginal ultrasound may be contraindicated or declined. These instances may include virgo intacta women, women with Müllerian anomalies (vaginal agenesis) and vaginal canalization defects, such as horizontal vaginal septum [9–11].

The scanning technique is similar to that employed with transvaginal assessment of the pelvis. As in transvaginal ultrasound, the probe should be sheathed and lubricated, then gently inserted, and advanced into the rectum. Transrectal ultrasound carries the same benefits as transvaginal ultrasound in the assessment of obese women over transabdominal ultrasound due to the proximity of the probe to the organs being assessed [11]. It has been demonstrated to be more accurate, specific, sensitive, and more reliable than transabdominal ultrasound in the assessment of the female pelvis [12]. Timor-Tritsch et al. compared transabdominal and transrectal ultrasound images obtained in 42 patients where transvaginal ultrasound was contraindicated. They found that in the majority of cases (31/42 studies) the transrectal images were clearer and more detailed when compared to the images obtained using transabdominal ultrasound. This group included four obese patients where the transabdominal ultrasound was insufficient to determine pelvic anatomy, whereas the transrectal ultrasound could demonstrate normal ovaries in two women with the other two women having ovarian cysts. Transrectal ultrasound carries comparable efficacy and sensitivity to transvaginal ultrasound and MRI when performing ultrasound for deep endometriosis or polycystic ovaries [9,12].

The transrectal approach has some perceived disadvantages. It may be viewed as an unorthodox approach to patients. However, after adequate counseling and explanation of the benefits in regards to image quality and information gained, most patients are amenable to the idea and find the process less traumatic and uncomfortable than envisaged. There are no reports of rectal or anal trauma in the literature [11,12]. Furthermore, the size of the probe is comparable to that of a colonoscope (13—14 mm diameter) and is smaller than an anoscope [10].

Transperineal approach

The transperineal approach is useful in obtaining images of the lower female genital tract and particularly the pelvic floor [10]. It is useful when transvaginal ultrasound is unavailable; however, it does not demonstrate any significant advantage in obese and normal-weight women.

In instances where ultrasound is unable to provide a diagnosis, MRI is the preferred modality in women of reproductive age.

Ultrasound settings

A modification of patient positioning, when performing a transabdominal assessment, together with advances in ultrasound technology has enabled improved image quality. These include:

- Differential tissue harmonic imaging that produces higher quality images than conventional methods. Harmonic beams are generated within tissue, with the intensity of the ultrasound beam increasing with depth of tissue traversed [3]. The maximum intensity of harmonic waves generated is directly proportional to the nonlinearity coefficient of the tissue, with fat carrying the highest nonlinearity coefficient [1]. Thus increasing adipose tissue generates increasing harmonic wave, which improves beam effective frequency, allowing for greater penetration of tissue and, thus, overall image quality [1,3,5].
- Compound imaging reduces the scanning angle and, thus, focuses the ultrasound beam to produce a superior image.
- Speckled reduction filters minimize noise artifacts [1,3,6].
- Pre- and postprocessing filters improve signal-to-noise ratio.
- "Penetrate mode" should be applied, when available, for obese patients, as it allows greater depth of penetration at lower frequencies [1,3].
- Newly developed 1 MHz crystal array and isotropic transducers allow real-time multiplanar imaging that may be beneficial in the obese population [8].
- Beam-forming algorithms can better model the pelvis in obese patients.
- Tissue aberration correction programs modify the speed at which the ultrasound beam travels through tissue, enabling greater penetration and improved image quality [1].

Ergonomic considerations

There are special ergonomic challenges that exist when performing gynecological ultrasound in obese patients, with increased risks to both the patient and practitioner. Ultrasound operators are at increased risk of musculoskeletal injuries, including repetitive strain injury, scanning in an awkward or uncomfortable position and needing to maintain transducer pressure for a prolonged period [1]. In fact, it has been shown that health-care professionals experience higher rates of musculoskeletal injuries when compared with workers in other industries, including construction, manufacturing, and retail, due to repetitive manual handling and heavy lifting [5].

Strategies that can be employed to minimize these ergonomic challenges include:

- An awareness of standard table weight restrictions and avoiding exceeding these as it increases the risk of patient injury and reduces the life span of the table.
- A table that has an adjustable height, preferably with voice-operated commands to raise and lower the table, optimizing scanning height and allowing for the ease of transfer of the patient on and off the table [1].
- Foot pedals that enable the sonographer to work from either side of the ultrasound table.
- Adequate task rotation ensuring bariatric patients are not scheduled sequentially and increasing the duration

of the visit to ensure sufficient time for consultation and for the ultrasound operator to rest afterward [1].
- Performing a transvaginal or transrectal ultrasound as this reduces the ergonomic stress on the ultrasound operator.
- The patient may be unable to tolerate a semirecumbent position due to aortocaval compression resulting in hypoxia and hypotension, thus placing the patient on an oblique lie or lateral decubitus position is beneficial [1].

Clinical applications

Miscarriage

Obesity significantly increases the risk of pregnancy-related complications, including miscarriage and recurrent miscarriage even in chromosomally normal pregnancies [13,14]. Postulated theories discussing the correlation between obesity and miscarriage are based around a dysfunction of the hypothalamic–pituitary axis that in turn affects embryo quality and development and endometrial receptivity [13]. Furthermore, obesity simulates a chronic inflammatory state that can reduce endometrial receptivity [13]. A dating pregnancy ultrasound is commonly offered to pregnant women between 6 and 8 weeks of gestation. This assists in determining gestational age of the pregnancy, pregnancy location, number of fetuses present, and viability. A transabdominal ultrasound in an obese woman may not provide the most detailed view (Fig. 18.6).

Polycystic ovarian syndrome and infertility

Obese women are at increased risk of chronic anovulation, with the leading cause being PCOS. PCOS is diagnosed using a combination of clinical, biochemical, and ultrasound-based factors. Women with PCOS are at increased risk of infertility and endometrial cancer [1,15,16]. Higher insulin levels associated with obesity exacerbate hyperandrogenism, due to the presence of multiple antral follicles, in half of the women with PCOS. Transvaginal ultrasound forms part of the assessment for PCOS. Sonographic features as documented by the Rotterdam Consensus (2003) include:

- Enlarged ovarian volume (≥ 10 mL^3), with the ovaries appearing more spherical.

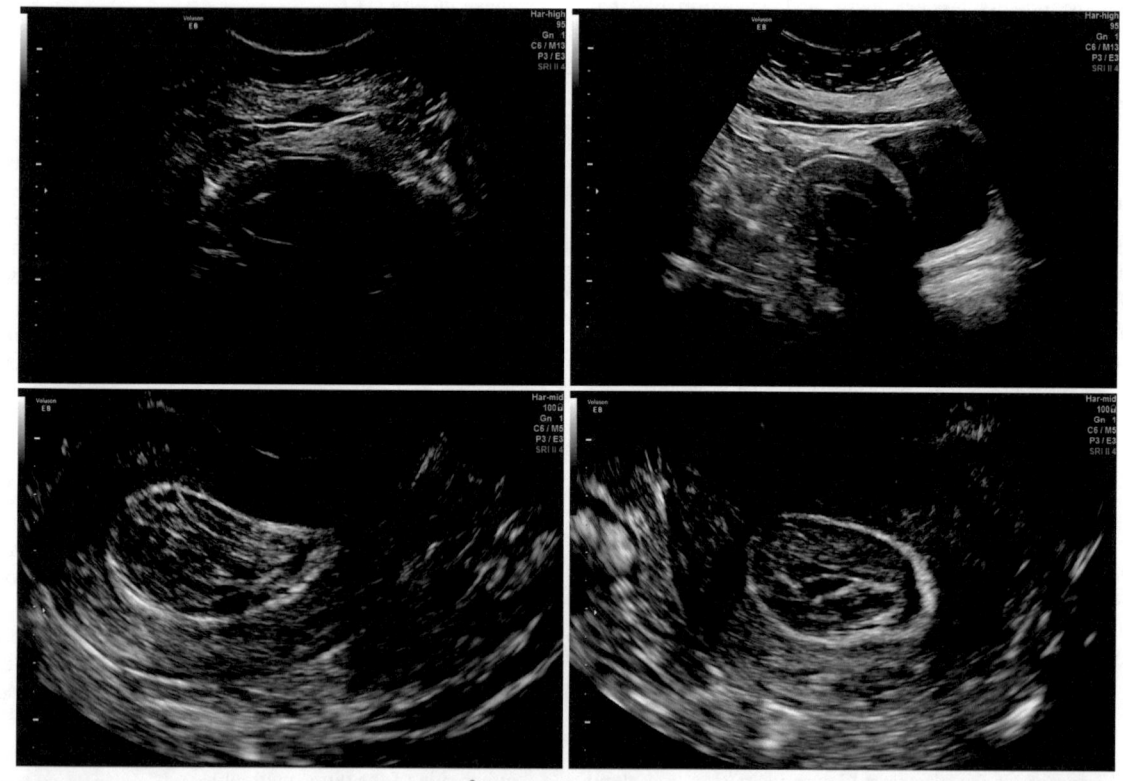

FIGURE 18.6 A 34-year-old female patient (BMI = 37 kg/m^2) at 6 weeks of gestational age as per the last menstrual period, presenting with vaginal bleeding. Transabdominal ultrasound (top images) showing possible intrauterine gestational sac with fetal pole compared to transvaginal ultrasound (bottom images) showing no definite gestational sac, suggestive of an incomplete miscarriage. *BMI*, Body mass index.

- Multiple peripherally arranged antral follicles (diameter 2–9 mm). The threshold for the number of antral follicles per ovary is being debated, with more recent literature suggesting the threshold should be ≥ 25 follicles rather than ≥ 12.
- An increased in central hyperechoic stroma.
- Increased stromal blood flow [15,17] (Fig. 18.7).

As mentioned, obesity is associated with infertility. Gynecology ultrasound is a useful tool in the investigation of female factor infertility. The ultrasound study makes an assessment of the following features:

- Ovarian reserve, by documenting the antral follicle count.
- Ovarian morphology and factors, which may inhibit ovulation, such as features of PCOS, ovarian cysts, and endometriomas.
- Uterine pathology such as adenomyosis, Müllerian anomalies, and intracavitary or endometrial pathologies (polyps, fibroids, adhesions), which may impair implantation.
  - Intracavitary pathology may be assessed using saline infusion sonohysterography [18].
- Tubal pathology such as a hydrosalpinx that can interfere with implantation or tubal blockage that can prevent fertilization.
  - In the absence of hydrosalpinges, tubal patency can be assessed through hysterosalpingo-contrast sonography [16].
- Extragenital pathology such as endometriosis of the rectouterine pouch, bowel, bladder, vagina, uterosacral ligaments, parametrium, and rectovaginal septum.
  - Endometriosis can be assessed through advanced or expert-guided transvaginal ultrasound as per the International Deep Endometriosis Analysis group [19,20] (Figs. 18.8 and 18.9).

Endometrial hyperplasia and endometrial cancer

The most significant risk factor for endometrial hyperplasia and endometrial cancer is obesity [1,16]. The incidence of endometrial cancer is more than doubled in obese women compared to those of normal weight [1]. Obesity leads to excess in peripheral unopposed estrogen that acts on the endometrium causing hyperplasia and, in turn, endometrial cancer. There are no

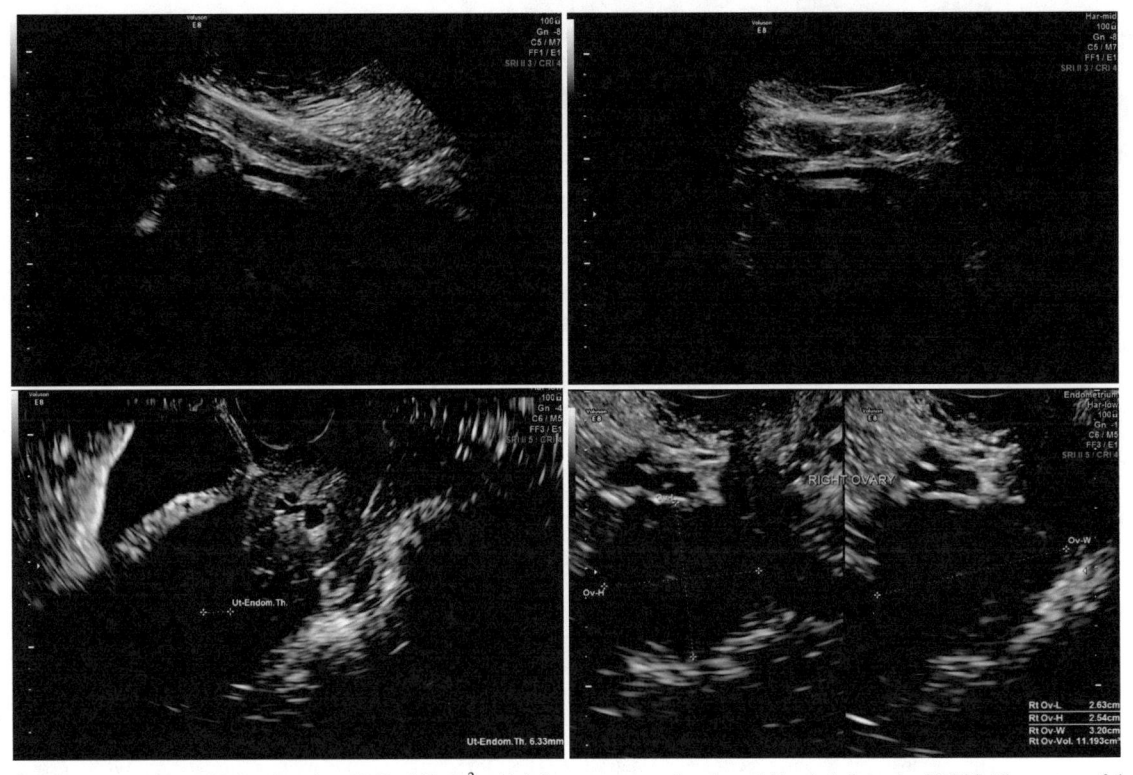

FIGURE 18.7 A 36-year-old female patient (BMI = 56 kg/m^2) with infrequent menstrual cycles and historical diagnosis of PCOS. The contour of the uterus was visible on transabdominal ultrasound; however, endometrial thickness could not be assessed and ovaries were unable to be visualized. Both the endometrial thickness and ovaries were seen on transvaginal ultrasound, with the right ovary demonstrating a volume of >10 mL^3. Taken together with her clinical history, her ultrasound findings are in line with a current PCOS diagnosis. *BMI*, Body mass index; *PCOS*, polycystic ovarian syndrome.

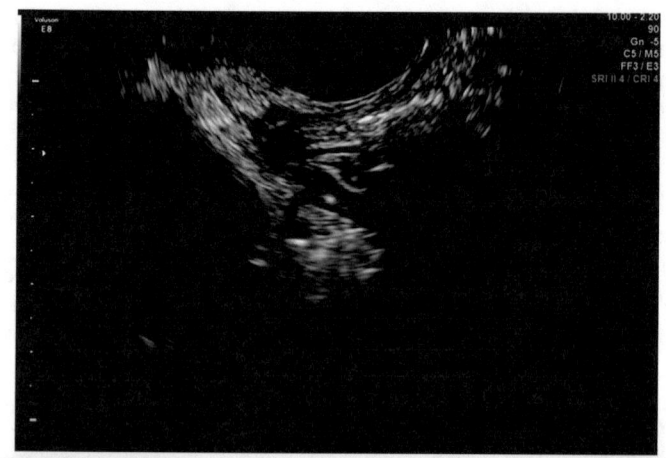

FIGURE 18.8 A 41-year-old female patient (BMI = 36 kg/m^2) presenting with a long-standing history of severe dysmenorrhea, dyspareunia, and dyschezia, also requiring assisted reproduction to conceive. Transvaginal ultrasound demonstrated a right ovarian endometrioma and bowel endometriosis nodule (small hypoechoic area to the left of the endometrioma). Pouch of Douglas obliteration was noted on ultrasound with a negative sliding sign. Her intraoperative findings correlated with ultrasound findings and this patient attended a surgical excision of her endometriosis. *BMI*, Body mass index.

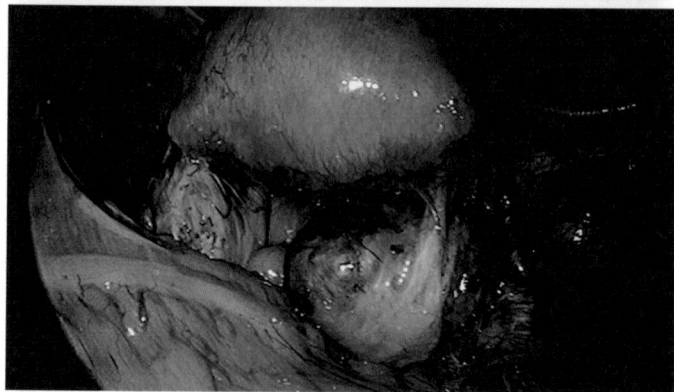

specific features of endometrial hyperplasia on ultrasound and the junctional zone remains intact [16]. Diagnosis is based on the histological assessment of an endometrial sample [16]. Despite this, Shokouhi demonstrated a high degree of sensitivity and specificity for endometrial hyperplasia in postmenopausal women.

Postmenopausal bleeding is a cardinal sign of endometrial cancer and all women presenting with this should be investigated. Ultrasound can be useful to facilitate diagnosis. In describing features of the endometrium, regardless of age, menopausal status, or indication for ultrasound, the International Endometrial Tumor Analysis group terms, definitions, and measurements should be used [21]. Sonographic features include:

- Endometrial thickness of ≥ 5 mm, though more recent metaanalyses suggest ≥ 4 mm, is abnormal [22]. Endometrial cancer is common if the thickness is ≥ 12 mm and almost all women with endometrial thickness ≥ 15 mm had cancer.
- Prominent and multiple vessels and global vascular pattern.
- Heterogenous endometrial mass.
  - Larger masses may cause disruption of the junctional zone.
- Moderate-to-high color score [16,23,24] (Fig. 18.10).

In the instance where the endometrium cannot be assessed sonographically, a sonohysterogram, hysteroscopy, or endometrial sample should be arranged [25].

Ovarian cancer

There is no clear association between ovarian cancer and obesity. To date, no screening tool is available for ovarian cancer, even in women who are at higher risk, including those with a family history or genetic marker positive [8]. Ultrasound, however, is the modality of choice to image the ovaries and characterize lesions [1]. The International Ovarian Tumor Analysis group's "Easy Descriptors," "Simple Rules," or Assessment of Different NEoplasias in the adneXa (ADNEX) model should be used [26–29].

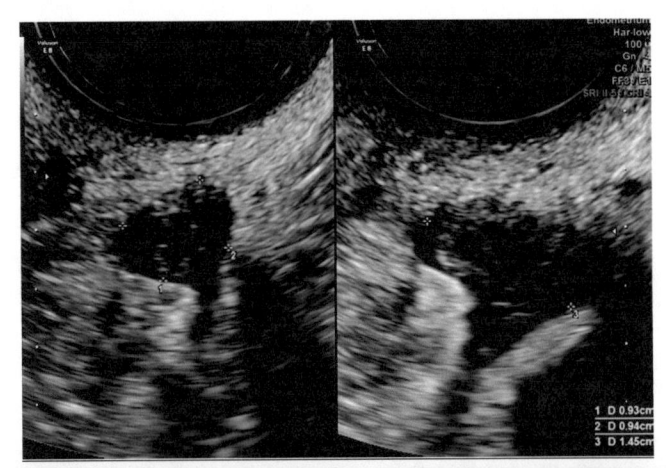

Obesity and gynecology ultrasound **Chapter | 18** **167**

FIGURE 18.9 A 36-year-old female patient (BMI = 45 kg/m^2) presenting with dysmenorrhea, dyspareunia, dyschezia, heavy menstrual bleeding, and primary infertility. Transvaginal ultrasound demonstrated a deep endometriosis nodule in the left uterosacral ligament and left hydroureter resulting in renal impairment. The bottom image is her intraoperative findings. This patient attended surgical excision of her endometriosis. *BMI*, Body mass index.

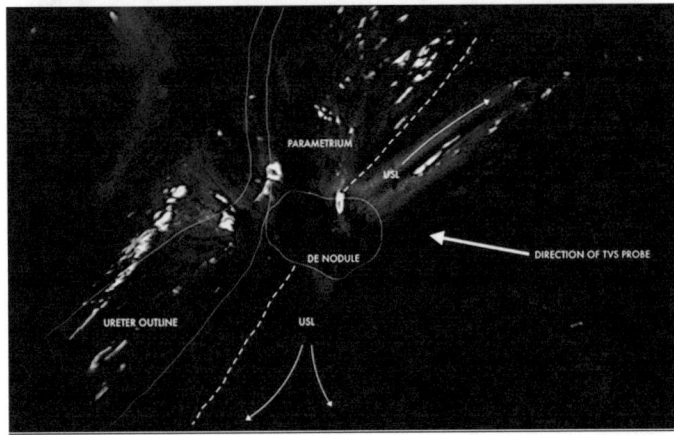

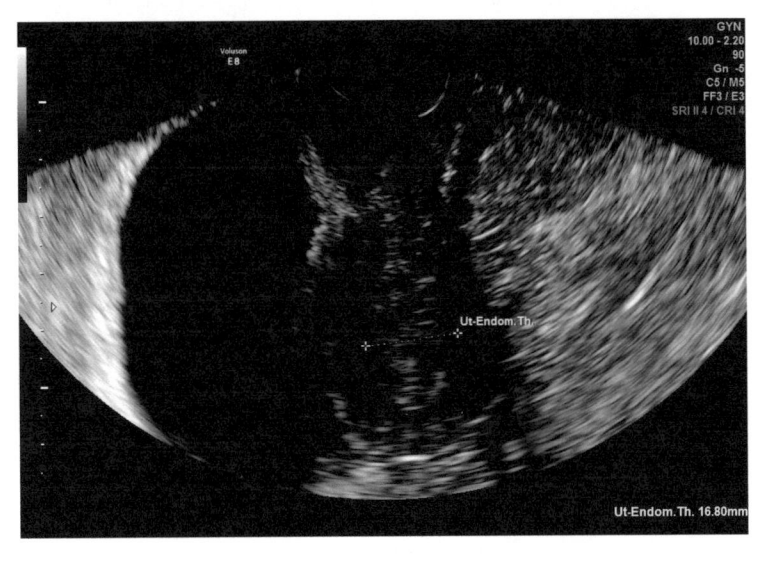

FIGURE 18.10 A 66-year-old female patient (BMI = 37 kg/m^2) referred with two episodes of postmenopausal bleeding. Transvaginal ultrasound demonstrates an axial uterus and thickened endometrium, measuring 16.8 mm (normal should be <5 mm in postmenopausal setting). Histological assessment confirmed grade 2 endometrial adenocarcinoma. *BMI*, Body mass index.

Cervical cancer

There is an increased incidence of cervical cancer in obese women. It is well documented that obesity leads to an underutilization of cervical screening, likely due to socioeconomic and psychological factors [1]. Thus obese women tend to present with more advanced disease. Cervical cancer is diagnosed and staged through a clinical examination, biopsies, and staging imaging. MRI is currently the imaging modality of choice to investigate for tumor burden and spread [1]; however, Epstein et al. have demonstrated pelvic ultrasound to be as accurate in the assessment of early-stage cervical cancer and supersede MRI when determining residual tumor burden and parametrial invasion.

Pelvic floor dysfunction

Pelvic floor dysfunction affects up to 60% of morbidly obese women [1]. Severity of symptoms and clinical findings is proportional to increasing weight. Pelvic floor assessment is clinical with ultrasound and MRI used as adjunctive tools. Perineal or translabial ultrasound is not greatly affected by obesity due to the closer proximity of organs [1], and instructional papers have been published to guide sonologists and clinicians when performing these [30].

Conclusion

Ultrasound is the imaging modality most affected by obesity. Patient positioning, optimizing ultrasound settings, and selecting the transvaginal or transrectal approach may overcome ergonomic stressors and improve image quality enabling a more meaningful study and taking advantage of ultrasound being a safe, economic, and easily accessible tool to gain valuable clinical information.

Acknowledgments

We would like to express our gratitude to the patients who have given their consent to use examples from their sonographic evaluations. We would also like to thank Ms. Mandy Ali of @me_myself_endo (Instagram) for her graphics of patient positioning.

References

[1] Glanc P, O'Hayon BE, Singh D, Bokhari SAJ, Maxwell CV. Challenges of pelvic imaging in obese women. Radiographics 2012;32:1839–62.

[2] Miller JC. Imaging and obese patients. In: Radiology rounds, vol. 3, 7. Massachusetts General Hospital Department of Radiology; 2005.

[3] Modica MJ, Kanal KM, Gunn ML. The obese emergency patient: imaging challenges and solutions. Radiographics 2011;31:811–23.

[4] Rochester NY. New ultrasound technology useful for imaging of overweight, obese patients. Carestream News Release 2015, November; <https://www.carestream.com/es/co/newsandevents/news-releases/2015/new-ultrasound-technology-useful-for-imaging-of-overweight-obese>.

[5] Reynolds A. Obesity and medical imaging challenges. Radiol Technol 2011;82:219–39.

[6] Benacerraf BR. A technical tip on scanning obese gravidae. Ultrasound Obstet Gynecol 2010;35:615–16.

[7] Uppot RN, Sahani DV, Hahn PF, Gervais D, Mueller PR. Impact of obesity on medical imaging and image-guided intervention. Am J Roentgenol 2007;188:433–40.

[8] Niazi M, Kamal MM, Malik N, Farooq M, Wahid N. Transabdominal vs transvaginal sonography – comparison in pelvic pathologies. J Rawalpindi Med Coll 2015;19:223–6.

[9] Alborzi S, Rasekhi A, Shomali Z, Madadi G, Alborzi M, Kazemi M, et al. Diagnostic accuracy of magnetic resonance imaging, transvaginal and transrectal ultrasonography in deep infiltrating endometriosis. Medicine 2018;97:e9536.

[10] Timor-Tritsch IE, Montaegudo A, Rebarber A, Goldsteins SR, Tsymbal T. Transrectal scanning: an alternative when transvaginal scanning is not feasible. Ultrasound Obstet Gynecol 2003;21:473–9.

[11] Guducu N, Sida G, Isci H, Yigiter AB, Dunder I. The utility of transrectal ultrasound in adolescents when transabdominal or transvaginal ultrasound is not feasible. J Pediatr Adolesc Gynecol 2013;26:265–8.

[12] Lee DE, Park SY, Lee SR, Jeong K, Chung HW. Diagnostic usefulness of transrectal ultrasound compared with transvaginal ultrasound assessment in young Korean women with polycystic ovary syndrome. J Menopausal Med 2015;21:149–54.

[13] Cavalcante MB, Manoel S, Peixoto AB, Araujo Junior E, Ricardo B. Obesity and recurrent miscarriage: a systematic review and meta-analysis. J Obstet Gynaecol Res 2019;45:30–8.

[14] Pettigrew R, Hamilton-Fairley D. Obesity and female reproductive function. Br Med Bull 1997;53:341–58.

[15] Zhu RY, Wong YC, Yong EL. Sonographic evaluation of polycystic ovaries. Best Pract Res Clin Obstet Gynaecol 2016;37:25–37.

[16] Campbell S. *Ultrasound evaluation in female infertility. Part 1, the ovary and the follicle*. Obstet Gynecol Clin North Am 2019;46:683–96.

[17] Lujan ME, Jarrett BY, Brooks ED, Reines JK, Peppin AK, Muhn N, et al. Updated ultrasound criteria for polycystic ovary syndrome: reliable threshold for elevated follicle population and ovarian volume. Hum Reprod 2013;28:1361–8.

[18] Valenzano M, Lijoi D, Mistrangelo E, Fortunato T, Costantini S, Ragni N. The value of sonohysterography in detecting intracavitary benign abnormalities. Arch Gynecol Obstet 2005;272:265–8.

[19] Leonardi M, Condous G. How to perform an ultrasound to diagnose endometriosis. Australas J Ultrasound Med 2018;21:61–9.

[20] Guerriero S, Condous G, van den Bosch T, Valentin L, Leone F, et al. Systematic approach to sonographic evaluation of the pelvis in women with suspected endometriosis including terms, definitions measurements: a consensus opinion from the International Deep Endometriosis Analysis (IDEA) group. Ultrasound Obstet Gynecol 2016;48:318–32.

[21] Leone F, Timmerman D, Bourne T, Valentin L, Epstein E, et al. Terms, definitions and measurements to describe the sonographic features of the endometrium and intrauterine lesions: a consensus

opinion from the International Endometrial Tumor Analysis (IETA) group. Ultrasound Obstet Gynecol 2010;35:103−12.
[22] Timmermans A, Opmeer B, Khan K, Bachmann LM. Endometrial thickness measurement for detecting endometrial cancer in women with postmenopausal bleeding: a systematic review and meta-analysis. Obstet Gynecol 2010;116:160−7.
[23] Epstein E, Van Holsbeke C, Mascilini F, Masback A, Kannisto P, Ameye L, et al. Gray-scale and color Doppler ultrasound characteristics of endometrial cancer in relation to stage, grade and tumor size. Ultrasound Obstet Gynecol 2011;38:586−93.
[24] Epstein E, Fischerova D, Valentin L, Testa AC, Franchi D, Saldkevicius P, et al. Ultrasound characteristics of endometrial cancer as defined by International Endometrial Tumor Analysis (IETA) consensus nomenclature: prospective multicenter study. Ultrasound Obstet Gynecol 2018;51:818−28.
[25] ACOG Committee Opinion. The role of transvaginal ultrasonography in evaluating the endometrium of women with postmenopausal bleeding, <https://www.acog.org/Clinical-Guidance-and-Publications/Committee-Opinions/Committee-on-Gynecologic-Practice/The-Role-of-Transvaginal-Ultrasonography-in-Evaluating-the-Endometrium-of-Women-With-Postmenopausal?IsMobileSet = false>; 2018.
[26] Timmerman D, Testa AC, Bourne T, et al. Simple ultrasound-based rules for the diagnosis of ovarian cancer. Ultrasound Obstet Gynecol 2008;31:681−90.
[27] Timmerman D, Ameye L, Fischerova D, et al. Simple ultrasound rules to distinguish between benign and malignant masses before surgery: prospective validation by IOTA group. Br Med J 2010;341:c6839.
[28] Van Calster B, Van Hoorde K, Valentin L, et al. Evaluating the risk of ovarian cancer before surgery using the ADNEX model to differentiate between benign, borderline, early and advanced stage invasive, and secondary metastatic tumours: prospective multicenter diagnostic study. Br Med J 2014;349:g5920.
[29] Ameye L, Timmerman D, Valentin L, et al. Clinically oriented three-step strategy for assessment of adnexal pathology. Ultrasound Obstet Gynecol 2012;40:582−91.
[30] Dietz H. Pelvic floor ultrasound. A review. Clin Obstet Gynecol 2017;60:58−81.

Further reading

Dueholm M, Hjorth IM. Structured imaging technique in the gynecologic office for the diagnosis of abnormal uterine bleeding. Best Pract Res Clin Obstet Gynaecol 2016;40:23−43.

Epstein E, Testa A, Gaurilcikas A, Di Legge A, Ameye L, et al. Early-stage cervical cancer: tumor delineation by magnetic resonance imaging and ultrasound — a European multicenter trial. Gynecol Oncol 2013;128:449−53.

Kayemba-Kays S, Pambou A, Heron A, Benosman SM. Polycystic ovary syndrome: pelvic MRI as alternative to pelvic ultrasound for the diagnosis in overweight and obese adolescent girls. Int J Pediatr Adolesc Med 2017;4:147−52.

Shokouhi B. Role of transvaginal ultrasonography in diagnosing endometrial hyperplasia in pre- and post-menopause women. Niger Med J 2015;56:353−6.

Chapter 19

Obesity and menstrual disorders

Jane J. Reavey[1], W. Colin Duncan[1], Savita Brito-Mutunayagam[1], Rebecca M. Reynolds[2] and Hilary O.D. Critchley[1]

[1]MRC Centre for Reproductive Health, The Queen's Medical Research Institute, The University of Edinburgh, Edinburgh, United Kingdom,
[2]University/BHF Centre for Cardiovascular Science, The Queen's Medical Research Institute, The University of Edinburgh, Edinburgh, United Kingdom

Introduction

An association between obesity and reproductive function has long been recognized. Hippocrates wrote, "People of such constitution cannot be prolific .... fatness and flabbiness are to blame. The womb is unable to receive the semen and they menstruate infrequently and little" [1].

The prevalence of obesity, defined as a body mass index (BMI) >30 kg/m^2, is rising worldwide and is a major health concern. Obesity increases the risk of developing type 2 diabetes, hypertension, ischemic heart disease, cancer (particularly colon, endometrial, and breast), cerebrovascular disease, and osteoarthritis. Obesity also increases risk of death; a recent metaanalysis, including data from 57 prospective studies with 894,576 participants, reported that for every increase in BMI of 5 kg/m^2, there was a 30% overall higher mortality with a 40% increase in vascular mortality; a greater than 50% increase in diabetic, renal, and hepatic mortality; a 10% increase in neoplastic mortality; and 20% increase in respiratory and other mortality [2].

Data from the Health Survey for England showed that the prevalence of obesity in 2016 was 26% in adult men and 27% in adult women [3]. This has increased from 15% in 1993 but overall has remained at a similar level since 2010. These rates are lower than in the United States where most recent data from 2013 to 2014 show the prevalence of obesity is 35% among men and 40.4% among women [4]. The rates of obesity in the general population are paralleled among pregnant women [5,6]. In the United Kingdom, approximately 20% of pregnant women are obese [7]. This problem was highlighted in the most recent MBRACE-UK report of maternal deaths between 2013 and 2015 where 34% of women who died had a BMI ≥ 30 kg/m^2 [8]. The risks of obesity during pregnancy have been extensively reviewed in Volume 1.

Childhood and adolescent obesity is also a growing concern. Twenty percent of 14-year olds in the United Kingdom are classified as obese and face an increased risk of multiple health problems [9]. Obesity affects virtually every organ system in the body, including the reproductive organs.

Abnormal uterine bleeding—Causes

In women of reproductive age, obesity influences the development and progression of menstrual problems. Obese women are reported to be three times more likely to suffer from menstrual abnormalities than women of a normal weight [10]. In addition, weight loss can restore menstruation to a normal pattern [10]. Abnormal uterine bleeding (AUB) may be classified using the PALM-COEIN paradigm (Fig. 19.1) [11]. This useful acronym describes the etiological basis of menstrual problems and may be used as a framework to describe the impact of obesity on the menstrual bleeding experience.

PALM-COEIN Classification

Polyps

A survey of premenopausal women with endometrial polyps found that 82% reported AUB [12]. In these women, obesity, particularly in combination with hypertension, was an important risk factor for polyp development [12]. Further evidence for an association between obesity and endometrial polyps comes from a retrospective study that found a significant increase in obesity prevalence in women with endometrial polyps compared to women without [13]. In addition, in an infertility setting, BMI was an independent risk factor for the development of endometrial polyps [14]. Obese women would

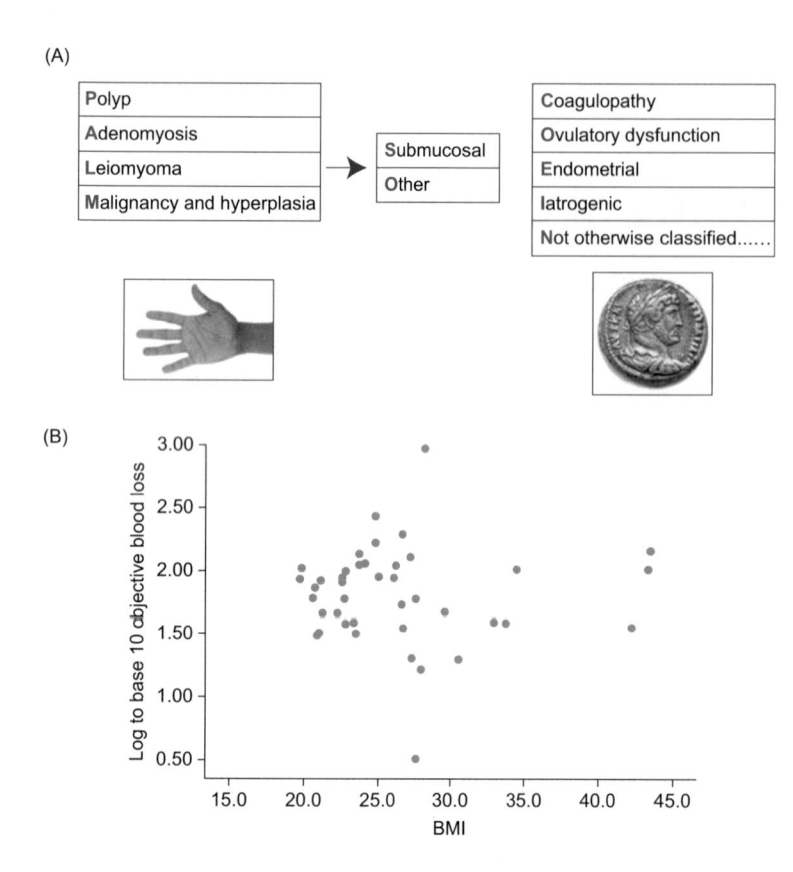

FIGURE 19.1 (A) The PALM-COEIN classification of abnormal uterine bleeding and (B) Magnitude of blood loss in relation to body mass index. *(A) - Reproduced with permission from Munro MG, Critchley HOD, Fraser IS, F.M.D. Committee. The two FIGO systems for normal and abnormal uterine bleeding symptoms and classification of causes of abnormal uterine bleeding in the reproductive years: 2018 revisions. Int J Gynaecol Obstet 2018;143 (3):393—408.*

therefore appear to be at an increased risk of developing endometrial polyps. One potential mechanism for this may be higher serum estrogen levels in women with obesity. Adipose tissue is an active endocrine organ that converts androgens to estrogens through aromatase activity [15]. There is evidence that both levels of estradiol and duration of unopposed estrogen increase with increasing body weight [16], which may have an augmented proliferative effect on the endometrium.

Malignancy and hyperplasia

Obesity also increases the risk of malignancy developing within an endometrial polyp [17]. The major risk factors for the development of endometrial cancer and endometrial hyperplasia are the same, and prominent among these is obesity [18,19]. It is estimated that 40% of all endometrial cancer is attributable to obesity [20], and in one study, it is reported that 86% of women with complex hyperplasia were obese [21]. In addition, BMI is predictive of endometrial thickness on ultrasound scan, and this is predictive of hyperplasia [22]. A recent systematic review of premenopausal women found a significant dose—response association between obesity and endometrial hyperplasia and cancer [23]. The risk of endometrial cancer varied from an almost 4-fold increase in women with a BMI ≥ 25 kg/m^2 (pooled OR 3.85, 95% CI 2.53—5.84) to an almost 20-fold increase in women with a BMI ≥ 40 kg/m^2 (pooled OR 19.79, 95% CI 11.18—35.03). Bariatric surgery may be beneficial in risk reduction. A significant beneficial effect on the rate of endometrial cancer development was found in obese women who underwent bariatric surgery compared to obese control women [24]. Furthermore, there is evidence from a recent prospective cohort study that bariatric surgery in women with BMI ≥ 40 kg/m^2 can reverse atypical hyperplasia [25].

Ovulatory dysfunction

Several studies have described associations between obesity and irregularity of the menstrual cycle as assessed by self-reported questionnaire. A large study carried out over three decades ago, including 26,638 women aged between 20 and 40 years, reported associations between menstrual cycle irregularity and anovulation with overweight and obesity [26]. Women with evidence of anovulatory cycles, defined as irregular cycles greater than 36 days, were more than 13.6 kg heavier than women with no menstrual cycle abnormalities, after adjusting for age and height.

More contemporary studies have reported similar findings. For example, a cross-sectional study in the United States, including 3941 women who described their menstrual cycle characteristics by questionnaire, found that women with BMI ≥ 35 kg/m^2 had increased risk of long cycles compared to women with BMI 22−23 kg/m^2 (OR 5.4, 95% CI 2.1−13.7) [27]. Similar findings were reported in a cross-sectional study of 726 Australian women aged 26−36 years who were not taking hormonal contraceptives and were not pregnant or breastfeeding [28]. Compared to nonobese women, obese women had at least twofold greater odds of having an irregular cycle, defined as ≥ 15 days between the longest and shortest cycle in the last 12 months. The findings are also applicable to other ethnic groups, for example, both a small study of 120 Mexican women [29] and another study of 322 Samoan women [30] found that increasing obesity was associated with increasing likelihood of amenorrhea and oligomenorrhea. A recent study of 4788 Korean women has also shown that women who are overweight (BMI 25−29.9 kg/m^2) or obese (BMI ≥ 30 kg/m^2) report a higher rate of menstrual irregularities compared to women with a normal BMI [31].

Generally, associations between obesity and menstrual irregularity appear to be similar when using BMI as a marker of general obesity, or when using markers of central obesity such as waist circumference [28,32], waist-to-hip ratio [28,30], or trunk fat [33]. Importantly, weight loss in infertile women restores normal menstrual cyclicity [34,35]. Interestingly, there is evidence that the association between menstrual disorders and obesity is not just to do with current levels of obesity but is also related to a prior history of obesity. Using data from the 1958 British birth cohort study in 5799 females, obesity in early adulthood at age 23 years, as well as obesity in childhood at age 7 years, both independently increased the risk of menstrual problems by age 33 years (OR = 1.75 and 1.59, respectively) after adjusting for other confounding factors [36]. The evidence thus strongly indicates that obese women are at increased risk of developing ovulatory dysfunction.

The endometrium

Heavy menstrual bleeding (HMB) is defined as excessive menstrual blood loss, which interferes with a woman's physical, social, emotional, and or material quality of life [37]. In clinical research, HMB has an objective definition wherein a total menstrual blood loss of 80 mL or greater is considered as HMB [38].

HMB affects up to 30% of premenopausal women [37,39] and is a source of considerable morbidity among women of reproductive age irrespective of BMI. Menstrual complaints are a significant burden to general practitioner as well as specialist health service resources. HMB has a significant socioeconomic impact with 1 in 20 women aged between 30 and 49 years attending their General Practioner each year due to menstrual problems [37]. Disorders of the menstrual cycle are one of the four most common reasons for general practitioner consultations [40]. Although the etiology underlying HMB of endometrial origin [41] remains to be fully defined, the body of knowledge concerning molecular and cellular mechanisms underpinning HMB of endometrial origin may lead to the development of novel therapeutic approaches [42,43].

A raised BMI is associated with earlier menarche [44] and menstrual irregularities during adolescence. There are minimal data available in the literature on the influence of obesity on menstrual blood loss. A raised BMI will certainly impact on endometrial function in the context of an increased risk of endometrial hyperplasia and endometrial carcinoma [45]. Raised circulating estrogen levels, as a consequence of peripheral conversion of androgens by adipose tissue aromatase enzyme, have been implicated in the increased proliferative activity of endometrial cells. Circulating adipokines have also been associated with increased angiogenesis as well as cell proliferation. HMB is a common complaint among those women who are premenopausal and who are subsequently diagnosed with endometrial cancer. It would therefore not be unlikely if a raised BMI was found to impact on the volume of menstrual blood loss. One retrospective study looking at menstrual disorders found that HMB appeared to be more prevalent in women who were overweight [26]. However, further research into the impact of obesity on menstrual blood loss is needed.

One clear effect of obesity is that the management of HMB among women with a raised BMI is a challenge. The treatments available for HMB may be limited although data showing treatment outcome in relation to BMI are lacking. Raised BMI is associated with poor efficacy of hormonal contraception [46,47] suggesting an effect of obesity on bioavailability or action of steroids.

Hysterectomy has additional complications in the presence of a raised BMI [48]. Data are sparse on the outcome of less invasive surgical approaches for the management of HMB, that is, endometrial ablation in obese women. One follow-up study of 44 patients who underwent endometrial ablation found a trend toward failure with this intervention in women with a BMI > 34 kg/m^2 [49]. However, a more recent retrospective cohort study of 666 women undergoing endometrial ablation found comparable rates of amenorrhoea and treatment failure in the obese and nonobese cohort of women [50].

Some options seem to be suited to obese women. The levonorgestrel-releasing intrauterine system (LNG-IUS) is considered a "first-time" treatment option for the management of HMB [37]. It also protects against endometrial hyperplasia in ovulatory dysfunction. A study among adolescent women undergoing bariatric surgery showed a high acceptance rate of this method for the management of menstrual complaints [51].

Polycystic ovary syndrome

It seems therefore that the most important endometrial effects of obesity are polyps and endometrial hyperplasia. Both of these are associated with ovulatory dysfunction that is clearly influenced by obesity. Anovulation leads to unopposed estrogen stimulation of the endometrium and the risk of hyperplasia increases with the duration of amenorrhea [52]. The ovulatory dysfunction associated with polycystic ovary syndrome (PCOS) is a risk factor for endometrial hyperplasia. Hyperplasia has been reported in one-third of anovulatory women with PCOS [22,52], and this is associated with increased endometrial thickness on a pelvic ultrasound scan [52]. PCOS is also associated with an increased incidence of endometrial polyps [14]. Women with PCOS have ovulatory dysfunction and are at increased risk of developing endometrial polyps and hyperplasia.

The prevalence of menstrual dysfunction in women with PCOS is 75%−85% [53]. In the United States the estimated cost of hormonally treating menstrual dysfunction and AUB in women with PCOS is $1.35 billion [53]. PCOS itself is therefore a significant cause of AUB. Importantly, however, women with PCOS are commonly overweight or obese (38%−66%) [10,54]. It is therefore difficult to separate the effects of obesity from the effects of PCOS. This is particularly problematic as obesity has a profound effect on the expression of the PCOS syndrome [10].

Population studies suggest that one in five women have polycystic ovaries (PCO) on a pelvic ultrasound scan [55,56] but only one-third of women with PCO on an ultrasound scan have PCOS [57]. Obesity affects the clinical manifestations of PCOS [58] and women with PCO on ultrasound scan in the absence of PCOS are more likely to develop PCOS with increasing weight gain [59]. It means that increasing obesity will increase the prevalence of PCOS [10]. The development of the symptoms of PCOS is predicted by weight and weight gain is attributed to the development of symptoms in more than one-third of women with PCOS [60]. As obesity rates rise, the public health significance of PCOS will therefore increase [61].

As women with PCOS gain weight, they are more likely to exhibit hyperandrogenism and menstrual cycle disturbances [10]. Thus obesity modifies the severity of PCOS and more women will experience menstrual problems as their weight increases [62]. Several studies have demonstrated that menstrual abnormalities are more prevalent in obese than nonobese women with PCOS [54]. Indeed, a history of weight gain frequently precedes the onset of oligomenorrhea and menstrual problems [54]. An important observation is that weight loss improved menstrual function in adolescent [63], infertile [64], and other adult [65,66] women with PCOS. Obesity will thus exaggerate PCOS and increase PCOS-related AUB.

Obesity in the absence of polycystic ovary

It is likely that increasing ovulatory dysfunction associated with PCOS is the major factor in obesity-related menstrual problems. However, obese women may suffer from perturbations of the hypothalamic pituitary ovarian axis, with menstrual cycle disturbance, in the absence of PCOS [10,67]. Obesity independently increases hyperandrogenism, hirsutism, and infertility [61]. This may be secondary to the altered body fat−dependent hormone milieu and the development of insulin resistance associated with weight gain [10]. Both weight loss and metformin administration are reported to benefit menstrual bleeding disorders in the absence of PCOS [46]. The presence of insulin resistance predicts a thicker endometrium on ultrasound scan [68], and BMI is positively associated with the thickness of the endometrium in the absence of PCOS. In addition, obesity has been independently associated with increased uterine blood flow as measured by Doppler uterine artery pulsatility index [55]. Therefore there may be a PCOS-independent effect of obesity although this remains to be established.

Most of the studies reporting associations between menstrual cycle irregularity have been conducted among women who are not currently pregnant and so may have included women with fertility problems where PCOS is overrepresented. We therefore investigated menstrual history among pregnant women attending an antenatal metabolic clinic [69]. Severely obese ($n = 249$) [BMI 44.3 (sd 4.1) kg/m^2] and normal weight ($n = 109$) [BMI 22.6 (1.6) kg/m^2] women recorded information about their menstrual cycles prior to pregnancy. All women had conceived naturally and none had received fertility treatment.

The percentage of women with regular menstrual cycles was similar in obese and lean (76% vs 80%, P = ns) and there was no difference in length of the menstrual cycle [28.7 (3.4) vs 28.9 (4.6) days, P = ns]. Likewise, there were no significant differences in the numbers of women with a history of PCOS. Thus in this group of fertile women, severe obesity did not impact on regularity of the menstrual cycle, suggesting there may be a subgroup of women who are less susceptible to the adverse influences of obesity. In accord with other studies the obese women had a significantly earlier age at menarche compared to normal-weight women [12.5 (1.5) versus 13.0 (1.3) years, $P = .0002$)] (all unpublished data).

Summary

As raised BMI is associated with menstrual disorders, the current epidemic of obesity is manifesting by increased rates of AUB. Further research is needed into the differential impact of obesity on endometrial and ovarian function. Studies should address the optimal lifestyle, medical, and surgical management of abnormal bleeding in obese women. As obesity is a risk for serious endometrial pathology, BMI as well as age should be taken into account in the decision for endometrial biopsy in women with AUB.

Acknowledgments

We thank Mrs. Sheila Milne for secretarial support and Mr. Ronnie Grant for assistance with the illustrations. We thank Miss N Marielle ten Brink for collating the menstrual cycle history data. She was supported by an Erasmus studentship and Tommy's. We also acknowledge the contribution from Dr. Jacqueline Maybin in the study involving women attending our local gynecology clinic with menstrual problems. Some of the data described herein was made possible through support from MRC Centre Grant (MR/N022556/1).

References

[1] Lloyd GER, editor. Hippocratic writings. London: Penguin Books; 1978.
[2] Whitlock G, Lewington S, Sherliker P, Clarke R, Emberson J, Halsey J, et al. Body-mass index and cause-specific mortality in 900 000 adults: collaborative analyses of 57 prospective studies. Lancet 2009;373(9669):1083—96.
[3] NHS. Statistics on obesity, physical activity and diet—England. Available from: <https://digital.nhs.uk/data-and-information/publications/statistical/statistics-on-obesity-physical-activity-and-diet/statistics-on-obesity-physical-activity-and-diet-england-2018>; 2018 (accessed 15.08.19).
[4] Flegal KM, Kruszon-Moran D, Carroll MD, Fryar CD, Ogden CL. Trends in obesity among adults in the United States, 2005 to 2014. JAMA 2016;315(21):2284—91.
[5] Heslehurst N, Ells LJ, Simpson H, Batterham A, Wilkinson J, Summerbell CD. Trends in maternal obesity incidence rates, demographic predictors, and health inequalities in 36,821 women over a 15-year period. BJOG 2007;114(2):187—94.
[6] Heslehurst N, Rankin J, Wilkinson JR, Summerbell CD. A nationally representative study of maternal obesity in England, UK: trends in incidence and demographic inequalities in 619 323 births, 1989-2007. Int J Obes (Lond) 2010;34(3):420—8.
[7] CMACE. Centre for Maternal and Child Enquiries. Maternal obesity in the UK: findings from a national project. London: CMACE; 2010. Available from: <https://www.publichealth.hscni. net/sites/default/files/Maternal%20Obesity%20in%20the%20UK. pdf> (accessed 15.08.19).
[8] M. Knight, M. Nair, D. Tuffnell, J. Shakespeare, S. Kenyon, J.J. Kurinczuk, editors, on behalf of MBRRACE-UK. Saving Lives, Improving Mothers' Care - Lessons learned to inform maternity care from the UK and Ireland Confidential Enquiries into Maternal Deaths and Morbidity 2013-15. Oxford: National Perinatal Epidemiology Unit, University of Oxford; 2017. Available from: <https://www.npeu.ox.ac.uk/mbrrace-uk/reports/confidential-enquiry-into-maternal-deaths> (accessed 4th May 2019).
[9] Fitzsimmons E, Pongiglione B. Prevalence and trends in overweight and obesity in childhood and adolescence. Findings from the Millennium Cohort Study, with a focus on age 14. In: Centre for Longitudinal Studies working paper 2017/16. London: UCL; 2017.
[10] Brewer CJ, Balen AH. The adverse effects of obesity on conception and implantation. Reproduction 2010;140(3):347—64.
[11] Munro MG, Critchley HOD, Fraser IS, F.M.D. Committee. The two FIGO systems for normal and abnormal uterine bleeding symptoms and classification of causes of abnormal uterine bleeding in the reproductive years: 2018 revisions. Int J Gynaecol Obstet 2018;143(3):393—408.
[12] Reslova T, Tosner J, Resl M, Kugler R, Vavrova I. Endometrial polyps. A clinical study of 245 cases. Arch Gynecol Obstet 1999;262(3-4):133—9.
[13] Serhat E, Cogendez E, Selcuk S, Asoglu MR, Arioglu PF, Eren S. Is there a relationship between endometrial polyps and obesity, diabetes mellitus, hypertension? Arch Gynecol Obstet 2014;290 (5):937—41.
[14] Onalan R, Onalan G, Tonguc E, Ozdener T, Dogan M, Mollamahmutoglu L. Body mass index is an independent risk factor for the development of endometrial polyps in patients undergoing in vitro fertilization. Fertil Steril 2009;91(4):1056—60.
[15] Stocco C. Tissue physiology and pathology of aromatase. Steroids 2012;77(1—2):27—35.
[16] Emaus A, Espetvedt S, Veierod MB, Ballard-Barbash R, Furberg AS, Ellison PT, et al. 17-Beta-estradiol in relation to age at menarche and adult obesity in premenopausal women. Hum Reprod 2008;23(4):919—27.
[17] Gregoriou O, Konidaris S, Vrachnis N, Bakalianou K, Salakos N, Papadias K, et al. Clinical parameters linked with malignancy in endometrial polyps. Climacteric 2009;12(5):454—8.
[18] Linkov F, Edwards R, Balk J, Yurkovetsky Z, Stadterman B, Lokshin A, et al. Endometrial hyperplasia, endometrial cancer and prevention: gaps in existing research of modifiable risk factors. Eur J Cancer 2008;44(12):1632—44.
[19] Epplein M, Reed SD, Voigt LF, Newton KM, Holt VL, Weiss NS. Risk of complex and atypical endometrial hyperplasia in relation to anthropometric measures and reproductive history. Am J Epidemiol 2008;168(6):563—70 discussion 571—6.
[20] Arnold M, Pandeya N, Byrnes G, Renehan PAG, Stevens GA, Ezzati PM, et al. Global burden of cancer attributable to high body-mass index in 2012: a population-based study. Lancet Oncol 2015;16(1):36—46.
[21] Horn LC, Schnurrbusch U, Bilek K, Hentschel B, Einenkel J. Risk of progression in complex and atypical endometrial hyperplasia: clinicopathologic analysis in cases with and without progestogen treatment. Int J Gynecol Cancer 2004;14(2):348—53.

[22] McCormick BA, Wilburn RD, Thomas MA, Williams DB, Maxwell R, Aubuchon M. Endometrial thickness predicts endometrial hyperplasia in patients with polycystic ovary syndrome. Fertil Steril 2011;95(8):2625−7.

[23] Wise MR, Jordan V, Lagas A, Showell M, Wong N, Lensen S, et al. Obesity and endometrial hyperplasia and cancer in premenopausal women: a systematic review. Am J Obstet Gynecol 2016;214(6):689.e1−689.e17.

[24] Winder AA, Kularatna M, MacCormick AD. Does bariatric surgery affect the incidence of endometrial cancer development? A systematic review. Obes Surg 2018;28(5):1433−40.

[25] MacKintosh ML, Derbyshire AE, McVey RJ, Bolton J, Nickkho-Amiry M, Higgins CL, et al. The impact of obesity and bariatric surgery on circulating and tissue biomarkers of endometrial cancer risk. Int J Cancer 2019;144(3):641−50.

[26] Hartz AJ, Barboriak PN, Wong A, Katayama KP, Rimm AA. The association of obesity with infertility and related menstrual abnormalities in women. Int J Obes 1979;3(1):57−73.

[27] Rowland AS, Baird DD, Long S, Wegienka G, Harlow SD, Alavanja M, et al. Influence of medical conditions and lifestyle factors on the menstrual cycle. Epidemiology 2002;13(6):668−74.

[28] Wei S, Schmidt MD, Dwyer T, Norman RJ, Venn AJ. Obesity and menstrual irregularity: associations with SHBG, testosterone, and insulin. Obesity (Silver Spring) 2009;17(5):1070−6.

[29] Castillo-Martinez L, Lopez-Alvarenga JC, Villa AR, Gonzalez-Barranco J. Menstrual cycle length disorders in 18- to 40-y-old obese women. Nutrition 2003;19(4):317−20.

[30] Lambert-Messerlian G, Roberts MB, Urlacher SS, Ah-Ching J, Viali S, Urbanek M, et al. First assessment of menstrual cycle function and reproductive endocrine status in Samoan women. Hum Reprod 2011;26(9):2518−24.

[31] Bae J, Park S, Kwon JW. Factors associated with menstrual cycle irregularity and menopause. BMC Womens Health 2018;18(1):36.

[32] De Pergola G, Tartagni M, d'Angelo F, Centoducati C, Guida P, Giorgino R. Abdominal fat accumulation, and not insulin resistance, is associated to oligomenorrhea in non-hyperandrogenic overweight/obese women. J Endocrinol Invest 2009;32(2):98−101.

[33] Douchi T, Kuwahata R, Yamamoto S, Oki T, Yamasaki H, Nagata Y. Relationship of upper body obesity to menstrual disorders. Acta Obstet Gynecol Scand 2002;81(2):147−50.

[34] Clark AM, Thornley B, Tomlinson L, Galletley C, Norman RJ. Weight loss in obese infertile women results in improvement in reproductive outcome for all forms of fertility treatment. Hum Reprod 1998;13(6):1502−5.

[35] Hollmann M, Runnebaum B, Gerhard I. Effects of weight loss on the hormonal profile in obese, infertile women. Hum Reprod 1996;11(9):1884−91.

[36] Lake JK, Power C, Cole TJ. Women's reproductive health: the role of body mass index in early and adult life. Int J Obes Relat Metab Disord 1997;21(6):432−8.

[37] NICE. NG88: heavy menstrual bleeding: assessment and management. National Institute for Health and Clinical Excellence (NICE); 2018. Available from: <https://www.nice.org.uk/guidance/ng88> (accessed 15.08.19).

[38] Warner PE, Critchley HO, Lumsden MA, Campbell-Brown M, Douglas A, Murray GD. Menorrhagia II: is the 80-mL blood loss criterion useful in management of complaint of menorrhagia? Am J Obstet Gynecol 2004;190(5):1224−9.

[39] RCOG. National heavy menstrual bleeding audit—final report. London: RCOG; 2014. Available from: <https://www.rcog.org.uk/globalassets/documents/guidelines/research--audit/national_hmb_audit_final_report_july_2014.pdf> (accessed 15.08.19).

[40] Palep-Singh M, Prentice A. Epidemiology of abnormal uterine bleeding. Best Pract Res Clin Obstet Gynaecol 2007;21(6):887−90.

[41] Munro MG, Critchley HO, Broder MS, Fraser IS. FIGO classification system (PALM-COEIN) for causes of abnormal uterine bleeding in nongravid women of reproductive age. Int J Gynaecol Obstet 2011;113(1):3−13.

[42] Critchley HO, Maybin JA. Molecular and cellular causes of abnormal uterine bleeding of endometrial origin. Semin Reprod Med 2011;29(5):400−9.

[43] Maybin JA, Critchley HO. Menstrual physiology: implications for endometrial pathology and beyond. Hum Reprod Update 2015;21(6):748−61.

[44] Tam CS, de Zegher F, Garnett SP, Baur LA, Cowell CT. Opposing influences of prenatal and postnatal growth on the timing of menarche. J Clin Endocrinol Metab 2006;91(11):4369−73.

[45] Dossus L, Rinaldi S, Becker S, Lukanova A, Tjonneland A, Olsen A, et al. Obesity, inflammatory markers, and endometrial cancer risk: a prospective case-control study. Endocr Relat Cancer 2010;17(4):1007−19.

[46] Lash MM, Armstrong A. Impact of obesity on women's health. Fertil Steril 2009;91(5):1712−16.

[47] Kulie T, Slattengren A, Redmer J, Counts H, Eglash A, Schrager S. Obesity and women's health: an evidence-based review. J Am Board Fam Med 2011;24(1):75−85.

[48] McMahon MD, Scott DM, Saks E, Tower A, Raker CA, Matteson KA. Impact of obesity on outcomes of hysterectomy. J Minim Invasive Gynecol 2014;21(2):259−65.

[49] Fakih M, Cherfan V, Abdallah E. Success rate, quality of life, and descriptive analysis after generalized endometrial ablation in an obese population. Int J Gynaecol Obstet 2011;113(2):120−3.

[50] Madsen AM, El-Nashar SA, Hopkins MR, Khan Z, Famuyide AO. Endometrial ablation for the treatment of heavy menstrual bleeding in obese women. Int J Gynaecol Obstet 2013;121(1):20−3.

[51] Hillman JB, Miller RJ, Inge TH. Menstrual concerns and intrauterine contraception among adolescent bariatric surgery patients. J Womens Health (Larchmt) 2011;20(4):533−8.

[52] Cheung AP. Ultrasound and menstrual history in predicting endometrial hyperplasia in polycystic ovary syndrome. Obstet Gynecol 2001;98(2):325−31.

[53] Azziz R, Marin C, Hoq L, Badamgarav E, Song P. Health care-related economic burden of the polycystic ovary syndrome during the reproductive life span. J Clin Endocrinol Metab 2005;90(8):4650−8.

[54] Gambineri A, Pelusi C, Vicennati V, Pagotto U, Pasquali R. Obesity and the polycystic ovary syndrome. Int J Obes Relat Metab Disord 2002;26(7):883−96.

[55] Polson DW, Adams J, Wadsworth J, Franks S. Polycystic ovaries—a common finding in normal women. Lancet 1988;1(8590):870−2.

[56] Farquhar CM, Birdsall M, Manning P, Mitchell JM, France JT. The prevalence of polycystic ovaries on ultrasound scanning in a population of randomly selected women. Aust N Z J Obstet Gynaecol 1994;34(1):67–72.

[57] Fauser BC, Tarlatzis BC, Rebar RW, Legro RS, Balen AH, Lobo R, et al. Consensus on women's health aspects of polycystic ovary syndrome (PCOS): the Amsterdam ESHRE/ASRM-Sponsored 3rd PCOS Consensus Workshop Group. Fertil Steril 2012;97(1):28–38.e25.

[58] Salehi M, Bravo-Vera R, Sheikh A, Gouller A, Poretsky L. Pathogenesis of polycystic ovary syndrome: what is the role of obesity? Metabolism 2004;53(3):358–76.

[59] Pettigrew R, Hamilton-Fairley D. Obesity and female reproductive function. Br Med Bull 1997;53(2):341–58.

[60] Laitinen J, Taponen S, Martikainen H, Pouta A, Millwood I, Hartikainen AL, et al. Body size from birth to adulthood as a predictor of self-reported polycystic ovary syndrome symptoms. Int J Obes Relat Metab Disord 2003;27(6):710–15.

[61] Teede H, Deeks A, Moran L. Polycystic ovary syndrome: a complex condition with psychological, reproductive and metabolic manifestations that impacts on health across the lifespan. BMC Med 2010;8:41.

[62] Welt CK, Gudmundsson JA, Arason G, Adams J, Palsdottir H, Gudlaugsdottir G, et al. Characterizing discrete subsets of polycystic ovary syndrome as defined by the Rotterdam criteria: the impact of weight on phenotype and metabolic features. J Clin Endocrinol Metab 2006;91(12):4842–8.

[63] Ornstein RM, Copperman NM, Jacobson MS. Effect of weight loss on menstrual function in adolescents with polycystic ovary syndrome. J Pediatr Adolesc Gynecol 2011;24(3):161–5.

[64] Clark AM, Ledger W, Galletly C, Tomlinson L, Blaney F, Wang X, et al. Weight loss results in significant improvement in pregnancy and ovulation rates in anovulatory obese women. Hum Reprod 1995;10(10):2705–12.

[65] Lass N, Kleber M, Winkel K, Wunsch R, Reinehr T. Effect of lifestyle intervention on features of polycystic ovarian syndrome, metabolic syndrome, and intima-media thickness in obese adolescent girls. J Clin Endocrinol Metab 2011;96(11):3533–40.

[66] Tang T, Glanville J, Hayden CJ, White D, Barth JH, Balen AH. Combined lifestyle modification and metformin in obese patients with polycystic ovary syndrome. A randomized, placebo-controlled, double-blind multicentre study. Hum Reprod 2006;21(1):80–9.

[67] Rachon D, Teede H. Ovarian function and obesity—interrelationship, impact on women's reproductive lifespan and treatment options. Mol Cell Endocrinol 2010;316(2):172–9.

[68] Iatrakis G, Tsionis C, Adonakis G, Stoikidou M, Anthouli-Anagnostopoulou F, Parava M, et al. Polycystic ovarian syndrome, insulin resistance and thickness of the endometrium. Eur J Obstet Gynecol Reprod Biol 2006;127(2):218–21.

[69] Denison FC, MacGregor H, Stirrat LI, Stevenson K, Norman JE, Reynolds RM. Does attendance at a specialist antenatal clinic improve clinical outcomes in women with class III obesity compared with standard care? A retrospective case-note analysis. BMJ Open 2017;7(5):e015218.

Chapter 20

Incontinence and pelvic organ prolapse in the obese woman

Clare F. Jordan[1] and Douglas G. Tincello[1,2]

[1]*University Hospitals of Leicester NHS Trust, Leicester, United Kingdom,* [2]*Department of Health Sciences, University of Leicester, Leicester, United Kingdom*

Introduction

This chapter will discuss the available literature relating to the impact of obesity on the prevalence and severity of pelvic floor dysfunction (PFD) in women (urinary and anal incontinence and uterovaginal prolapse). The evidence for symptom resolution after weight loss will be presented, and recent data on the efficacy and safety of urogynecology surgery will be examined. Throughout this chapter, unless specifically indicated, the definitions of overweight and obesity will be the standard WHO definitions, based on body mass index (BMI): overweight BMI 25–29 kg/m^2, obese BMI 30–35 kg/m^2, and morbid obesity BMI > 35 kg/m^2.

The International Continence Society publishes standardization documents from time to time, with terminology and definitions of common symptoms, signs, and diagnoses relating to incontinence [1]. Those definitions will be used throughout this chapter. Urinary symptoms related to continence in women can be divided into storage symptoms, voiding symptoms, or postvoiding symptoms (Table 20.1). It is now recognized that these symptoms do not correlate well with the underlying cause of incontinence or findings during urodynamic testing. While women with the isolated symptom of stress urinary incontinence (SUI) are very likely to have urethral sphincter weakness, the relationship between other storage symptoms and diagnosis is less clear. Thus the symptom syndrome "overactive bladder" (OAB) is now defined as "urgency, with or without urge incontinence, usually with frequency and nocturia" [1].

Incidence and prevalence

UI is a cause of significant morbidity and cost, estimated at over £500 million representing 1% of the health-care budget in the United Kingdom [2], €400 billion in Europe [3], and between $25 and $50 billion in the United States [4,5]. Incontinence becomes more prevalent with increasing age [6] (Fig. 20.1). It has recently been estimated that 11% of the global population suffers from OAB and 8% from UI [7]. Previous population studies such as the EPINCONT study in Norway reported prevalence rates of any form of UI as high as 25%, with SUI being the most common type in their local population [8]. Known risk factors for UI include parity, vaginal childbirth [9], large babies [10], perineal trauma, operative delivery [11], increasing age [12], and prior hysterectomy. Certain modifiable lifestyle factors—current or former cigarette smoking (>20 a day) as well as intake of tea—have been shown to be positively associated with UI [13]. Particular medications (e.g., diuretics and doxazosin) are known to affect lower urinary tract function and may cause UI. Detailed analysis of data demonstrating the relationship between urinary symptoms and obesity will be discussed in the following sections.

Normal bladder function and causes of incontinence in women

The bladder has two functions: to store urine and to then void the stored urine. During the storage phase of micturition, the detrusor muscle of the bladder is quiescent and accommodates increasing urine volumes with no increase in pressure. As bladder capacity is reached, sensory afferents (stretch receptors) in the bladder wall are triggered to give rise to increasing sensation of bladder filling. The sphincter mechanism (discussed later) is closed. Cortical inhibition of the spinal voiding reflex arc, learned during toilet training in infancy, allows delay of micturition until socially convenient. During

TABLE 20.1 Selected symptoms and signs of urinary incontinence.

	Definition
Storage symptoms	
Increased daytime frequency	Patient who considers he/she voids too often by day
Nocturia	Waking at night one or more times to void
Urgency	A sudden compelling desire to pass urine, which is difficult to defer
Urinary incontinence	Any involuntary leakage of urine
Stress urinary incontinence	Involuntary leakage on effort or exertion or on sneezing and coughing
Urge urinary incontinence	Involuntary leakage accompanied by or immediately preceded by urgency
Mixed urinary incontinence	Involuntary leakage associated with urgency and also with exertion, effort, sneezing or coughing
Nocturnal enuresis	The loss of urine occurring during sleep
Voiding symptoms	
Slow stream	Reported by the individual as her perception of reduced urine flow, usually compared to previous performance or in comparison to others
Splitting or spraying	... of the stream may be reported
Intermittent stream	When the individual describes urine flow that stops and starts, on more than one occasion, during micturition
Hesitancy	Difficulty in initiating micturition resulting in a delay in the onset of voiding after the individual is ready to pass urine
Terminal dribble	A prolonged final part of micturition, when the flow has slowed to a trickle or dribble
Postmicturition symptoms	
Incomplete emptying	A self-explanatory term for a feeling experienced by the individual after passing urine
Postmicturition dribble	The involuntary loss of urine immediately after he or she has finished passing urine, usually after rising from the toilet in case of a women

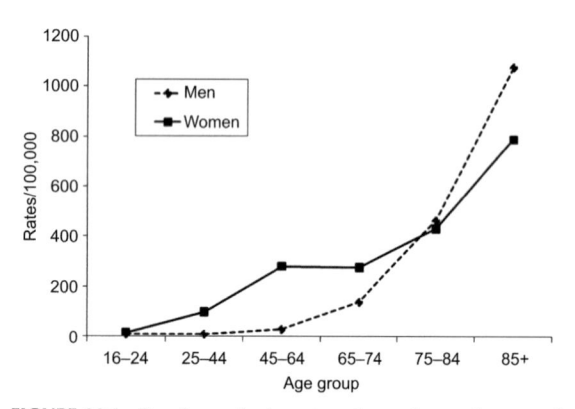

Source: Definitions from the International Continence Society Standardisation Document [1].

FIGURE 20.1 Prevalence of urinary incontinence by age in men and women. *Data from Hospital Episode Statistics, UK Department of Health.*

voiding, cortical inhibition is removed and a coordinated relaxation of the pelvic floor and urethral sphincters occurs synchronously with detrusor contraction. Detrusor contraction is mediated by muscarinic cholinergic nerves of the parasympathetic nervous system, and urethral sphincter tone is maintained by noradrenergic neurons of the sympathetic nervous system and somatic fibers from the pudendal nerves.

In women, continence is achieved by a combination of the ligaments supporting the bladder neck, the urethral sphincter itself and the pelvic floor muscles. The urethral sphincter mechanism in women is a functional system, including the internal (smooth muscle) and external (striated muscle) sphincters, together with the muscles of the pelvic floor, and the ligaments supporting the urethra (pubourethral ligament). In the normal situation the ligaments and pelvic floor maintain the urethra above the urogenital hiatus and thus within the abdomen. Increases in abdominal pressure are transmitted equally to the bladder and bladder neck (Fig. 20.2A).

Urodynamic stress incontinence

As mentioned previously, SUI is both a symptom and a sign [1]. In isolation, SUI is a reasonably good predictor of the presence of an incompetent urethral sphincter, leading to incontinence. Urethral sphincter weakness in most cases is due to hypermobility, where the pelvic floor and ligaments cannot retain the urethra in position, and it falls through the urogenital hiatus during increases in abdominal pressure, leading to loss of pressure transmission to the urethra and hence leakage of urine (Fig. 20.2).

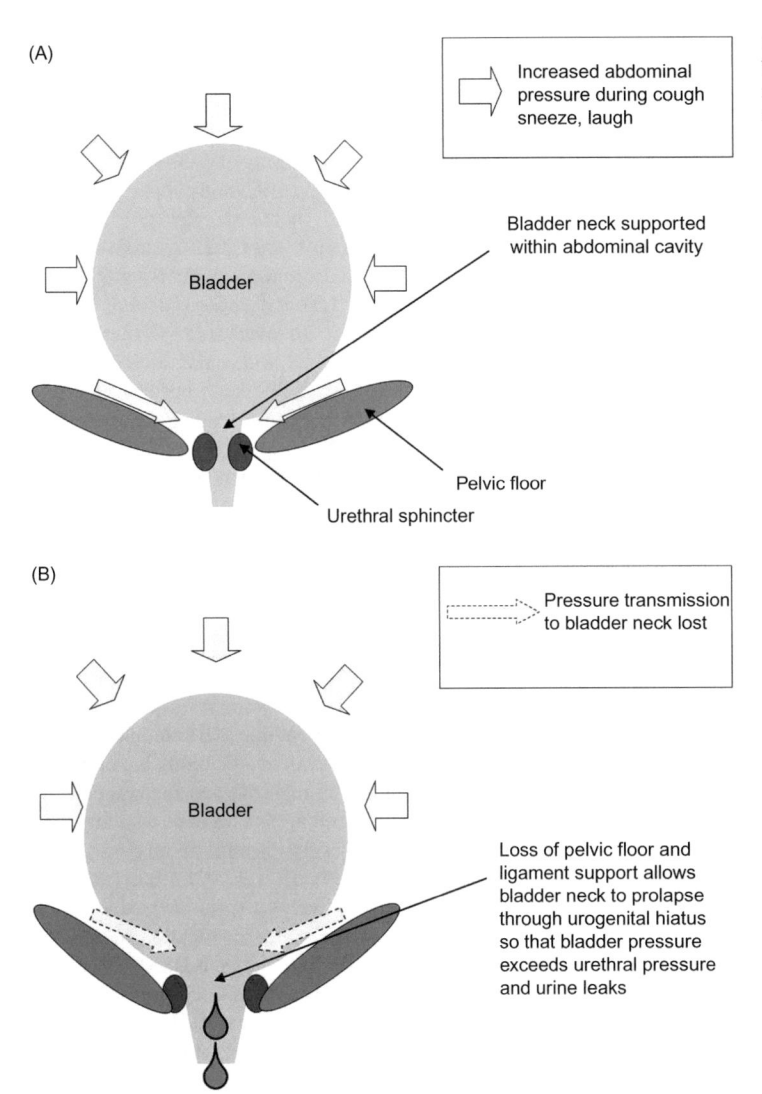

FIGURE 20.2 Schematic diagram of (A) normal bladder neck position and (B) bladder neck displacement through the urogenital hiatus causing stress incontinence.

Intrinsic sphincter deficiency is less common and occurs where urethral closure pressure is low without any urethral mobility. Urethral incompetence can be demonstrated during urodynamic testing by involuntary leakage associated with increased abdominal pressure and no detrusor muscle contraction, and it is termed urodynamic stress incontinence (USI) [1].

Detrusor overactivity

Detrusor overactivity (DO) is a condition characterized by the urodynamic observation of involuntary detrusor contractions during the filling phase of micturition [1]. DO is a poorly understood condition of uncertain etiology. Women with DO will often complain of symptoms of OAB but may not be incontinent unless the urethral sphincter function is compromised, or the detrusor contractions are of very high pressure amplitude and overcome urethral resistance. It should be remembered that the relationship between demonstrable DO and OAB syndrome is not absolute, and patients with OAB may demonstrate normal urodynamic investigations.

Obesity and urinary incontinence

Bearing in mind the changes in bladder neck movement and of the abdominal pressure transmission as mentioned previously, it seems most likely that obesity will affect the risk of developing USI rather than OAB/DO. There exist some urodynamic data in obese women before and after surgery, which have been recently reviewed in a systematic review [14] to confirm this theory. Abdominal and intravesical pressures are higher in obese women [15,16], and successful weight loss is associated with reduced urethral mobility and intravesical pressure. Bariatric surgery (BS) (techniques to reduce stomach capacity) has been shown to reduce not only the weight dramatically but also the bladder pressure by almost 50% (17−10 cm H_2O) [17]. A recent study found that obese women had more severe incontinence as defined by a lower Valsalva leak point pressure, in keeping with the theory that obesity increases the pressure load upon the bladder and bladder neck [18].

In terms of urinary symptoms, there is now a large body of evidence confirming that urinary symptoms are more prevalent in overweight and obese women. A recent metaanalysis of prospective cohort studies assessed the relationship between overweight/obesity and the risk of UI in young and middle-aged women. This showed a substantial risk increase (68%) of UI with any excess weight in young to mid-aged women. When this was broken down into BMI categories, the risk of developing UI increased by about a third in women in the overweight category (35%) and nearly doubled in women with obesity (95%) [16]. The Leicestershire Incontinence Study, funded by the UK Medical Research Council, surveyed over 12,000 community-dwelling women aged 40 years or over with validated urinary symptom questionnaires [19]. Over 7000 respondents were also sent a food questionnaire. The multivariate analysis showed an increased risk of new onset SUI both for overweight [odds ratio (OR) 1.25, 95% confidence intervals (CI) 0.94, 1.67] and obese (OR 1.74; CI 1.22, 2.48) women. A questionnaire study of 1336 Swedish women revealed a 12% prevalence of weekly or more UI [20]. Among those women, obesity was a major risk factor (OR 4.1; CI 2.6, 6.6), but overweight women also had increased risk (OR 1.8; CI 1.3, 2.5). More recently, obesity has been confirmed as an independent risk factor for SUI by other groups from the United States [21−23], Central America [24], Korea [25], and Taiwan [26] and the large EPINCONT study from Scandinavia [13] (OR 1.4; CI 1.2, 1.5 for overweight; OR 1.7; CI 1.6, 2.0 for obese) indicating that this effect is independent of ethnic origin. Interestingly, Lawrence et al. [21] found that obesity and diabetes were independent but additive risk factors for stress incontinence. Although there is a clear association between obesity and USI, it should be remembered that large epidemiological research studies often identify women with relatively mild and not bothersome symptoms. Gerten et al. [27] compared a cohort of women seeking BS for weight loss and found the impact of their incontinence to be less than the women who attended a urogynecology clinic, as assessed by the Incontinence Impact Questionnaire and Urogenital Distress Inventory. However, among women who underwent surgery for incontinence, obese women reported more distress and bother from their symptoms than women of normal BMI [15].

The association between obesity and OAB/DO appears as strong. The Leicestershire MRC study reported similar ORs for OAB as for SUI in overweight (OR 1.24; CI 0.98, 1.63) and obese (OR 1.46; CI 1.02, 2.09) women [19]. The EPINCONT study found similar risks for overweight (OR 1.1; CI 0.9, 1.3) and obese (OR 1.5; CI 1.2, 1.9) women [13]. A study of over 3000 women in the United States found not only a higher risk of OAB in the obese women (OR 2.67; CI 2.20, 3.22) but also that this risk was further increased in obese women who also had type 2 diabetes [21]. This finding agrees with other work suggesting diabetes to be an independent risk factor for OAB [28,29]. A comprehensive systematic review presents all the evidence confirming the association, and readers are directed to this for a full discussion of the evidence [30].

Obesity and fecal incontinence

Fecal incontinence (FI) is also common in obese women. A survey of 256 women attending a BS seminar in the United States reported a prevalence of flatus incontinence of 87%, liquid incontinence of 80%, and incontinence of solid stool of 19% [31]. Among 551 women who attended a urogynecology clinic, every 5-point increase in BMI carried increased risk of FI (OR 1.21; CI 1.05, 1.40) and of constipation (OR 1.13; CI 0.98, 1.31) [32]. Sileri et al. [33] found severe FI in 25% and constipation in 20% of 139 patients (93 women) who underwent BS. A 2011 systematic review of BS research (men and women) identified 13 studies that reported the rates of FI between 16% and 68%, in each case higher than the rates reported for nonobese individuals [34]. This review also reported constipation rates between 17% and 29%.

Obesity and prolapse

The increased intraabdominal pressure associated with obesity will theoretically lead to increased pressure upon the pelvic floor, uterine and vaginal ligaments, and connective tissue supports and, therefore, increases the risk of developing symptomatic prolapse. Although many authors have examined the obstetric and gynecology factors associated with prolapse, there are few studies that examine obesity. A large, community-based epidemiological survey of 17,000 women found a trend between increasing obesity and prevalence of prolapse [35]. A survey of over 21,000 women attending menopause clinics in Italy found a 5% prevalence of prolapse, and the risk of prolapse was higher in overweight (OR 1.4; CI 1.2, 1.7) and obese (OR

1.6; CI 1.3, 1.9) women [36]. Similar risks are reported for overweight women (OR 1.9; CI 1.2, 3.1) [37] in a study of 5000 women from Sweden.

A large US study analyzed data from 16,608 women in a hormone replacement trial followed for 5 years [38]. They found that overweight and obese women had increased risks of progression of all prolapse types (assessed by pelvic examination). The excess risks for anterior vaginal prolapse were 32% and 48% in overweight and obese women, respectively, for posterior prolapse 37% and 58% and for uterine prolapse 43% and 69%. In this study, loss of weight was associated with no regression of prolapse and a suggestion that uterine prolapse may worsen.

Weight loss and the effects upon continence and prolapse

Given that obesity is associated with incontinence and prolapse, in theory due to increased abdominal pressure, it makes sense that weight loss should reduce the severity and impact of these symptoms. Weight loss is a simple intervention with minimal side effects, so are there data that confirm that relief of PFD symptom occurs?

Two systematic reviews, of BS studies and nonrandomized weight reduction intervention studies, were published in 2008 and 2009 after analyzing the literature [30,39]. Six studies from BS were reviewed; in patients (predominantly women) who lost weight (typically 50% of excess weight or a fall in BMI of 15–20 points), urinary continence cure rates were between 30% and 64%. Two studies reported urinary continence outcomes after weight reduction programs: loss of 5% of body weight (a typical target for such programs) was associated with 50% or more improvement of symptom and reduction in leakage severity assessed by pad testing (median reduction of 19 g).

A pilot trial of 48 women randomized to a weight control program, or delayed enrollment, followed for 6 months demonstrated a median weight loss of 16 kg in the intervention group (BMI from 35 to 28 kg/m^2) that was associated with significant reductions in mean incontinence episodes (60% vs 15%) [40]. The trial team subsequently completed a larger study (the PRIDE study) randomizing 338 women to a 6-month weight control program or an education program [41]. Mean weight loss was 8% of body weight at baseline, and there were improvements in weekly incontinence episodes of 47.4% (CI 39.9, 54.0) in the intervention group compared to 28.1% (CI 12.6, 40.9) in the control group. This difference was dominated by changes in stress incontinence symptoms. Secondary analysis of the data showed that larger weight loss (up to 10% of baseline) was associated with greater improvements in symptoms. Women losing up to 10% weight were more likely to achieve at least 70% of symptom relief than those losing less weight (compared to controls): for 5%–10% loss of weight, OR 2.4 (CI 1.1, 5.1) and for more than 10% loss, OR 3.3 (CI 1.7, 6.4) [42]. These beneficial effects of weight loss persisted up to 18 months after the intervention [43]. Similar improvements in urinary symptoms were reported in three recent cohort studies with follow-up between 3 months and 5 years [44–46].

A more recent systematic review [47] determined the effect of BS on the prevalence of PFDs, specifically on UI, pelvic organ prolapse (POP), and FI. This meta-analysis confirmed that patients undergoing BS will benefit from a >60% reduction in UI risk. BS improves POP but has no significant benefit in FI.

Data on improvements in anal incontinence after weight loss are more sparse. Burgio et al. [48] found that the prevalence of incontinence to solids or liquids fell from 19.4% to 9.1% at 6 months and to 8.6% at 12 months after BS. A secondary analysis of the patients in the PRIDE study showed that 16% of women had at least monthly solid or liquid incontinence. The majority of these women (33 of 55) had improvement in their symptoms during the study follow-up [49]. A recent cohort study of 160 women in Israel also reported improvement in anorectal symptoms 3–6 months after BS [45].

Data on whether weight loss alters prolapse severity are also scarce. As mentioned previously, a large prospective study found that weight loss had only minimal effects upon anatomical prolapse [38]. A study from Egypt assessed 400 women and found vaginal prolapse in 65% of the sample, although symptoms were much less common [50]. Among the identified risk factors, a history of significant weight loss actually was associated with an increased risk of prolapse. Cuicchi et al. [51] reported pelvic floor symptoms from 100 obese women before and after BS. POP symptoms were reported by 56 women, and 15 women had documented anatomical prolapse. After surgery, 74% of the affected women had resolution of their prolapse symptoms. The Israel study [45] demonstrated short-term improvements in prolapse symptoms and a 5-year follow-up study of over 400 women showed an inverse relationship between the amount of weight loss after BS and prevalence of POP and UI symptoms [46]. The larger the amount of weight lost, the lower the likelihood of POP and UI symptoms, including stress and urgency UI, up to 9 years following surgery (mean, 5 years). POP symptoms/prevalence was reduced by 50% with weight loss following BS. "The strength of association persisted irrespective of age, pre BS weight, time since surgery, or other potential confounders."

Weight-loss programs or BS appears to be an effective way of reducing or even curing symptoms of UI. A target weight loss between 5% and 10% seems the optimal target to balance achievability with significant and meaningful reductions in urinary symptoms [42].

Continence and prolapse surgery in the obese woman

Surgery is a commonly used treatment for women with stress incontinence and also for prolapse if vaginal pessaries and pelvic floor exercises have failed. Surgery in the obese women may not only be technically more difficult, with higher rates of complications, but may also carry a higher failure rate for both continence and prolapse. Several studies address this issue, all of which are retrospective reviews with variable length of follow-up. A study of 242 women having retropubic tension-free vaginal tape (TVT) showed the subjective cure rate at 6 months to be 85%, 95%, and 89% in women of normal BMI, the overweight, and the obese women, respectively [52], with no difference in complication rate. Similar data were reported from a study of 285 Korean women, 45 of whom were obese and 159 overweight [53]. Rafii et al. [54] compared the outcomes after TVT in 187 women: 39 with BMI > 30 kg/m^2 and 62 with BMI between 25 and 30 kg/m^2. After follow-up of at least 6 months, the objective cure and subjective cure rates were statistically not different between 82% and 93%. Persistent urge UI (a complication of continence surgery) was more common in women with BMI > 30 kg/m^2 (17.9% compared to 6.4% or 3.4%, $P = .02$).

More recent retrospective studies of other continence procedures have demonstrated similar cure rates, irrespective of BMI for the transobturator tapes (TOT) [55–57] and fascial slings [58]. One study reported 10-year outcomes after TVT insertion and reported cure rates of 61% in women with BMI < 30 kg/m^2, compared to 50% in women with BMI > 30 kg/m^2, although no statistical comparisons were done [59]. In this cohort the reoperation rate was 2% in the women with low BMI and 4.5% in those with BMI of >30 kg/m^2.

Only one study reported any difference in either short- or long-term complications. An analysis of 31 morbidly obese women (BMI > 40 kg/m^2) compared to 52 women with BMI < 30 kg/m^2 having TVT with a mean follow-up of 18.5 months reported similar cure rates but a statistically significant excess of short-term complications (overall 48.4% vs 38.5%), although actual numbers of clinically significant events (chronic retention, wound hematoma, deep vein thrombosis, pneumonia or cardiac events) were rare in either group [60].

The systematic review by Greer et al. [14] reviewed all the data available from surgical reports at the time (2007–08). Pooled data from seven studies of TVT confirmed the low rate of bladder perforation in both obese and nonobese women, which was actually less in the obese group (1.2% vs 6.6%; OR 0.277; CI 0.098, 0.782). Other complications, including new urgency, appeared similar in obese or nonobese women. The authors were able to conduct a metaanalysis of cure rate using data from 453 obese and 1186 nonobese women [14]. Pooled cure was 81% in the obese and 85% in the nonobese women (OR 0.576; CI 0.426–0.779).

Thus overall, the data suggest that mid-urethral sling surgery for stress incontinence (TVT and TOT) is slightly less effective in obese women than in those of normal body weight, at least in the medium term (2 years). There is limited evidence about long-term outcome, so this needs to be confirmed by more long-term studies. The complication rate and adverse event profile appear similar regardless of BMI.

For prolapse surgery, data on the relationship between obesity and the outcomes and complications are extremely limited. Five-year follow-up of women having prolapse surgery revealed no association between obesity and the risk of prolapse recurrence [61]. Similarly, obesity was found not to be a risk factor for the development of prolapse after colposuspension [62]. A recent secondary analysis of data from a trial of prophylactic colposuspension with sacrocolpopexy found that although operating time was longer in the obese women, there were no differences in outcomes for incontinence, prolapse symptoms, or satisfaction [63]. A small Spanish study reported higher failure rates of prolapse surgery at 1 year in women with higher BMI, with 17 of 69 women having surgical failure overall. BMI was greater in the women whose surgery failed (29.6 ± 2.03 vs 27.1 ± 3.32) [64].

We were able to identify only one study comparing surgical complications after prolapse surgery by different BMI [65]. Adverse events were rare; the need for blood transfusion (OR 2.46; CI 1.38, 4.39) and the incidence of long-term urinary retention (OR 2.20; CI 1.21, 4.03) were more common in the women of normal BMI.

So, overall, there are few data reporting on the relationship between the success of prolapse surgery, or complications, and obesity. It appears that obesity carries neither excess risk of surgical failure, nor complications, although data from larger studies with longer follow-up are needed to be confident of this conclusion.

Conclusion

Obesity and being overweight are associated with a clearly increased risk of UI, both for stress incontinence and for OAB. In addition, obesity is associated with a higher prevalence of symptoms of anal incontinence and POP. Weight loss (either by diet and exercise or by BS) is associated with large improvements in urinary, anal, and prolapse symptoms. Achieving a target weight loss between 5% and 10% of baseline weight will bring about complete resolution of UI, anal incontinence, and prolapse symptoms in up to 70% of women. On this basis, weight loss should be considered the first-line management in the obese or overweight woman with pelvic floor symptoms, and the prospect of cure of these symptoms is likely to be a major source of motivation to comply with weight-loss strategies.

Where surgery is deemed necessary, women should be advised that both continence and prolapse surgery appear equally safe in the obese patient but that the cure rate of mid-urethral tapes for UI is slightly compromised. Further information is required to confirm whether the long-term failure rate after prolapse surgery is greater in the obese woman.

References

[1] Abrams P, Cardozo L, Fall M. The standardisation of terminology of lower urinary tract function: report from the standardisation sub-committee of the International Continence Society. Neurourol Urodyn 2002;21:167−78.

[2] Turner DA, Shaw C, McGrother CW, Dallosso HM, Cooper NJ, The Leicestershire MRC Incontinence Study Team. The cost of clinically significant urinary storage symptoms for community dwelling adults in the UK. BJU Int 2004;93:1246−52.

[3] Irwin DE, Mungapen L, Milsom I, Kopp Z, Reeves P, Kelleher C. The economic impact of overactive bladder syndrome in six Western countries. BJU Int 2009;103:202−9.

[4] Onukwugha E, Zuckerman IH, McNally D, Coyne KS, Vats V, Mullins CD. The total economic burden of overactive bladder in the United States: a disease-specific approach. Am J Manage Care 2009;15:S90−7.

[5] Kannan H, Radican L, Turpin RS, Bolge SC. Burden of illness associated with lower urinary tract symptoms including overactive bladder/urinary incontinence. Urology 2009;74:34−8.

[6] McGrother CW, Donaldson MM, Shaw C. Storage disorder of the bladder: prevalence, incidence and need for services in the UK. BJU Int 2004;93:763−9.

[7] Irwin DE, Kopp ZS, Agatep B, Milsom I, Abrams P. Worldwide prevalence estimates of lower urinary tract symptoms, overactive bladder, urinary incontinence and bladder outlet obstruction. BJU Int 2011;108:1132−8.

[8] Hannestad YS, Rortveit G, Sandvik H, Hunskaar S, Norwegian EPINCONT study. A community-based epidemiological survey of female urinary incontinence: The Norwegian EPINCONT Study. J Clin Epidemiol 2000;53:1150−7.

[9] MacLennan AH, Taylor AW, Wilson DH, Wilson D. The prevalence of pelvic floor disorders and their relationship to gender, age, parity and mode of delivery. BJOG 2000;107:1460−70.

[10] Glazener CM, Herbison GP, MacArthur C. New postnatal urinary incontinence: obstetric and other risk factors in primiparae. BJOG 2006;113:208−17.

[11] Boyles SH, Li H, Mori T, Osterweil P, Guise JM. Effect of mode of delivery on the incidence of urinary incontinence in primiparous women. Obstet Gynecol 2009;113:134−41.

[12] Swithinbank LV, Donovan JL, du Heaume JC. Urinary symptoms and incontinence in women: relationship between occurrence, age, and perceived impact. Br J Gen Pract 2002;49:897−900.

[13] Hannestad YS, Rortveit G, Daltveit AK, Hunskaar S. Are smoking and other lifestyle factors associated with female urinary incontinence? The Norwegian EPINCONT Study. Br J Obstet Gynaecol 2005;110:247−54.

[14] Greer WJ, Richter HE, Bartolucci AA, Burgio KL. Obesity and pelvic floor disorders: a systematic review. Obstet Gynecol 2008;112:341−9.

[15] Richter HE, Kenton K, Huang L. The impact of obesity on urinary incontinence symptoms, severity, urodynamic characteristics and quality of life. J Urol 2010;183:622−8.

[16] Lamerton TJ, Torquati L, Brown WJ. Overweight and obesity as major, modifiable risk factors for urinary incontinence in young to mid-aged women: a systematic review and meta-analysis. Obes Rev 2018;19:1735−45.

[17] Sugarman H, Windsor A, Bessos M, Kellum J, Reines H, DeMaria E. Effects of surgically induced weight loss on urinary bladder pressure, sagittal abdominal diameter and obesity co-morbidity. Int J Obest Relat Metab Disord 2008;22:230−5.

[18] El-Hefnawy AS, Wadie BS. Severe stress urinary incontinence: objective analysis of risk factors. Maturitas 2011;68:374−7.

[19] Dallosso HM, McGrother CW, Matthews RJ, Donaldson MM, The Leicestershire MRC Incontinence Study Team. The association of diet and other lifestyle factors with overactive bladder and stress incontinence: a longitudinal study in women. BJU Int 2003;92:69−77.

[20] Uustal Fornell E, Wingren G, Kjolhede P. Factors associated with pelvic floor dysfunction with emphasis on urinary and fecal incontinence and genital prolapse: an epidemiological study. Acta Obstet Gynecol Scand 2004;83:383−9.

[21] Lawrence JM, Lukacz ES, Liu IL, Nager CW, Luber KM. Pelvic floor disorders, diabetes, and obesity in women: findings from the Kaiser Permanente Continence Associated Risk Epidemiology Study. Diabetes Care 2007;30:2536−41.

[22] Markland AD, Richter HE, Fwu CW, Eggers P, Kusek JW. Prevalence and trends of urinary incontinence in adults in the United States, 2001 to 2008. J Urol 2011;186:589−93.

[23] Hawkins K, Pernarelli J, Ozminkowski RJ. The prevalence of urinary incontinence and its burden on the quality of life among older adults with medicare supplement insurance. Qual Life Res 2011;20:723−32.

[24] Lopez M, Ortiz AP, Vargas R. Prevalence of urinary incontinence and its association with body mass index among women in Puerto Rico. J Womens Health 2009;18:1607−14.

[25] Ham E, Choi H, Seo JT, Kim HG, Palmer MH, Kim I. Risk factors for female urinary incontinence among middle-aged Korean women. J Womens Health 2009;18:1801−6.

[26] Hsieh CH, Hsu CS, Su TH, Chang ST, Lee MC. Risk factors for urinary incontinence in Taiwanese women aged 60 or over. Int Urogynecol J Pelvic Floor Dysfunct 2007;18:1325−9.

[27] Gerten KA, Richter HE, Burgio KL, Wheeler TL, Goode PS, Redden DT. Impact of urinary incontinence in morbidly obese women versus women seeking urogynecologic care. Urology 2007;70:1082—5.

[28] Liu RT, Chung MS, Lee WC. Prevalence of overactive bladder and associated risk factors in 1359 patients with type 2 diabetes. Urology 2011;78:1040—5.

[29] Uzun H, Zorba OU. Metabolic syndrome in female patients with overactive bladder. Urology 2012;79:72—5.

[30] Hunskaar S. A systematic review of overweight and obesity as risk factors and targets for clinical intervention for urinary incontinence in women. Neurourol Urodyn 2008;27:749—57.

[31] Wasserberg N, Haney M, Petrone P. Fecal incontinence among morbid obese women seeking for weight loss surgery: an underappreciated association with adverse impact on quality of life. Int J Colorectal Dis 2008;23:493—7.

[32] Erekson EA, Sung VW, Myers DL. Effect of body mass index on the risk of anal incontinence and defecatory dysfunction in women. Am J Obstet Gynecol 2008;198:596—604.

[33] Sileri P, Franceschilli L, Cadeddu F. Prevalence of defaecatory disorders in morbidly obese patients before and after bariatric surgery. J Gastrointest Surg 2012;16:62—6.

[34] Poylin V, Serrot FJ, Madoff RD. Obesity and bariatric surgery: a systematic review of associations with defecatory dysfunction. Colorectal Dis 2011;13:e92—e103.

[35] Mant J, Painter R, Vessey M. Epidemiology of genital prolapse: observations from the Oxford Family Planning Association Study. Br J Obstet Gynaecol 1997;104:579—85.

[36] Progetto Menopausa Italia Study Group. Risk factors for genital prolapse in non-hysterectomized women around menopause. Results from a large cross-sectional study in menopausal clinics in Italy. Eur J Obstet Gynecol Reprod Biol 2000;93:135—40.

[37] Miedel A, Tegerstedt G, Maehle-Schmidt M, Nyren O, Hammarstrom M. Nonobstetric risk factors for symptomatic pelvic organ prolapse. Obstet Gynecol 2009;113:1089—97.

[38] Kudish BI, Iglesia CB, Sokol RJ. Effect of weight change on natural history of pelvic organ prolapse. Obstet Gynecol 2009;113:81—8.

[39] Subak LL, Richter HE, Hunskaar S. Obesity and urinary incontinence: epidemiology and clinical research update. J Urol 2009;182:S2—7.

[40] Subak LL, Whitcomb E, Shen H, Saxton J, Vittinghoff E, Brown JS. Weight loss: a novel and effective treatment for urinary incontinence. J Urol 2005;174:190—5.

[41] Subak LL, Wing R, West DS. Weight loss to treat urinary incontinence in overweight and obese women. N Engl J Med 2009;360:481—90.

[42] Wing RR, Creasman JM, West DS. Improving urinary incontinence in overweight and obese women through modest weight loss. Obstet Gynecol 2010;116:284—92.

[43] Wing RR, West DS, Grady D. Effect of weight loss on urinary incontinence in overweight and obese women: results at 12 and 18 months. J Urol 2010;184:1005—10.

[44] Knepfler T, Valero E, Triki E, Chilinsteva N, Koensgen S, Rohr S. Bariatric surgery improves pelvic floor disorders. J Visc Surg 2016;153:95—9.

[45] Leshem A, Shimonov M, Amir H, Gordon D, Groutz A. Effects of bariatric surgery on female pelvic floor disorders. Urology 2017;105:42—7.

[46] Gabriel I, Tavakkoli A, Minassian VA. Pelvic organ prolapse and urinary incontinence in women after bariatric surgery: 5-year follow-up. Female Pelvic Med Reconstr Surg 2018;24:120—5.

[47] Lian W, Zheng Y, Huang H, Chen L, Cao B. Effects of bariatric surgery on pelvic floor disorders in obese women: a meta-analysis. Arch Gynecol Obstet 2017;296:181—9.

[48] Burgio KL, Richter HE, Clements RH, Redden DT, Goode PS. Changes in urinary and fecal incontinence symptoms with weight loss surgery in morbidly obese women. Obstet Gynecol 2007;110:1034—40.

[49] Markland AD, Richter HE, Burgio KL, Myers DL, Hernandez AL, Subak LL. Weight loss improves fecal incontinence severity in overweight and obese women with urinary incontinence. Int Urogynecol J 2011;22:1151—7.

[50] Gomman HM, Nossier SA, Fotohi EM, Kholeif AE. Prevalence and factors associated with genital prolapse: a hospital-based study in Alexandria (Part I). J Egypt Public Health Assoc 2001;76:313—35.

[51] Cuicchi D, Lombardi R, Cariani S, Leuratti L, Lecce F, Cola B. Clinical and instrumental evaluation of pelvic floor disorders before and after bariatric surgery in obese women. Surg Obes Relat Dis 2013;9:69—75. Available from: https://doi.org/10.1016/j.soard.2011.08.013.

[52] Mukherjee K, Constantine G. Urinary stress incontinence in obese women: tension-free vaginal tape is the answer. BJU Int 2001;88:881—3.

[53] Ku JH, Oh JG, Shin JW, Kim SW, Paick JS. Outcome of midurethral sling procedures in Korean women with stress urinary incontinence according to body mass index. Int J Urol 2006;13:379—84.

[54] Rafii A, Darai E, Haab F, Samain E, Levardon M, Deval B. Body mass index and outcome of tension-free vaginal tape. Eur Urol 2003;43:288—92.

[55] Tchey DU, Kim WT, Kim YJ, Yun SJ, Lee SC, Kim WJ. Influence of obesity on short-term surgical outcome of the transobturator tape procedure in patients with stress urinary incontinence. Int Neurourol J 2010;14:13—19.

[56] Rechberger T, Futyma K, Jankiewicz K, Adamiak A, Bogusiewicz M, Skorupski P. Body mass index does not influence the outcome of anti-incontinence surgery among women whereas menopausal status and ageing do: a randomised trial. Int Urogynecol J 2010;21:801—6.

[57] Liu PE, Su CH, Lau HH, Chang RJ, Huang WC, Su TH. Outcome of tension-free obturator tape procedures in obese and overweight women. Int Urogynecol J 2011;22:259—63.

[58] Haverkorn RM, Williams BJ, Kubricht WS, Gomelsky A. Is obesity a risk factor for failure and complications after surgery for incontinence and prolapse in women? J Urol 2011;185:987—92.

[59] Aigmueller T, Trutnovsky G, Tamussino K. Ten-year follow-up after the tension-free vaginal tape procedure. Am J Obstet Gynecol 2011;205.

[60] Skriapas K, Poulakis V, Dillenburg W. Tension-free vaginal tape (TVT) in morbidly obese patients with severe urodynamic stress incontinence as last option treatment. Eur Urol 2006;49:544—50.

[61] Clark AL, Gregory T, Smith VJ, Edwards R. Epidemiologic evaluation of reoperation for surgically treated pelvic organ prolapse and urinary incontinence. Am J Obstet Gynecol 2003;189:1261—7.

[62] Auwad W, Bombieri L, Adekanmi O, Waterfield M, Freeman R. The development of pelvic organ prolapse after colposuspension: a prospective, long-term follow-up study on the prevalence and predisposing factors. Int Urogynecol J Pelvic Floor Dysfunct 2006;17:389—94.

[63] Bradley CS, Kenton KS, Richter HE. Obesity and outcomes after sacrocolpopexy. Am J Obstet Gynecol 2008;199:690—8.

[64] Diez-Calzadilla NA, March-Villalba JA, Ferrandis C. Risk factors in the failure of surgical repair of pelvic organ prolapse. Actas Urol Esp 2011;35:448—53.

[65] Nam KH, Jeon MJ, Hur HW, Kim SK, Bai SW. Perioperative and long-term complications among obese women undergoing vaginal surgery. Int J Gynaecol Obstet 2010;108:244—6.

Chapter 21

Urinary and fecal incontinence in obese women

Vasilios Pergialiotis[1,2] and Stergios K. Doumouchtsis[1,3,4,5]

[1]*Laboratory of Experimental Surgery and Surgical Research N.S. Christeas, National and Kapodistrian University of Athens, Athens, Greece,*
[2]*3rd Department of Obstetrics and Gynecology, Attikon Hospital, National and Kapodistrian University of Athens, Athens, Greece,* [3]*Department of Obstetrics and Gynaecology, Epsom and St Helier University Hospitals NHS Trust, London, United Kingdom,* [4]*St George's, University of London, London, United Kingdom,* [5]*American University of the Caribbean School of Medicine, Coral Gables, FL, United States*

Introduction—epidemiology

Urinary incontinence

Urinary incontinence (UI) is a common disorder that affects approximately 25% of the general population according to large epidemiological studies [1,2]. According to the findings of a metaanalysis that accumulated data from approximately 230,000 people (men and women), the prevalence of the disease peaks at 35 years of age and remains constant thereafter, although the severity of symptoms seems to progress following the completion of the seventh decade of life [3]. Several factors have been implicated in the course of UI, including parity, operative vaginal delivery, length of labor, obesity, chronic cough, depression, anxiety, poor health status, lower urinary tract symptoms, previous hysterectomy, and smoking. As expected, the association of the various factors with the risk of developing UI becomes more significant as the symptoms of the disorder progress to more severe forms.

UI is associated with substantial economic burden. In a study from 2014 the total national cost for treating overactive bladder in the United States was estimated at 65.9 billion dollars, with projected costs at 76.2 billion dollars in 2015 and 82.6 billion dollars in 2020 [4]. Internationally, the burden remains still significant given the increasing number of people referred to physicians for treatment [5]; however, the heterogeneity of cost estimates around the globe renders a cost-effectiveness analysis of the various treatment alternatives impossible given the risks of bias [6].

On the other hand, data related to quality of life (QoL) are more homogeneous and permit proper interpretation of findings and show that the impact of the disorder can be detrimental irrespective of the patients' age [7,8]. Specifically, patients with UI suffer from depression, anxiety, and sexual dysfunction [9–11]. The symptomatology is worse when obesity is copresent as women with a body mass index (BMI) greater than 30 kg/m^2 tend to have more severe symptoms [12].

Anal incontinence

Anal incontinence (AI) is the complaint of involuntary loss of flatus or feces. In other words, AI refers to the involuntary loss of gas (flatus incontinence) that may be combined or not with fecal incontinence (FI). The latter is defined as the involuntary loss of solid or liquid stool. Epidemiological data related to AI are more scarce compared to those available for UI. This is primarily owed to the emotional consequences of these disorders, which have a significant impact on the patient's self-esteem and QoL as well as to infrequent screening from health-care providers. Response rates of participants differ significantly when compared to those of patients with UI. Specifically, only 60%–70% of participants respond to FI-related questionnaires, whereas the actual response rates for UI exceed 80% [3,13]. Nevertheless, the actual incidence of the disease is estimated to range from 7% to 15% in community-dwelling men and women. AI is more prevalent among institutionalized individuals with data referring to nursing homes reporting an actual incidence between 50% and 70% [14,15]. This fact is primarily the result of the increased prevalence of dementia in these populations, a disorder that has a direct negative impact on anal sphincter control.

Several risk factors have been related to AI, including operative vaginal delivery, obstetric anal sphincter injury, advanced age, decreased physical ability, neurological diseases, obesity, and intestinal motility problems (primarily diarrhea).

The impact of AI on patients' QoL can be catastrophic as the majority of them lack self-esteem and progressively diminish their social interactions to the minimum together with hiding their problem from the relatives and in several occasions even from their general practitioner.

Epidemiological data in obese populations

Obesity aggravates the severity of UI symptoms. In the Norwegian EPICONT (Epidemiology of Incontinence in the County of Nord-Trøndelag) study, which included 34,755 women, researchers observed that obesity had a significant impact on UI (stress, urgency, and mixed) [16]. Previous studies suggested that for each increase in BMI units, the odds ratio (OR) of developing UI increases by 1.6 [17] and women with BMI that exceeds 35 kg/m^2 have a prevalence that peeks at 67.3% [18]. The detrimental effect of obesity on UI was also confirmed by a randomized trial that revealed that a reduction of 5% in body weight seems to improve the symptoms by reducing their frequency by approximately 50% [19].

Although UI is more prevalent in postmenopausal women, young as well as middle-aged women are also affected by obesity. In a recent metaanalysis that was based on eight studies, overweight status was associated with an increased risk of developing UI by approximately 35%, while the risk nearly doubled among obese women (relative risk 1.95, 95% CI 1.58, 2.42) [20]. In this specific population the effect of unit-increase in weight on UI ranged between 1% and 10%; however, the outcome measures that were used (waist circumference, weight, and BMI) varied significantly, hence precluding definitive conclusions. The impact of obesity was more significant in the case of stress UI (risk ratio 2.45) and less significant in urgency incontinence (risk ratio 1.90).

As previously stated, the effect of age on the prevalence of incontinence becomes minimal after the completion of the fifth decade of life [3]; however, the severity of symptoms becomes more pronounced after 70 years of age. Several factors contribute to this, including chronic ischemia of the lower urinary tract as well as the higher prevalence of abdominal obesity. The impact of this latter risk factor on the prevalence of UI of elderly women is demonstrated in a subcohort of 471 women who were retrieved in the EXERNET trial's database [21]. The findings of this study revealed that obesity significantly increased the incidence of stress, urgency, and mixed UI as well as nocturia and also decreased the rates of "never wet" incidents.

The impact of obesity on AI is not as widely evaluated as data are limited in this field. A small number of articles, however, point the fact that incontinent women have increased BMIs compared to the general population and that central obesity may confer to AI. In a previous cohort of obese and overweight patients that was based on the Program to Reduce Incontinence by Diet and Exercise, randomized trial researchers not only observed that the prevalence of FI was 16%, but they also observed a marginal protective effect of each increase in BMI units (OR 0.9, 95% CI 0.1, 1.0, $P = .01$) [22]. The single most important dietary factor that was associated with FI was a low intake of dietary fiber. As the symptoms become worse, a tendency toward seeking medical advice is observed [23].

To date, data concerning the cumulative effect of age on symptom severity in obese patients are lacking and, given the fact that the majority of information is based on community-dwelling individuals, generalization of findings cannot be extrapolated. During childhood the impact of obesity on functional defecation disorders (including functional constipation and functional nonretentive FI) seems to be important; however, its effect on AI, in particular, does not seem to be affected by the patient's BMI, although data are extremely limited to draw safe conclusions (only 17 children with FI were available in the most recent systematic review) [24]. The most relevant study was based on 2106 community-dwelling women older than 40 years of whom 511 reported some form of FI [25]. A positive correlation of age with FI rates was observed after subgrouping women according to their decade of life and the multivariable analysis revealed that age (per 5 years) increased the odds of developing FI by approximately 10%, although the effect was not statistically significant (OR 1.1, 95% CI 1, 1.2, $P = .15$). However, substratification according to the presence of obesity was not present in this study; hence, it remains unknown whether there is an additive effect of obesity and age on the risk of developing AI.

Pathophysiology of incontinence in the obese population

Urinary incontinence

The pathophysiology of UI in the obese population has been investigated in several studies in the international literature. In 2008 Hunskaar performed a systematic review of the literature, investigating the factors that predispose overweight and obese women to develop UI [26]. The findings of this study suggested that obesity increases the intraabdominal pressure, thus weakening the pelvic musculature and innervation. Concurrently, the increase in intraabdominal pressure leads to altered urodynamic results, including increased maximal intravesical peak pressures during cough, although it does not seem to alter the urethral function (obese patients have similar Valsalva leak point pressures and comparable maximal urethral closure pressures) [27–29].

The aggravation of symptoms of UI that is observed in excess and morbid obesity is the result of an increasing intraabdominal pressure that seems to reach a value of approximately 9 mmHg (a result that significantly differs from the nonexistent intraabdominal pressure in healthy controls) [30,31]. This, however, is significantly smaller compared to that expected in patients with intraabdominal hypertension (>12 mmHg) and, thus, its effects are rather subtle (chronic defects in pelvic vasculature, innervation, and musculature as explained earlier) than associated with organ failure [30,32].

Anal incontinence

Research evidence on AI is scarcer and the results of individual studies have not been reproduced and validated in their majority; however, several pathophysiological mechanisms have been proposed and these deserve further investigation. Altered stool consistency is among the proposed factors that contribute to FI as obese patients report altered bowel habits that are accompanied by unformed stools [33]. The incidence of fecal urgency in this population reaches 38.5% of cases and FI rates are comparable to those of studies conducted in community-dwelling women (17%). The findings of this study were confirmed in a recent prospective case-matched study that compared obese patients to age- and sex-matched nonobese patients with FI [34]. This latter study reported that the rates of FI were comparable between the two groups, although stool inconsistency seemed to be more prevalent among obese patients. Significant differences were also observed in anorectal manometry results with obese patients having higher upper- and lower part resting pressures, higher intraabdominal pressure during effort, and increased maximum tolerable volume. Similar to those, were the findings of an earlier study, which also showed that baseline anal resting and squeeze pressures are increased in obese women with FI, suggesting that the threshold for leakage lowers as the pressure increases [35].

Outcomes of incontinence procedures in obese women

Urinary incontinence

The efficacy of surgical procedures has been evaluated in nonobese as well as obese incontinent patients. The majority of available evidence is based on the outcomes of patients treated with mid-urethral slings, while evidence concerning the efficacy of bulking agents in obese women is lacking. Only one study on a cohort of obese male patients who underwent surgical bulking procedures for UI is available [36]. Its findings reveal that the group that was initially treated with bulking agents had the highest rates of subsequent interventions (40.1% during the first year and 52.9% for the entire follow-up after surgery) compared to slings (10.4% and 15.5%) and artificial urinary sphincter (2.3% and 20%). Evidence concerning bladder neck suspension procedures is limited to case reports; hence, definitive conclusions cannot be drawn to guide clinical practice.

On the other hand, several studies investigated the efficacy of slings in overweight and obese women. Current data suggest that transobturator slings should be preferred compared to single-incision slings as the latter option is associated with worse postoperative incontinence symptoms [37]. Either way, the impact of BMI on outcomes following midurethral sling placement is negative according to the findings of a recent systematic review that summarized data from 2846 women [38]. Specifically, objective cure rates are lower in overweight and obese patients compared to healthy controls, although subjective cure rates do not seem to differ. The severity of obesity also negatively influences the results of the operation as morbidly obese patients are twice as likely to report a failure following a mid-urethral sling operation [39]. In the long term, obese women undergoing sling procedures have worse outcomes according to the findings of a recent study that followed up patients for a period of 5 years [40]. Specifically, obese patients had worse objective cure rates (65.9% compared to 87.4% in nonobese) as well as subjective cure rates (53.6% vs 76.7%). The incidence of urinary urgency incontinence was comparable in both groups; however, bothersome symptoms were more likely to persist in obese women (58.9% vs 42.1%).

Anal incontinence

AI is primarily treated with behavioral treatment that aims to reduce stool inconsistency. Various treatment alternatives have been proposed, including bowel training, biofeedback, antidiarrheal drugs, and bulk laxatives (in cases of chronic constipation). Neither of these methods has been evaluated in obese populations. Sphincteroplasty remains the cornerstone of treatment in cases of damaged anal sphincter. A study that included 15 obese and 64 nonobese women, followed-up for a median period of 64 months [41], showed that although the risk of complications was comparable between the two groups, improvement was less evident in obese patients. Perianal bulking has also been used and showed promising results;

however, data in obese populations are unavailable and considering the limited efficacy of bulking agents in the field of UI, they cannot be yet recommended for the treatment of FI.

Incontinence symptoms following weight loss

Urinary incontinence

The largest randomized trial investigating the impact of a 6-month weight-loss program on outcomes of UI was published in 2009 and included 338 overweight and obese women with at least 10 UI episodes per week [42]. This study found that weight reduction in the intervention group was approximately 8.0 kg (compared to 1.5 kg in the control group), and this was accompanied by a significant reduction in incontinence episodes at 6 months (47% vs 28%). The difference was significant, however, only for cases with stress UI, whereas patients with urgency UI did not benefit from the results of the program.

Taking into account the modest weight loss that accompanies behavioral programs of weight loss, one can assume that the impact of surgical procedures would be more drastic. Outcomes concerning the impact of bariatric surgery on UI are widely available in the international literature and a recent systematic review summarized them by accumulating evidence from 33 cohort studies, including a total of 2910 patients followed up for a median period of 12 months [43]. Despite the low quality of evidence that was available and the fact that outcomes referred to mixed populations (men and women), bariatric surgery was associated with the improvement or resolution of UI in approximately 55% of cases. Stress UI was less likely to be treated (47% of cases) compared to urgency UI (53% of cases). Worsening and new onset UI was observed in approximately 3% of cases. Another systematic review that included studies reporting on the impact of bariatric surgery in women with UI showed that incontinence-specific QoL scores were improved by 14%, while the proportion of women who were cured from any type of UI reached 58% [44]. On the other hand, the authors noted that the significant heterogeneity among included studies as well as the short follow-up period (median 12 months) limits the findings of this study.

The pathophysiology behind the improvement of UI symptoms following bariatric procedures has been described since 1992, when Bump et al. observed that weight reduction leads to statistically significant changes in intravesical pressure, magnitude of bladder pressure increases during coughing, bladder-to-urethra pressure transmission with cough, and urethral axial mobility [45].

Anal incontinence

The impact of bariatric surgery on AI symptoms is rather disappointing. The most recent systematic review summarized evidence from 20 studies that included 3684 patients with pelvic floor dysfunction [46]. Outcomes on FI were available in nine of these studies and revealed a modest effect with a reduction of the odds of FI by approximately 20%; however, the change was not significant (OR 0.80, 95% CI 0.53, 1.21) and this observation may be attributed to the relatively small number of enrolled patients. The pathophysiology behind this association remains an assumption and relies on the actual etiology of FI in obese women, which is primarily owed to stool inconsistency. In addition, a study that investigated functional anorectal parameters in 46 obese women with pelvic floor disorders revealed that bariatric surgery had no impact on internal and external anal sphincter size and on mean anorectal angle during squeeze and during defecation [47]. Similar observations were also reported in a previous systematic review from 2011, which included six studies with reported outcomes prior to, and after bariatric surgery [48]. In the latter study the authors noted that although FI seemed to improve after Roux-en-Y procedures, the effects of the operation on diarrhea were unclear.

Conclusion

Obese women are more prone to develop urinary and AI compared to normal-weight women. Increased intraabdominal pressure seems to play a major role in the process of the development of UI with a direct negative impact on pelvic structures. On the other hand, evidence in the field of AI is less clear but suggests that stool inconsistency is more important than the increase in intraabdominal pressure. Consequently, weight loss results in important changes in urinary tract function, whereas its impact on anorectal manometry and functional parameters is minimal if any. Concerning continence procedures, the vast majority of available evidence is based on mid-urethral slings with transobturator tapes exhibiting a superior effect, compared to single incision tapes. Long-term outcomes are less clear; however, obesity is associated with aggravated severity of symptoms postoperatively. Evidence around surgical treatment of FI is limited and does not suffice to reach conclusions for clinical practice.

References

[1] Ebbesen MH, Hunskaar S, Rortveit G, Hannestad YS. Prevalence, incidence and remission of urinary incontinence in women: longitudinal data from the Norwegian HUNT study (EPINCONT). BMC Urol 2013;13:27. Available from: https://doi.org/10.1186/1471-2490-13-27.

[2] Hannestad YS, Rortveit G, Sandvik H, Hunskaar S. A community-based epidemiological survey of female urinary incontinence: the Norwegian EPINCONT study. Epidemiology of Incontinence in the County of Nord-Trøndelag. J Clin Epidemiol 2000;53:1150–7.

[3] Minassian VA, Drutz HP, Al-Badr A. Urinary incontinence as a worldwide problem. Int J Gynaecol Obstet 2003;82:327–38.

[4] Coyne KS, Wein A, Nicholson S, Kvasz M, Chen CI, Milsom I. Economic burden of urgency urinary incontinence in the United States: a systematic review. J Manag Care Pharm 2014;20:130–40. Available from: https://doi.org/10.18553/jmcp.2014.20.2.130.

[5] Milsom I, Coyne KS, Nicholson S, Kvasz M, Chen CI, Wein AJ. Global prevalence and economic burden of urgency urinary incontinence: a systematic review. Eur Urol 2014;65:79–95. Available from: https://doi.org/10.1016/j.eururo.2013.08.031.

[6] Zwolsman S, Kastelein A, Daams J, Roovers JP, Opmeer BC. Heterogeneity of cost estimates in health economic evaluation research. A systematic review of stress urinary incontinence studies. Int Urogynecol J 2019;30:1045–59. Available from: https://doi.org/10.1007/s00192-018-3814-0.

[7] Ko Y, Lin SJ, Salmon JW, Bron MS. The impact of urinary incontinence on quality of life of the elderly. Am J Manag Care 2005;11:S103–11.

[8] Kwon BE, Kim GY, Son YJ, Roh YS, You MA. Quality of life of women with urinary incontinence: a systematic literature review. Int Neurourol J 2010;14:133–8. Available from: https://doi.org/10.5213/inj.2010.14.3.133.

[9] Lai HH, Shen B, Rawal A, Vetter J. The relationship between depression and overactive bladder/urinary incontinence symptoms in the clinical OAB population. BMC Urol 2016;16:60. Available from: https://doi.org/10.1186/s12894-016-0179-x.

[10] Lai HH, Rawal A, Shen B, Vetter J. The relationship between anxiety and overactive bladder or urinary incontinence symptoms in the clinical population. Urology 2016;98:50–7. Available from: https://doi.org/10.1016/j.urology.2016.07.013.

[11] Duralde ER, Rowen TS. Urinary incontinence and associated female sexual dysfunction. Sex Med Rev 2017;5:470–85. Available from: https://doi.org/10.1016/j.sxmr.2017.07.001.

[12] Pace G, Silvestri V, Guala L, Vicentini C. Body mass index, urinary incontinence, and female sexual dysfunction: how they affect female postmenopausal health. Menopause 2009;16:1188–92. Available from: https://doi.org/10.1097/gme.0b013e3181a6b491.

[13] Bharucha AE, Dunivan G, Goode PS, et al. Epidemiology, pathophysiology, and classification of fecal incontinence: state of the science summary for the National Institute of Diabetes and Digestive and Kidney Diseases (NIDDK) workshop. Am J Gastroenterol 2015;110:127–36. Available from: https://doi.org/10.1038/ajg.2014.396.

[14] Bliss DZ, Harms S, Garrard JM, et al. Prevalence of incontinence by race and ethnicity of older people admitted to nursing homes. J Am Med Dir Assoc 2013;14:451.e451–7. Available from: https://doi.org/10.1016/j.jamda.2013.03.007.

[15] Nelson R, Furner S, Jesudason V. Fecal incontinence in Wisconsin nursing homes: prevalence and associations. Dis Colon Rectum 1998;41:1226–9. Available from: https://doi.org/10.1007/bf02258218.

[16] Hannestad YS, Rortveit G, Daltveit AK, Hunskaar S. Are smoking and other lifestyle factors associated with female urinary incontinence? The Norwegian EPINCONT Study. BJOG 2003;110:247–54. Available from: https://doi.org/10.1046/j.1471-0528.2003.02327.x.

[17] Brown JS, Seeley DG, Fong J, Black DM, Ensrud KE, Grady D. Urinary incontinence in older women: who is at risk? Study of Osteoporotic Fractures Research Group. Obstet Gynecol 1996;87:715–21.

[18] Schreiber Pedersen L, Lose G. Prevalence of urinary incontinence among women and analysis of potential risk factors in Germany and Denmark. Acta Obstet Gynecol Scand 2017;96:939–48. Available from: https://doi.org/10.1111/aogs.13149.

[19] Subak LL, Whitcomb E, Shen H, Saxton J, Vittinghoff E, Brown JS. Weight loss: a novel and effective treatment for urinary incontinence. J Urol 2005;174:190–5. Available from: https://doi.org/10.1097/01.ju.0000162056.30326.83.

[20] Lamerton TJ, Torquati L, Brown WJ. Overweight and obesity as major, modifiable risk factors for urinary incontinence in young to mid-aged women: a systematic review and meta-analysis. Obes Rev 2018;19:1735–45. Available from: https://doi.org/10.1111/obr.12756.

[21] Moreno-Vecino B, Arija-Blázquez A, Pedrero-Chamizo R, et al. Associations between obesity, physical fitness, and urinary incontinence in non-institutionalized postmenopausal women: the elderly EXERNET multi-center study. Maturitas 2015;82:208–14. Available from: https://doi.org/10.1016/j.maturitas.2015.07.008.

[22] Markland AD, Richter HE, Burgio KL, Bragg C, Hernandez AL, Subak LL. Fecal incontinence in obese women with urinary incontinence: prevalence and role of dietary fiber intake. Am J Obstet Gynecol 2009;200:566.e1–6. Available from: https://doi.org/10.1016/j.ajog.2008.11.019.

[23] Richter HE, Burgio KL, Clements RH, Goode PS, Redden DT, Varner RE. Urinary and anal incontinence in morbidly obese women considering weight loss surgery. Obstet Gynecol 2005;106:1272–7. Available from: https://doi.org/10.1097/01.AOG.0000187299.75024.c4.

[24] Koppen IJ, Kuizenga-Wessel S, Saps M, et al. Functional defecation disorders and excessive body weight: a systematic review. Pediatrics 2016;138. Available from: https://doi.org/10.1542/peds.2016-1417.

[25] Varma MG, Brown JS, Creasman JM, et al. Fecal incontinence in females older than aged 40 years: who is at risk? Dis Colon Rectum 2006;49:841–51. Available from: https://doi.org/10.1007/s10350-006-0535-0.

[26] Hunskaar S. A systematic review of overweight and obesity as risk factors and targets for clinical intervention for urinary incontinence in women. Neurourol Urodyn 2008;27:749–57. Available from: https://doi.org/10.1002/nau.20635.

[27] Fuganti PE, Gowdy JM, Santiago NC. Obesity and smoking: are they modulators of cough intravesical peak pressure in stress urinary incontinence? Int Braz J Urol 2011;37:528–33. Available from: https://doi.org/10.1590/s1677-55382011000400013.

[28] Richter HE, Kenton K, Huang L, et al. The impact of obesity on urinary incontinence symptoms, severity, urodynamic characteristics and quality of life. J Urol 2010;183:622–8. Available from: https://doi.org/10.1016/j.juro.2009.09.083.

[29] Swenson CW, Kolenic GE, Trowbridge ER, et al. Obesity and stress urinary incontinence in women: compromised continence mechanism or excess bladder pressure during cough? Int Urogynecol J 2017;28:1377–85. Available from: https://doi.org/10.1007/s00192-017-3279-6.

[30] Wilson A, Longhi J, Goldman C, McNatt S. Intra-abdominal pressure and the morbidly obese patients: the effect of body mass index. J Trauma 2010;69:78–83. Available from: https://doi.org/10.1097/TA.0b013e3181e05a79.

[31] Lambert DM, Marceau S, Forse RA. Intra-abdominal pressure in the morbidly obese. Obes Surg 2005;15:1225–32. Available from: https://doi.org/10.1381/096089205774512546.

[32] Rogers WK, Garcia L. Intraabdominal hypertension, abdominal compartment syndrome, and the open abdomen. Chest 2018;153:238–50. Available from: https://doi.org/10.1016/j.chest.2017.07.023.

[33] Pares D, Vallverdu H, Monroy G, et al. Bowel habits and fecal incontinence in patients with obesity undergoing evaluation for weight loss: the importance of stool consistency. Dis Colon Rectum 2012;55:599–604. Available from: https://doi.org/10.1097/DCR.0b013e3182446ffc.

[34] Brochard C, Venara A, Bodere A, Ropert A, Bouguen G, Siproudhis L. Pathophysiology of fecal incontinence in obese patients: a prospective case-matched study of 201 patients. Neurogastroenterol Motil 2017;29. Available from: https://doi.org/10.1111/nmo.13051.

[35] Ellington DR, Polin MR, Szychowski JM, Deng L, Richter HE. The effect of obesity on fecal incontinence symptom distress, quality of life, and diagnostic testing measures in women. Int Urogynecol J 2013;24:1733–8. Available from: https://doi.org/10.1007/s00192-013-2103-1.

[36] Chughtai B, Sedrakyan A, Isaacs AJ, et al. National study of utilization of male incontinence procedures. Neurourol Urodyn 2016;35:74–80. Available from: https://doi.org/10.1002/nau.22683.

[37] Lau HH, Enkhtaivan S, Su TH, Huang WC. The outcome of a single-incision sling versus trans-obturator sling in overweight and obese women with stress urinary incontinence at 3-year follow-up. J Clin Med 2019;8. Available from: https://doi.org/10.3390/jcm8081099.

[38] Xia Z, Qian J, Chen Y, Liao B, Luo D. Does body mass index influence the outcome of midurethral sling procedures for stress urinary incontinence? Int Urogynecol J 2017;28:817–22. Available from: https://doi.org/10.1007/s00192-016-3181-7.

[39] Elshatanoufy S, Matthews A, Yousif M, et al. Effect of morbid obesity on midurethral sling efficacy for the management of stress urinary incontinence. Female Pelvic Med Reconstr Surg 2018;. Available from: https://doi.org/10.1097/spv.0000000000000594.

[40] Brennand EA, Tang S, Birch C, Murphy M, Ross S, Robert M. Five years after midurethral sling surgery for stress incontinence: obesity continues to have an impact on outcomes. Int Urogynecol J 2017;28:621–8. Available from: https://doi.org/10.1007/s00192-016-3161-y.

[41] Hong KD, DaSilva G, Dollerschell JT, Wexner SD. Suboptimal results after sphincteroplasty: another hazard of obesity. Tech Coloproctol 2014;18:1055–9. Available from: https://doi.org/10.1007/s10151-014-1195-z.

[42] Subak LL, Wing R, West DS, et al. Weight loss to treat urinary incontinence in overweight and obese women. N Engl J Med 2009;360:481–90. Available from: https://doi.org/10.1056/NEJMoa0806375.

[43] Lee Y, Yu J. The impact of bariatric surgery on urinary incontinence: a systematic review and meta-analysis. BJU Int 2019;. Available from: https://doi.org/10.1111/bju.14829.

[44] Purwar B, Cartwright R, Cavalcanti G, Digesu GA, Fernando R, Khullar V. The impact of bariatric surgery on urinary incontinence: a systematic review and meta-analysis. Int Urogynecol J 2019;30:1225 37. Available from: https://doi.org/10.1007/s00192-018-03865-x.

[45] Bump RC, Sugerman HJ, Fantl JA, McClish DK. Obesity and lower urinary tract function in women: effect of surgically induced weight loss. Am J Obstet Gynecol 1992;167:392–7. Available from: https://doi.org/10.1016/s0002-9378(11)91418-5 discussion 397–399.

[46] Montenegro M, Slongo H, Juliato CRT, Minassian VA, Tavakkoli A, Brito LGO. The impact of bariatric surgery on pelvic floor dysfunction: a systematic review. J Minim Invasive Gynecol 2019;26:816–25. Available from: https://doi.org/10.1016/j.jmig.2019.01.013.

[47] Cuicchi D, Lombardi R, Cariani S, Leuratti L, Lecce F, Cola B. Clinical and instrumental evaluation of pelvic floor disorders before and after bariatric surgery in obese women. Surg Obes Relat Dis 2013;9:69–75. Available from: https://doi.org/10.1016/j.soard.2011.08.013.

[48] Poylin V, Serrot FJ, Madoff RD, et al. Obesity and bariatric surgery: a systematic review of associations with defecatory dysfunction. Colorectal Dis 2011;13:e92–e103. Available from: https://doi.org/10.1111/j.1463-1318.2011.02584.x.

Chapter 22

Role of obesity in cancer in women

Joanna M. Cain
Department of Obstetrics and Gynecology, University of Massachusetts Medical School, Worcester, MA, United States

Overview

The development of cancer in an individual is stimulated by genetic components, environmental exposures, and lifestyle exposures, all impacting the cellular microenvironment and initiating carcinogenesis. The profusion of nutrients and products of adipocytes and other environmental disturbances affect the balance of that microenvironment and increase the risks for cancer. Adipose tissue functions as a homeostatic "endocrine organ" for energy via multiple hormones, including estrogen, leptin, and adiponectin as well as many chemokines. The links between obesity and cancer have been documented for hormonally influenced cancers such as endometrial and breast cancer, and links to a broader range of cancers are emerging [1]. For certain cancers the changes to the systemic and microenvironment from obesity and hypernutrition are not only a significant risk factor for carcinogenesis but also impact the success and outcome of treatment. Obesity also impacts survival as women with a BMI of greater than 40 kg/m^2 have a 1.6-fold higher risk for death from cancer [1].

The impact of the global epidemic of obesity, particularly in developing areas such as Asia (including China), the Middle East, Oceania, and other countries, is significant as the need for treatment for cancers in men and women as well as other associated obesity-linked diseases such as diabetes and hypertension escalates [2]. This epidemic has been facilitated by the counterintuitive effect of positive developments, including the increase in global food supply and the growth of economies and individual wealth facilitating better nutrition [3]. The food supply itself however, while adequate, may not have a range of healthy options to choose from affecting the obesity crisis. The cost to economies in loss of life and cost of health care with increasing obesity in the population is significant.

In the United States the proportion of adults 20 or over categorized as overweight or obese is over 70% [4]. The percentage varies by gender, region, age, racial, and ethnic status interacting with and compounding other differences in risk for cancer in these diverse groups. In the United States, non-Hispanic black individuals had the highest prevalence of obesity, followed by Hispanics, non-Hispanic white, and non-Hispanic Asian individuals leading to higher burdens of cancer and affecting outcomes [5].

The increase in obesity and concomitant increase in cancer burden globally argue for a preventive focus on nutrition and weight management [6]. Of particular concern is the growing epidemic of cancers in obese young people (24–29) that had previously been seen predominantly only in the elderly, including endometrial, colorectal, gallbladder, kidney, pancreatic, and multiple myeloma [7]. Assuring that the next generation is not harmed by supporting public health initiatives and research to control obesity is a key strategy for all nations.

Epidemiologic evidence for links between obesity and cancer

Cancers that have been linked to obesity with variable but consistent relationships include esophageal (with greater increases for higher BMI), gastric, liver, kidney (renal cell), pancreatic, colorectal, gallbladder, multiple myeloma and meningioma, and potentially some thyroid and ovarian cancers. The link between obesity and breast and endometrial cancers has a long trail of evidence as discussed in the next section. A study of colorectal cancer risk in Asian women showed not only a link but also increased risk based on Waist Circumference for colon cancer but not for rectal cancers [8].

There are gender differences for different cancer types. For example, obese women have a 4.8- to 5.3-fold increase in kidney and overall intestinal cancers, which is higher than in men, while obese men have a higher 2.6- to 4.5-fold increase in pancreas and liver cancer [9] and a higher risk for colorectal cancer [10]. The contribution of obesity as a unique risk factor for the development of cancer differs by site as well. As an example, up to 54% of gallbladder cancer in women is linked to obesity [11]. The International Agency for Research on Cancer working group on Body Fatness in 2016 noted sufficient evidence for links between obesity and postmenopausal breast, colon, endometrial, esophageal, gallbladder, kidney, liver, meningioma, multiple myeloma, ovary, pancreas, stomach, and thyroid [12]. Others have proposed links between advanced prostate cancer, mouth, pharynx, and larynx, although the relationship is less clear [13].

Cancers unique to or more common in women

The most striking example of the interaction between obesity and cancer is endometrial cancer where not only more than 50% is associated with obesity but also there is evidence for a striking dose/response curve from 1.5-fold increase for overweight women to 7.1-fold increase for Class 3 obesity [14–16].

The primary mechanism is likely linked to the increase in aromatase, conversion of androgens to estrogens, and then the hormonal influence on proliferation of endometrial cells. However, other mechanisms proposed for nongynecologic cancers may also play a role such as chronic inflammation and production of inflammatory cytokines, increases in insulin-like growth factors, and their associated cellular and genetic impacts. Not surprisingly, weight reduction decreases this risk as shown by the reduction of risk with bariatric surgery [17] of up to a 60% reduction.

Breast cancer also shows a strong association with obesity. An analysis of the Women's Health Initiative Clinical Study population [18] showed that there was an increased risk for invasive breast cancer for overweight and above women compared to normal-weight women with the highest risks in those with grade 2 and 3 obesity (HR 1.58) and again a dose–response relationship was found. Like endometrial cancer, this is primarily in the hormone receptor positive disease cell types. However, for this study duration, obese women who lost weight did not have a corresponding drop in risk, unlike the impact seen with endometrial cancer and bariatric surgery.

It is also not clear when, during the life cycle, obesity most impacts the risk for breast cancer. Obesity during the trial may have been lifelong and the results reflect critical windows at reproductive age or earlier childhood as well. In fact, premenopausal obese women may actually have a protective effect from the body fat and a decreased risk in that life stage [15,19]. Whether obesity in adolescence during breast development or premenopausal is a greater or lesser risk than postmenopausal is difficult to study but will be important to understand for prevention. In addition, high body fat proportions, even if the BMI is normal, may increase the risk for breast cancer [20]. The type and distribution of body fat is an area of intensive research and may hold keys to future prevention strategies.

Potential mechanisms for oncogenesis

There are many potential pathways that are engaged and could result in carcinogenesis with obesity. Insulin resistance and inflammation are markers of disturbance of the microenvironment and have been suggested as more accurate risk predictors than BMI alone [21]. Including waist to hip ratios as well as BMI may be helpful in assessment of risk [22]. Associated conditions, in particular, metabolic syndrome and diabetes, have a significant impact on outcomes in breast cancer [23,24]. The mechanisms that underpin this relationship are still obscure. However, the cross talk between adipocytes and macrophages is of particular interest, as molecular mechanisms such as inflammatory cytokines, TSC1–TSC2 complex-mTOR cross talk, increased aromatase activity, and other interactions with leptin and insulin resistance have been associated with adipocyte–macrophage breast cancer cell clusters [25].

Another component of the altered inflammatory environment may come from inhibition of natural killer (NK) cell activity [26]. Michelet et al. have shown obesity interfering with NK and CD8 + T cells through paralyzing their cellular metabolism by transport of lipids into the mitochondria. Blocking this accumulation appeared to restore cytotoxicity, suggesting that further research on these various pathways may provide additional avenues for cancer treatment specific to obesity-linked cancers.

The type and location of the adipose deposits also play a role in the hormones and cytokines secreted, with white fat and visceral fat implicated in poor survival and treatment outcomes with colon, esophageal, and renal cancers [27]. In addition, Tangen et al. [28] noted that a high percentage of visceral fat was a larger contributor to estradiol production compared to subcutaneous fat in postmenopausal endometrial cancer patients. This group also noted that a high percentage of visceral fat in endometrial cancer was independently associated with reduced disease specific survival [29]. The association of the increasing fat mass with increasing fasting insulin levels and hyperinsulinemia implicates insulin as well as estradiol as an important biologic link associating trunk adiposity with breast cancer risk [30]. Furthermore, the importance of body composition rather than simply BMI has been highlighted by Iyengar et al. [20] who looked at normal-weight women with higher whole body fat and trunk fat mass and found higher risk for breast cancer. High body fat levels were associated with elevated levels of insulin, C-reactive protein, interleukin 6, leptin, and triglycerides as well as lower levels of high-density lipoprotein cholesterol and sex hormone−binding globulin.

The functional causal mechanisms supported by obesity may differ between location with chronic secretion and inflammation being posited for elevation of gallbladder cancer risks, reflux esophagitis and inflammation for esophageal cancer risk, associated hypertension with renal cancer risk, and endogenous estrogen with breast and endometrial cancer risks. Ultimately the actions affecting oncogenesis are carried out at the cellular level with inflammatory mechanisms a common component.

One pathway, as seen with breast and endometrial cancers, comes through the downstream influence of increased aromatase activity and the stimulation from circulating estrogens as noted previously with visceral fat [31,32]. However, these varying pathways interact, and other downstream linkages between obesity and cancer are being explored to tease out the critical interactions, including those leading to insulin resistance (Table 22.1). One characteristic of the obesity−cancer relationship appears to be the significant ongoing cross talk between adipose tissue components (adipocytes, stem cells, macrophages, stromal cells, recruited inflammatory cells, and extracellular matrix), their products, and the cancer cells themselves leading to a microenvironment that not only promotes carcinogenesis but also promotes metastasis [27,33]. It may be that the more disturbed the environment is with insulin resistance and more exposure to growth promoting secretome of adipocytes, the more aggressive and potentially metastatic the resulting tumor [34].

Many genes have emerged as potential bridges between obesity and cancer, and much research is needed to fully explore the gene regulation of the various components involved. Of growing interest is the role of micro RNAs [35,36] in regulation of carcinogenic events for multiple cancers. Their association with other obesity-related phenomena such as adipocyte differentiation or metabolic syndrome makes their role important as a bridge to cancers. In a study of miR-10b and breast cancer (more prominent with ductal cancers), the levels correlated with tumor grade as well as showing different regulation in primary versus metastatic tumor deposits [35].

Clinical implications for prevention and treatment of cancer in obese patients

Weight reduction by whatever means seems a clear path for prevention lifelong, given the risk windows where obesity interacts with oncogenesis may differ for various cancers from adolescence through postmenopausal life stages. The resulting prevention of disruption of the microenvironment and prevention of a proinflammatory state locally and systemically will reduce certain cancers such as endometrial and is likely to reduce others as well in addition to reduction in high blood pressure, metabolic syndrome, and diabetes [37,38]. Adoption of

TABLE 22.1 Cell pathway examples related to oncogenesis in obesity.

Cell type/ microenvironment	Specific mediators	Results/outcomes
Adipose tissue	Adipokines: Leptin Adiponectin Aromatase ASC	Induced IL6 and other cytokines, angioneogenesis REDUCED in obesity/acts against leptin Increased estradiol, stimulates multiple signaling pathways, receptors for cell growth
Adipose tissue: inflammatory environment	Cytokines: MCP-1 IL6, IL8	Recruits proinflammatory macrophages Increases acute phase proteins, recruits lymphocytes, increases B lymphocytes, increases aromatase expression (IL6), recruits leukocytes (IL8)
Tumor-associated adipose tissue	Transfer of triglycerides to cancer cells	Increase free fatty acids/potentiated metastases [14]
Macrophages	TNF alpha Multiple cytokines and growth factors	Variable, may contribute to cancer cell proliferation, stimulate angioneogenesis, modulation of extracellular matrix
Insulin receptors/ multiple cells		Association with adiponectin

ASC, Adipose stem cells; *IL*, interleukin; *MCP-1*, monocyte chemotactic protein 1.

a low fat dietary program has shown promise for increase in overall survival with breast cancer, including potentially fewer deaths from cancers [39].

Medications that address the role of the insulin pathway, for example metformin, have been shown to have a positive role in lowering cancer incidence in diabetics [40,41]. In addition, progesterones have a role in endometrial cancer prevention and estrogen agonists/antagonists in breast cancer to reduce endocrine stimulation. Health providers need to be aware of the links of obesity to multiple cancers, not just endocrine-related cancers, and have a low threshold for the evaluation of new symptoms and signs. This also requires an awareness of the elevated risks of those with normal body mass index but high body fat and truncal fat depositions. Oncologists need to be mindful of changing recommendations for the treatment of obese patients, particularly accuracy, and need for modification of dosing of chemotherapeutic agents to treat the cancers [42].

Health-care providers have been vexed by the complex mix of individual psychological, environmental, and decision/motivational factors that hinder weight-reduction. There is a paucity of effective tools or interventions to consistently and effectively address this health issue at present other than bariatric surgery [43]. The bias in the global as well as medical culture that patients are the ones accountable and lack the willpower to address the obesity can result in failure of patients to seek help and failure of the health providers to address the issue. Survey findings from the NEJM Catalyst group [43] show providers hold patients accountable, while limiting their own responsibility, which demonstrably has failed to provide positive results. Finding a means to engage the population with effective and sufficient support will likely require engaging a broader health team with nutritionists and counselors as well as the educational and motivational support from all levels of the health-care team. In addition, individual health providers can advocate for policies within their practice, locally and nationally, to support more active lifestyles, availability of healthy food options, and healthy diets.

Globally, public health advocates and governmental entities have reasons to address the problem from multiple fronts. The evidence is emerging that encouragement of healthy physical activity lifelong and promotion of a balanced diet [44,45] may be the most important actions that governments and nongovernmental organizations can take. Assuring that food production focuses on healthy food distribution and an environment (including school environments) that promotes physical activity may do more than any other mechanisms for reduction in the burden of obesity-related cancers. Taxes, subsidies, media campaigns, social marketing, education of health professionals and educators, education of leaders and communities may all have a role dependent on the local circumstances. The overall reduction in costs to economies from the reduction of multiple diseases and disabilities linked to obesity makes it worth governmental interventions [46].

Summary and ongoing needs for research

There is a critical need to understand the cellular cross talk that results in the oncogenic environment stimulating cancers in obese individuals. This not only offers a mechanistic understanding but also opens doors to potential targets for prevention, treatment of the cancer, and reduction of recurrence and metastatic disease. In addition, increasing understanding of basic genetic and neuronal information about homeostasis, particularly weight regulation and energy balance, may provide additional means to prevent the obesity epidemic and oncogenic outcomes [47].

The role of weight loss on control of cancer is another area that is particularly difficult to research, given the multiplicity of treatments and factors impacting each type of cancer. However, the role that weight loss during or after treatment plays in control and prevention of recurrence is one that both cancer survivors and their healthcare providers are anxious to understand.

Finally, there must be public health advocacy and leadership globally to address policies and implementation of programs to promote healthy foods and lessen the overproduction and consumption of foods associated with obesity. Along with this, examination of our living and working environments to enhance their promotion of physical activity for all levels of abilities will make a lasting impact on the prevention of the component of cancers attributable to obesity. Continued research into evidence-based pathways to reduce individual obesity and control appetite will additionally assist in the prevention of these cancers. Increasing health information, support, and positive intervention options for health professional interaction with patients struggling with obesity is needed. Expanding global knowledge of the benefits of addressing obesity and the risks for individuals and populations in failing to do so, including cancer risks, is an important advocacy role for every health provider.

References

[1] Calle EE, Kaaks R. Overweight, obesity and cancer: epidemiological evidence and proposed mechanisms. Nat Rev Cancer 2004;4:579—91.

[2] Arnold M, Leitzmann M, Freisling H, Bray F, Romieu I, Renehan A, et al. Obesity and cancer: an update of the global impact. Cancer Epidemiol 2016;41:8—15.

[3] Swinburn BA, Sacks G, Hall KD, et al. The global obesity pandemic: shaped by global drivers and local environments. Lancet 2011;378:804—14.

[4] National Center for Health Statistics. Health, United States, 2015: with special feature on racial and ethnic health disparities. Hyattsville, MD: National Center for Health Statistics; 2016.

[5] Ogden CL, Carroll MD, Kit BK, Flegal KM. Prevalence of childhood and adult obesity in the United States 2011-2012. JAMA 2014;311(8):806−14.

[6] Sung H, Siegel R, Torre LA, Pearson-Stuttard J, et al. Global patterns in excess body weight and the associated cancer burden. CA Cancer J Clin 2019;69:88−112.

[7] Sung H, Siegel RL, Rosenberg PS, Jemal A. Emerging cancer trends among young adults in the USA: analysis of a population-based cancer registry. Lancet Public Health 2019; http://dx.doi.org/10.1016/S2468-2667(18)30267-6.

[8] Wong TS, Chay WY, Tan MH, Chow KY, Lim WY. Reproductive factors, obesity and risk of colorectal cancer in a cohort of Asian women. Cancer Epidemiol 2018;58:33−43.

[9] Font-Burgada J, Sun B, Karin M. Obesity and cancer: the oil that feeds the flame. Cell Metab 2016;23:48−60.

[10] Ma Y, Yang Y, Wang F, et al. Obesity and risk of colorectal cancer: a systematic review of prospective studies. PLoS One 2013;8: e53916.

[11] Arnold M, Pandeya N, Byrnes G, et al. Global burden on cancer attributable to high body mass index in 2012: a population-based study. Lancet Oncol 2015;16(1):36−46.

[12] Lauby-Secretan B, Scoccianti C, Loomis D, et al. Body fatness and cancer—viewpoint of the IARC Working Group. N Engl J Med 2016;375:794−8.

[13] World Cancer Research Fund/American Institute for Cancer Research. Continuous update project report 2018. Body fatness and weight gain and the risk of cancer. London: World Cancer Research Fund International; 2018. Wcrf.org/sites/default/files/Body-fatness-and-weight-gain.pdf [accessed 03.07.19].

[14] Onstad MA, Schmandt RE, Lu KH. Addressing the role of obesity in endometrial cancer risk, prevention and treatment. J Clin Oncol 2016;34:4225−30.

[15] Renehan AG, Tyson M, Egger M, et al. Body-mass index and incidence of cancer: a systematic review and meta-analysis of prospective observational studies. Lancet 2008;371:569−78.

[16] Setiawan VW, Yang HP, Pike MC, et al. Type I and II endometrial cancers: have they different risk factors? J Clin Oncol 2013;31:2607−18.

[17] Upala S, Sanguankeo A. Bariatric surgery reduces risk of endometrial cancer. Surg Obes Relat Dis 2015;11:1410.

[18] Neuhouser ML, Aragaki AK, Prentice RL, Manson JE, et al. Overweight, obesity, and postmenopausal invasive breast cancer risk. JAMA Oncol 2015;1:611−21.

[19] Laudisio D, Muscogiuri G, Barrea L, Savastano S, Colao A. Obesity and breast cancer in premenopausal women: current evidence and future perspectives. Eur J Obstet Gynecol Reprod Biol 2018;230:217−21.

[20] Iyengar NM, Arthur R, Manson JE, et al. Association of body fat and risk of breast cancer in postmenopausal women with normal body mass index: a secondary analysis of a randomized clinical trial and observational study. JAMA Oncol 2019;5: 155−163.

[21] Sahakyan KR, Somers VK, Rodriguez-Escudero JP, et al. Normal weight central obesity: implications for total and cardiovascular mortality. Ann Intern Med 2015;163:827−35.

[22] Ando S, Gelsomino L, Panza S, Giordano C, Bonofiglio D, Brone I, et al. Obesity, leptin and breast cancer: epidemiological evidence and proposed mechanisms. Cancers 2019;11:62. Available from: https://doi.org/10.3390/cancers11010062.

[23] Berrino F, Villarini A, Traina A, et al. Metabolic syndrome and breast cancer prognosis. Breast Cancer Res Treat 2014;147:159−65.

[24] Erickson K, Patterson RE, Flatt SW, et al. Clinically defined type 2 diabetes mellitus and prognosis in early-stage breast cancer. J Clin Oncol 2011;29:54−60.

[25] Engin AB, Engin A, Gonul II. The effect of adipocyte-macrophage cross talk in obesity related breast cancer. J Mol Endocrinol 2019; https://doi.org/10.1530/JME-18-0252. Pii:JME-18-0252.R1, [Epub ahead of print].

[26] Michelet X, Dyck L, Hogan A, Lofthus RM, et al. Metabolic reprogramming of NK cells in obesity limits antitumor responses. Nat Immunol 2018;19:1330−40.

[27] Himbert C, Delphan M, Shcerer D, Bowers LW, Hursting S, Ulrich C. Signals from the adipose microenvironment and the obesity-cancer link—a systematic review. Cancer Prev Res 2017;10:494−506.

[28] Tangen IL, Fasmer KE, Konings GF, Jochems A, et al. Blood steroids are associated with prognosis and fat distribution in endometrial cancer. Gynecol Oncol 2019;152:46−52.

[29] Mauland KK, Eng O, Ytre-Hauge S, Tangen IL, Berg A, Salvesen HB, et al. High visceral fat percentage is associated with poor outcome in endometrial cancer. Oncotarget 2017;8:105184−95.

[30] Hvidtfeldt UA, Gunter MJ, Lange T, et al. Quantifying mediating effects of endogenous estrogen and insulin the relation between obesity, alcohol consumption and breast cancer. Cancer Epidemiol Biomarkers Prev 2012;21:1203−12.

[31] Liu L, Wang L, Zheng J, Tang G. Leptin promotes human endometrial carcinoma cell proliferation by enhancing aromatase (P450arom) expression and estradiol formation. Eur J Obstet Gynecol Reprod Biol 2013;170:198−201.

[32] Rajapaksa G, Thomas C, Gustafsson JA. Estrogen signaling and unfolded protein response in breast cancer. J Steroid Biochem Mol Biol 2016;163:45−50.

[33] Amor S, Iglesias-de la Cruz MC, Ferrero E, Garcia-Villar O, Barrios V, Fernandez N, et al. Peritumoral adipose tissue as a source of inflammatory and angiogenic factors in colorectal cancer. Int J Colorectal Dis 2016;31:365−75.

[34] Rosendahl AH, Bergqvist M, Lettiero B, Kimbung S, Borgquist S. Adipocytes and obesity-related conditions jointly promote breast cancer cell growth and motility: associations with CAP1 for prognosis. Front Endocrinol 2018;9:689. Available from: https://dx.doi.org/10.3389/fendo.2018.00689.

[35] Meerson A, Eliraz Y, Yehuda H, Knight B, et al. Obesity impacts the regulation of miR-10b and its targets in primary breast tumors. BMC Cancer 2019;19:86.

[36] Ali AS, Ali S, Ahmad A, Bao B, Philip PA, Arkar FH. Expression of microRNAs: potential molecular link between obesity, diabetes and cancer. Obes Rev 2011;12:1050−62.

[37] Iyengar NM, Gucalp A, Dannenberg AJ, Hudis CA. Obesity and cancer mechanisms. Tumor microenvironment and inflammation. J Clin Oncol 2016;34:4270−6.

[38] Clement K, Viguerie N, Poitou C, et al. Weight loss regulates inflammation-related genes in white adipose tissue of obese subjects. FASEB J 2004;18:2277−86.

[39] Chlebowski RT, Aragaki AK, Anderson GL, Simon MS, et al. Association of low-fat dietary pattern with breast cancer overall survival: a secondary analysis of the Women's Health Initiative Randomized clinical trial. JAMA Oncol 2018;4(10): e181212.

[40] Libby G, Donnelly LA, Donnan PT, et al. New users of metformin are at low risk of incident cancer: a cohort study among people with type 2 diabetes. Diabetes Care 2009;32:1620–5.

[41] Morales DR, Morris AD. Metformin in cancer treatment and prevention. Ann Rev Med 2015;66:17–29.

[42] Griggs JJ, Mangu PB, Anderson H, Balaban EP, et al. Appropriate chemotherapy dosing for obese adult patients with cancer. American Society of Clinical Oncology clinical practice guideline. J Clin Oncol 2012;30:1553–61.

[43] Volpp KG, Mohta NS. Patient engagement survey: the failure of obesity efforts and the collective nature of solutions. N Engl J Med 2018; Catalyst.NEJM.org. Insights report.

[44] World Health Organization. Governance: development of a draft global action plan to promote physical activity 2018-2030. Geneva: World Health Organization; 2018. who.int/ncds/governance/physical_activity_plan/en/ [accessed 03.07.20].

[45] Hawkes C, Smith TG, Jewell J, et al. Smart food policies for obesity prevention. Lancet 2015;385:2410–21.

[46] Nyberg ST, Batty GD, Penti J, et al. Obesity and loss of disease-free years owing to major non-communicable diseases: a multicohort study. Lancet Public Health 2018;3:3490–7.

[47] Lowell BB. New neuroscience of homeostasis and drives for food, water, and salt. N Engl J Med 2019;380:459–71.

Chapter 23

Obesity and breast cancer

Chiara Benedetto, Emilie Marion Canuto and Fulvio Borella
Department of Surgical Sciences, Sant'Anna Hospital, University of Torino, Torino, Italy

Epidemiology

Breast cancer (BC), the most prevalent female cancer, is on increase and it has been calculated that *33% of BCs in postmenopause are due to obesity* [1–3]. Currently, BC is responsible for 15% of all cancer deaths in women, with 626,679 deaths per year worldwide [3].

The association between obesity and BC in relation to *menopausal status* has been extensively studied.

A large UK population–based cohort study reported a linear association between obesity and the overall risk of BC in postmenopause, with a hazard ratio (HR) of 1.05 (99% confidence interval, CI 1.03–1.07), for each body mass index (BMI) 5 kg/m^2 increase.

Conversely, there was an inverse correlation between obesity and BC in premenopause, with an estimated HR of 0.89 (99% CI 0.86–0.92) for each BMI 5 kg/m^2 increase [4]. However, a metaanalysis, focused on premenopausal women, reported that differences were observed when data were stratified by *ethnicity* and other anthropometric parameters, such as *waist-to-hip ratio* (WHR) and *height*, were taken into account. Indeed, for each 5 kg/m^2 increase in BMI, there was a 7% and 5% reduction in BC risk in Caucasian and African women, respectively, while there was a 5% increase in Asian women. As to WHR, there was a statistically significant relative risk (RR) increase of 8% for each 0.1 U increment and the strongest association was observed in Asian women (19%; RR = 1.19, 95% CI: 1.15–1.24), while small but significant effects of 5% and 6% were found in Caucasian and African women, respectively. As to height, an overall, significant 3% higher risk was demonstrated for each 10 cm increase. When adjusted for ethnicity, the increased risk was 12% for African women and 2% and 3% for Asian and Caucasian women, respectively [5].

As to the relationship between obesity and the *status of BC hormone receptors*, a large prospective cohort study showed that obesity increases the risk of developing hormone receptor positive (estrogen receptor, ER + ; progesterone receptor, PR +) BC in women aged ≥ 65 years (HR 1.25, CI 1.16–1.34 per 5 kg/m^2 increase in BMI) but not in women aged ≤ 49 years (HR 0.79, CI 0.68–0.91 per 5 kg/m^2 increase in BMI) [6]. These findings were confirmed by a large metaanalysis of 89 epidemiologic studies [7]. The relationship between obesity and hormone receptor negative BC risk is more complex. In premenopausal women, obesity is associated with an increased risk of hormone receptor negative BC (RR 1.06, 95% CI 0.71–1.60) as well as triple negative BC [ER, PR, and human epidermal growth factor (EGF) receptor 2—HER2 negative] (OR: 1.43; 95% CI: 1.23–1.65) [8], while it decreases the risk of hormone receptor positive BC (RR: 0.78; 95% CI: 0.67–0.92) [7]. In postmenopausal women, there is an association between obesity and an increased risk of hormone receptor negative BC in those who never used *hormone replacement therapy* (*HRT*) (multivariate HR 1.59, CI 1.08–2.34) but not in past–current users [6,9].

Another study reported that the overall risk of obesity-dependent BC is lower for women on HRT (RR: 1.18; 95% CI: 0.98–1.42 vs 1.42; 95% CI: 1.30–1.55) than in never users, suggesting that hormonal therapy is a confounding factor in the obesity–cancer relationship [7]. According to a metaanalysis on the effect of HRT on the BC risk, obesity attenuates the absolute and relative excess BC risk associated with HRT. More specifically, there was a poor relationship between the incidence of ER − disease and BMI among never users of HRT, while the ER + disease incidence increased as did the BMI. Thus ER + disease accounted for almost all of the associations reported on BC risk with BMI in never users and of the HRT-associated excess risk in users [9].

To date, no association between obesity and risk of specific *BC subtypes* has been demonstrated [5].

The *impact of weight loss* on breast cancer (BC) risk has been widely investigated. Intentional weight loss is associated with a lower BC risk in observational studies. In particular, in bariatric surgery observational trials, a weight loss of approximately 30% was associated with a reduction in BC risk of up to 80% [10].

Pathogenetic mechanisms

Visceral fat is involved in BC carcinogenesis and is considered as a tumor-promoting microenvironment [3]. The different relationship between obesity and BC risk in pre- versus postmenopausal women suggests that the underlying biology may differ according to the menopausal status. Indeed, the development of BC in obese women may be influenced by various factors, such as endogenous sex hormones, hyperinsulinemia, insulin-like growth factor 1 (IGF-1), hyperglycemia, adipokines, chronic inflammation, and the microbiome (Fig. 23.1) [2].

Sex hormones

Estrogen levels are higher in obese women than in nonobese women. Indeed, the adipose tissue is an important source of estrogens produced through the peripheral conversion of circulating androgens into *estradiol* by the aromatase enzyme. Furthermore, obese women have a reduced hepatic synthesis of *sex hormone−binding globulin (SHBG)*, which leads to higher levels of the bioavailable fraction of both estradiol and testosterone [12].

In vitro, estrogens have been shown to have mitogenic and mutagenic effects and to promote proliferation, genetic instability, and DNA damage in both normal and neoplastic mammary epithelial cells [2,11].

However, the risk of developing BC in obese women seems to be correlated not only with an increase in estrogen but also in androgen levels. In fact, the *testosterone* levels in obese women are higher than those in nonobese women and elevated serum levels of androgens are associated with an increased BC risk, in both premenopausal and postmenopausal women, implying that androgens in obese women play a role in BC pathogenesis [11−14].

Hyperinsulinemia

Visceral obesity is closely related to metabolic syndrome, insulin resistance, and hyperinsulinemia [15]. About 80% of diabetic women are obese: a high BMI is the single strongest independent risk factor for type 2 diabetes mellitus (T2DM) and it is associated with many metabolic abnormalities that lead to insulin resistance. Moreover, abdominal obesity, assessed by waist circumference or waist−hip ratio, predicts T2DM risk, whatever the BMI is [15,16].

Hyperinsulinemia may promote carcinogenesis via two main mechanisms: directly, by promoting cell growth, and indirectly, through the *IGF-1* axis. Insulin and IGF receptors are overexpressed in cancer cells, which may also express hybrid receptors capable of binding to both insulin and IGF-1 [17]. Hyperinsulinemia leads to increased concentrations of circulating bioactive IGF-1, due to the suppression of the insulin-like growth factor−binding proteins 1 and 2 and to the hepatic activation of the growth hormone receptor causing an increased secretion of GH that, in turn, stimulates IGF-1 [17]. The binding of insulin and IGF-1 with their receptor triggers mitogenesis, activation of antiapoptotic mechanisms (e.g., phosphatidylinositol-3-kinase PI3K and Ras pathways), angiogenesis and lymphangiogenesis, promoting carcinogenesis and neoplastic spread [11,15]. Indeed, a pooled individual data analysis of 17 prospective studies reports that high concentrations

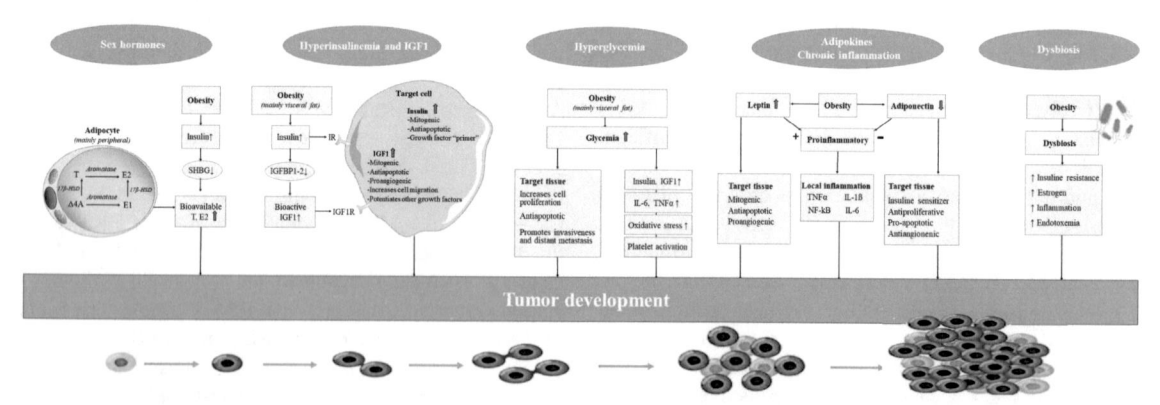

FIGURE 23.1 Schematic representation of the possible pathogenetic mechanisms linking obesity with breast cancer. *Δ4A*, Δ4-Androsetenedione; *17β-HSD*, 17β-hydroxysteroid dehydrogenases; *E1*, estrone; *E2*, estradiol; *IGF-1*, insulin-like growth factor 1; *IGF-1R*, IGF-1 receptor; *IGFBP*, insulin-like growth factor−binding protein; *IL*, interleukin; *IR*, insulin receptor; *NF-κB*, nuclear factor κB; *SHBG*, sex hormone−binding globulin; *T*, testosterone; *TNFα*, tumor necrosis factor α. *Modified from Renehan AG, Zwahlen M, Egger M. Adiposity and cancer risk: new mechanistic insights from epidemiology. Nat Rev Cancer 2015;15(8):484 [11].*

of circulating IGF-1 are associated with the risk of developing BC, specifically ER + tumors [18].

Moreover, evidence suggests a direct relationship between circulating IGF-1 levels and the risk of developing chemotherapy resistance [2]. An excess of insulin not only acts synergistically with the increased IGF-1 but also raises the bioavailability of estradiol and testosterone through the activation of the insulin−sex hormone axis, leading to reduced SHBG-circulating levels and an increased aromatase enzyme activity. In addition, insulin enhances the pulsatility of the hypothalamic−pituitary−ovarian axis, provoking an [17,19,20] increased ovarian androgen synthesis.

Hyperglycemia is yet another factor, linked to an excess of visceral fat, which influences tumor development. Unlike normal cells, tumor cells produce energy mainly through the less efficient aerobic glycolysis (a mechanism known as the "Warburg effect") and require a high glucose intake. In this context, hyperglycemia promotes proliferative, antiapoptotic, and metastatic processes [17,21]. Hyperglycemia induces the overexpression of glucose transporters (GLUT1 and GLUT2), protein kinase C α, peroxisome proliferator-activated receptor α and β, and EGF in cancer cells, increasing cell proliferation [21]. Elevated glucose levels promote metastasis and increased invasiveness due to the epithelial to mesenchymal transition process [17]. Furthermore, hyperglycemia acts indirectly on BC cells by increasing (1) the circulating insulin and IGF levels; (2) inflammatory cytokines, such as interleukin-6 (IL-6) and tumor necrosis factor α (TNFα); (3) oxidative stress; and (4) platelet activation [22]. Hyperglycemia also alters the epigenetic regulation of neoplastic cells, creating a "hyperglycemic memory," a condition associated with a chronic activation of oncogenic pathways, even if blood glucose levels return within the normal range [23].

Adipokines

Adipokines are a family of polypeptides synthesized by adipocytes that include over 100 different molecules: leptin and adiponectin are the most commonly studied in the context of carcinogenesis. Both of them have opposite biological effects.

Leptin is a potent proinflammatory agent and its systemic concentrations are proportional to the amount of body fat. It exerts several activities potentially relevant for the process of carcinogenesis: it is mitogenic, antiapoptotic, immunosuppressive, and proangiogenic by itself and in synergy with vascular endothelial growth factor expression. The binding of leptin to its long-form receptor (LRb) activates several intracellular downstream signaling pathways, such as the PI3K/Akt, mitogen-activated protein kinase (MAPK), and the Janus kinase 2/signal transducer and the activator of transcription 3 (JAK2/STAT3) pathways that are involved in the control of cell survival, proliferation, differentiation, migration, and invasion [11,24].

Although various epidemiological studies have reported conflicting results as to the association between leptin concentrations and BC, a metaanalysis demonstrated a positive association between leptin levels and BC risk [25].

Moreover, leptin and its receptor have been found to be associated with higher grade BCs, distant metastasis, and poor prognosis [26−29] and some studies have suggested that leptin has a potential role as a biomarker for BC risk [30,31].

Adiponectin has a potent antiinflammatory activity and, although it is the most abundant adipokines predominantly secreted by the visceral adipose tissue, its circulating levels are inversely correlated with body fat. This may be explained by the finding that, unlike leptin, adiponectin is produced only by mature adipocytes [11], which in obese humans have been shown to represent only about 20% of the total number of cells observed in the dysfunctional obese adipose tissue [32].

Adiponectin influences various cellular processes involved in carcinogenesis as it reduces fatty acid and protein synthesis, cellular growth, proliferation, and DNA mutagenesis and increases apoptosis. These effects are obtained both indirectly, by sensitizing cells to insulin and inhibiting inflammation, and, directly, by sequestering growth factors at the prereceptor level or by activating 5′AMP-activated protein kinase and inhibiting the extracellular-signal-regulated kinase 1 (ERK1) and ERK2, PI3K−AKT, WNT−β-catenin, nuclear factor-κB, and JAK2/STAT3 pathways [11,25].

A systematic review and metaanalysis has shown that low adiponectin concentrations are associated with an increased BC risk [33]. An inverse association between serum adiponectin concentrations and BC recurrence was specifically observed in ER/PR patients [34].

Chronic inflammation

Obesity is a state of chronic, low-grade inflammation, which may play an important role in tumor development and progression. In obese women, visceral fat, higher leptin, and estrogen levels are associated with an increase in proinflammatory molecules, such as IL-1β, IL-6, TNFα, and prostaglandin E2, all of which promote carcinogenesis [34].

High levels of *IL-1β* are associated with cancer cell growth, invasion, and angiogenesis. Both positive and negative hormone receptor BC express IL-1β receptors [35].

IL-6 is expressed by adipocytes, BC, and stromal cells and can play both pro- and antiinflammatory roles in BC pathogenesis. On the one hand, IL-6 is associated with epithelial–mesenchymal transformation and cell migration. IL-6, as well as TNFα, may also increase estrogen levels both systemically and in tumor and adipose tissue through the activation of the main enzyme complexes involved in estrogen synthesis. On the other, IL-6 can inhibit cell proliferation in early-stage positive hormone receptor BC, improving prognosis.

The expression of *IL-1β, IL-6,* and *TNFα* promotes T-regulatory lymphocytes chemotaxis and their accumulation, which inhibit the cytotoxic activity of $CD8^+$ T cells. This immunitary mechanism is associated with poor BC prognosis [36,37].

Microbiome

The human microbiome has been proven to play a fundamental role in some diseases, including cancer. *Microbial alterations (dysbiosis)* have been observed more in BC patients than in healthy women. It is also known that an increase in calorie intake can lead to dysbiosis, resulting in alterations in the carbohydrate and lipid metabolism, insulin resistance, perturbations in endocrine systems, and in a state of chronic inflammation [38,39]. The gut microbiota may induce the transformation of chemical compounds derived from the host diet into obesogenic and diabetogenic molecules that play a role in carcinogenesis. In murine models a high-fat diet can induce dysbiosis (e.g., lower abundance of *Clostridium leptum*, *Enterococcus* and *Nitrospira* species) and an increase in intestinal wall permeability, resulting in endotoxemia, systemic inflammation, and increased visceral fat [40]. Alterations in the gut microbiota may also influence the production of estrogen metabolites and, therefore, increase the estrogen circulating levels. Indeed, dysbiosis, obesity, and high levels of estrogens may act synergistically to increase the risk of BC [40,41].

Although intestinal dysbiosis has been implicated in obesity and may influence the risk of BC, future research is needed to better understand the relationship between obesity, the microbiome, and BC [38].

Diagnosis

Even if the sensitivity of mammography is similar in obese and nonobese women, *obesity* may *negatively impact BC diagnosis* [42]. Indeed, obese women are usually less aware of the importance of a healthy lifestyle, including the benefit of mammography screening. A metaanalysis of 16 studies showed that obese women aged ≥ 40 years had taken *less* advantage of *access to mammography* than nonobese women [43] and data from a retrospective cohort study suggest that the main cause of poor compliance was pain during the procedure [44].

Some *psychosocial factors* may contribute to the creation of barriers between patients and physicians, which can adversely affect the diagnosis and treatment of obese women with BC. Indeed, women with a low socioeconomic status generally tend to be more obese than women with higher status and this may hamper their accessing medical care.

In addition, on the one hand, obesity often leads to fear, shame, fatalism, alienation, low self-esteem, and embarrassment that can contribute to lower adherence to screening and/or medical care; on the other, some healthcare providers may mirror the societal stigma toward obesity and this may negatively impact their perceptions and decision-making [42].

Therapy

Obesity poses specific challenges in surgical, radio-, chemo-, and endocrine therapy (Fig. 23.2).

Surgery

Obese patients may make *anesthesia* riskier as it may be more difficult to intubate them and maintain ventilation support [42].

Recent data from the American College of Surgeons National Surgery Quality Improvement Program (ACS-NSQIP) database on 7202 women, who had *unilateral mastectomy without reconstruction*, reported an association between obesity and an increase in both minor and major surgical complications, in particular bleeding complications and surgical site infections [45].

A systematic review and metaanalysis of 29 studies on women who underwent *breast reconstruction following mastectomy* showed that obese women are at greater risk, in both implant-based and autologous reconstruction, of (1) surgical complications (RR 2.36, 95% CI 2.22–2.52; $P < .00001$), including wound dehiscence, hematoma, seroma, and flap failure or necrosis; (2) medical complications (RR 2.89, 95% CI 2.50–3.35; $P < .00001$) such as deep venous thrombosis, pulmonary embolism, urinary tract infections, myocardial infarction, pneumonia; and (3) reoperation [46].

Another metaanalysis of 14 studies, assessing the effect of obesity on the outcomes of *free autologous breast reconstruction, either immediate or delayed*, reported an almost threefold increase in the prevalence of complications, including recipient site and donor site complications overall, donor site wound infection and seroma, abdominal bulge/hernia, skin flap necrosis, recipient site delayed wound healing and partial flap failure, in obese compared with nonobese patients. A BMI ≥ 40 was identified as the threshold value at which the prevalence of complications became prohibitively high [47].

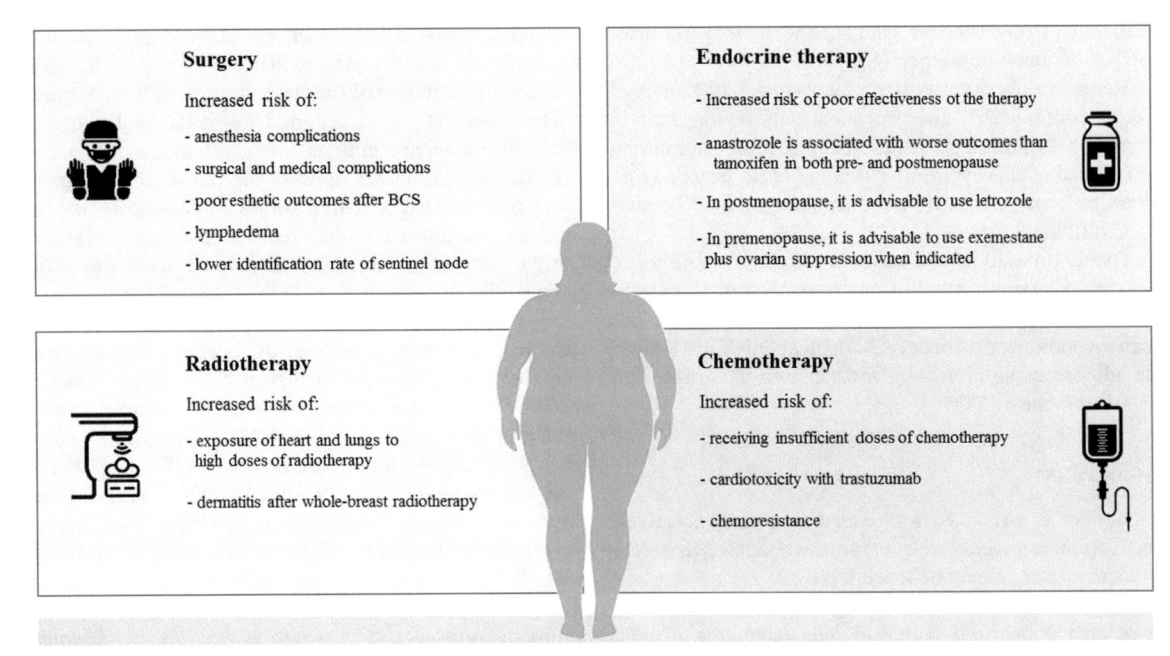

FIGURE 23.2 Summary of the most relevant challenges in the treatment of BC in obese women. *BC*, Breast cancer; *BCS*: breast-conserving surgery.

Data on patients that have *breast-conserving surgery followed by radiation therapy* are controversial, that is, some series reported comparable local control rates in obese and normal-weight women, while others observed higher local recurrence rates in obese women [48–50].

In general, the cosmetic outcome with breast conservation (lumpectomy and radiation) is poorer in obese than in nonobese women. In particular, patients with a BMI of >30 kg/m² have more postoperative breast asymmetry and deformity [42].

A recent retrospective review of 1566 patients with operable BC at a single institution showed that BC-conserving surgery was more common in obese patients."However, obese patients had an overall higher incidence of surgical site infections (12% vs 6%, $P < .001$), return to emergency departments after discharge (5.2% vs 2.5%, $P = .004$), and hospital readmission within 30 days of surgery (4% vs 2%, $P = .017$)" [51].

Sentinel node mapping has significantly reduced the number of women requiring axillary lymph node dissection thus decreasing the incidence of lymphedema, which is generally more common in obese than in nonobese patients [46]. However, sentinel node mapping is more difficult in obese women, as node identification rates are lower [42].

Radiotherapy

Obese patients with large breasts may receive *increased radiotherapy doses* also to critical organs, such as the heart and/or lungs, especially when treated in the supine position. Prone whole-breast radiation and hypofractionated radiotherapy may minimize this toxicity [42].

Furthermore, high BMI (>25 kg/m²) and large breast size have been associated with an increased risk of dermatitis after whole-breast radiotherapy [52].

Chemotherapy

Chemotherapy in obese women may pose some clinical challenges, given the frequent presence of comorbidities and the delicate balance between efficacy and toxicity.

It has been demonstrated that reductions from standard dose and dose intensity of chemotherapy may have a negative impact on disease-free and overall survival. Despite the importance of full weight–based cytotoxic chemotherapy dosing (both intravenous and oral), obese patients are more likely to receive *insufficient chemotherapy doses* as compared to normal-weight women. Indeed, some oncologists still base doses on ideal body weight, adjusted ideal body weight, or an empirical value of body surface area (BSA) rather than the use of actual body weight to calculate BSA. Concerns about toxicity or overdosing in obese patients, based on the use of actual body weight, are not evidence-based as there is no proof that either short- or long-term toxicity is increased in obese patients receiving full weight–based doses. On these premises the American Society of Clinical Oncology recommends that full weight–based chemotherapy doses be used in the treatment of obese cancer patients and toxicity be managed as in nonobese women [53].

Recently, obesity has been associated with a higher risk of *cardiotoxicity* after treatment with *trastuzumab* in women with HER-2-positive BC, thus close monitoring and effective management of cardiac risk factors (e.g., blood pressure, cholesterol, and smoking) should be considered in these patients [54].

Overall, obesity is considered a factor of *resistance to anticancer therapy*. Obesity has been proven to modify the pharmacokinetics of chemotherapy drugs and it has been hypothesized to induce biological modifications of the adipose tissue promoting resistance to the drugs used for BC treatment [55].

Endocrine

As obesity is associated with elevated aromatase activity and serum estrogen levels in postmenopause, therefore endocrine therapy may be less effective.

The current literature data report that *anastrozole* is associated with *worse outcomes* than tamoxifen in both post- and premenopausal obese patients.

This does not apply to letrozole versus tamoxifen in obese patients in postmenopause. Therefore, in *obese postmenopausal women*, it is advisable to use *letrozole*, rather than anastrozole as it is a more potent inhibitor of aromatase.

In *obese premenopausal women* the current standard of care from data of the TEXT (Tamoxifen and Exemestane Trial) and SOFT (Suppression of Ovarian Function Trial) is to use *exemestane* plus ovarian suppression when indicated [56].

Prognosis

Both premenopausal and postmenopausal obese women with BC have a *worse disease-free and overall survival*, despite appropriate local and systemic therapies, than do nonobese women.

A large metaanalysis of 82 follow-up studies of women with BC reported a RR for total mortality of 1.41 (95% CI, 1.29–1.53) and a RR for BC-specific mortality of 1.35 (95% CI, 1.24–1.47) for obese versus normal-weight patients. For each 5 kg/m² increment of BMI before and ≥ 12 months after diagnosis, increased risks of 17% and 8% for total mortality and 18% and 29% for BC mortality were observed [57].

The impact of *postdiagnosis weight loss*, resulting from changes in caloric intake, physical activity, or other interventions, on BC outcomes is under investigation. Data on the long-term effects of a low-fat diet on BC risk and postdiagnosis outcomes in the WHI (Women's Health Initiative) randomized prevention trial of dietary fat reduction versus no dietary change showed a significantly reduced risk of developing BC (HR 0.68) and deaths after BC diagnosis (HR 0.65) in the intervention arm. Women in the intervention are reported lower fat intake; higher fruit, vegetable, and grain intake; and a modest amount of weight loss. Although these results are encouraging, as the intervention was started before BC diagnosis, it does not necessarily mean that dietary modification and/or weight loss after BC diagnosis would have the same effects [58].

Preclinical and clinical evidence suggest that *metformin* may have antitumor activity in BC as it improves many of the potential physiologic mediators of the effect of obesity on BC (e.g., glucose, insulin, leptin, and C-reactive protein) whatever the baseline insulin or BMI is. Moreover, metformin is associated with modest weight loss. The largest clinical trial on metformin in BC, the National Cancer Institute of Cancer Clinical Trials Group (NCIC CTG) MA.32, investigating the effects of metformin versus placebo on early BC outcomes, is still ongoing [59].

To date, the safety and the impact of BC outcomes of *bariatric surgery* and *approved weight loss medications* are unknown.

References

[1] Ferlay J, Ervik M, Lam F, Colombet M, Mery L, Piñeros M, et al. Cancer base no. 11. Cancer Today (powered by GLOBOCAN 2018) 2018;.

[2] Benedetto C, Salvagno F, Canuto EM, Gennarelli G. Obesity and female malignancies. Best Pract Res Clin Obstet Gynaecol 2015;29 (4):528–40.

[3] Sung H, Siegel RL, Torre LA, Pearson-Stuttard J, Islami F, Fedewa SA, et al. Global patterns in excess body weight and the associated cancer burden. CA Cancer J Clin 2019;69(2):88–112.

[4] Bhaskaran K, Douglas I, Forbes H, dos-Santos-Silva I, Leon DA, Smeeth L, et al. Body-mass index and risk of 22 specific cancers: a population-based cohort study of 5·24 million UK adults. Lancet 2014;384(9945):755–65.

[5] Amadou A, Ferrari P, Muwonge R, Moskal A, Biessy C, Romieu I, et al. Overweight, obesity and risk of premenopausal breast cancer according to ethnicity: a systematic review and dose-response meta-analysis. Obes Rev 2013;14(8):665–78.

[6] Ritte R, Lukanova A, Berrino F, Dossus L, Tjønneland A, Olsen A, et al. Adiposity, hormone replacement therapy use and breast cancer risk by age and hormone receptor status: a large prospective cohort study. Breast Cancer Res 2012;14(3):R76.

[7] Munsell MF, Sprague BL, Berry DA, Chisholm G, Trentham-Dietz A. Body mass index and breast cancer risk according to postmenopausal estrogen-progestin use and hormone receptor status. Epidemiol Rev 2014;36(1):114–36.

[8] Pierobon M, Frankenfeld CL. Obesity as a risk factor for triple-negative breast cancers: a systematic review and meta-analysis. Breast Cancer Res Treat 2013;37(1):307–14.

[9] Collaborative Group on Hormonal Factors in Breast Cancer. Type and timing of menopausal hormone therapy and breast cancer risk:

individual participant meta-analysis of the worldwide epidemiological evidence. Lancet 2019;394(10204):1159–68.
[10] Christou NV, Lieberman M, Sampalis F, Sampalis JS. Bariatric surgery reduces cancer risk in morbidly obese patients. Surg Obes Relat Dis 2008;4(6):691–5.
[11] Renehan AG, Zwahlen M, Egger M. Adiposity and cancer risk: new mechanistic insights from epidemiology. Nat Rev Cancer 2015;15(8):484.
[12] Baglietto L, English DR, Hopper JL, MacInnis RJ, Morris HA, Tilley WD, et al. Circulating steroid hormone concentrations in postmenopausal women in relation to body size and composition. Breast Cancer Res Treat 2009;115(1):171–9.
[13] Yager JD, Davidson NE. Estrogen carcinogenesis in breast cancer. N Engl J Med 2006;354(3):270–82 19.
[14] Argolo DF, Hudis CA, Iyengar NM. The impact of obesity on breast cancer. Curr Oncol Rep 2018;20(6):47.
[15] Giovannucci E, Harlan DM, Archer MC, Bergenstal RM, Gapstur SM, Habel LA, et al. Diabetes and cancer: a consensus report. Diabetes Care 2010;33:1674–85.
[16] Zheng Y, Ley SH, Hu FB. Global aetiology and epidemiology of type 2 diabetes mellitus and its complications. Nat Rev Endocrinol 2018;14(2):88.
[17] Wojciechowska J, Krajewski W, Bolanowski M, Kręcicki T, Zatoński T. Diabetes and cancer: a review of current knowledge. Exp Clin Endocrinol Diabetes 2016;124(05):263–75.
[18] Endogenous Hormones and Breast Cancer Collaborative Group, Key TJ, Appleby PN, Reeves GK, Roddam AW. Insulin-like growth factor 1 (IGF1), IGF binding protein 3 (IGFBP3), and breast cancer risk: pooled individual data analysis of 17 prospective studies. Lancet Oncol 2010;11(6):530–42.
[19] Clayton PE, Banerjee I, Murray PG, Renehan AG. Growth hormone, the insulin-like growth factor axis, insulin and cancer risk. Nat Rev Endocrinol 2011;7:11–24.
[20] Upadhyay J, Farr O, Perakakis N, Ghaly W, Mantzoros C. Obesity as a disease. Med Clin North Am 2018;102(1):13–33.
[21] Ryu TY, Park J, Scherer PE. Hyperglycemia as a risk factor for cancer progression. Diabetes Metab J 2014;38:330–6.
[22] Samuel SM, Varghese E, Varghese S, Büsselberg D. Challenges and perspectives in the treatment of diabetes associated breast cancer. Cancer Treat Rev 2018;70:98–111.
[23] Siebel AL, Fernandez AZ, El-Osta A. Glycemic memory associated epigenetic changes. Biochem Pharmacol 2010;80(12):1853–9.
[24] Andò S, Gelsomino L, Panza S, Giordano C, Bonofiglio D, Barone I, et al. Obesity, leptin and breast cancer: epidemiological evidence and proposed mechanisms. Cancers 2019;11(1):62.
[25] Niu J, Jiang L, Guo W, Shao L, Liu Y, Wang L. The association between leptin level and breast cancer: a meta-analysis. PLoS One 2013;8(6):e67349.
[26] Ishikawa M, Kitayama J, Nagawa H. Enhanced expression of leptin and leptin receptor (OB-R) in human breast cancer. Clin Cancer Res 2004;10:4325–31.
[27] Garofalo C, Koda M, Cascio S, Sulkowska M, Kanczuga-Koda L, Golaszewska J, et al. Increased expression of leptin and the leptin receptor as a marker of breast cancer progression: possible role of obesity-related stimuli. Clin Cancer Res 2006;12:1447–53.
[28] Jarde T, Caldefie-Chezet F, Damez M, Mishellany F, Penault-Llorca F, Guillot J, et al. Leptin and leptin receptor involvement in cancer development: a study on human primary breast carcinoma. Oncol Rep 2008;19:905–11.
[29] Miyoshi Y, Funahashi T, Tanaka S, Taguchi T, Tamaki Y, Shimomura I, et al. High expression of leptin receptor mRNA in breast cancer tissue predicts poor prognosis for patients with high, but not low, serum leptin levels. Int J Cancer 2006;118:1414–19.
[30] Pan H, Deng LL, Cui JQ, Shi L, Yang YC, Luo JH, et al. Association between serum leptin levels and breast cancer risk: an updated systematic review and meta-analysis. Medicine 2018;97: e11345.
[31] Gui Y, Pan Q, Chen X, Xu S, Luo X, Chen L. The association between obesity related adipokines and risk of breast cancer: a meta-analysis. Oncotarget 2017;8:75389–99.
[32] Tchoukalova YD, Sarr MG, Jensen MD. Measuring committed preadipocytes in human adipose tissue from severely obese patients by using adipocyte fatty acid binding protein. Am J Physiol Regul Integr Comp Physiol 2004;287(5):R1132–40.
[33] Tworoger SS, Eliassen AH, Kelesidis T, Colditz GA, Willett WC, Mantzoros CS, et al. Plasma adiponectin concentrations and risk of incident breast cancer. J Clin Endocrinol Metab 2007;92:1510–16.
[34] Oh SW, Park CY, Lee ES, Yoon YS, Lee ES, Park SS, et al. Adipokines, insulin resistance, metabolic syndrome, and breast cancer recurrence: a cohort study. Breast Cancer Res 2011;13:R34.
[35] Deng T, Lyon CJ, Bergin S, Caligiuri MA, Hsueh WA. Obesity, inflammation, and cancer. Annu Rev Pathol 2016;11:421–49.
[36] Perrier S, Caldefie-Chezet F, Vasson MP. IL-1 family in breast cancer: potential interplay with leptin and other adipocytokines. FEBS Lett 2009;583:259–65.
[37] Dethlefsen C, Hojfeldt G, Hojman P. The role of intratumoral and systemic IL-6 in breast cancer. Breast Cancer Res Treat 2013;138:657–64.
[38] Agurs-Collins T, Ross S, Dunn BK. The many faces of obesity and its influence on breast cancer risk. Front Oncol 2019;9:765.
[39] Djuric Z. Obesity-associated cancer risk: the role of intestinal microbiota in the etiology of the host proinflammatory state. Transl Res 2017;179:155–67.
[40] Rogers CJ, Prabhu KS, Vijay-Kumar M. The microbiome and obesity-an established risk for certain types of cancer. Cancer J 2014;20(3):176–80.
[41] Shapira I, Sultan K, Lee A, Taioli E. Evolving concepts: how diet and the intestinal microbiome act as modulators of breast malignancy. ISRN Oncol 2013;693920.
[42] Lee K, Kruper L, Dieli-Conwright CM, Mortimer JE. The impact of obesity on breast cancer diagnosis and treatment. Curr Oncol Rep 2019;21(5):41.
[43] Maruthur NM, Bolen S, Brancati FL, Clark JM. Obesity and mammography: a systematic review and meta-analysis. J Gen Intern Med 2009;24(5):665–77.
[44] Feldstein AC, Perrin N, Rosales AG, Schneider J, Rix MM, Glasgow RE. Patient barriers to mammography identified during a reminder program. J Women's Health 2011;20(3):421–8.
[45] Garland M, Hsu FC, Clark C, Chiba A, Howard-McNatt M. The impact of obesity on outcomes for patients undergoing mastectomy using the ACS-NSQIP data set. Breast Cancer Res Treat 2018;168(3):723–6.
[46] Panayi AC, Agha RA, Brady AS, Orgill DP. Impact of obesity on outcomes in breast reconstruction: a systematic review and meta-analysis. J Reconstr Microsurg 2018;34(05):363–75.

[47] Schaverien MV, Mcculley SJ. Effect of obesity on outcomes of free autologous breast reconstruction: a meta-analysis. Microsurgery 2014;34(6):484–97.

[48] Ewertz M, Jensen JB, Gunnarsdóttir KA, Højris I, Jakobsen EH, Nielsen D, et al. Effect of obesity on prognosis after early-stage breast cancer. J Clin Oncol 2011;29(1):25–31.

[49] Bergom C, Tracy K, Meena B, Hina S, Prior P, Rein LE, et al. The association of local-regional control with high body mass index in women undergoing breast conservation therapy for early stage breast cancer. Int J Radiat Oncol Biol Phys 2016;96(1):65–71.

[50] Warren LEG, Ligibel JA, Chen YH, Truong L, Catalano PJ, Bellon BJ. Body mass index and locoregional recurrence in women with early-stage breast cancer. Ann Surg Oncol 2016;23(12):3870–9.

[51] Burkheimer E, Starks L, Khan M, Oostendorp L, Melnik MK, Chung MH, et al. The impact of obesity on treatment choices and outcomes in operable breast cancer. Am J Surg 2019;217(3):474–7.

[52] Ross KH, Gogineni K, Subhedar PD, Lin JY, McCullough LE. Obesity and cancer treatment efficacy: existing challenges and opportunities. Cancer 2019;125(10):1588–92.

[53] Griggs JJ, Mangu PB, Temin S, Lyman GH. Appropriate chemotherapy dosing for obese adult patients with cancer: American Society of Clinical Oncology clinical practice guideline. J Clin Oncol 2012;30:1553e61.

[54] Mantarro S, Rossi M, Bonifazi M, D'Amico R, Blandizzi C, Vecchia C La. Risk of severe cardiotoxicity following treatment with trastuzumab: a meta-analysis of randomized and cohort studies of 29,000 women with breast cancer. Intern Emerg Med 2016;11:123–40.

[55] Vaysse C, Muller C, Fallone F. Obesity: an heavyweight player in breast cancer's chemoresistance. Oncotarget 2019;10(35):3207–8.

[56] Jiralesrspong S, Goodwin P. Obesity and breast cancer prognosis: evidence, challenges and opportunities. J Clin Oncol 2016;34(35):4203–16.

[57] Chan DSM, Vieira AR, Aune D, Bandera EV, Greenwood DC, McTiernan A, et al. Body mass index and survival in women with breast cancer: systematic literature review and meta-analysis of 82 follow-up studies. Ann Oncol 2014;25:1901–14.

[58] Chlebowski RT, Aragaki AK, Anderson GL, Thomson CA, Manson JE, Simon MS, et al. Low-fat dietary pattern and breast cancer mortality in the Women's Health Initiative randomized controlled trial. J Clin Oncol 2017;35(25):2919.

[59] DiSipio T, Rye S, Newman B, Hayes S. Incidence of unilateral arm lymphoedema after breast cancer: a systematic review and meta-analysis. Lancet Oncol 2013;14:500–15.

Chapter 24

Obesity and female malignancies

Ketankumar B. Gajjar[1] and Mahmood I. Shafi[2]

[1]Department of Gynaecological Oncology, Nottingham University Hospitals NHS Trust, Nottingham, United Kingdom, [2]Nuffield Health, Cambridge, United Kingdom

Introduction

The prevalence of obesity is rapidly increasing worldwide. The proportion of overweight and obese children and adults has steadily increased [1,2]. The estimated age-standardized prevalence of obesity in 2014 was 10.8% among men, 14.9% among women, and 5.0% among children [3,4]. Overall, about 3% of all cancers are linked to obesity, while cancers linked to obesity comprise approximately 51% of newly diagnosed cases among women [3,5]. The particular malignancies that obesity and overweight have an association with are increased risks of endometrial cancer (1.5-fold), postmenopausal breast cancer (twofold), ovarian cancer, and possibly cervical cancer (Table 24.1) [6–8]. It is estimated that 20%–30% of deaths from cancer in women could be attributed to overweight and obesity [9,10]. Furthermore, obesity has an impact on the screening, diagnosis, and treatment of female malignancies. There is evidence from a systematic review suggesting that 15%–30% of weight loss in women is associated with a reduced risk of cancer [11].

Epidemiology

Obesity and endometrial cancer

Obesity accounts for about 40% cases of endometrial cancer in the developed world [12]. A linear increase in the risk of endometrial cancer with increasing weight and body mass index (BMI) has been observed [13]. There is convincing and consistent evidence from both case–control and cohort studies that overweight and obesity are strongly associated with type I (estrogen-dependent) endometrial cancer [2,7,8]. Overweight and obese women have two to four times greater risk of developing endometrial cancer than do women of a healthy weight, regardless of their menopausal status, and extremely obese women are about seven times as likely to develop type I endometrial cancer [14,15]. Obesity in the menopause produces a state of excess estrogen production. This is due to the peripheral conversion, in the adipose tissue, of androgens secreted from the adrenal glands and ovaries into estrone, by the enzyme aromatase. Prolonged unopposed estrogen exposure will lead to a continuous spec-

TABLE 24.1 Relative risk of cancer incidence in relation to body mass index (BMI) [6].

	Floating absolute risk (floated CI)			
		BMI		
Cancer site	25–27.4	27.5–29.9	≥30	Trend (95% CI) per 10 U
Endometrium	1.21 (1.11–1.32)	1.43 (1.29–1.58)	2.73 (2.55–2.92)	2.89 (2.62–3.18)
Breast (premenopausal)	0.93 (0.82–1.05)	0.99 (0.84–1.16)	0.79 (0.68–0.92)	0.86 (0.73–1.00)
Breast (postmenopausal)	1.10 (1.04–1.16)	1.21 (1.13–1.29)	1.29 (1.22–1.36)	1.40 (1.31–1.49)
Ovary	0.99 (0.91–1.08)	1.13 (1.02–1.25)	1.12 (1.02–1.23)	1.14 (1.03–1.27)
Cervix	0.94 (0.75–1.19)	0.79 (0.57–1.10)	1.02 (0.80–1.31)	1.04 (0.79–1.38)

CI, Confidence interval.

Obesity and Gynecology. DOI: https://doi.org/10.1016/B978-0-12-817919-2.00024-3
© 2020 Elsevier Inc. All rights reserved.

trum of change from proliferative endometrium through endometrial hyperplasia/polyps to endometrial carcinoma (Figs. 24.1 and 24.2). Progesterone containing intrauterine contraceptive devices are a good contraceptive choice for obese women given their high efficacy irrespective of weight and the ability of the levonorgestrel intrauterine system (Mirena) to prevent endometrial hyperplasia in obese anovulatory women.

Obesity and breast cancer

Obesity seems to increase the risk of breast cancer only among postmenopausal women who do not use hormonal replacement therapy (HRT) (30%−50% increased risk) [16−18]. Obese women are also at higher risk of dying from breast cancer after the menopause [19]. Weight gain during adulthood has been found to be the most consistent and strongest predictor of postmenopausal breast cancer risk [20−22]. It is estimated that 11,000−18,000 deaths per year from breast cancer in US women over age 50 might be avoided if women could maintain a BMI under 25 kg/m^2 throughout their adult lives [19]. Women with central obesity have a greater breast cancer risk than those whose fat is distributed over the hips, buttocks, and lower extremities [23]. In addition, adult weight gain has been associated with a higher risk of postmenopausal breast cancer than the actual BMI level [24,25]. Among postmenopausal HRT users, there is no significant difference in breast cancer risk between obese women and women of a healthy weight [17,26]. Paradoxically, premenopausal obese women have a lower risk of developing breast cancer than do women of a healthy weight [17,27,28].

Both the increased risk of developing breast cancer and dying from it after the menopause seem to be due to increased levels of endogenous estrogen in obese women [29]. Estrogen levels in postmenopausal women are 50%−100% higher among obese women, compared to lean women [30], leading to a more rapid growth of estrogen-responsive/sensitive breast tumors. Breast cancer also seems to be detected later in obese women, leading to a poorer outcome. This is because the detection of a breast tumor is more difficult in overweight women [28].

Obesity and ovarian cancer

Evidence from more than 10 prospective studies and more than 20 case−control studies indicates a positive dose−response relationship between BMI and the risk of epithelial ovarian cancer. Higher BMI is associated with a slight increase (a 5-U increase in BMI is associated with a 10% increase) in the risk of ovarian cancer, particularly in women who have never used HRT [6,31]. Among nonusers of HRT, overweight women had a relative risk of about 1.1, whereas that for obese women was about 1.2 compared with normal-weight women. There was no association among users of HRT [31]. Among different histological subtypes of ovarian cancer, obesity was positively associated with clear cell tumors but less correlated with invasive endometrioid or mucinous tumors [32].

Only a few studies focused on the association between obesity and ovarian cancer survival and the evidence is conflicting. A recent metaanalysis found that women with ovarian cancer with obesity during early adulthood or before diagnosis had worse survival [33]. However, the effect of obesity on surgical morbidity in primary ovarian cancer after optimal primary tumor debulking remains controversial. Neither peri- or postoperative morbidity was affected by BMI [34,35].

In patients with ovarian cancer treated with carboplatin-based chemotherapy, obesity makes carboplatin dosing more challenging. Incorporating a weight adjustment may cause greater grade 3 and 4 systemic side effects (see paper below) whilst not incorporating a weight adjustment may lead to an increased risk of disease progression if the dose is not high enough. Women with obesity have been shown to have a lower relative decrease in their platelet counts and haemoglobin levels. There was a trend toward increased risk for disease progression in women with a BMI > 30 [36].

Obesity and cervical cancer

Studies of the association between BMI and cervical cancer are limited and inconclusive. While some studies reported cervical cancer to be associated with elevated BMI (two- to threefold increased risk) [6,13,37], others found a lower relative risk [38]. No association was observed in a cohort study of Swedish women [39]. The increased risk among overweight and obese women was

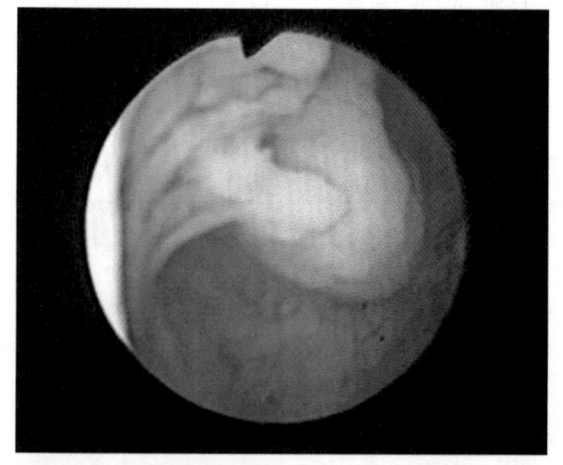

FIGURE 24.1 Hysteroscopy showing an endometrial polyp associated with a thickened malignant endometrium.

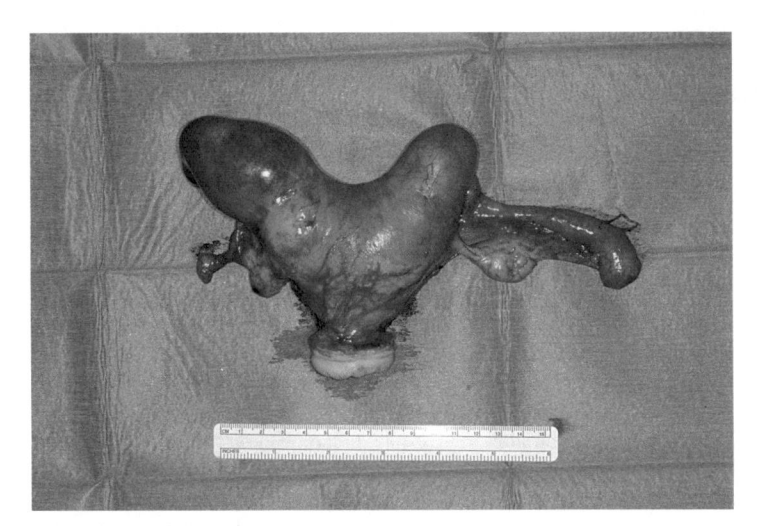

FIGURE 24.2 A surgical specimen following a total abdominal hysterectomy and bilateral salpingo-oophorectomy performed for the treatment of high-grade endometrial carcinoma. Note the presence of a bulky enlarged uterus with endometrial cancer filling and distending both cornua and the right fallopian tube.

mainly for cervical adenocarcinoma, with a smaller increased risk for squamous cell carcinoma [40]. Differential screening and health behavior, where obese women might be less likely to go for screening on a regular basis than women of normal weight, could partly explain the observed increased risk.

A retrospective study found that women with BMI > 35 with cervical cancer had a higher risk of both all-cause death (HR 1.26, 95% confidence interval (CI) 1.10−1.45) and disease-specific death (HR 1.24, 95% CI 1.06−1.47) than their normal-weight counterparts [41]. Both treatment-related and biological factors may contribute to decreased disease-specific survival in morbidly obese patients with cervical cancer.

Mechanisms relating obesity to female malignancies

Obesity affects the production of peptides [e.g., insulin and insulin-like growth factor 1 (IGF1), sex hormone−binding globulin (SHBG)] and steroid hormones (i.e., estrogen, progesterone, and androgens). It is likely that prolonged exposure to high levels of estrogen and insulin associated with obesity may contribute to the development of female malignancies [2,42,43].

Circulating levels of estrone and estradiol are directly related to amount of adipose tissue in postmenopausal women. [42]. Obesity in the menopause produces a state of excess estrogen production. Adipose tissue cells express various steroid hormone−metabolizing enzymes and are an important source of circulating estrogens, especially in postmenopausal women. This is due to the peripheral conversion of androgens secreted from the adrenal glands and ovaries into estrone, by the enzyme aromatase in the fat cells. The situation is aggravated by the fact that increased body fat is associated with decreased circulating levels of both progesterone and SHBG. With lower SHBG, there is a higher circulating level of free active estrogens.

Excess weight, increased plasma triglyceride levels, and low levels of physical activity can all raise circulating insulin levels, leading to chronic hyperinsulinemia that has been associated with cancers of the breast [44,45] and the endometrium [2]. The carcinogenic effects of hyperinsulinemia could be directly mediated by insulin receptors in the target cells or might be due to related changes in endogenous female sex hormones synthesis and bioavailability. Insulin also promotes the synthesis and biological activity of IGF1. Both insulin and IGF1 can act as growth factors that promote cell proliferation and inhibit apoptosis [46,47]. Endometrial cancer risk is inversely related to blood levels of IGF-binding protein 1 and 2 (IGFBP1 and IGFBP2), which reduce the amount of bioavailable IGF1 [48]. There is an increased risk of breast cancer in women with increased serum levels of insulin or IGF1 [49−51], especially in premenopausal women. The increase in blood levels of insulin and IGF1 results in reduced hepatic synthesis and blood concentrations of SHBG, increasing the bioavailability of estradiol [2,52,53]. Proteins secreted by adipose tissue (adipokines) also contribute to the regulation of immune response (leptin), inflammatory response (tumor necrosis factor α, interleukin-6, and serum amyloid A), vasculature and stromal interactions, and angiogenesis (vascular endothelial growth factor 1), as well as extracellular matrix components (type VI collagen).

Avoiding weight gain lowers the risk of endometrial and postmenopausal breast cancers [54]. However, there is limited evidence that intentional weight loss will affect cancer risk [55,56]. Physical activity among postmenopausal women at a level of walking about 30 min/day was associated with a 20% reduction in breast cancer risk, mainly among women who were of normal weight. The protective effect of physical activity was not found among overweight or obese women with breast cancer [57].

Effect of obesity on management of female malignancies

Obese patients have a poorer outcome compared to lean patients. For instance, obesity is associated both with reduced likelihood of survival and increased likelihood of recurrence among patients with breast cancer, regardless of menopausal status and after adjustment for stage and treatment (Table 24.2) [6,19,58]. The poorer outcomes in obese women probably reflect a true biological effect of adiposity on survival, a delayed diagnosis in heavier women, and a higher rate of treatment-associated complications. Heavier women are less likely to receive mammography or cervical screening [59]. For women who self-detect their breast cancers, nonlocalized disease is more common with a high BMI [60].

Clinical examination of the obese women with female malignancies can be difficult. Manual handling of these patients can also be challenging. Special hospital beds and operating tables should be available. The best way to ensure a safe and successful treatment is adequate preoperative evaluation, preparation, and counselling. Assistance with proper dosing and monitoring of medications should be considered. Prescriptions must take into account the concepts of total body weight as well as ideal body weight, as certain doses (e.g., corticosteroids, penicillin, and cephalosporins) are calculated based on ideal body weight, while others are calculated based on total body weight (e.g., heparins). For the women who are moderately or severely obese receiving adjuvant radiotherapy for endometrial cancer, a wider planning target margin is required to reduce the magnitude of setup error [61].

A high BMI increases the risk of perioperative complications and mortality, particularly in the presence of comorbidities. The risks are increased in case of morbid obesity (BMI > 35). Obese women should receive careful counseling about the increased risk of complications and technical difficulties that may be encountered during surgery [62,63]. Preoperative evaluation should include a cardiovascular and respiratory assessment. Obese patients are at much higher risk for postoperative complications given the more frequent comorbidities, such as diabetes, hypertension, coronary artery disease, sleep apnea, hypoventilation, and osteoarthritis of the knees and hips. Thus respiratory or cardiac failure, venous thromboembolism, aspiration, wound infection and dehiscence, and postoperative asphyxia are all more common in obese patients. Control of the airway is critical in obese surgical patients. Planned admission to high-dependency units is often advisable. Ventilation may be aided by the use of noninvasive positive pressure ventilation units, particularly if the patient has a history of sleep apnea. Venous access can be problematic in obese patients. Doppler ultrasound scans could assist the safe placement of intravenous lines.

Women with gynecological malignancies can be managed in a standard fashion, and in most instances there is no need to compromise surgical treatment where this is indicated [64]. The route of any surgical intervention needs to be considered, as abdominal procedures are more of an issue than vaginal surgery. The abdominal wall anatomy is distorted by the overhanging skin and fat (panniculus). Obesity is recognized as a potential limiting factor in the application of laparoscopic surgery because of a higher rate of failed entry, hindered manipulation, and poor views. Obesity may not allow steep Trendelenburg because of unacceptably high peak inspiratory pressure. In addition, obesity may prevent adequate mobilization of the small bowel out of the pelvis to allow for proper pelvic visualization. Nevertheless, laparoscopic and robotic

TABLE 24.2 Relative risk of cancer mortality in relation to body mass index (BMI) [6].

Cancer site	Floating absolute risk (floated CI) BMI			Trend (95% CI) per 10 U
	25−27.4	27.5−29.5	≥30	
Endometrium	1.09 (0.82−1.45)	1.21 (0.85−1.71)	2.28 (1.81−2.87)	2.46 (1.78−3.39)
Breast (premenopausal)	1.05 (0.67−1.64)	0.91 (0.49−1.70)	0.64 (0.34−1.21)	0.68 (0.37−1.24)
Breast (postmenopausal)	1.26 (1.07−1.47)	1.22 (0.99−1.49)	1.49 (1.27−1.75)	1.36 (1.12−1.66)
Ovary	0.93 (0.84−1.03)	1.02 (0.89−1.16)	1.16 (1.04−1.30)	1.17 (1.03−1.33)
Cervix	0.77 (0.51−1.17)	0.61 (0.34−1.11)	1.15 (0.79−1.70)	1.53 (0.95−2.47)

CI, Confidence interval.

surgery has additional benefits for the obese: they have less postoperative ileus, fewer wound infections, and they mobilize more quickly than those undergoing laparotomy [65]. Obesity presents problems with laparotomy incision placement and closure. Adequate wound antisepsis is necessary, as obese women are at increased risk of wound infection and wound failure. Possible etiologies include decreased oxygen tension, immune impairment, and tension and secondary ischemia along suture lines [66]. Access to the pelvis can be challenging and there is a higher incidence of intraoperative complications due to problems with access or distorted anatomy. Difficulty with hemostasis, particularly among women when removing the cervix and suturing the vaginal vault, requires experience to manage. Good assistance, retraction, and lighting are essential.

The use of regional anesthesia (spinal or epidural) is to be encouraged because apart from the anesthetic benefits, it will help with postoperative pain control. However, regional anesthesia may be difficult or even impossible due to high BMI leading to difficulty locating spine. The availability of an experienced anesthetist may minimize technical failure of regional anesthesia. A recent large multisite retrospective cohort study found that in severe obesity, bariatric surgery was associated with a lower risk of incident cancer, particularly obesity-associated cancers, such as postmenopausal breast cancer, endometrial cancer, and colon cancer [67]. In another study, bariatric surgery—induced weight loss resulted in significant beneficial changes in circulating biomarkers of insulin resistance, inflammation, and reproductive hormones, in endometrial morphology, and in molecular pathways that are implicated in endometrial carcinogenesis. These results may have important implications for screening, prevention, and treatment of endometrial cancer [68].

Future directions

Further research to define the causal role of obesity in gynecological malignancies is needed. Obesity-associated dysregulation of adipokines is likely to contribute not only to tumorigenesis and tumor progression but also to metastatic potential. It will also be important to develop successful intervention strategies, both at the individual and community levels, for weight loss and maintenance. The tobacco control experience has taught us that policy and environmental changes are crucial to achieving changes in individual behavior. Future trials may involve studies on the effect of dietary changes on weight gain and cancer risk, the effect of patterns of physical activity (the intensity, frequency, and duration of various sorts of physical activity) in relation to weight gain and cancer risk, and the combined effects of changes in diet and physical activity on obesity and female cancer risk.

Clinicians should be aware that preventing/treating obesity should be considered part of cancer prevention. Further research is needed to clarify the mechanism and role of bariatric surgery to lower the risk of incident cancer in severe obesity.

References

[1] Zaninotto P, Head J, Stamatakis E, Wardle H, Mindell J. Trends in obesity among adults in England from 1993 to 2004 by age and social class and projections of prevalence to 2012. J Epidemiol Community Health 2009;63(2):140–6.

[2] Kaaks R, Lukanova A, Kurzer MS. Obesity, endogenous hormones, and endometrial cancer risk: a synthetic review. Cancer Epidemiol Biomarkers Prev 2002;11(12):1531–43.

[3] NCD Risk Factor Collaboration (NCD-RisC). Trends in adult body-mass index in 200 countries from 1975 to 2014: a pooled analysis of 1698 population-based measurement studies with 19·2 million participants. Lancet 2016;387(10026):1377–96.

[4] Ng M, Fleming T, Robinson M, et al. Global, regional, and national prevalence of overweight and obesity in children and adults during 1980–2013: a systematic analysis for the Global Burden of Disease Study 2013. Lancet 2014;384(9945):766–81.

[5] Polednak AP. Trends in incidence rates for obesity-associated cancers in the US. Cancer Detect Prev 2003;27(6):415–21.

[6] Reeves GK, Pirie K, Beral V, et al. Cancer incidence and mortality in relation to body mass index in the Million Women Study: cohort study. BMJ 2007;335(7630):1134.

[7] Kalliala I, Markozannes G, Gunter MJ, et al. Obesity and gynaecological and obstetric conditions: umbrella review of the literature. BMJ 2017;359:j4511.

[8] Kyrgiou M, Kalliala I, Markozannes G, et al. Adiposity and cancer at major anatomical sites: umbrella review of the literature. BMJ 2017;356:j477.

[9] Allison DB, Fontaine KR, Manson JE, Stevens J, VanItallie TB. Annual deaths attributable to obesity in the United States. JAMA 1999;282(16):1530–8.

[10] Banegas JR, López-García E, Gutiérrez-Fisac JL, Guallar-Castillón P, Rodríguez-Artalejo F. A simple estimate of mortality attributable to excess weight in the European Union. Eur J Clin Nutr 2003;57(2):201–8.

[11] Birks S, Peeters A, Backholer K, O'Brien P, Brown W. A systematic review of the impact of weight loss on cancer incidence and mortality. Obes Rev 2012;13(10):868–91.

[12] Bergström A, Pisani P, Tenet V, Wolk A, Adami HO. Overweight as an avoidable cause of cancer in Europe. Int J Cancer 2001;91(3):421–30.

[13] Calle EE, Rodriguez C, Walker-Thurmond K, Thun MJ. Overweight, obesity, and mortality from cancer in a prospectively studied cohort of U.S. adults. N Engl J Med 2003;348(17):1625–38.

[14] Setiawan VW, Yang HP, Pike MC, et al. Type I and II endometrial cancers: have they different risk factors? J Clin Oncol 2013;31(20):2607–18.

[15] Dougan MM, Hankinson SE, Vivo ID, Tworoger SS, Glynn RJ, Michels KB. Prospective study of body size throughout the life-course and the incidence of endometrial cancer among

premenopausal and postmenopausal women. Int J Cancer 2015;137(3):625–37.
[16] Galanis DJ, Kolonel LN, Lee J, Le Marchand L. Anthropometric predictors of breast cancer incidence and survival in a multi-ethnic cohort of female residents of Hawaii, United States. Cancer Causes Control 1998;9(2):217–24.
[17] van den Brandt PA, Spiegelman D, Yaun S-S, et al. Pooled analysis of prospective cohort studies on height, weight, and breast cancer risk. Am J Epidemiol 2000;152(6):514–27.
[18] Friedenreich CM. Review of anthropometric factors and breast cancer risk. Eur J Cancer Prev 2001;10(1):15–32.
[19] Petrelli JM, Calle EE, Rodriguez C, Thun MJ. Body mass index, height, and postmenopausal breast cancer mortality in a prospective cohort of US women. Cancer Causes Control 2002;13(4):325–32.
[20] Kawai M, Minami Y, Kuriyama S, et al. Adiposity, adult weight change and breast cancer risk in postmenopausal Japanese women: the Miyagi Cohort Study. Br J Cancer 2010;103(9):1443–7.
[21] Ahn J, Schatzkin A, Lacey JV, et al. Adiposity, adult weight change, and postmenopausal breast cancer risk. Arch Intern Med 2007;167(19):2091–102.
[22] Eliassen AH, Colditz GA, Rosner B, Willett WC, Hankinson SE. Adult weight change and risk of postmenopausal breast cancer. JAMA 2006;296(2):193–201.
[23] Kaaks R, Van Noord PA, Den Tonkelaar I, Peeters PH, Riboli E, Grobbee DE. Breast-cancer incidence in relation to height, weight and body-fat distribution in the Dutch "DOM" cohort. Int J Cancer 1998;76(5):647–51.
[24] Feigelson HS, Jonas CR, Teras LR, Thun MJ, Calle EE. Weight gain, body mass index, hormone replacement therapy, and postmenopausal breast cancer in a large prospective study. Cancer Epidemiol Biomarkers Prev 2004;13(2):220–4.
[25] Schairer C, Lubin J, Troisi R, Sturgeon S, Brinton L, Hoover R. Menopausal estrogen and estrogen-progestin replacement therapy and breast cancer risk. JAMA 2000;283(4):485–91.
[26] Lahmann PH, Lissner L, Gullberg B, Olsson H, Berglund G. A prospective study of adiposity and postmenopausal breast cancer risk: the Malmö Diet and Cancer Study. Int J Cancer 2003;103(2):246–52.
[27] Trentham-Dietz A, Newcomb PA, Storer BE, et al. Body size and risk of breast cancer. Am J Epidemiol 1997;145(11):1011–19.
[28] Cui Y, Whiteman MK, Flaws JA, Langenberg P, Tkaczuk KH, Bush TL. Body mass and stage of breast cancer at diagnosis. Int J Cancer 2002;98(2):279–83.
[29] Toniolo PG, Levitz M, Zeleniuch-Jacquotte A, et al. A prospective study of endogenous estrogens and breast cancer in postmenopausal women. J Natl Cancer Inst 1995;87(3):190–7.
[30] Huang Z, Hankinson SE, Colditz GA, et al. Dual effects of weight and weight gain on breast cancer risk. JAMA 1997;278(17):1407–11.
[31] Collaborative Group on Epidemiological Studies of Ovarian Cancer. Ovarian cancer and body size: individual participant meta-analysis including 25,157 women with ovarian cancer from 47 epidemiological studies. PLoS Med 2012;9(4):e1001200.
[32] Olsen CM, Nagle CM, Whiteman DC, Purdie DM, Green AC, Webb PM. Body size and risk of epithelial ovarian and related cancers: a population-based case-control study. Int J Cancer 2008;123(2):450–6.

[33] Yang HS, Yoon C, Myung SK, Park SM. Effect of obesity on survival of women with epithelial ovarian cancer: a systematic review and meta-analysis of observational studies. Int J Gynecol Cancer 2011;21(9):1525–32.
[34] Skírnisdóttir I, Sorbe B. Prognostic impact of body mass index and effect of overweight and obesity on surgical and adjuvant treatment in early-stage epithelial ovarian cancer. Int J Gynecol Cancer 2008;18(2):345–51.
[35] Fotopoulou C, Richter R, Braicu EI, et al. Impact of obesity on operative morbidity and clinical outcome in primary epithelial ovarian cancer after optimal primary tumor debulking. Ann Surg Oncol 2011;18(9):2629–37.
[36] Wright JD, Tian C, Mutch DG, et al. Carboplatin dosing in obese women with ovarian cancer: a Gynecologic Oncology Group study. Gynecol Oncol 2008;109(3):353–8.
[37] Poorolajal J, Jenabi E. The association between BMI and cervical cancer risk: a meta-analysis. Eur J Cancer Prev 2016;25(3):232–8.
[38] Wolk A, Gridley G, Svensson M, et al. A prospective study of obesity and cancer risk (Sweden). Cancer Causes Control 2001;12(1):13–21.
[39] Törnberg SA, Carstensen JM. Relationship between Quetelet's index and cancer of breast and female genital tract in 47,000 women followed for 25 years. Br J Cancer 1994;69(2):358–61.
[40] Lacey JV, Swanson CA, Brinton LA, et al. Obesity as a potential risk factor for adenocarcinomas and squamous cell carcinomas of the uterine cervix. Cancer 2003;98(4):814–21.
[41] Frumovitz M, Jhingran A, Soliman PT, Klopp AH, Schmeler KM, Eifel PJ. Morbid obesity as an independent risk factor for disease-specific mortality in women with cervical cancer. Obstet Gynecol 2014;124(6):1098–104.
[42] Key TJ, Appleby PN, Reeves GK, et al. Body mass index, serum sex hormones, and breast cancer risk in postmenopausal women. J Natl Cancer Inst 2003;95(16):1218–26.
[43] Gajjar K, Martin-hirsch P, Martin F. CYP1B1 and hormone-induced cancer. Cancer Lett. 2012;324(1):13–30.
[44] Kaaks R. Nutrition, hormones, and breast cancer: is insulin the missing link? Cancer Causes Control 1996;7(6):605–25.
[45] Stoll BA. Oestrogen/insulin-like growth factor-I receptor interaction in early breast cancer: clinical implications. Ann Oncol 2002;13(2):191–6.
[46] Prisco M, Romano G, Peruzzi F, Valentinis B, Baserga R. Insulin and IGF-I receptors signaling in protection from apoptosis. Horm Metab Res 1999;31(2–3):80–9.
[47] Khandwala HM, McCutcheon IE, Flyvbjerg A, Friend KE. The effects of insulin-like growth factors on tumorigenesis and neoplastic growth. Endocr Rev 2000;21(3):215–44.
[48] Lukanova A, Zeleniuch-Jacquotte A, Lundin E, et al. Prediagnostic levels of C-peptide, IGF-I, IGFBP-1, -2 and -3 and risk of endometrial cancer. Int J Cancer 2004;108(2):262–8.
[49] Cust AE, Stocks T, Lukanova A, et al. The influence of overweight and insulin resistance on breast cancer risk and tumour stage at diagnosis: a prospective study. Breast Cancer Res Treat 2009;113(3):567–76.
[50] Pichard C, Plu-Bureau G, Neves-E Castro M, Gompel A. Insulin resistance, obesity and breast cancer risk. Maturitas 2008;60(1):19–30.
[51] Schairer C, Hill D, Sturgeon SR, et al. Serum concentrations of IGF-I, IGFBP-3 and c-peptide and risk of hyperplasia and cancer

of the breast in postmenopausal women. Int J Cancer 2004;108(5):773−9.

[52] Tchernof A, Després JP. Sex steroid hormones, sex hormone-binding globulin, and obesity in men and women. Horm Metab Res 2000;32(11−12):526−36.

[53] Kokkoris P, Pi-Sunyer FX. Obesity and endocrine disease. Endocrinol Metab Clin North Am 2003;32(4):895−914.

[54] Vainio H, Kaaks R, Bianchini F. Weight control and physical activity in cancer prevention: international evaluation of the evidence. Eur J Cancer Prev 2002;11(Suppl. 2):S94−100.

[55] Ziegler RG, Hoover RN, Nomura AM, et al. Relative weight, weight change, height, and breast cancer risk in Asian-American women. J Natl Cancer Inst 1996;88(10):650−60.

[56] Trentham-Dietz A, Newcomb PA, Egan KM, et al. Weight change and risk of postmenopausal breast cancer (United States). Cancer Causes Control 2000;11(6):533−42.

[57] McTiernan A, Kooperberg C, White E, et al. Recreational physical activity and the risk of breast cancer in postmenopausal women: the Women's Health Initiative Cohort Study. JAMA 2003;290(10):1331−6.

[58] Chlebowski RT, Aiello E, McTiernan A. Weight loss in breast cancer patient management. J Clin Oncol 2002;20(4):1128−43.

[59] Wee CC, McCarthy EP, Davis RB, Phillips RS. Screening for cervical and breast cancer: is obesity an unrecognized barrier to preventive care? Ann Intern Med 2000;132(9):697−704.

[60] Reeves MJ, Newcomb PA, Remington PL, Marcus PM, MacKenzie WR. Body mass and breast cancer. Relationship between method of detection and stage of disease. Cancer 1996;77(2):301−7.

[61] Martra F, Kunos C, Gibbons H, et al. Adjuvant treatment and survival in obese women with endometrial cancer: an international collaborative study. Am J Obstet Gynecol 2008;198(1):89.e1−8.

[62] Peña MM, Taveras EM. Preventing childhood obesity: wake up, it's time for sleep!. J Clin Sleep Med 2011;7(4):343−4.

[63] Haslam D, Sattar N, Lean M. ABC of obesity. Obesity—time to wake up. BMJ 2006;333(7569):640−2.

[64] Papadia A, Ragni N, Salom EM. The impact of obesity on surgery in gynecological oncology: a review. Int J Gynecol Cancer 2006;16(2):944−52.

[65] Lamvu G, Zolnoun D, Boggess J, Steege JF. Obesity: physiologic changes and challenges during laparoscopy. Am J Obstet Gynecol 2004;191(2):669−74.

[66] DeMaria EJ, Carmody BJ. Perioperative management of special populations: obesity. Surg Clin North Am 2005;85(6):1283−9.

[67] Schauer DP, Feigelson HS, Koebnick C, et al. Bariatric surgery and the risk of cancer in a large multisite cohort. Ann Surg 2019;269(1):95−101.

[68] MacKintosh ML, Derbyshire AE, McVey RJ, et al. The impact of obesity and bariatric surgery on circulating and tissue biomarkers of endometrial cancer risk. Int J Cancer 2019;144(3):641−50.

Chapter 25

Challenges in gynecological surgery in obese women

Chu Lim[1] and Tahir A. Mahmood[1]

[1]*Obstetrics and Gynaecology, Victoria Hospital, Kirkcaldy, United Kingdom,* [2]*Department of Obstetrics and Gynaecology, Victoria Hospital, Kirkcaldy, United Kingdom*

Introduction

Obesity is a worldwide health problem that increases morbidity and mortality. In the United States the prevalence of obesity is roughly 35% in adults [1]. It is estimated that over the past 5 years the incidence of obesity has nearly doubled. Worldwide in 2008, 35% of adults over the age of 20 were overweight and 11% were obese. According to data published by the WHO, there were 300 million obese women in 2008 [2]. In women, it is associated with an increased risk of death and morbid conditions (including hypertension, diabetes mellitus, obstructive sleep apnea, and hypercholesterolemia) as well as a number of gynecological diseases, such as menstrual problems, infertility, prolapse, and urinary incontinence. They are also at increased risk of developing malignancies such as endometrial and postmenopausal breast cancer.

Body mass index (BMI) is used as a way to classify individuals according to their weight. It is calculated by dividing the person's weight in kilograms by the square of their height in meters (kg/m^2). Obesity is defined as having a body mass of 30 or greater, and it can be further subdivided: class I obesity is defined as a BMI of 30 to less than 35, class II obesity is defined as a BMI of 35 to less than 40, and class III obesity is defined as a BMI of 40 or greater.

Obesity is associated with various conditions, including diabetes mellitus, hypertension, hypercholesterolemia, heart disease, asthma, and arthritis, all these contribute to increased morbidity and mortality. It is a known risk factor for a number of gynecological diseases, such as menstrual problems, infertility, prolapse, urinary incontinence, and endometrial cancer. Women who are 9–22 kg above their healthy body weight have a threefold increase in having endometrial cancer, rising to ninefold if they are 22 kg over their ideal health weight.

Based on the data and expert opinion, The American College of Obstetricians and Gynecologists in recent committee opinion [3] had made the following recommendations:

1. Gynecologic surgeons should have the knowledge to counsel obese women on the risks specific to this group.
2. As with all patients, evidence demonstrates that, in general, vaginal hysterectomy is associated with better outcomes and fewer complications than laparoscopic or abdominal hysterectomy.

Indications for surgery

The first consideration is to ensure that surgery is appropriately indicated. Many gynecological conditions can be treated without surgery, and weight loss alone will improve conditions such as stress incontinence and menstrual disorders. Clinical examination of the obese person can be limited, and skilled ultrasonography is helpful. Conservative therapies, such as the levonorgestrel-releasing intrauterine system [4] for menstrual dysfunction, bladder retraining and physiotherapy for urinary problems, and pessaries for prolapse, should readily be considered for women who are obese.

Obese women should receive careful counseling about the increased risk of complications and technical difficulties that may be encountered during surgery. They should be weighed and assessed, and a weight loss program should be offered, providing they are willing to change [4]. Weight loss and exercise are, clearly, useful options, but they take time and motivation. It is clearly the doctor's duty to help them understand the problem from a medical point of view and grasp how the doctor is working to reduce the risks as far as is possible. There is case to offer bariatric surgery if conservative treatment had failed, and there are other significant comorbidities associated with the obesity.

Risk of obese women undergoing surgery

"Obesity paradox" was noted in a prospective, multiinstitutional, risk-adjusted cohort study of 118,707 patients who underwent nonbariatric general surgery. The mortality and morbidity risk was found to be highest in women who are underweight and morbidly obese extremes and the lowest rates in the overweight and moderately obese [5]. In the absence of metabolic syndrome (specifically, hypertension and diabetes), the overall mortality and composite morbidity of the overweight and moderately obese has been shown to be lower than that of normal-weight patients. Obese patients with metabolic syndrome who undergo general, vascular, and orthopedic surgery are at increased risk of perioperative morbidity and mortality compared with normal-weight patients [6].

Wound complications, surgical site infections, and venous thromboembolism are the main cause of morbidity in obese women who underwent open abdominal surgery. Every effort should be made to offer all patients, regardless of BMI, the least invasive procedure in order to decrease complications, length of hospital stay, and postoperative recovery time. In a Cochrane review of studies on hysterectomy, the risk of wound complication and surgical site infections is less when the procedure is done through a vaginal or laparoscopic approach compared with an open route [7]. Conversion rate to open procedure is higher in obese women but tends to decrease over time with surgical experience [8].

Physiological changes in the obese patients

Cardiovascular disease is the leading cause of both morbidity and mortality in the obese patient. There is an increased risk of ischemic heart disease, hypertension, and heart failure. A Scottish health survey showed that the prevalence of cardiovascular disease was 37% in adults with a BMI more than 30 compared to only 10% in adults with a BMI of less than 25 [9]. Hypertension is common in obese patients with 60% of obese patients having mild-to-moderate hypertension and 5%–10% having severe hypertension [10].

Cardiac arrhythmias are more common in obese patients and may be caused by a number of factors, including hypoxia, electrolyte disturbance, myocardial hypertrophy, and fatty infiltration of the conducting system [11–13].

The association between obesity and cardiomyopathy is well recognized [14,15]. Autopsy studies have shown that there is a 20%–55% increase in cardiac diameter, ventricle size, and cardiac weight for the obese patient compared to the nonobese patient [16].

Obesity is associated with a decrease in functional residual capacity (FRC), expiratory reserve volume, and total lung capacity [17,18]. This is caused by reduced chest wall compliance because of the excess body weight around the ribs and under the diaphragm. Oxygen consumption and carbon dioxide production are increased in the obese patient because of the metabolic activity of the increased amount of adipose tissue. Furthermore, FRC is reduced in the obese patients when lying in supine position with an impaired tolerance for the Trendelenburg position for the laparoscopic surgery. The FRC is further compromised by anesthesia to levels lower than closing capacity resulting in airway closure and hypoxemia [19,20].

Obese patients are at an increased risk of gastric acid aspiration during surgery because of large gastric volume, increased predisposition to reflux, lower gastric pH, and delayed gastric emptying [21,22].

Obese patients have altered pharmacokinetics due to smaller total body water, increased adipose tissue and increased blood volume, and altered renal blood flow [12]. All these changes are important in understanding challenges faced by the anesthetists in order to optimize their techniques and may be a limiting factor for total duration of surgery as well.

Preoperative evaluation

Obese patients are at higher risk of coronary artery disease (CAD), hypertension, diabetes mellitus, obstructive sleep apnea, and venous thromboembolism. Morbid obesity is associated with left ventricular enlargement and systolic and diastolic dysfunction, even in the absence of overt cardiac disease [23]. The risk of atrial fibrillation is similarly increased. Preoperative consultation with an anesthetist should be considered for the obese patient. The anesthetist will consider whether tracheal intubation and airway management will be difficult due to adipose tissue in the neck and limited neck/cervical spine movement. The use of regional anesthesia is to be encouraged: in addition to the anesthetic benefits, this helps with postoperative analgesia and reduces the risk of thromboembolism by half [24]. In practice, regional anesthesia may prove difficult or even impossible, but the availability of an experienced anesthetist will minimize technical failure.

Obese patients with metabolic syndrome undergoing noncardiac surgery are at increased risk of cardiovascular complications; a 12-lead electrocardiogram is recommended at preoperative evaluation together with and other tests based on physical examination findings. In obese patients with diabetes mellitus, blood glucose evaluation and counseling the woman on the importance of euglycemia to improve postoperative wound healing are important.

Obstructive sleep apnea can be associated with postoperative respiratory complications (pneumonia, postoperative hypoxemia, and unplanned reintubation). Specialist investigations will be required if the condition is suspected from a history of daytime somnolence, morning headaches, nocturnal wakening, and partner reports of loud snoring and apneic episodes during sleep. It is unknown whether screening for obstructive sleep apnea improves postoperative outcomes [25]. Routine pulmonary function tests on the obese patient are not recommended.

Thorough abdominal and bimanual pelvic examination must be carried to guide the route of surgery. In obese patients, this may be difficult, and preoperative imaging (e.g., ultrasound scan, magnetic resonance imaging) may help to determine the best route of surgery in these cases, and an examination under anesthesia may provide more guidance.

Equipment and general considerations

Clear and early communication is needed between staff involved in the management of these women, as time is needed to plan personnel and the availability of resources such as equipment. Outside bariatric surgery departments, hospitals may be ill-equipped to provide care for severely obese patients. Evidence from the United States suggests that this is becoming a clinical risk and medicolegal issue as patients may receive suboptimal treatment and staff can sustain injury when attempting to mobilize patients with extreme obesity [26,27]. In addition, a paper from the United Kingdom highlighted the patient safety incidents associated with obesity [28]. The National Patient Safety Agency (NPSA) and the Health and Safety Executive (HSE) have both recognized the potential risks to bariatric patients since 2007, producing reports and recommendations for NHS trusts [29,30].

All staff should undergo appropriate manual handling training to protect both themselves and patients [23]. Every operating table, trolley, and bed should be labeled with its maximum weight capacity. Special hospital beds should be available that can accommodate the weight and enable movement of the patient without manual handling. A standard operating table can usually support a body weight of 130–160 kg, and most theaters have tables available for supporting a body weight of up to 300 kg. Width extensions can be used on some tables to prevent the patient's body overhanging the edges of the table. All equipment should be electronically operated. Royal College of Anaesthetists in the United Kingdom proposed "obesity packs" with specific equipment and guidelines [23]. Obese people are at risk of slipping off the table during position changes and, therefore, they must be well secured to the table. All pressure points should be well padded, as there is a risk of nerve injury and of rhabdomyolysis of the gluteal muscles leading to renal failure [31] among the morbidly obese.

Anesthetic challenges

One of the main challenges in the anesthetic management of the obese patient is the maintenance of adequate oxygenation [32]. As a good practice, perhaps two experienced anesthetists may pair up to support each other during complex procedures. During anesthesia, obese patients in supine position require a 15% higher minute ventilation to maintain normocarbia [32]. The increased weight on the chest wall in the supine position and peumoperitoneum adds to the increased inspiratory resistance, and requires higher minute ventilation, and pressures to maintain satisfactory oxygenation. In Trendelenburg position the steeper the head-down position, and the higher the pneumoperitoneum pressures, the greater the problem becomes. An imbalance between perfusion and ventilation within the lung tissue is exacerbated by these factors, resulting in increased difficulty for the anesthetist to maintain oxygenation for these patients especially in prolonged and complex surgery.

Thromboprophylaxis

It is well recognized that the risk of perioperative deep vein thrombosis and pulmonary embolism is higher among obese people than among those of normal weight [24]. Venous stasis is more pronounced and postoperative mobility reduced. Obesity is also associated with increased levels of fibrinogen and factor VIII. Although rarely used in obese women, hormone replacement therapy and contraceptives containing estrogen should be stopped 4 weeks before surgery. Appropriately sized antithromboembolic stockings should be used and mechanical devices such as intermittent pneumatic compression are recommended. Low-molecular-weight heparin, such as enoxaparin 40 mg daily, starting a minimum of 2 hours postoperatively, is advised unless there is a risk of bleeding. This should be continued throughout the hospital stay and for 1 week after surgery [24]. Treatment should be extended to 4 weeks in cases of pelvic surgery for malignancy. If appropriate, hormone treatment can be restarted, once the additional risk of thrombosis has passed after 4 weeks.

Intraoperative challenges

Laparoscopic surgery

Laparoscopy in the obese patient is more technically challenging than in the normal-weight patient and should be undertaken by those who have adequate laparoscopic surgical experience. Obesity used to be considered a relative contraindication to laparoscopic surgery because of higher rate of failed entry, hindered manipulation, and poor views

[33] Excessive weight on the thorax can cause difficulty in ventilation, particularly in steep head-down positions.

With the improvement in anesthetics, surgical skills, and equipment, laparoscopic approach had been shown to be feasible in obese women with the added advantage of reduced recovery time and hospital stay, combined with the reduced risk of thromboembolic disease and wound infection. Eltabbakh et al. showed that obese women with endometrial cancer can be safely managed with laparoscopic surgery [34]. Heinberg et al. compared total laparoscopic hysterectomy in obese versus nonobese women and demonstrated that total laparoscopic surgery can be safely performed in obese patients with similar complications rates to those with a normal BMI [35].

Gaining access into the intraperitoneal cavity can be challenging in the obese patient. It is important to recognize that obesity per se brings anatomical changes as the distance between the skin and peritoneum is increased in obese women making the placement of the Veress needle in the peritoneal cavity more difficult. If the Veress needle is used, the 150 mm length may help achieve peumoperitoneum and avoid preperitoneal insufflation; longer ancillary trocars (up to 150 mm) also may be useful.

The Royal College of Obstetricians and Gynaecologists has published a guideline for preventing entry-related gynecological laparoscopic injuries [27], and in this guideline, they have recommended the use of the transumbilical open technique or entry at Palmer's point for in morbidly obese patient to gain access to the intraperitoneal cavity [36].

Placement of ancillary trocars can be more challenging because of the suboptimal visualization of the inferior epigastric vessels. It is very important to visualize the inferior epigastric vessels prior to placement of secondary lateral ports. Therefore incision should be made extremely lateral to the edge of the rectus sheath, ensuring avoidance of injury to the pelvic sidewall vessels. The increased thickness of the abdominal wall is more likely to result in inadvertent displacement of the lateral ports when instruments are being removed or replaced, and therefore the use of cuffed ports, which will not become displaced during the procedure, should also be considered.

Exposure can be difficult when operating in the pelvis of an obese patient. Operating in the pelvis requires the Trendelenburg position that may cause difficulty in ventilating the patient. A higher pneumoperitoneal pressure may be required, but the higher pressure may hamper the ability to provide adequate ventilation. The omental fat and limited manipulation of instruments also pose difficulty.

Postoperative hernias are more likely in obese patients. Closure of any port size at least 10 mm or greater often presents the greatest challenge; a port closure technique that affords laparoscopic visualization may be useful in this situation. It could therefore be recommended to use the smallest ports feasible when operating on obese women reducing the risk of a postoperative hernia

Open abdominal surgery

Abdominal surgery in the obese patient may be challenging. As the amount of subcutaneous tissue increases, retraction and adequate exposure are difficult. Compared with abdominal surgery in normal-weight women, abdominal hysterectomy has been associated with longer operative times in obese women [37]. Access to the pelvis can be challenging, and there is a higher incidence of intraoperative complications due to problems with access or distorted anatomy. Good assistance, retraction, and lighting are essential. Flexible illuminators are available that can provide good light in deep cavities, and long instruments are all helpful

Obese women are at increased risk of wound infection and wound failure [38]. Possible etiologies include decreased oxygen tension, immune impairment, and tension and secondary ischemia along suture lines [39]. Antibiotics are an essential part of any strategy for reducing wound infection; lowest infection rates are associated with antibiotic administration before the incision is made [40]. A recent investigation of maternal obesity in cesarean delivery has highlighted a decreased antibiotic tissue concentration in the obese patient using standard prophylactic antibiotic recommendations; this may account for the increased surgical site infection rate in obese patients [41]. When prophylactic antibiotics are warranted, a higher or weight-dependent dose of antibiotics should be considered.

Postoperative issues

Morbidly obese individuals are often admitted to a high-dependency unit postoperatively due to comorbidity. Adequate thromboprophylaxis is vital, venous thromboembolisms occur in 5%−12% of obese patients who undergo surgery [42]. All women should be fitted with thromboembolic-deterrent stockings, advised on rehydration and early mobilization. Standard prophylaxis of subcutaneous low-molecular-weight heparin will not be adequate, and larger doses are often necessary. This dose will depend on the individual's BMI and clinical condition.

Adequate analgesia is crucial to allow early ambulation; local anesthetic infusions can be used. Many obese patients have degenerative joint disease, so early ambulation may prove challenging. Patients who are not fully ambulatory before surgery may benefit from extended venous thromboembolism prophylaxis.

Respiratory morbidity is more common in the obese patient. Postoperative hypoxemia, which is experienced more frequently in the obese patient, due to reduced FRC and atelectasis, can be improved with supplemental oxygen, semirecumbent positioning, and chest physiotherapy.

Medicolegal implication

A report from the Royal College of Surgeons in 2012, derived following the search of NHS Litigation Authority (NHSLA) database since 2004, that there was only one claim from an obese patient directly relating to compromised care and inadequate management secondary to obesity. However, 29 claims have been registered by NHS staff for injuries sustained while being involved in the care of an "obese" patient, predominantly owing to manual handling issue [43]. Furthermore, a paper from the United States in 2009 showed that a disproportionate number of staff manual handling injuries and lost workdays resulted from working with morbidly obese patients [44].

The NPSA reported 25 incidents in the 2009–10 period where "delays or limited access to appropriate equipment had the potential to cause harm to bariatric patients" and highlighted care of obese patients as a signal item, an emerging issue from national review of serious patient safety incidents [45]. Data from the United States suggest that patients are becoming aware of this inequality of access to imaging due to their size [46]. There are also legal cases in England where delay in access to MRI owing to patient weight may have had an effect on the outcome, and legal claims are being considered.

The small number of legal cases registered with the NHSLA over the past few years is unlikely to represent the true "injury" rates to both patients and staff [47], and yet more evidence from the United States suggests that obesity is an increasing factor in legal claims [48].

Conclusion

Obese patients commonly have comorbid conditions (e.g., obstructive sleep apnea, CAD, poorly controlled hypertension, or a difficult airway) that can complicate intraoperative and postoperative care management. It is essential doctors caring for overweight patients understand the physiological abnormalities that are associated with obesity. Doctors should consider the need for a particular procedure and the likely benefits and risks. They should have a frank discussion with the woman and involve her in decision-making. A comprehensive preparation and thorough preoperative assessment are essential. Appropriate planning of infrastructure upgrading to allow safe management of morbidly obese patients in an appropriate, safe, and adequately equipped environment is vital in the current upward trend of obesity in the world.

References

[1] Ogden CL, Carroll MD, Kit BK, et al. Prevalence of childhood and adult obesity in the United States, 2011-2012. J Am Med Assoc 2014;311:806–14.

[2] Mahmood T, Arulkumaran S, editors. Obesity – A ticking time bomb for reproductive health. London: Elsevier Insights; 2013. ISBN:978-0-12-416045-3.

[3] COMMITTEE OPINION Number 619. January 2015 (Reaffirmed 2019) Committee on Gynecologic Practice. American College of Obstetricians and Gynaecologists, 2019.

[4] Logue J, Thompson L, Romanes F, Wilson DC, Thompson J, Sattar N. Management of obesity: summary of SIGN guidelines. BMJ 2010;340:154.

[5] Mullen JT, Moorman DW, Davenport DL. The obesity paradox: body mass index and outcomes in patients undergoing nonbariatric general surgery. Ann Surg 2009;250:166–72.

[6] Glance LG, Wissler R, Mukamel DB, Li Y, Diachun CA, Salloum R, et al. Perioperative outcomes among patients with the modified metabolic syndrome who are undergoing noncardiac surgery. Anesthesiology 2010;113:859–72.

[7] Nieboer TE, Johnson N, Lethaby A, Tavender E, Curr E, Garry R, et al. Surgical approach to hysterectomy for benign gynaecological disease. Cochrane Database Syst Rev 2009;(3). Available from: https://doi.org/10.1002/14651858.CD003677.pub4 Art. No: CD003677.

[8] Wattiez A, Soriano D, Cohen SB, Nervo P, Canis M, Botchorishvili R, et al. The learning curve of total laparoscopic hysterectomy: comparative analysis of 1647 cases. J Am Assoc Gynecol Laparosc 2002;9:339–45.

[9] Lean ME. Obesity and cardiovascular disease: the wasted years. Br J Cardiol 1999;6:269–73.

[10] Alexander JK. Obesity and cardiac performance. Am J Cardiol 1964;14:860–5.

[11] Hubert HB, Feinleib M, McNamara PM, Castelli WB. Obesity as an independent risk factor for cardiovascular disease: a 26-year follow-up of participants in the Framingham heart study. Circulation 1983;67:968–77.

[12] Shenkman Z, Shir Y, Brodsky JB. Perioperative management of the obese patient. Br J Anaesth 1993;70:349–59.

[13] Valensi P, Thi BN, Lormeau B, Paries J, Attali JR. Cardiac autonomic function in obese patients. Int J Obes Relat Metab Disord 1995;19:113–18.

[14] Duflou J, Virmani R, Rabin I, Burke A, Farb A, Smialek J. Sudden death as a result of heart disease in morbid obesity. Am Heart J 1995;130:306–13.

[15] Murphy PG. Obesity. In: Hemmings Jr HC, Hopkins PM, editors. Foundations of anaesthesia. Basic and clinical sciences. London: Mosby; 2000. p. 703–11.

[16] DeDivitiis O, Fazio S, Petitto M, Maddalena G, Contaldo F, Mancini M. Obesity and cardiac function. Circulation 1981;64:477–82.

[17] Ray C, Sue D, Bray G, Hansen JE, Wasserman K. Effects of obesity on respiratory function. Am Rev Respir Dis 1983;128:501–6.

[18] Biring MS, Lewis MI, Liu JI, Mohsenifar Z. Pulmonary physiologic changes of morbid obesity. Am J Med Sci 1999;318:293–7.

[19] Luce JM. Respiratory complications of obesity. Chest 1980;78:626–31.

[20] Ray C, Sue D, Bray G, Hansen JE, Wasserman K. Effects of obesity on respiratory function. Am Rev Respir Dis 1983;128:501–6.

[21] Vaughan RW, Bauer S, Wise L. Volume and pH of gastric juice in obese patients. Anesthesiology 1975;43:686–9.

[22] Lam AM, Grace DM, Penny F, Vezina WC. Prophylactic intravenous cimetidine reduces the risk of acid aspiration in morbidly obese patients [abstract]. Anesthesiology 1983;59:A242.

[23] Association of Anaesthetists of Great Britain and Ireland. Perioperative management of the morbidly obese patient. London: AAGBI; 2007.

[24] National Institute for Health and Clinical Excellence. Venous thromboembolism: reducing the risk of thromboembolism (deep vein thrombosis and pulmonary embolism) in patients admitted to hospital. Clinical guideline CG92. London: NICE; 2009.

[25] Balk EM, Moorthy D, Obadan NO, Patel K, Ip S, Chung M, et al. Diagnosis and treatment of obstructive sleep apnea in adults. Comparative Effectiveness Review No. 32. (Prepared by Tufts Evidence-based Practice Center under Contract No. 290-2007-10055-1). AHRQ Publication No. 11-EHC052-EF. Rockville, MD: Agency for Healthcare Research and Quality; 2011.

[26] Tizer K. Extremely obese patients in the healthcare setting: patient and staff safety. J Ambul Care Manage 2007;30.134–41.

[27] Randall SB, Poriesw J, Pearson A, Drake DJ. Expanded Occupational Safety and Health Administration 300 log as metric for bariatric patient-handling staff injuries. Surg Obes Relat Dis 2009;5:463–8.

[28] Booth CM, Moore CE, Eddleston J, et al. Patient safety incidents associated with obesity: a review of reports to the National Patient Safety Agency and recommendations for hospital practice. Postgrad Med J 2011;87:694–9.

[29] National patient Safety Agency. Risk of harm to bariatric patients from delays in treatment – signal. London: NpSA; 2010.

[30] Hignett S, Chipchase S, Tetley A, Griffiths P. Risk assessment and process planning for bariatric patient handling pathways. Bootle: HSE; 2007.

[31] Bostanjian D, Anthone GJ, Hamoui N, Crookes PF. Rhabdomyolysis of gluteal muscles leading to renal failure: a potentially fatal complication of surgery in the morbidly obese. Obes Surg 2003;13:302–5. Available from: https://doi.org/10.1381/096089203764467261.

[32] Sprung J, Whalley DG, Falcone T, Warner DO, Hubmayr RD, Hammel J. The impact of morbid obesity, pneumoperitoneum, and posture on respiratory system mechanics and oxygenation during laparoscopy. Anesth Analg 2002;94:1345–50.

[33] Irvine LM, Shaw RW. The effects of patient obesity in gynaecological practice. Curr Opin Obstet Gynecol 2003;13:179–84.

[34] Eltabbakh GH, Shamonki MI, Moody JM, Garafano LL. Hysterectomy for obese women with endometrial cancer: laparoscopy or laparotomy? Gynecol Oncol 2000;78:329–35.

[35] Heinberg EM, Crawford BL, Weitzen SH, Bonilla DJM. Total laparoscopic hysterectomy in obese versus non obese patient. Obstetrics Gynecol 2004;103:674–80.

[36] RCOG. Preventing entry-related gynecological laparoscopic injuries _ RCOG Green TopGuidelineNo49. London: RCOG; 2008.

[37] Khavanin N, Lovecchio FC, Hanwright PJ, Brill E, Milad M, Bilimoria KY, et al. The influence of BMI on perioperative morbidity following abdominal hysterectomy. Am J Obstet Gynecol 2013;208:449.e1–6.

[38] Incisions for gynaecology surgery. In: Rock A, Jones HW, editors. TeLinde's operative gynaecology. 10th ed. Philadelphia, PA: Lippincott; 2003.

[39] Demaria EJ, Carmody BJ. Perioperative management of special populations obesity. Surg Clin North Am 2005;85:1283–9. Available from: https://doi.org/10.1016/j.suc.2005.09.002.

[40] Classesn DC, Evans RS, Pestotnik SL, Horn SD, Menlove RL, Burke JP. The timing of prophylactic administration antibiotics and the risk of surgical wound infection. N Engl J Med 1992;326:281–6. Available from: https://doi.org/10.1056/NEJM199201303260501.

[41] Pevzner L, Swank M, Krepel C, Wing DA, Chan K, Edmiston Jr. CE. Effects of maternal obesity on tissue concentrations of prophylactic cefazolin during cesarean delivery. Obstet Gynecol 2011;117:877–82.

[42] Wilson AT, Reilly CS. Anaesthesia and the obese patient. Int J Obes 1993;17:427–35.

[43] Butcher K, Morgan J, Norton S. Inadequate provision of care for morbidly obsess patients in UK Hospital. Ann R Coll Surg Engl (Suppl) 2012;94:338–41.

[44] Randall SB, Pories WJ, Pearson A, Drake DJ. Expanded Occupational Safety and Health Administration 300 log as metric for bariatric patient-handling staff injuries. Surg Obes Relat Dis 2009;5:463–8.

[45] National patient Safety Agency. Risk of harm to bariatric patients from delays in treatment – signal. London: NPSA; 2010.

[46] Kaminsky J, Gadaleta D. A study of discrimination within the medical community as viewed by obese patients. Obes Surg 2002;12:14–18.

[47] Medical Litigation Consultants. The role of the medical expert, <http://www.medlit.info/guests/mmpcanadian/medlit.htm> (cited March 2012).

[48] LaCaille RA, DeBerard MS, LaCaille LJ, et al. Obesity and litigation predict workers' compensation costs associated with interbody cage lumbar fusion. Spine J 2007;7:266–72.

Chapter 26

Laparoscopic and robotic surgery in obese women

Manou Manpreet Kaur and Thomas Ind
The Royal Marsden Hospital, London, United Kingdom

Introduction

Health care is facing a growing number of challenges at multiple levels related to the rising prevalence of obesity. Clinicians are now more frequently required to surgically manage obese (and often older) patients, having both benign- and/or malignant conditions. Whereas obesity was once considered to be associated with high-income countries, the differences between low- and middle-income countries are narrowing as Western civilization is expanding. Worldwide obesity has risen almost threefold since 1975 with 39% of adults aged 18 years or older being overweight in 2016 and at least 2.8 million people die annually as a result of being overweight or obese. There are a larger number of obese individuals than underweight [1].

The World Health Organization (WHO) uses a classification system to define obesity (Table 26.1) [3]. According to the most recent Health Survey for England (HSE) nearly two-thirds (64%) of adults were classed as "overweight" [body mass index (BMI) of at least 25 kg/m^2 but less than 30 kg/m^2] or "obese" (BMI > 30 kg/m^2) in 2017. Men were more likely to be *overweight* compared to women (40% vs 31%), whereas women were more likely to be *obese* (30% vs 27%) compared to men. This figure demonstrates nearly twice as many obese women compared to figures from 1993 at 16.4% (Fig. 26.1). Furthermore, 5% of women were "morbidly obese" or in the obesity III category according to WHO classification system (BMI > 40 kg/m^2) in 2017. Younger generations are becoming obese at earlier ages and are staying obese for longer. Thirty percent of children aged 2–15 years in England were overweight or obese and 17% of these were obese [2].

In the surgical literature the class III obesity sometimes has a further stratification, describing "morbid obesity" as BMI of 40–44.9 kg/m^2, "superobesity" as BMI 45–59.9 kg/m^2, and "super-super obesity" as BMI > 60 kg/m^2 [4,5].

In addition to measuring obesity by calculation of BMI, waist circumference (WC) is often considered to evaluate the distribution of body fat (BF) around the abdomen (central obesity), in which a man is obese with a WC > 102 cm and a woman is obese with a WC of > 88 cm. This tool has been recommended by the NICE guidelines on obesity, published in 2014 as well, and is useful to identify individuals with a BMI of < 35 kg/m^2, but who have a high health risk due to central obesity [2,3].

In the 2017 HSE survey, 49% of women had a WC > 88 cm, compared to 26% in 1993 (Fig. 26.1) [6].

TABLE 26.1 Health risk from body mass index (BMI) and waist circumference as illustrated by Health Survey for England [2].

BMI	WHO BMI classification	WC low (<80)	WC high (80–88)	WC very high (>88)
18.5–24.9	Normal	No increased risk	No increased risk	Increased risk
25.0–29.9	Overweight	No increased risk	Increased risk	High risk
30.0–34.9	Obese I	Increased risk	High risk	Very high risk
35.0–39.9	Obese II	Very high risk	Very high risk	Very high risk
>40.0	Obese III (also known as "morbid obese")	Very high risk	Very high risk	Very high risk

BMI in kg/m^2; *WC*, waist circumference in cm.

Obesity and Gynecology. DOI: https://doi.org/10.1016/B978-0-12-817919-2.00026-7
© 2020 Elsevier Inc. All rights reserved.

The American Association of Clinical Endocrinologists (AACE) characterize obesity by BF percentage as >35% in females and >25% in males. This characterization also enables to identify those who have a normal weight obesity (i.e., BMI < 25 kg/m^2, but BF percentage >35) with a similar obesity-related risk profile [2,7] (Fig. 26.2).

Obesity is associated with increased risk for conditions such as diabetes; cardiovascular disease; asthma; and some malignancies, such as colon carcinoma, breast carcinoma, and, especially, endometrial carcinoma in women. The latter is the sixth most common cancer in women worldwide and the most common gynecological malignancy in the developed world. Having a BMI of 30 and <35 kg/m^2 is associated with 2.6-fold increased risk to develop endometrial cancer while this risk increases to 6.3-fold in morbidly obese women. This is etiologically associated with the peripheral conversion of androstenedione to estrone (E1) and suppression of sex hormone—binding globulin production from the liver allowing more free bioavailable estrogens (Figs. 26.3 and 26.4). If the current trend in this progressive rise in obesity continues, obesity could overtake smoking as the biggest preventable cause of cancer among UK women by 2040. It has been demonstrated that the risk for poor health increases sharply with an increasing BMI. Public Health England estimated that the overall cost of obesity to be £27 billion in 2017. The report also estimated that the NHS spent £6.1 billion on overweight and obesity-related ill-health in 2014—15, whereas it is projecting these costs to reach £9.7 billion by 2050 with wider costs to society being estimated to reach £49.9 billion per year [1,2].

Public Health England commissioned the "Whole System Obesity" (WSO) programme in 2015 as an effort to develop a practical guide to help councils create a Whole Systems Approach (WSA) in their local area on creating the right environment for change and collaborative working across the local system with emphasis on tackling obesity to build up a stronger local economy and reduce social care costs. This program is currently ongoing with an update given in July 2018 [6].

Apart from the core anatomical and surgical challenges linked to obesity, gynecological surgeons are also confronted with additional issues associated with the perioperative management of obese women. These are more pronounced in the category of patients who are morbidly obese (BMI > 40 kg/m^2). This subgroup of patients can be seen as a vulnerable population and is often subject to physician bias when decisions are made to perform surgery. In this chapter, we aim to review the literature and

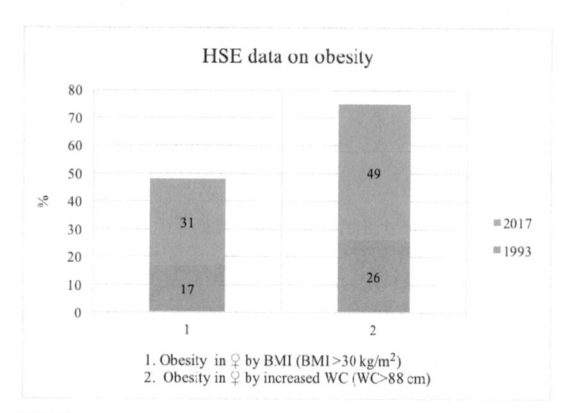

FIGURE 26.1 Increase in obesity in England according to HSE data [2]. *HSE*, Health Survey for England.

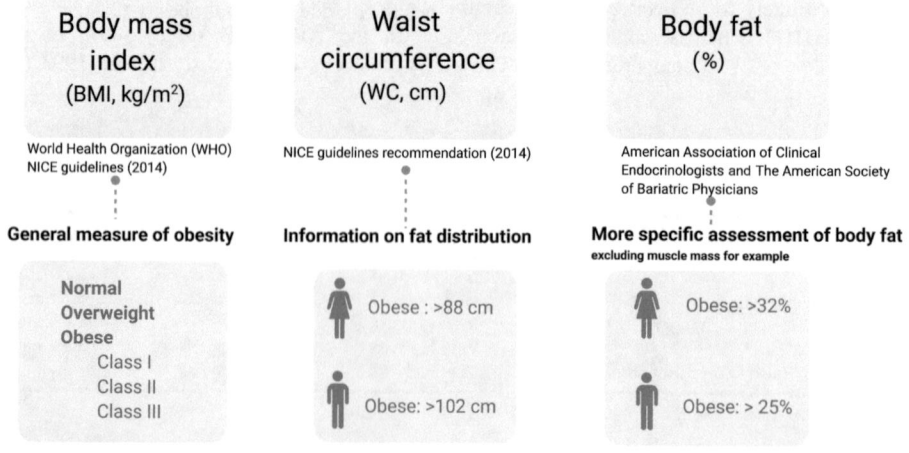

FIGURE 26.2 Different classification systems to measure obesity.

FIGURE 26.3 Increased peripheral aromatase activity and its effect on endometrium.

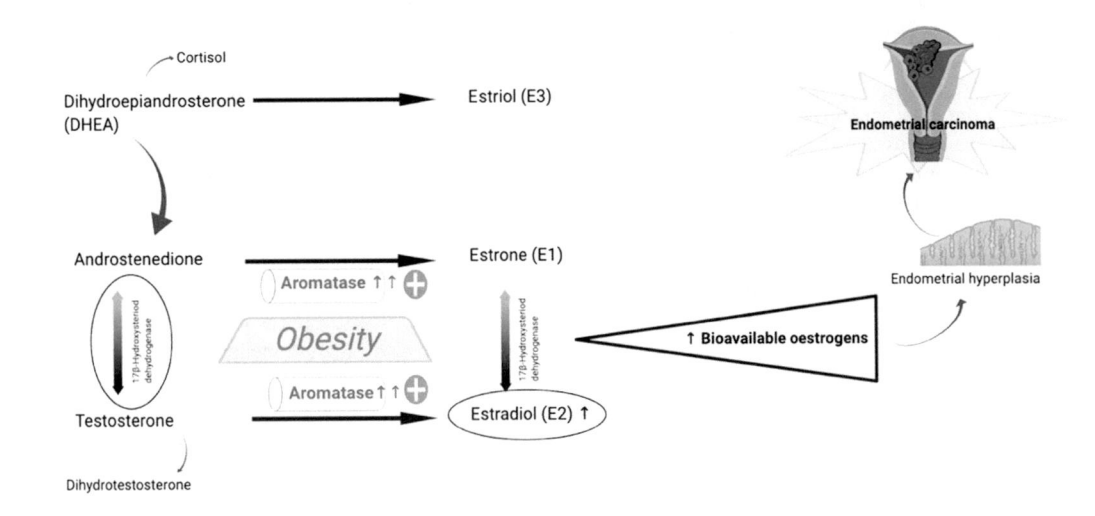

17β-hydroxysteriod dehydrogenase and aromatase enzyme activity is increased by adipocytes. 17β-hydroxysteriod dehydrogenase also catalyses E1 (biologically less active) to E2 and androstenedione to testosterone. This leads to increased concentration of bioavailable free circulating oestrogens.

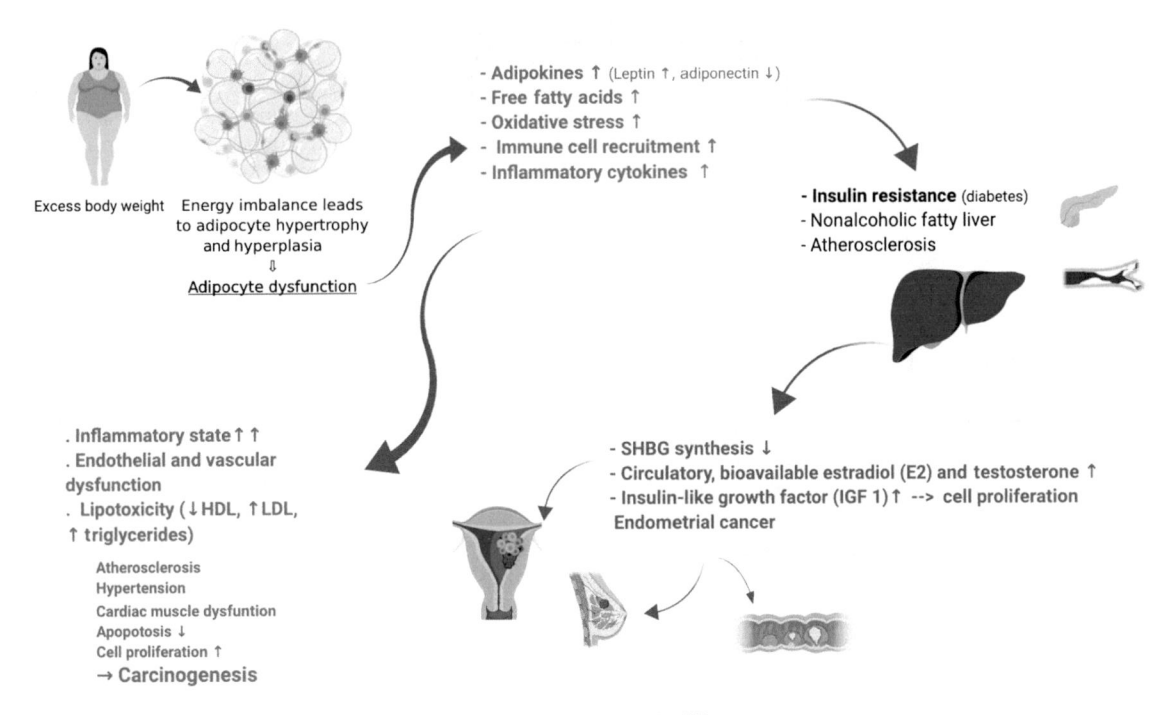

FIGURE 26.4 Obesity induced adipocyte dysfunction and the associated changes in different organ systems.

present the current state of art regarding the evidence of the role of (laparoscopic and) robotic surgery in the population of obese women in gynecology and to discuss the solutions for the emerging challenges perioperatively that this entity brings with it alongside the medical condition for which a woman is treated.

Physiological changes in (obese) surgical patient

Lipotoxicity is the state when there is an imbalance in the fat-regulated homeostasis, leading to cell dysfunction, increased inflammation, and insulin resistance by the metabolically active overabundance "visceral" fat depots. Lipotoxicity associated with centripetal obesity has important implications for the physiological function of various organ systems [8,9] (Table 26.2). Presence of macrophages together with expression of some inflammatory factors has been shown to be more frequent in omental fat ("visceral" obesity) than in subcutaneous fat (peripheral obesity) for example [10].

Cardiovascular alterations

Central "visceral" adipose tissue amassing leads to several cardiovascular and hemodynamic changes with associated physiological abnormalities. Cardio-metabolic variations occur causing antagonistic heart redesigning and ventricular malfunction due to an increase in blood volume, stroke volume, arterial pressure, and left ventricular wall stress. Cardiovascular hypertrophy and pulmonary hypertension lead to worsening of the cardiac function, and form fundamental forerunners of heart failure and arrhythmias in obese individuals [9,11–13].

Respiratory function

Truncal adiposity leads to a decrease in respiratory muscle endurance and chest wall compliance (30% lower in class III obese individuals), resulting in rapid, shallow breathing pattern in the background of increased overall airway resistance (68% higher in class III obese individuals) [14]. The latter leads to an increase in peak inspiratory pressure. The mechanism for these physiological changes is the increased intraabdominal pressure (IAP) in combination with cephalad shift of the diaphragm. As the functional residual capacity (FRC) and expiratory reserve volume (ERV) drop hereby, the mismatch in ventilation perfusion promotes alveolar collapse and atelectasis at the lung bases. The decrease in chest wall compliance can be as high as 60% after pneumoperitoneum is realized [14–16].

Steep Trendelenburg positioning in combination with CO_2-pneumoperitoneum results in a greater arterial partial pressure of CO_2 ($PaCO_2$). The end-tidal CO_2 ($EtCO_2$), however, remains relatively constant and therefore leads to an elevated $PaCO_2$–$EtCO_2$ gradient, which on its turn reflects increased dead space [13,14]. Apart from being associated with obesity hypoventilation syndrome with hypercapnia and disordered breathing, obesity is also a well-established risk factor for developing obstructive sleep apnea (OSA) through mechanical and biochemical changes. It is estimated to be as high as 31% in individuals with $BMI > 35$ kg/m^2 and 48% in adults with $BMI > 50$ kg/m^2. Obesity increases risk of asthma and may aggravate asthma-like symptoms. Pulmonary function tests (PFTs) are valuable tools to assess the respiratory function and can include a maneuver known as maximal voluntary ventilation, which can aid to identify the exponential drop in FRC and ERV with BMI values even <30 kg/m^2 [17–19]. Abdelbadee et al. [20] showed

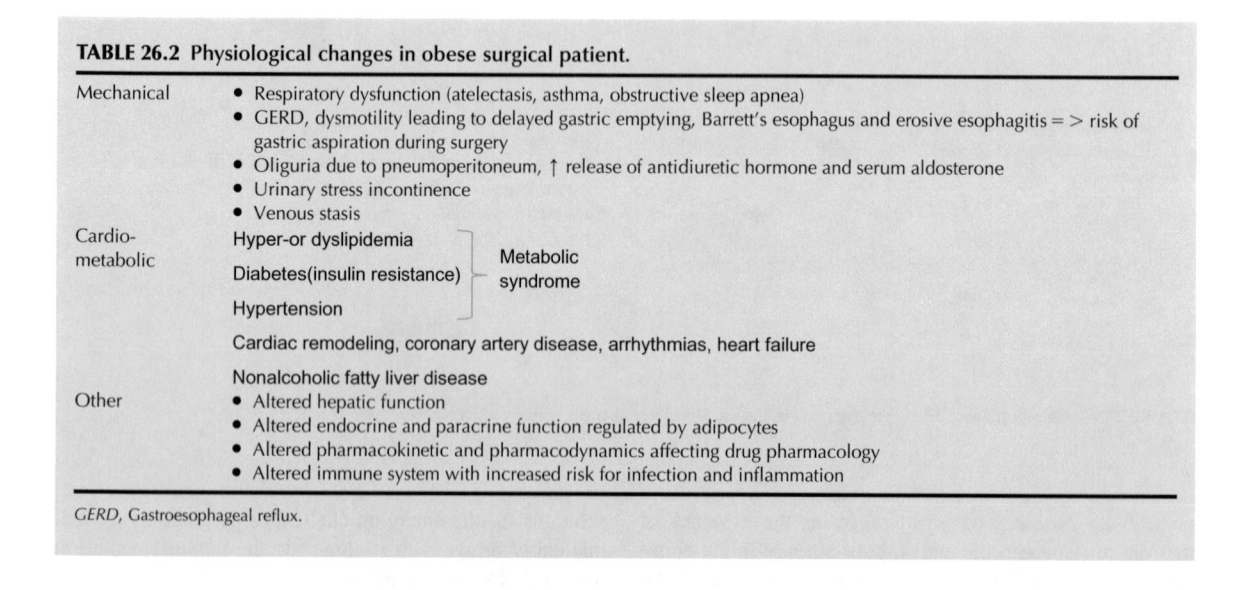

TABLE 26.2 Physiological changes in obese surgical patient.

Mechanical	• Respiratory dysfunction (atelectasis, asthma, obstructive sleep apnea) • GERD, dysmotility leading to delayed gastric emptying, Barrett's esophagus and erosive esophagitis = > risk of gastric aspiration during surgery • Oliguria due to pneumoperitoneum, ↑ release of antidiuretic hormone and serum aldosterone • Urinary stress incontinence • Venous stasis
Cardio-metabolic	Hyper-or dyslipidemia ⎫ Diabetes(insulin resistance) ⎬ Metabolic syndrome Hypertension ⎭ Cardiac remodeling, coronary artery disease, arrhythmias, heart failure
Other	Nonalcoholic fatty liver disease • Altered hepatic function • Altered endocrine and paracrine function regulated by adipocytes • Altered pharmacokinetic and pharmacodynamics affecting drug pharmacology • Altered immune system with increased risk for infection and inflammation

GERD, Gastroesophageal reflux.

that visceral fat volume at L2–L3 best predicts the highest pulmonary tolerance in obese patients.

Effects of CO_2 pneumoperitoneum

Class III obese patients have a 2–3 times higher IAP than that of nonobese individuals. High aqueous solubility and diffusibility of CO_2 enables it to be eliminated through the lungs as it is systemically absorbed across the peritoneum. When intraoperative ventilation is impaired, hypercapnia and acidosis can occur. Hypercapnia can cause cardiac arrhythmias and vasoconstriction of pulmonary vessels, whereas in the presence of acidosis, hypercapnia exerts a depressive effect on myocardial contractility and through stimulation of autonomic nervous system, it can lead to tachycardia [13]. Overall, absorption and excretion of CO_2 in morbidly obese patients appear to be similar to that of nonobese patients. The estimated CO_2 absorption during laparoscopy ranges from 38 to 42 mL/min, which represents 30% increase in the CO_2 load. Appropriate ventilator adaptations commonly avoid hypercarbia [14,18,21,22].

Gastrointestinal changes

The mechanical effects of obesity (e.g., lower esophageal sphincter abnormalities, increased intragastric pressure, larger acidic gastric contents) and altered secretion of adipokines in obese individuals form a risk factor for several digestive diseases, such as gastroesophageal reflux, delayed gastric emptying, Barrett's esophagus, and erosive esophagitis. These predispose the obese patient population to develop gastric aspiration during minimally invasive abdominopelvic surgeries. H2 antagonists and/or prokinetic drugs can be considered to be given as prophylaxis prior to surgery [9,12,21,22].

As obese patients in BMI class III often have preexisting liver conditions, understanding that an increased IAP can reduce the portal venous flow is important, as this can lead to hepatic hypoperfusion and acute hepatocyte injury. A transient elevation of liver enzymes can be seen occasionally [13,21].

Renal function

A decreased urine output during minimally invasive surgeries has been well documented and the amount of oliguria is directly correlated with the level of increased IAP. Pneumoperitoneum can have a direct pressure effect on the renal vasculature, resulting in reduced renal blood flow, whereas intraoperative release of antidiuretic hormone, plasma renin activity, and serum aldosterone may also diminish urine output [11,13]. In class III obese subjects, it can decrease up to 64%. This transient oliguria does not cause harmful effect on the overall kidney function, which is reflected by unchanged levels of serum creatinine and blood urea for example [13].

Venous stasis

Increased IAP during minimally invasive abdominopelvic surgery can reduce the peak femoral systolic velocity by 43% and increase the femoral vein cross-sectional area by 52% [23]. Sequential compression devices have been shown to partially reverse these adverse effects of pneumoperitoneum on femoral venous flow, compared to a complete reversal in nonobese population. Combining these compression devices with prophylactic antithrombotic agents seems necessary to prevent deep venous thrombosis in this subset of patients [13,15].

Drug pharmacology

The altered pharmacokinetics and pharmacodynamics must be considered when managing obese patients, which can lead to impaired drug penetration and/or absorption, altered metabolism, and clearance of drugs administered. These are resulting from a combination of different factors: the greater ratio in body adipose tissue to body water content, altered gut permeability, increased blood volume with altered distribution, including not only hepatic- and renal perfusion but also a modified hepatic- and renal function (steatohepatitis, cholesterol gall bladder stones, initially increased and then decreased GFR due to chronically elevated intraglomerular pressure) [11,15,24].

Immune system

Positive energy balance in obesity leads to adipose tissue hypertrophy and complex adaptive changes in adipocytes, their blood supply, and immunological milieu, which over time causes apoptosis and chronic hypoxia. This leads to an altered secretion of adipokines, increased production of inflammatory markers such as TNF-α and IL-6, an impaired chemotaxis and macrophage differentiation, which can have a negative association with increased infection risk in obese individuals, often manifesting as urinary tract infections or skin- and soft tissue infections (Fig. 26.4). Broad-spectrum prophylactic antibiotics should be considered, especially in morbid obese individuals [11,12,21].

Benefits of minimally invasive surgery

Technological features

Laparoscopic (keyhole) surgery is widely used in gynecology and when compared to open surgery, it offers the advantages of smaller incisions; lower infection rates; reduced postoperative pain and quicker mobilization; decreased blood loss, which then indirectly have a positive impact on health care and society by a shorter hospital stay, faster recovery, and quicker return to workplace hereby also reducing a significant amount of societal financial burden. In addition to this, the minimally invasive approach enhances the surgeon's visualization of the operating field with improved tissue approximation [5,25].

The development of 3D high-definition video technology has been introduced to improve the spatial distance determination in laparoscopy, but it requires polarized glasses to be worn, whereas the angle of the video monitor to the surgeon is critical to obtain the optimal 3D image [26]. European Association for Endoscopic Surgery (EAES) recently published a consensus report on the use of 3D laparoscopic imaging systems by screening through 9967 abstracts and suggests that this technology not only reduces the operative time but also acknowledges that future robust clinical research is required to solidify the potential benefit of 3D laparoscopy. Furthermore, they found a significant reduction in complication rate using 3D approach and in 69 box trainer or simulator studies, more than 50% concluded, trainees were significantly faster and made fewer errors [27].

Laparoscopic surgery is being further revolutionized by the introduction of 4K (and 8K) ultrahigh-definition video technology and new developments in advanced sealing and suturing devices, instruments with 6 degrees of freedom, ergonomic platforms with armrests, and sophisticated camera holders [26].

Lately, the advantages of laparoscopic surgery have been expanded by the introduction of narrow band imaging (NBI) and fluorescence imaging (FI). In NBI, light is reflected from tissue when illuminated with light in the blue and green regions of the visible spectrum. NBI enhances the visualization of superficial blood vessels and, to a lesser extent, the peritoneal architecture. Fluorescent imaging is created when fluorophores in the tissues are excited by being illuminated by blue- or near infrared (NIR) light, which results in emission of light at a longer wave length. "Autofluorescence" imaging exposes the existing differences in biochemical tissue composition by reflecting the signal from fluorophores naturally present in the tissues. In "vector-enhanced" FI, a signal is created by an externally administered fluorophore/drug, which accumulates in the target tissue. Indocyanine green (ICG) with NIR is increasingly used for the detection of sentinel node in cervical and endometrial carcinomas. The sentinel node dissection approach has been formally included in the 2014 National Comprehensive Cancer Network Endometrial Cancer Guidelines. 5-Aminolevulinic acid—enhanced FI or stand-alone blue light autofluorescence laparoscopy and NBI aim to improve the detection of small and nonpigmented endometriosis lesions, and peritoneal implants in ovarian epithelial carcinomas. A small number of studies have reported the use of image-enhanced minimally invasive surgery (MIS) to identify the ureters when ICG or methylene blue is used with NIR FI [28].

Robotic-assisted laparoscopic surgery (RAS) using the da Vinci (Intuitive Surgical Inc., Sunnyvale, California, United States) is a progressively growing innovation as an enhancement for minimally invasive gynecological surgery since its first FDA-approved gynecological cases were conducted in 2005. This relatively novel surgical technique allows the surgeon to execute the procedure from a computer console containing a 3D high-definition vision system, consisting of two high-definition screens where two images are fused by mirror technology, which makes it better than an operative microscope. The system is situated away from the patient and is operated via remote-controlled mechanical arms attached to a robotic platform at the surgical table. Its benefits on patient outcomes are well documented consisting of less blood loss and need of transfusions, reduced analgesic requirements, shorter hospital stay, quicker return to normal activities, and fewer early readmissions. Since recently, new robotic technologies are making their way into the market, which potentially will further refine and/or develop new equipment; some of them have launched their end products such as Senhance (TransEnterix Surgical), which enables digital laparoscopy assisted with Robotic Technology and Versius (CMR Surgical Ltd) robotic system, while others are awaited such as from Medronic and the joint venture of Ethicon and Google (Verb) [19,29].

In the data published by Zakhari et al. [30] on over 10,000 women undergoing RAS or laparoscopic surgery for endometrial carcinoma, it was found that the RAS group had higher comorbidity and was more often obese and more likely to undergo staging. This is one of the examples reflecting a positive change in gynecological surgery offering MIS to this challenging population, despite their high comorbid status and increasing severity of obesity.

Lim et al. [31] reported similar favorable data from their large cohort of hysterectomies for both benign and malignant diseases by different routes: abdominal, vaginal, laparoscopic, and robotic-assisted laparoscopy, that the robotic cohort encountered fewer intraoperative complications compared to vaginal and open approaches, and fewer postoperative complications when compared to all the other cohorts, despite consisting of more "complex" pathology (e.g., pelvic adhesions, obesity, and large uteri).

Furthermore, RAS enhances the surgical field of vision by providing the surgeon with a stable, 3D view and an increased instrumental degree of freedom using the Endowrist Instruments consisting of seven degrees of movement, hereby mimicking the surgeon's hand and wrist to allow working around corners by improving dexterity. Furthermore, it eliminates the fulcrum effect of straight stick surgery (i.e., laparoscopy) and provides tremor filtration allowing more precise movements to enable microsurgical dissection, which are a result of up to 10 times reduced movements of surgical instruments [26,32]. An additional advantage for the surgeon is the possibility to get trained and/or train utilizing a dual console, whereas it is well established that the learning curve to perform a task is much shorter [32,33].

Single-incision laparoscopic [34–36] and single-incision robotic-assisted [37,38] procedures by the da Vinci platform have been reported to be feasible and safe although this approach does offer technical challenges for the surgeon as it limits the triangulation of instruments. This limitation is less in the robotic platform. The utilization of this technique both in laparoscopy and RAS is merely in descriptive context by small case series and needs yet to be established at a larger scale.

Transvaginal natural orifice transluminal endoscopy has been described by some authors to perform not only diagnostic evaluation of the pelvis but also a total hysterectomy for example [39–43]. This approach warrants further validation as the practice has been limited to only a very few centers worldwide. There is no data on its feasibility in obese patients.

Indications

The gynecological conditions managed by MIS are endless and range from a diagnostic evaluation of the pelvis to salpingectomy, ovarian cystectomy, oophorectomy, excision of endometriosis, (sub)total hysterectomy, sacrocolpopexy, myomectomy, abscess drainage, pelvic (and/or paraaortic) lymph node dissection with or without sentinel lymph node(s), transabdominal cerclage as the current mainstay.

The first laparoscopy in pregnant patient was a cholecystectomy in 1991. There is little research on MIS in specifically obese pregnant women, but literature has demonstrated that laparoscopic procedures in pregnant patients can be executed safely during any trimester without any risk to mother or fetus. However, a careful preoperative, intraoperative, and postoperative planning is mandatory to optimize both maternal and fetal outcomes. General principles apply for the obese pregnant patient requiring a laparoscopy with regard to most of the precautions undertaken to prevent comorbidity, such as thromboembolic prophylaxis [44,45]. The incidence of adnexal masses during pregnancy is around 2%, and the current evidence supports close observation when ultrasound features are reassuring with a normal CA 125, LDH, and the patient is asymptomatic. Surgery can be necessary if an ovarian torsion occurs, which with a delay in diagnosis may lead to peritonitis, spontaneous abortion, and/or preterm delivery. Progesterone therapy is warranted if the corpus luteum is removed in the first trimester [45].

The feasibility of laparoscopic management of obese women diagnosed with tubal ectopic pregnancy (with or without hemoperitoneum) when compared with nonobese has been reported with no difference in complication rate, including a similar conversion rate [46,47]. A case report has reported the feasibility to perform a total hysterectomy with bilateral salpingoophorectomy on a patient with a BMI of 98 kg/m^2 using the da Vinci robotic system [48].

Minimally invasive surgery and obesity: laparoscopy–robotics

Most of the studies evaluating the outcomes of obesity after MIS are nongynecological in nature and the large majority of the existing data on obese women undergoing MIS comes from retrospective gynecological oncology patients. When compared to open abdominal surgery, MIS clearly offers a great benefit for obese patient population by reducing the rates of (wound) infections, abdominal wall hernias, postoperative ileus, and susceptibility to thromboembolic events, which can be life-threatening [49–55].

Corrado et al. [56] demonstrated ($N = 655$), comparing laparoscopic surgery with RAS in obese patients for endometrial cancer, that the RAS group had a lower conversion rate compared to laparoscopic group with the same BMI (0.8% vs 3.7%, $P = .024$). The conversion rate was higher for patients with BMI ≥ 50 kg/m^2 in laparoscopic group, whereas in the RAS group, the conversion rate significantly reduced as the BMI increased. Furthermore, the RAS group was able to perform twice as more pelvic lymphadenectomies and had a significantly shorter postoperative hospital stay. The operating times and total blood loss was, on the other hand, longer in the RAS arm [56]. Others have reported similar lower conversion rates to laparotomy in the RAS group of all BMIs (including class III obese patients) compared to laparoscopic group [50,51,57,58]. In the surgical treatment of endometrial cancer in obese patients, for example, RAS approach is becoming the mainstay surgical modality [19,57,59–61].

Operative exposure and access to deep pelvic structures can be limited in open surgery in obese patients, which is also the case in MIS. The feasibility to perform MIS in obese patient population had to be proven in the past as it was once considered as a contraindication due to the many challenges faced in this subgroup of patients, such as difficulty to gain intraabdominal entry, creating and maintaining a pneumoperitoneum, intolerance to a Trendelenburg position, poor vision of operative field due to excess amount of intraabdominal fat, problematic manipulation of instruments and trocars due to increased thickness of the abdominal wall, and a large panniculus changing anatomic landmarks. In addition to this, minimally invasive abdominopelvic surgery in obese patients has been associated with a higher complication rate, especially ureteric injuries, longer operating times, and conversion to open surgery [49,62,63]. RAS has been proposed to overcome some of these challenges due to its advantages of three-dimensional stereoscopic vision, wristed instruments improving dexterity and tremor canceling software improving surgical precision, whereas the resistance caused by an enlarged abdominal wall with movements of laparoscopic instruments is eliminated.

Intraabdominal visceral fat is a predictor for early laparotomic conversion in obese subjects with endometrial cancer [64]. A prospective randomized trial showed a high conversion rate of 25.8% despite having low median BMIs (28 and 29 kg/m^2) in women undergoing laparoscopic treatment for endometrial cancer [55]. A study of 497 patients evaluated laparoscopic surgical feasibility in nonobese (58%), obese (33%), and morbidly obese (9%) patients undergoing gynecologic oncological procedures and found that although the category of morbidly obese patients had longer operative times, conversion to laparotomy was comparable as were the low intraoperative- and severe postoperative complications rates [53]. Similar results have been echoed by several other authors showing a lower conversion rate even in obese patients [33,49,58,59,61,65].

A recent systematic review and metaanalysis on conversions and complications in laparoscopic and robotic hysterectomy for endometrial cancer in obese patients reported a similar conversion rate in both arms for BMI $\geq$ 30 kg/m^2 but a statistically significant higher conversion rate in the laparoscopic arm for BMI $\geq$ 40 kg/m^2, and it found that this was attributed to positional intolerance in 31% of laparoscopic cases versus 6% in the robotic arm. The authors concluded that robotic approach may offer benefit over laparoscopy when obesity is encountered, which is estimated to affect over 50% of the endometrial cancer patients [58]. Others have reported conversion rates in the obese RAS group of as low as 0.6% [66]. The proposed advantages of RAS in the obese subgroup of patients have been suggested by others as well with at least a similar conversion rate to laparoscopic groups [12,19,50,59,60,67].

Obtaining and then maintaining a pneumoperitoneum can be challenging to achieve, due to not only the increased abdominal wall thickness and preperitoneal fat but also respiratory dysfunction in the obese while requiring a Trendelenburg position. A metaanalysis showed a lower pulmonary morbidity for laparoscopic surgery in obese patients when compared to open procedures (1.6% vs 3.6%) [68]. Wysham et al. [66] reported that only 3% of patients with a BMI $\geq$ 37 kg/m^2, who underwent a robotic-assisted hysterectomy, had pulmonary complications, with an association tied only to age. It has been demonstrated that robotic-assisted hysterectomy procedure for example is feasible with less Trendelenburg position and/or lower levels of pneumoperitoneum at pressures of 10–12 mmHg [69]. Angioli et al. [70] solidified this finding in their trial demonstrating a linear relationship between surgical field visualization and the IAP in conventional laparoscopy, whereas in the RAS arm, there was no significant difference noted in the surgical field visualization at lower pressures, even when decreasing it down to 10–5 mmHg.

A great wealth of information confirms that minimally invasive gynecological surgery in obese patients, both by laparoscopy and robotic-assisted approach, is feasible, safe and significantly improves patient outcomes when compared to open surgery, making it cost-effective. The literature stays heterogenous about the extent obesity has on surgical parameters, such as operating time, but is consistent on higher rate of surgery-related comorbidity in this specific group of women, especially the morbidly obese. This makes it important to ensure a careful multidisciplinary preoperative planning as most technical challenges can be overcome by well-trained and dedicated surgical teams at all levels of the provided perioperative patient care.

Alternatives for class III (morbidly) obese patients

Although MIS has been shown to be feasible in morbidly obese patients, in some occasions, when on balance, the comorbidity profile is significant apart from morbid obese status for example, it may be considered to treat these patients with nonsurgical modalities. In benign conditions, progesterone-releasing intrauterine device (IUD) such as Mirena/Levosert or less invasive treatments such as (outpatient) endometrial ablation can offer adequate symptom control for menorrhagia. Similarly, other nonsurgical management options can be considered, not only to treat menorrhagia but also to alleviate symptoms associated with fibroids, endometriosis, and prolapse, such as gonadotropin-releasing hormone analogues, ulipristal acetate, uterine artery embolization, and/or pessaries. Megestrol acetate can be used to treat atypical endometrial hyperplasia in combination with a Mirena IUD, while in exceptional cases the patient may be referred to have a bariatric procedure in order to prevent a high peripheral aromatization to estrone, which can be the mainstay of endometrial cancer in obese women.

Cost-effectiveness

There is a lack of level-1 evidence in the literature addressing the exact role of RAS [71]. At present, cost-effectiveness studies have shown laparoscopy to be the least expensive MIS approach for a hysterectomy when compared to robotic and open approach [60]. The subgroup of obese patients is associated with higher costs in all the three arms of surgical modalities, open, laparoscopic, and robotic assisted [49,72,73].

Zakhari et al. [30] demonstrated the higher cost associated with RAS, showing a difference in median charges of $38,161 versus $31,476 in those undergoing RAS versus laparoscopic surgery ($P < .0001$). This is consistent with other reports from the literature suggesting about 10%–15% higher cost of RAS [60,74,75]. The open laparotomy approach has been shown to be the highest in overall cost [59,76,77]. A systematic review, including studies from 2008 to 2016 by Iavazzo and Gkegkes [33], found reports suggesting RAS to be 1.6 times more expensive than conventional laparoscopic surgery. However, when a subanalysis of different costs is performed, Venkat et al. [78] demonstrated that despite the overall higher cost of RAS in endometrial cancer at $64,266 versus $55,130 for laparoscopy, the reimbursement to the hospital was statistically not different: $13,003 for RAS and $10,245 for laparoscopic approach. Equally, there were no differences in reimbursement to the surgeon ($P = .74$) or anesthesiologist ($P = .84$) between the two surgical modalities. Furthermore, robotic hysterectomy costs less than a laparotomy when societal costs are included, whereas the costs of robotic surgery decrease with increasing procedural volume and, additionally, are positively correlating to the use of virtual reality simulator [33,67,79]. This was demonstrated by an excellent economic analysis of robotic-assisted hysterectomies in 10,906 endometrial cancer patients, where the cost differential between laparoscopic and robotic approach decreased with increasing hospital volume from $2471 for the first 5–15 cases to $924 for more than 50 cases. When data was analyzed on the basis of surgeon volume, RAS for endometrial cancer was $1761 more expensive than laparoscopy for those who had performed <5 cases and this differential declined to $688 for >50 RAS procedures when compared with laparoscopic hysterectomy [75]. Similarly, the per protocol subanalysis comparing laparoscopic to RAS for hysterectomy in the RCT ($N = 122$) published by Lönnerfors et al. [67] in 2016, showed a similar cost in both approaches ($7059 vs $7016 for RAS). Ind et al. [80] equally reported in their prospective evaluation that introducing a robot can lead to a reduction in overall cost by reducing the total number of open procedures. Similar results were reported by Leitao et al. [59], especially for the morbidly obese subgroup of patients.

A prospective RCT by Deimling et al. [81] demonstrated the noninferiority of RAS with regards to operating times when compared to conventional laparoscopy performed by surgeon experienced in both techniques (73.9 minutes in RAS vs 74.9 minutes in conventional laparoscopy). This study is one of the very scarce studies reporting on the total operative time from surgeon incision to surgeon stop, including the robot docking time. This study echoed the findings by Mäenpää et al. [61] who showed a shorter total operative time in RAS group compared to traditional laparoscopy for treatment for endometrial cancer with no conversion cases. Furthermore, a few authors have emphasized to recognize the difficulty to evaluate a true cost-effectiveness in RAS due to bias created by reports, including procedures performed during the initial learning curve of the surgeons, the underreported proportion type II error to identify a difference when one truly exists, lack of recognition on the trend seen in RAS being performed in more complex patients (including more obese), misclassification of conversions (sometimes performed to retrieve specimen), lack of attention given to surgeon's physical fatigue during difficult straight stick laparoscopy, and its effects on surgeon's performance [67,81–83].

Different ways to reduce the overall costs with RAS have been proposed by several authors. Collaboration between different specialties and hospitals can identify the cost-effective use of robotic systems and push manufacturers to reduce the purchase and procedure costs. A correct selection of surgical candidates such as obese (and elderly) patients with comorbidities, cases requiring a microsurgical and/or fine dissection such as severe endometriosis can minimize the costs because of the advantages at intraoperative surgical level and postoperative shorter stay in hospital with less analgesia requirements. Implementation of robotics in high-volume centers with improved training of dedicated robotic teams and careful utilization of robotic instruments by multiple surgical specialties are another ways to maximize cost-effectiveness [51,76,77,84].

Complications

Some authors have demonstrated that despite longer operative times in the morbidly obese patients, the increased BMI is not an independent risk factor for bowel or urinary tract injury [12,53]. Others, however, have reported an increased risk of adverse events in obese women undergoing gynecological MIS [49,62,85]. According to the current data in the literature, the proportions of complications such as organ/vessel injury, venous thromboembolism (VTE), and blood transfusion are low and not appreciably different in both laparoscopic and robotic approaches [58]. Complications such as wound breakdown, respiratory challenges, cardiac complications, and difficult intubations are associated with obesity. The surgical outcome worsens as the level of obesity increases; however, minimally invasive techniques lead to fewer complications compared to open procedures [86].

About 50% of MIS-related injuries are known to occur during the initial entry steps, which in the obese patient population can be challenging [87–89]. Major complications such as bowel perforation or major abdominal vessel injury at this stage have an overall reported incidence of 0.04%–0.5% and 0.04%–1.0%, respectively [87,90]. Waggling of the Veress needle must be avoided, as it can enlarge a 1.6 mm accidental puncture injury to an injury of up to 1 cm in viscera or blood vessels. Injury to retroperitoneal space or structures can occur when excessive force is used to insert the Veress needle or during retroperitoneal dissection. Visual entry trocars do not avoid visceral and vascular injury but do have the advantage of minimizing the size of entry wound and reducing the force necessary for insertion [91]. Type 1 injuries are caused by damage to major blood vessels or the bowel in a normal location during entry, whereas type 2 injuries are a result of damage to abdominal wall vessels (e.g., inferior epigastric vessels and superficial vessels) and to the bowel adherent to the abdominal wall. About 30%–50% of the bowel injuries and 13%–15% of vascular injuries are not detected immediately intraoperatively. Persistent pyrexia, tachycardia, or ileus with high inflammatory parameters and/or peritonitis signs should raise the suspicion of bowel injury. Most intraoperative bowel lesions can be sutured by a single layer, interrupted serosubmucosal 3-0 Vicryl or 4-0 PDS suture, but some may require a partial excision with a primary reanastomosis or occasionally a temporary stoma [87].

Urinary tract injuries are seldom entry related and unlike major vessel and iatrogenic bowel injuries, they are rarely associated with mortality. The etiology can range from inadequate knowledge of pelvic anatomy or of energy devices used, imprecise application of stapling devices, lack of surgical skill to open peritoneum and/or dissect retroperitoneally, or severely distorted anatomy due to dense pelvic adhesions. Ureteric injuries often are discovered in the late postoperative phase (e.g., urinoma appearing several days postoperatively). Postprocedure cystoscopy with intravenous injection of indigo carmine is increasingly used to demonstrate ureteric patency; however, this can still provide a false negative result in cases of ureteric injury secondary to thermal damage or where there is a partial obstruction. During repair of a bladder injury, a cystoscopy with insertion of a double-J catheter can be helpful to avoid additional injury to the trigone of the bladder [9,11,12,53,92].

The risk of port site hernias is lower when bladeless trocars are utilized and when a transverse incision is made. The incidence is low at <3% in ports of ≤ 8 mm and increases to 86% when left unclosed in ports ≥ 10 mm [93].

Case reports have reported postoperative vision defect or vision loss, attributed to ischemic optic neuropathy due to high venous pressure and interstitial edema resulting in decreased blood flow during steep Trendelenburg position, but interestingly, no such cases have been reported following gynecologic surgery. The intraocular pressure that increases during insufflation of the abdomen and due to the Trendelenburg position is reported to reach a plateau after 30–60 minutes [94,95].

Preoperative preparation

A comprehensive evaluation and counseling of the patient prior to her surgery is essential along with a meticulous theatre preparation and postoperative care plan on the ward to achieve an optimal surgical outcome. Preoperative assessment and surgical planning are excellent tools herein and can be further supported by the development of guidelines for obese surgical patient, which are tailored to the resources of the individual units and which should have a timely review to highlight any challenges or deficiencies to be dealt with.

General considerations

A correct indication to perform surgery for the most benefit of the patient is imperative, whereas the woman should be actively involved in the decision-making process. She should be fully informed about the additional risks imposed by obesity, which also includes discussing specific medical conditions that may further increase her surgical risk in the presence of obesity. Patient should understand that certain technical and practical difficulties can be encountered by the teams involved in her perioperative care, which may lead to a different care plan than initially stipulated; for example, a need to place a central line or conversion to laparotomy. Preoperative counseling may include discussion on weight loss, referral to a dietician or bariatric surgical specialist. Hereby, it is important for all the team members to be sensitive and responsive to their needs and treat them with respect and dignity. Communication with the patient's primary care physician/general practitioner and/or other relevant specialists is advisable to obtain an identification and stratification of the potential health risks, which could manifest perioperatively. This will also be helpful to outline an optimal postoperative clinical pathway.

All patients should have their height and weight recorded and BMI calculated. Baseline bloods should include full blood count, serum electrolytes, renal- and liver function tests, and glucose ($\pm$ HbA1c).

Pulmonary and cardiovascular assessment

Patients with cardiopulmonary dysfunction and/or class III obesity must be seen by the anesthesiologist [11,15]. An electrocardiogram (ECG) is recommended for all obese patients but is obligatory in patients with one or more risk factors for coronary heart disease (obesity is an independent risk factor), as it can highlight ischemia, arrhythmias, and cardiac hypertrophy [11]. Heart rate and mean arterial blood pressure commonly increase during minimally invasive abdominal surgery. An ECG evidence of right atrial hypertrophy, history of marked exertional dyspnea or morning headaches may indicate the presence of OSA, even with a low STOP-BANG score, need a prompt referral. Continuous positive airway pressure (CPAP) may be required and, if so, should ideally start preoperatively. If patients are CPAP tolerant and have been adequately treated for sleep-disordered breathing, they generally do not require high dependency care and can even be suitable for day surgery [15]. There is controversial evidence in the literature suggesting routine execution of PFTs in otherwise healthy obese patients. Patients with poor exercise capacity (less than four metabolic equivalents, e.g., unable to climb a flight of stairs, perform activities of daily life without symptoms) require a chest X-ray and an echocardiogram with a review by cardiologist [21]. Cardiopulmonary exercise test (CPET) can be considered to predict who is at high risk for postoperative complications but may require availability of recumbent bikes. An arterial blood gas analysis should be considered in the presence of respiratory wheeze at rest, O_2 saturation <95% on air, serum bicarbonate concentration >27 mmol/L, a forced vital capacity <3 L or forced expiratory volume in 1 second <1.5 L. An arterial pCO_2 > 6 kPa indicates a degree of respiratory failure and consequently likelihood of increased anesthetic risk [15]. Airway assessment is valuable as obesity is associated with 30% higher chance of difficult/failed intubation; a neck circumference >60 cm indicates a 35% probability of difficult laryngoscopy [11,15].

Additional considerations

It is recommended to screen for type 2 diabetes mellitus if the obese patient is >45 years old. Extrapolation from bariatric literature suggests that surgical risks are significantly increased with a HbA1C of >8.0% or in uncontrolled diabetes. Such patients may benefit from delayed surgery for medical optimization [15,21].

Peripheral vascular access should be evaluated as some may require invasive monitoring; any infections, including skin conditions such as folliculitis, should be treated preoperatively. In the setting of treatment of benign conditions, an endometrial sampling is recommended in women over 40 years old if they are having abnormal uterine bleedings as there is increased risk of hyperplasia [12].

Obese patients undergoing surgery for longer than 45 minutes and not at risk of major bleeding have a moderate (3%) risk for VTE and require prophylaxis with intraoperative sequential compression devices or perioperative pharmacological administration of low-molecular-weight heparins in "benign" disease and <40 years age; both mechanical and pharmacological prophylaxes are required in the case of malignancy, age >40 years, pulmonary dysfunction, and personal history of VTE or thrombophilia [12,96]. Dose adjustments will be required for patients weighing >100 kg [15]. The AACE, The Obesity Society (TOS), and the American Society for Metabolic and Bariatric Surgery (ASMBS) suggest that transdermal and oral estrogen therapy should be discontinued one cycle prior to surgery in premenopausal women and 3 weeks prior to surgery in postmenopausal women [92]. It is essential that if thromboembolic device stockings are used in the obese, they are fitted correctly to avoid vascular occlusion. Routine use of IVC filter is not supported by the current evidence [97,98]. In cases where it is considered, removal should take place within 3 months [21].

There is no consensus on the role of routine bowel preparation. According to some reports, it does not seem to improve the operative field (even in BMI > 50 kg/m^2) and that in addition, it can exacerbate underlying cardiovascular and renal dysfunction, precipitating postoperative complications [11,99]. However, some authors recommend a low-residue diet for 5–7 days preoperatively in low-complexity procedures to additionally giving a mechanical bowel preparation or an enema the night before/a few hours prior to surgery in cases where there is meticulous dissection required in pouch of Douglas and pararectally, such as deep infiltrating endometriosis [100–102].

The value of preoperative nutritional screening has been suggested by the guidelines of AACE, TOS, and ASMBS, as a good nutrition is important to decrease morbidity and mortality postoperatively due to surgical stress leading to protein malnutrition, ureagenesis, and accelerated protein breakdown in obese patients. Deficiencies in iron, thiamine, vitamin B12, and vitamin D are common in this patient group and should ideally be corrected preoperatively [21,103].

An up-front liaison with the theatre coordinator is recommended about the requirement of appropriate equipment, such as a suitable operating table, trolley and hospital bed (for postoperative patient transfer), stirrups, extra-long instruments, etc. (see next). A standard operating table can accommodate a body weight of 130–160 kg and most theatres will have tables available to support a body weight up to 300 kg and average stirrups can support a weight of 500 lb (227 kg) (Allen Medical Systems, Inc.) [5,96]. Patient's dignity is important, so suitable size of theatre gowns should be available. National Patient Safety Agency highlighted a clinical governance issue that many of the incidents related to obesity reported in a 2011 review were involving inadequate provision of suitable equipment [104]. Surgical teams should have training on managing obese patients, which also includes a safe transportation and moving of class III obese patients, whereas also in trouble shooting to any equipment failures.

> **Key points—preoperative preparation**
> - A correct indication to perform a minimally invasive surgery procedure is paramount, whereas a thoroughly explained informed consent form should be completed with an active involvement of the obese woman in decision-making, hereby understanding the additional potential risks and side effects related to her obesity profile.
> - The care providers need to ensure they remain sensitive and responsive to the needs of the obese patient and a liaison with patient's primary care provider (GP or other specialists involved) can be of significant value.
> - Where possible a preoperative optimization of patient's comorbidity issues should be organized to ensure a pathway leading to a favorable postoperative outcome in the long run.
> - All class III obese individuals and patients with known or suspected cardiopulmonary disease should be seen by the anesthesiologist and appropriate additional investigations should be completed before scheduling surgery. An electrocardiogram is essential in all classes of obesity.
> - A cardiopulmonary exercise test can be considered to help identify individuals who could be high risk for postoperative complications.
> - Airway and peripheral vascular access should be evaluated; nutritional screening could be considered to improve postoperative morbidity and mortality.
> - Endometrial sampling is recommended in ♀ ≥ 40 years presenting with presumed benign uterine conditions.
> - There is no consensus on the role of routine bowel preparation; a low-residue diet 5–7 days prior to surgery is often considered in low-complexity procedures.
> - A timely, up-front liaison with the theatre team is advised to ensure access to the required suitable equipment and armamentarium for the surgery to take place.

Intraoperative considerations

Theatre staff, operating room, and equipment

It is crucial that the surgical team is familiar with the additional requirements and bariatric instruments potentially being utilized. A team brief before the start of a theatre list and a consecutive discussion about the case with all the team members involved during the WHO surgical checklist will help ensure the presence of appropriate equipment for both anesthetic and/or surgical teams, which will prevent a slow setup and out-of-reach equipment causing delay. Extra time requirement, potential need for additional personnel and seniority of the team leader should be taken into consideration. A reduced operative time by an experienced surgeon will help limit the intraoperative morbidity. In addition to weight-bearing equipment such as operating table, other equipment requirements may consist of a suitable blood pressure measuring cuff, CPAP or other high-flow oxygen delivery device, raised step for anesthetist, gel pads and padding for pressure points (risk of nerve injury and rhabdomyolysis of the gluteal muscles leading to renal failure), extra pressure reducing mattresses, long spinal and epidural needles, portable ultrasound machine, readily available difficult airway equipment, extra-large compression stockings and intermittent compression devices, additional arm-boards and width table extensions with attachments for positioning [5,15].

Patient positioning

An optimal surgical positioning required for a good surgical access into the female pelvis needs to be balanced with minimum risks to the patient such as hemodynamic instability, impaired ventilation, or musculoskeletal injury. The upper airway should be accessible at all times and there must be a plan for tracheal intubation if required [15,102]. A Mayo stand, placed a few inches above the nose of patient, will protect patient's face, especially when ports are placed superior to the umbilicus and for example in RAS, the robotic camera can come into contact of the face. There are commercially available "face shields" as well [11,12,35,102]. Obese patients are a higher risk to develop pressure ulcers and nerve injuries. Prolonged compression for >6 hours can result in permanent injuries with ulnar- and sciatic neuropathies occurring most frequently [22,100]. An anti–slide gel or foam pads (commercially available such Pink Pad—Pigassi Patient Positioning System, NINOMED Safe-T-Secure) and surgical bean bags can be utilized to prevent patient sliding on table and minimize the risk of nerve injuries. Placing the arms in extended position should be avoided due to the risk of brachial plexus injuries. With both arms tucked at the patient's sides, the hands and elbows should be padded. It should be ensured that all intravenous access sites are adequately secured and accessible. The literature does not support the use of shoulder braces due to evidence of compression and stretch injuries of brachial plexus. A padded strap across the chest instead is recommended instead [12,100]. In addition to intermittent pneumatic compression devices, padded stirrups with extra padding around pressure points such as ankles and knees to minimize the risk of VTE [15,100,102].

Lithotomy is the most commonly used position in pelvic surgery and is defined as a supine position of the body with lower extremities separated, flexed, and supported in stirrups. Whereas in normal BMI patients the recommendation is to move the patient down the table until the buttocks lie just beyond the edge of the lower table break, it is advisable in obese patients to position the buttocks of the patient slightly lower than the edge of the table as when Trendelenburg position will be carried out the body will shift cephalad with the weight of the panniculus [12]. The position of the buttocks is one of the key elements of a smooth surgical procedure as a too high position will make it difficult to gain access vaginally with a Sims speculum for example, whereas, on the other hand, if the buttocks are overhanging, it will have a risk of patient slipping caudally. Correct use of stirrups with simultaneous elevation of the legs will avoid dislocation of the hip joint, minimize risks to other extremity, help avoid rotational stress on the lumbar spine and maintain the limbs in a symmetrical aligned position. The angle of flexion and external rotation will depend on the procedure being performed [101,102,105].

Lithotomy position is combined with a Trendelenburg position, which is a head-down position (classically a 45-degree tilt) to move abdominopelvic viscera superiorly and expand the surgical working area. In cases where access is required from both abdominal and perineal aspects, Lloyd-Davies position is used, which is defined as a supine position of the body with hips flexed at 15 degrees as the basic angle and a 30-degree head-down tilt. The key difference between lithotomy and Lloyd-Davies is the degree of hip and knee flexion [105].

Trendelenburg positioning can have deleterious effects on cardiopulmonary function in obese patients as described earlier by a combination of factors: a further reduction of the already decreased FRC and respiratory compliance, a further increase in respiratory resistance, whereas the oxygenation is impaired. Use of CPAP at 10 cm H_2O can help prevention of atelectasis [11,14,16,66]. Afors et al. [100] described in their recommendations for obese patients undergoing MIS to have a specific sequence in establishing Trendelenburg to minimize these effects in which a 15 mmHg pneumoperitoneum is established followed by a gradual 30-degree Trendelenburg position. The surgeon can then displace the bowel beginning from the caecum, followed by the last ileal loop above the sacral promontory and subsequently the pressure can be dropped to 12 mmHg after which the degree of Trendelenburg is adjusted to be maintained at a position where the bowel can stay reclined out of the pelvis from the point of pelvic brim [100].

A "tilt test" before draping is recommended in RAS because the robot is stationary and sliding will put tension on the port sites. The degree of Trendelenburg should be decreased to the level that is necessary to operate safely and lower abdominal pressures can be considered to aid the tolerance of Trendelenburg. In the event of an elevated CO_2 level in the middle of the surgery, the abdomen needs to be desufflated, the robotic arms undocked, and position should be taken out of Trendelenburg while the $etCO_2$ levels improve.

The Foley lap lift technique has been described using a 14-French Foley catheter inserted through the abdominal wall and tied to a retractor on the foot of the bed, which helps to reduce the pneumoperitoneum pressure [22]. The panniculus can be taped or weighted away from surgical field (laterally to the thighs for example); however, care should be taken to prevent skin trauma and/tissue necrosis [12,22].

Key points—patient positioning

- Upper airways should be accessible at all times; plan for unexpected tracheal intubation.
- Pressure areas should be protected with appropriate wrapping to avoid pressure ulcers and/or compression nerve injuries. Specialized anti–slide gel or foam pads can prevent sliding of the patient on surgical table when in Trendelenburg position, whereas a padded strap across the chest is highly recommended instead of shoulder supports.
- (Extra) Large intermittent pneumatic compression devices should be used.
- Class III obesity: Position the buttocks slightly lower than the edge of the table (cephalad shift of the body with weight of panniculus when moved to Trendelenburg).
- A gradual Trendelenburg position toward 30 degrees with a pneumoperitoneum at 15 mmHg is recommended, hereby stopping at the level of Trendelenburg, which allows the last ileal loop to be displaced above the sacral promontory with then also dropping the pressure to 12 mmHg if possible.
- A tilt test prior to draping is highly recommended in robotic-assisted laparoscopic surgery to prevent unrecognized tension on port sites if sliding occurs after docking.
- If an elevated CO_2 level is encountered in the middle of the surgery, the abdomen needs to be desufflated with neutralizing patient's position, and in the case of RAS a dedocking is required.

Medication

Multifactorial increased risk of surgical site infections requires adequately dosed antibiotic prophylaxis and abdominal skin preparation. Altered pharmacokinetics warrant an increased dose (e.g., standard dose of 2 g cefazolin needs to be replaced by 3 g cefazolin if patient is >120 kg) and an additional dose is required if operating time >4 hours or if blood loss is exceeding 1500 mL. Depending upon the complexity of the surgical procedure, the intravenous antibiotic prophylaxis can be continued for 24 hours. Larger gastric volume (>25 mL) and acidic gastric contents (pH < 2.5) with delayed emptying and a higher risk of aspiration in obese patients suggest to consider administration of antacids and prokinetics a few hours prior to surgery, whereas a nasogastric tube should always be inserted regardless of the port placement, especially in robotic-assisted cases [11,92].

Surgical technical diligence

Entry techniques

Access to the abdominal cavity forms a challenge in obese patients when performing MIS. The abdominal wall anatomy is distorted due to pannus and as a result, the umbilicus is displaced caudally about 3–6 cm below the aortic bifurcation. It is important to identify the bony landmarks such as the ischial spines, the costovertebral edge, and the xiphoid process. Different entry techniques exist but none has been found to be superior according to the most recent Cochrane Review by Ahmad et al. [90]. For a safe entry, it is advisable that the surgeons use an entry technique they are most familiar with, whereas it will also be dependent upon the equipment available in each unit.

If a Veress needle is used, the incision should be made in the base of the umbilicus, which can further include an umbilical stalk elevation (USE) by isolating the umbilical base after making a skin incision and lifting the umbilical stalk upward with a clamp [92,106]. The USE technique should not be confused with grasping the skin to elevate the abdominal wall as the latter will simply increase the distance from the skin to fascia, leading to a higher preperitoneal insufflation. At the base of the umbilicus, there is the shortest distance between the skin and the peritoneum as the anterior and the posterior rectus sheath are here directly attached with the peritoneum without any presence of subcutaneous fat that lowers the risk of failed entry due to preperitoneal insufflation that can lead to surgical emphysema. The Veress should have a vertical insertion (90-degree angle) in class III obesity patients who have an abdominal wall thickness often greater than 11 cm. Mostly a standard Veress needle should be sufficient as the average distance between the umbilical base and parietal peritoneum is about 6 cm (± 3 cm). An angle of 70 degrees is recommended in patients with class I and class II obesity [57,87,91,100,106,107].

> **Key points—entry techniques**
> - Abdominal wall anatomy is often distorted: umbilicus can be 3–6 cm below the aortic bifurcation.
> - RCOG recommendation: open Hasson technique in obese.
> - However, Cochrane Review: no entry technique found to be superior: the technique with which the surgeon is the most familiar should be used.
> - Umbilical stalk elevation can be helpful to avoid preperitoneal insufflation with a Veress needle.
> - A vertical insertion (90-degree angle) of Veress is recommended in class III and a 70-degree angle in class I and class II obese individuals.
> - Palmer's point entry or Lee–Huang point entry offer an alternative, especially when dealing with large pelvic masses and/or pelvic adhesions are encountered.
> - High-flow continuous insufflator is preferred in obese patients.

The Royal College of Obstetricians and Gynecologists recommends the open (Hasson) technique or an entry at the Palmer's point for morbidly obese patients, although it must be understood that the Hasson approach requires more dissection and there is risk of CO_2 leakage due to larger opening, especially in RAS with the da Vinci Xi [57,92]. A balloon trocar in laparoscopy can be helpful in this situation [5]. Due to a lower amount of subcutaneous fat at the Palmer's point, this forms an excellent alternative entry mode. In this which an incision is made in the midclavicular line, 3 cm below the left subcostal margin to insert the Veress needle perpendicular to the skin. Gastric decompression is paramount in this approach and this entry approach should not be performed in patients with previous splenic or gastric surgery, hepatosplenomegaly, portal hypertension, and gastropancreatic masses [48,91,107,108]. Lee–Huang point, located in the middle upper abdomen between xiphoid process and the umbilicus, can be an alternative entry option in patients who have had surgery in left upper quadrant, present with large pelvic pathology (e.g., fibroids, ovarian cysts), and have a failed umbilical/Palmer's approach [107]. Direct optical trocar entry as an alternative to Veress at Palmer's point has been described with a lower rate of failed entry, but is not recommended as standard practice [90,91]. Transuterine and vaginal posterior fornix entry to create CO_2 pneumoperitoneum have been described but also lack robust evidence to be recommended [91,109]. High-flow continuous insufflators deliver a rapid and continuous insufflation of large volumes of warm CO_2 (45 L/min) with the ability to maintain the abdominal volume as they constantly measure the intraabdominal pressure and should be preferred in the obese surgical patient.

Port placement

Depending upon the patient's habitus, the trocars may need to be inserted more laterally (away from the edge of rectus sheath), and cephalad because of the extent of panniculus and difficulty of visualizing the inferior epigastric vessels. As in nonobese patients, it is important that the trocars are placed with the patient in a neutral horizontal position because an insertion at 45 degrees with patient in Trendelenburg will actually mean an angulation of 75 degrees from the horizontal plane of patient's spine, hereby increasing the risk of major vessel injury. The tip of the secondary trocars should be visualized during placement and insertion should be perpendicular to the abdominal wall and underlying peritoneum. However, the tip of the lateral port trocar is often pointed directly at the external iliac vessels and, therefore, once the peritoneum is penetrated, the direction of insertion should be changed medially and caudally away from these vessels. An uncontrolled thrust of the trocar into the abdomen after an unexpected loss of resistance can be prevented by consciously balancing the force of the agonist muscles that produce the forward thrust with the antagonist muscles that stop it, whereas an extension of the index finger on the inserting trocar shaft can limit the depth of insertion as well. The speed should be slow, also in trocars with self-retracting blades or extending shields as they will have time to deploy. Trocars with intraabdominal and extraabdominal stabilizers are preferred as it can be otherwise challenging to replace a dislodged trocar. Extra-long trocars and instruments can be required. Reinsertion of a trocar can be aided by placing a laparoscopic instrument (e.g., blunt grasper) through the port and into the incision with locating the fascial and peritoneal incisions by gentle probing. Once it is located, the sleeve can be slid over this instrument. If a larger port is required to replace a smaller one, a blunt trocar can be used or there are trocars available with blades which can be retracted into blunt conical blade guards before reinsertion of the sleeve [12,35,87,101,108].

In RAS, it can be useful to assess the feasibility of the procedure by starting with a 5 mm Palmer's point laparoscope prior to draping and docking. The estimated rate of adhesions is 15% with a history of previous laparoscopy, 20%–28% when a low transverse laparotomy was performed in the past, whereas it is estimated at 50%–60% in the case of midline laparotomy [89,91]. The trocars should be inserted fully beyond the black guide marker as they withdraw themselves slightly when the cart is attached. Additional adjustments can take place after docking using the camera control. Noncutting trocars have the advantage of lower incidence of bleeding and less postoperative pain as the conical tip displaces the vessels instead of transecting them. It is very important that the position of the trocar is correct around the "remote center." This is the point in the trocar around which the arms within the trocar should move to provide an optimal balance between the maneuverability and the force exerted to the abdominal wall, reducing postoperative pain. The ports should be 'burped' after placement to create more intraabdominal working space. This will be required when patient's skin is puckered/indented with the trocars digging in. "Burping" can be achieved by lifting the trocar away from the abdominal wall, tenting it outward. To prevent clashing between the robotic arms, 8–10 cm space surrounding each arm should be ensured which may require rotating the arms to let the elbows face away from each other. A general principle that always should be remembered in both laparoscopic and RAS is to have a low threshold to place an additional assistant port if there is additional need for retraction [57,92,102].

All ports ≥10 mm require a formal deep fascia closure to prevent postoperative hernias and devices such as Endoclose or Berci needle can be helpful to achieve a closure under direct vision. These can also be used to ligate a bleeding superficial epigastric vessel in the abdominal wall [101].

> **Key points—port placement**
> - Trocars should be placed with patient in neutral, horizontal position and insertion should be perpendicular to the abdominal wall to avoid dislodging and prevent postoperative pain.
> - Extra-long trocars can be required in morbidly obese; trocars with intra- and/or extraabdominal stabilizers should be preferred. Low threshold to insert an additional trocar when needed.
> - An assessment of feasibility of the procedure can be done by inserting a 5 mm port in Palmer's point, especially in robotic-assisted laparoscopic surgery (RAS) as to avoid consumption of time and costs associated with draping and docking.
> - When RAS in obese patients, insert trocars beyond the black guide marker—adjust after docking to ensure the remote center on the trocar is correctly positioned (causes the least trauma), whereas "burp" the ports (creates more work space).
> - RAS: Ensure 8–10 cm space between each robotic arm to avoid clashing.
> - All ports ≥10 mm require a deep fascia closure.
> - High-flow continuous insufflator is preferred in obese patients.

Exposure

As fatty tissue can obscure the surgical view despite a decent Trendelenburg positioning, strategies such as suspending the bowel or ovaries to the anterior abdominal wall or manoeuvers to pack the small bowel to keep it displaced in upper abdomen can be considered. In the case of the rectosigmoid, this can be achieved by fixating the epiploic appendices to the anterolateral abdominal wall with a stitch. The sigmoid reflection and caecum can also be mobilized from their lateral peritoneal attachments that can help to reflect the large bowel out of the pelvis. A fan retractor (Endo Paddle Retract, Covidien) can decrease operating times and can allow lower abdominal pressure to operate with. An intrauterine manipulator is essential in many surgical indications but some may require the use of a manipulator without interference with uterine tissue, such as the McCartney tube that is placed around the cervix to push the uterus cranially. An endoloop applied to a divided round ligament and drawn out suprapubically can also help antevert the uterus and allow access to the pouch of Douglas, posterior cervix, and vagina. Similarly a prolene suture can be run along the peritoneal edge of overhanging perivascular fat and is drawn out suprapubically to give traction to the bladder superiorly [101]. As the ureters are often not visible transperitoneally in obese, a retroperitoneal dissection is advised to identify these which then on its turn can help to identify the origin of the uterine artery to minimize blood loss and the associated vision into surgical field. Using of a 30-degree scope can be considered in some cases. Adequate hemostasis is essential, not only for surgical vision but also to avoid the risk of postoperative pelvic infection. Active hemostatic agents can be used to stop diffuse oozing, whereas insertion of a drain at the end of the procedure can be considered [22,26,57,92,100,101].

> **Key points—exposure and surgical considerations**
> - Consider suspension of ovaries and/or large bowel epiploic appendices to anterolateral abdominal wall to optimize exposure.
> - Intrauterine manipulator can be the key element to ensure a successful procedure, whereas one should know about the extrauterine manipulation techniques when interference with intrauterine tissue is required to be avoided.
> - Identification of the ureters may require a retroperitoneal dissection in obese and a 30-degree scope can be helpful, whereas a diligent hemostasis should be the norm with a good knowledge of the available hemostatic agents.
> - A strategized and standardized approach toward a particular procedure is advisable.
>
> *(Continued)*

> **(Continued)**
> - Advanced devices with multiple functions can aid to achieve a better postoperative outcome.
> - A longer operating time directly correlates with the postoperative outcome in the obese patient; an experienced surgeon should perform the procedure when dealing with morbid obesity.
> - High-flow continuous insufflator is preferred in obese patients.

Surgical considerations

Minimization of instrument changes increases the chances of an optimal procedure in the obese. It is advisable to strategize the surgical approach and standardization where possible should be aimed at. Advanced devices with multiple functions to adequately seal/coagulate and ligate the tissue will allow reduction in operating time, total blood loss and, consequently, a better postoperative outcome. Skeletalization of vessels prior to ligation is advisable, whereas the ligation should be as distal as possible to the place where the coagulation was carried out to ensure adequate tissue at a pedicle is available to be grasped. As in obese patients the length of operative time directly correlates with the postoperative outcome, an experienced surgeon should perform the procedure, which may allow limited time for training of junior surgeons, and needs the procedure that may allow limited time for training of junior surgeons and need understanding.

Postoperative considerations

Routine postoperative care supported by enhanced recovery protocol can help early identification of potential surgery-related comorbidity such as postoperative infections. Early mobilization is vital. If possible, restricting the patient with urinary catheter, intravenous infusion or other devices should be avoided. Semirecumbent positioning and use of incentive spirometry and bilevel positive airway machines in patients with BMI $> 40 \text{ kg/m}^2$ and OSA, along avoidance of narcotic analgesics can reduce the respiratory complication [11,12]. Where possible, narcotic analgesia should be avoided. On the other hand, optimal analgesia is essential and can be ensured with nonopioids as the presence of pain can restrict deep respiration, thereby preventing good ventilation and mobilization. Intramuscular route of drug administration is to be avoided owing to unpredictable pharmacokinetics in the obese. Regional anesthesia in the early postoperative phase can be considered in some surgical cases. Depending upon the comorbidity profile and complexity of the surgery performed, some patients may require admission into the intensive care unit [15,21].

The BOLD database (Bariatric Outcomes Longitudinal Database) demonstrated that the majority of VTE (73%) occur after discharge [98]. An extended duration of VTE prophylaxis is therefore mandatory in the obese surgical patient population. Low-molecular-weight heparins (LMWH) are preferred over unfractionated heparin and dose should be adjusted to patient's body weight (0.5 mg/kg OD with a target Factor Xa level of 0.2–0.4 IU/mL) [4,11,21,98].

An association with obesity and acute kidney injury postoperatively has been reported. Apart from the known risk factors for AKI in obese patients (hypertension, diabetes, etc.), long operative time gives an overall incidence of AKI 7% and is secondary to rhabdomyolysis. Preoperative exposure to nephrotoxic agents (e.g., antibiotics), avoiding long operating times and intraopertaive hypotension and ensuring adequate fluid administration are recommended. Urine output of 1 mL/kg/h is based on lean body mass and is a reliable predictor of fluid replacement in obese patients [11].

The Association of Anaesthetists and the British Association of Day Surgery do not see morbid obesity as a contraindication for day case surgery provided there is an appropriate case selection and there are appropriate resources available, including additional time for anesthesia and surgery. They point out that obese patients in fact benefit from short-duration anesthetic techniques as well as from an early mobilization [110].

Conclusion

With the progressively increasing obesity rates worldwide, not only will we be challenged to broaden our surgical armamentarium with further advancement of equipment, but we will also have to form multidisciplinary integrated teams in standardized ways to accommodate this subset of patients for their perioperative requirements and postoperative care needs. This requires adequate knowledge around the physiological and anatomical changes associated with obesity.

Laparoscopy in obese patients can be challenging but outweighs these challenges due to its feasibility and the numerous benefits associated with MIS approach, such as better visualization of surgical field, decreased blood loss, quicker recovery, less postoperative pain, and lower complication rate. It is recommended that especially complex procedures are performed by skilled surgeons in a specialized center.

When it comes to determine the place of RAS in gynecologic MIS, the benefits of RAS have been documented heterogeneously with often a large emphasis on the associated higher costs. There is a lack of level-1 evidence when it comes to define the exact role of RAS in gynecology. Despite the progressive diffusion of RAS giving especially the obese patients the benefits to undergo MIS in a less challenging way and with much lower conversion rates when compared to laparoscopy, the higher purchase- and maintenance cost is at present the preeminent issue preventing its acceptance in routine practice for complex procedures. The growing body of competition in the market of robotic surgical technologies will potentially help reduce some of the fixed costs associated with RAS, whereas there is also a strong need to perform high-quality methodological cost analyses, including a broader cost perspective to include not only the purchase costs but also the complete resource use from hospitalization (including microcosting such as staffing costs, total annual number of RAS procedures, learning curve to achieve the expert level), and outside the health-care system (such as return to work, rehabilitation, long-term societal outcomes). As perpetual biotechnological innovation is driving physicians to adapt a personalized surgical approach offered to their patients. Until we reach further evidence on different MIS modalities, we know that RAS in obese women is feasible, safe, reproducible and could be a valid alternative to laparoscopic approach. It is likely that continued improvements in laparoscopic and robotic surgical systems, including options such as tele surgery, in the foreseeable future will further revolutionize the field of MIS.

Conclusions on comparisons between laparoscopic and robotic surgical approaches should be made with caution given the difficulty to compare two technologies and poor data quality around the outcomes at present. The route of surgical approach should be guided by the skill and comfort level of the surgeon that promises the highest level of safety and efficiency as both laparoscopy and RAS lead to a better outcome for the obese patient. RAS is ultimately an additional tool in the field of MIS and its role should be defined in an institutional environment that supports safe use of other surgical approaches in an integrated MIS structure.

References

[1] WHO. Obesity and overweight. Fact Sheet February. February 16, 2018;2018(2018):16; <https://www.who.int/en/news-room/fact-sheets/detail/obesity-and-overweight>.

[2] England HS. Health survey England 2017 adult and child overweight and obesity, <https://files.digital.nhs.uk/3F/6971DC/HSE17-Adult-Child-BMI-rep.pdf>; 2018, December 4.

[3] World Health Organisation. Global strategy on diet, physical activity and health. What is overweight and obesity? <https://www.who.int/dietphysicalactivity/childhood_what/en/>; June 2016.

[4] Bazurro S, Ball L, Pelosi P. Perioperative management of obese patient. Curr Opin Crit Care 2018;24(6):560–7.

[5] Biswas NHP. Surgical risk from obesity in gynaecology. Obstetrician Gynaecol 2011;13:87–91.

[6] GOV.UK. Public Health Matters Implementing the whole systems approach to obesity, <https://publichealthmatters.blog.gov.uk/2018/07/11/implementing-the-whole-systems-approach-to-obesity/>; November 2018.

[7] Poirier P, Alpert MA, Fleisher LA, et al. Cardiovascular evaluation and management of severely obese patients undergoing surgery: a science advisory from the American Heart Association. Circulation 2009;120(1):86–95.

[8] Ortega-Loubon C, Fernandez-Molina M, Singh G, Correa R. Obesity and its cardiovascular effects. Diab/Metab Res Rev 2019;35(4):e3135.

[9] Cullen A, Ferguson A. Perioperative management of the severely obese patient: a selective pathophysiological review. Can J Anaesth 2012;59(10):974–96.

[10] Harman-Boehm I, Bluher M, Redel H, et al. Macrophage infiltration into omental versus subcutaneous fat across different populations: effect of regional adiposity and the comorbidities of obesity. J Clin Endocrinol Metab 2007;92(6):2240–7.

[11] Leonard KL, Davies SW, Waibel BH. Perioperative management of obese patients. Surgical Clin North Am 2015;95(2):379–90.

[12] Louie M, Toubia T, Schiff LD. Considerations for minimally invasive gynecologic surgery in obese patients. Curr Opin Obstet Gynecol 2016;28(4):283–9.

[13] Nguyen NT, Wolfe BM. The physiologic effects of pneumoperitoneum in the morbidly obese. Ann Surg 2005;241 (2):219–26.

[14] Sprung J, Whalley DG, Falcone T, Warner DO, Hubmayr RD, Hammel J. The impact of morbid obesity, pneumoperitoneum, and posture on respiratory system mechanics and oxygenation during laparoscopy. Anesth Analg 2002;94(5):1345–50.

[15] Members of the Working Party, Nightingale MPM CE, Shearer E, Redman JW, Lucas JMC DN, Fox WTA, et al. Peri-operative management of the obese surgical patient 2015: Association of Anaesthetists of Great Britain and Ireland Society for Obesity and Bariatric Anaesthesia. Anaesthesia 2015;70(7):859–76.

[16] Tomescu DR, Popescu M, Dima SO, Bacalbasa N, Bubenek-Turconi S. Obesity is associated with decreased lung compliance and hypercapnia during robotic assisted surgery. J Clin Monit Comput 2017;31(1):85–92.

[17] Masa JF, Pepin JL, Borel JC, Mokhlesi B, Murphy PB, Sanchez-Quiroga MA. Obesity hypoventilation syndrome. Eur Respir Rev 2019;28(151).

[18] Pierce AM, Brown LK. Obesity hypoventilation syndrome: current theories of pathogenesis. Curr Opin Pulm Med 2015;21 (6):557–62.

[19] Beck TL, Schiff MA, Goff BA, Urban RR. Robotic, laparoscopic, or open hysterectomy: surgical outcomes by approach in endometrial cancer. J Minim Invasive Gynecol 2018;25(6):986–93.

[20] Abdelbadee AY, Paspulati RM, McFarland HD, et al. Computed tomography morphometrics and pulmonary intolerance in endometrial cancer robotic surgery. J Minim Invasive Gynecol 2016;23 (7):1075–82.

[21] Ortiz VE, Kwo J. Obesity: physiologic changes and implications for preoperative management. BMC Anesthesiol 2015;15:97.

[22] Scheib SA, Tanner 3rd E, Green IC, Fader AN. Laparoscopy in the morbidly obese: physiologic considerations and surgical techniques to optimize success. J Minim Invasive Gynecol 2014;21 (2):182–95.

[23] Nguyen NT, Cronan M, Braley S, Rivers R, Wolfe BM. Duplex ultrasound assessment of femoral venous flow during laparoscopic and open gastric bypass. Surg Endosc 2003;17(2):285–90.

[24] Smit C, De Hoogd S, Bruggemann RJM, Knibbe CAJ. Obesity and drug pharmacology: a review of the influence of obesity on pharmacokinetic and pharmacodynamic parameters. Expert Opin Drug Metab Toxicol 2018;14(3):275–85.

[25] Al Sawah E, Salemi JL, Hoffman M, Imudia AN, Mikhail E. Association between obesity, surgical route, and perioperative outcomes in patients with uterine cancer. Minim Invasive Surg 2018;2018:5130856.

[26] Rassweiler JJ, Teber D. Advances in laparoscopic surgery in urology. Nat Rev Urol 2016;13(7):387–99.

[27] Arezzo A, Vettoretto N, Francis NK, et al. The use of 3D laparoscopic imaging systems in surgery: EAES consensus development conference 2018. In: Surgical endoscopy. 2018.

[28] Vogell ABH, Ware M, Wright VJ, Georgakoudi I, Schnelldorfer T. Novel imaging technologies in laparoscopic gynecologic surgery—a systematic review. ASME J Med Diagnostics 2017;1 (1):010801–010801-6.

[29] Gueli Alletti S, Rossitto C, Cianci S, et al. The Senhance surgical robotic system ("Senhance") for total hysterectomy in obese patients: a pilot study. J Robotic Surg 2018;12(2):229–34.

[30] Zakhari A, Czuzoj-Shulman N, Spence AR, Gotlieb WH, Abenhaim HA. Laparoscopic and robot-assisted hysterectomy for uterine cancer: a comparison of costs and complications. Am J Obstet Gynecol 2015;213(5) 665.e661-667.

[31] Lim PC, Crane JT, English EJ, et al. Multicenter analysis comparing robotic, open, laparoscopic, and vaginal hysterectomies performed by high-volume surgeons for benign indications. Int J Gynaecol Obstet 2016;133(3):359–64.

[32] Rashid TG, Kini M, Ind TE. Comparing the learning curve for robotically assisted and straight stick laparoscopic procedures in surgical novices. Int J Med Robot Comput Assist Surg 2010;6 (3):306–10.

[33] Iavazzo C, Gkegkes ID. Cost-benefit analysis of robotic surgery in gynaecological oncology. Best Pract Res Clin Obstet Gynaecol 2017;45:7–18.

[34] Moulton LJ, Jernigan AM, Michener CM. Postoperative outcomes after single-port laparoscopic removal of adnexal masses in patients referred to gynecologic oncology at a large academic center. J Minim Invasive Gynecol 2017;24(7):1136–44.

[35] So KA, Lee JK, Song JY, et al. Tissue injuries after single-port and multiport laparoscopic gynecologic surgeries: a prospective multicenter study. Exp Ther Med 2016;12(4):2230–6.

[36] You SH, Huang CY, Su H, Han CM, Lee CL, Yen CF. The power law of learning in transumbilical single-port laparoscopic subtotal hysterectomy. J Minim Invasive Gynecol 2018;25(6):994–1001.

[37] Iavazzo C, Minis EE, Gkegkes ID. Single-site port robotic-assisted hysterectomy: an update. J Robotic Surg 2018;12 (2):201–13.

[38] Matanes E, Lauterbach R, Boulus S, Amit A, Lowenstein L. Robotic laparoendoscopic single-site surgery in gynecology: a systematic review. Eur J Obstet Gynecol Reprod Biol 2018;231:1–7.

[39] Baekelandt JF, De Mulder PA, Le Roy I, et al. Hysterectomy by transvaginal natural orifice transluminal endoscopic surgery versus laparoscopy as a day-care procedure: a randomised controlled trial. BJOG 2019;126(1):105–13.

[40] Liu J, Kohn J, Fu H, Guan Z, Guan X. Transvaginal natural orifice transluminal endoscopic surgery for sacrocolpopexy: a pilot study of 26 cases. J Minim Invasive Gynecol 2019;26(4):748−53.

[41] Naval S, Naval R, Naval S. Transvaginal Natural Orifice Transluminal Endoscopic Surgery Hysterectomy Aided by Transcervical Instrumental Uterine Manipulation. J Minim Invasive Gynecol 2019;26(7). Available from: https://doi.org/10.1016/j.jmig.2019.05.004 (video article).

[42] Chen Y, Li J, Zhang Y, Hua K. Transvaginal single-port laparoscopy sacrocolpopexy. J Minim Invasive Gynecol 2018;25(4):585−8.

[43] Baekelandt J. Transvaginal natural orifice transluminal endoscopic surgery: a new approach to ovarian cystectomy. Fertil Steril 2018;109(2):366.

[44] Dizon AM, Carey ET. Minimally invasive gynecologic surgery in the pregnant patient: considerations, techniques, and postoperative management per trimester. Curr Opin Obstet Gynecol 2018;30(4):267−71.

[45] Pearl JP, Price RR, Tonkin AE, Richardson WS, Stefanidis D. SAGES guidelines for the use of laparoscopy during pregnancy. Surg Endosc 2017;31(10):3767−82.

[46] Hsu S, Mitwally MF, Aly A, Al-Saleh M, Batt RE, Yeh J. Laparoscopic management of tubal ectopic pregnancy in obese women. Fertil Steril 2004;81(1):198−202.

[47] Takacs P, Goebel M, Medina C. Laparoscopic management of ectopic pregnancy in obese women. Int J Gynaecol Obstet 2007;97(3):200−1.

[48] Stone P, Burnett A, Burton B, Roman J. Overcoming extreme obesity with robotic surgery. Int J Med Robot Comput Assist Surg 2010;6(4):382−5.

[49] Armfield NR, Janda M, Obermair A. Obesity in total laparoscopic hysterectomy for early stage endometrial cancer: health gain and inpatient resource use. Int J Qual Health Care 2019;31(4):283−8.

[50] Boggess JF, Gehrig PA, Cantrell L, et al. A comparative study of 3 surgical methods for hysterectomy with staging for endometrial cancer: robotic assistance, laparoscopy, laparotomy. Am J Obstet Gynecol 2008;199(4) 360.e361-369.

[51] Iavazzo C, Gkegkes ID. Robotic assisted hysterectomy in obese patients: a systematic review. Arch Gynecol Obstet 2016;293(6):1169−83.

[52] Marra AR, Puig-Asensio M, Edmond MB, Schweizer ML, Bender D. Infectious complications of laparoscopic and robotic hysterectomy: a systematic literature review and meta-analysis. Int J Gynecol Cancer 2019;29(3):518−30.

[53] Peng J, Sinasac S, Pulman KJ, Zhang L, Murphy J, Feigenberg T. The feasibility of laparoscopic surgery in gynecologic oncology for obese and morbidly obese patients. Int J Gynecol Cancer 2018;28(5):967−74.

[54] Stephan JM, Goodheart MJ, McDonald M, et al. Robotic surgery in supermorbidly obese patients with endometrial cancer. Am J Obstet Gynecol 2015;213(1):49.e41−8.

[55] Walker JL, Piedmonte MR, Spirtos NM, et al. Laparoscopy compared with laparotomy for comprehensive surgical staging of uterine cancer: Gynecologic Oncology Group Study LAP2. J Clin Oncol 2009;27(32):5331−6.

[56] Corrado G, Vizza E, Cela V, et al. Laparoscopic versus robotic hysterectomy in obese and extremely obese patients with endometrial cancer: a multi-institutional analysis. Eur J Surg Oncol 2018;44(12):1935−41.

[57] Menderes G, Gysler SM, Vadivelu N, Silasi DA. Challenges of robotic gynecologic surgery in morbidly obese patients and how to optimize success. Curr Pain Headache Rep 2019;23(7):51.

[58] Cusimano MC, Simpson AN, Dossa F, et al. Laparoscopic and robotic hysterectomy in endometrial cancer patients with obesity: a systematic review and meta-analysis of conversions and complications. Am J Obstet Gynecol 2019;221(5):410−28. e19.

[59] Leitao MM, Narain WR, Boccamazzo D, et al. Impact of robotic platforms on surgical approach and costs in the management of morbidly obese patients with newly diagnosed uterine cancer. Ann Surg Oncol 2016;23(7):2192−8.

[60] Ind T, Laios A, Hacking M, Nobbenhuis M. A comparison of operative outcomes between standard and robotic laparoscopic surgery for endometrial cancer: a systematic review and meta-analysis. Int J Med Robot Comput Assist Surg 2017;13(4).

[61] Mäenpää MM, Nieminen K, Tomas EI, Laurila M, Luukkaala TH, Mäenpää JU. Robotic-assisted vs traditional laparoscopic surgery for endometrial cancer: a randomized controlled trial. Am J Obstet Gynecol 2016;215(5):588.e581−7.

[62] Gunderson CC, Java J, Moore KN, Walker JL. The impact of obesity on surgical staging, complications, and survival with uterine cancer: a Gynecologic Oncology Group LAP2 ancillary data study. Gynecol Oncol 2014;133(1):23−7.

[63] Eddib A, Danakas A, Hughes S, et al. Influence of morbid obesity on surgical outcomes in robotic-assisted gynecologic surgery. J Gynecol Surg 2014;30(2):81−6.

[64] Palomba S, Zupi E, Russo T, et al. Presurgical assessment of intraabdominal visceral fat in obese patients with early-stage endometrial cancer treated with laparoscopic approach: relationships with early laparotomic conversions. J Minim Invasive Gynecol 2007;14(2):195−201.

[65] Uccella S, Bonzini M, Palomba S, et al. Impact of obesity on surgical treatment for endometrial cancer: a multicenter study comparing laparoscopy vs open surgery, with propensity-matched analysis. J Minim Invasive Gynecol 2016;23(1):53−61.

[66] Wysham WZ, Kim KH, Roberts JM, et al. Obesity and perioperative pulmonary complications in robotic gynecologic surgery. Am J Obstet Gynecol 2015;213(1):33.e31−7.

[67] Lönnerfors C, Reynisson P, Persson J. A randomized trial comparing vaginal and laparoscopic hysterectomy vs robot-assisted hysterectomy. J Minim Invasive Gynecol 2015;22(1):78−86.

[68] Antoniou SA, Antoniou GA, Koch OO, Kohler G, Pointner R, Granderath FA. Laparoscopic versus open obesity surgery: a meta-analysis of pulmonary complications. Digestive Surg 2015;32(2):98−107.

[69] Gould C, Cull T, Wu YX, Osmundsen B. Blinded measure of Trendelenburg angle in pelvic robotic surgery. J Minim Invasive Gynecol 2012;19(4):465−8.

[70] Angioli R, Terranova C, Plotti F, et al. Influence of pneumoperitoneum pressure on surgical field during robotic and laparoscopic surgery: a comparative study. Arch Gynecol Obstet 2015;291(4):865−8.

[71] Korsholm M, Sorensen J, Mogensen O, Wu C, Karlsen K, Jensen PT. A systematic review about costing methodology in robotic surgery: evidence for low quality in most of the studies. Health Econ Rev 2018;8(1):21.

[72] Ind TEJ, Marshall C, Hacking M, Chiu S, Harris M, Nobbenhuis M. The effect of obesity on clinical and economic outcomes in

robotic endometrial cancer surgery. Robotic Surg (Auckl) 2017;4:33–7.
[73] Suidan RS, He W, Sun CC, et al. Impact of body mass index and operative approach on surgical morbidity and costs in women with endometrial carcinoma and hyperplasia. Gynecol Oncol 2017;145(1):55–60.
[74] Chan JK, Gardner AB, Taylor K, et al. Robotic versus laparoscopic versus open surgery in morbidly obese endometrial cancer patients – a comparative analysis of total charges and complication rates. Gynecol Oncol 2015;139(2):300–5.
[75] Wright JD, Ananth CV, Tergas AI, et al. An economic analysis of robotically assisted hysterectomy. Obstet Gynecol 2014;123(5):1038–48.
[76] Moawad GN, Abi Khalil ED, Tyan P, et al. Comparison of cost and operative outcomes of robotic hysterectomy compared to laparoscopic hysterectomy across different uterine weights. J Robotic Surg 2017;11(4):433–9.
[77] Wu CZ, Klebanoff JS, Tyan P, Moawad GN. Review of strategies and factors to maximize cost-effectiveness of robotic hysterectomies and myomectomies in benign gynecological disease. J Robotic Surg 2019;13(5):635–42.
[78] Venkat P, Chen LM, Young-Lin N, et al. An economic analysis of robotic versus laparoscopic surgery for endometrial cancer: costs, charges and reimbursements to hospitals and professionals. Gynecol Oncol 2012;125(1):237–40.
[79] Bogani G, Multinu F, Dowdy SC, et al. Incorporating robotic-assisted surgery for endometrial cancer staging: analysis of morbidity and costs. Gynecol Oncol 2016;141(2):218–24.
[80] Ind TE, Marshall C, Hacking M, et al. Introducing robotic surgery into an endometrial cancer service—a prospective evaluation of clinical and economic outcomes in a UK institution. Int J Med Robot Comput Assist Surg 2016;12(1):137–44.
[81] Deimling TA, Eldridge JL, Riley KA, Kunselman AR, Harkins GJ. Randomized controlled trial comparing operative times between standard and robot-assisted laparoscopic hysterectomy. Int J Gynaecol Obstet 2017;136(1):64–9.
[82] Madueke-Laveaux OS, Advincula AP. Robot-assisted laparoscopy in benign gynecology: advantageous device or controversial gimmick? Best Pract Res Clin Obstet Gynaecol 2017;45:2–6.
[83] Moss EL, Sarhanis P, Ind T, Smith M, Davies Q, Zecca M. The impact of obesity on surgeon ergonomics in robotic and straight stick laparoscopic surgery J Minim Invasive Gynecol 2019;18S1553-4650(19)30313-9. Available from: https://doi.org/10.1016/j.jmig.2019.07.009.
[84] Xie Y. Cost-effectiveness of robotic surgery in gynecologic oncology. Curr Opin Obstet Gynecol 2015;27(1):73–6.
[85] Bijen CB, de Bock GH, Vermeulen KM, et al. Laparoscopic hysterectomy is preferred over laparotomy in early endometrial cancer patients, however not cost effective in the very obese. Eur J Cancer 2011;47(14):2158–65 (Oxford, England: 1990).
[86] Orekoya O, Samson ME, Trivedi T, Vyas S, Steck SE. The impact of obesity on surgical outcome in endometrial cancer patients: a systematic review. J Gynecol Surg 2016;32(3):149–57.
[87] Alkatout I. Complications of laparoscopy in connection with entry techniques. J Gynecol Surg 2017;33(3):81–91.
[88] Sivaraman A, Sanchez-Salas R, Prapotnich D, et al. Robotics in urological surgery: evolution, current status and future perspectives. Actas Urol Esp 2015;39(7):435–41.
[89] Sanchez AM, Medina LG, Husain FZ, Otelo, R. Complications of robotic surgical access. 3rd ed.; 2018.
[90] Ahmad G, Baker J, Finnerty J, Phillips K, Watson A. Laparoscopic entry techniques. Cochrane Database Syst Rev 2019;1:Cd006583.
[91] Vilos GA, Ternamian A, Dempster J, Laberge PY. Laparoscopic entry: a review of techniques, technologies, and complications. J Obstet Gynaecol Canada 2007;29(5):433–47.
[92] Mikdachi H, Schreck A. Robotic surgery in the obese patient: tips and tricks for the benign gynecologist. Int J Gynecol Clin Pract 2018;5(1).
[93] O'Sullivan OE, O'Reilly BA, Hewitt M. Tips and tricks for robotic surgery. Cham: Springer; 2017.
[94] Grosso A, Ceruti P, Morino M, Marchini G, Amisano M, Fioretto M. Comment on the paper by Mondzelewski and Colleagues: "Intraocular pressure during robotic-assisted laparoscopic procedures utilizing steep Trendelenburg positioning.". J Glaucoma 2015;24(6):399–404 Journal of glaucoma.2017;26(4):e166-e167.
[95] Mondzelewski TJ, Schmitz JW, Christman MS, et al. Intraocular pressure during robotic-assisted laparoscopic procedures utilizing steep Trendelenburg positioning. J Glaucoma 2015;24(6):399–404.
[96] Committee on Gynecologic Practice. Committee opinion no. 619: gynecologic surgery in the obese woman. Obstet Gynecol 2015;125(1):274–8.
[97] Birkmeyer NJ, Finks JF, English WJ, et al. Risks and benefits of prophylactic inferior vena cava filters in patients undergoing bariatric surgery. J Hosp Med 2013;8(4):173–7.
[98] Li W, Gorecki P, Semaan E, Briggs W, Tortolani AJ, D'Ayala M. Concurrent prophylactic placement of inferior vena cava filter in gastric bypass and adjustable banding operations in the Bariatric Outcomes Longitudinal Database. J Vasc Surg 2012;55(6):1690–5.
[99] Siedhoff MT, Clark LH, Hobbs KA, Findley AD, Moulder JK, Garrett JM. Mechanical bowel preparation before laparoscopic hysterectomy: a randomized controlled trial. Obstet Gynecol 2014;123(3):562–7.
[100] Afors K, Centini G, Murtada R, Castellano J, Meza C, Wattiez A. Obesity in laparoscopic surgery. Best Pract Res Clin Obstet Gynaecol 2015;29(4):554–64.
[101] Hackethal A, Brennan D, Rao A, et al. Consideration for safe and effective gynaecological laparoscopy in the obese patient. Arch Gynecol Obstet 2015;292(1):135–41.
[102] Lim PC, Kang E. How to prepare the patient for robotic surgery: before and during the operation. Best Pract Res Clin Obstet Gynaecol 2017;45:32–47.
[103] Mechanick JI, Youdim A, Jones DB, et al. Clinical practice guidelines for the perioperative nutritional, metabolic, and nonsurgical support of the bariatric surgery patient—2013 update: cosponsored by American Association of Clinical Endocrinologists, The Obesity Society, and American Society for Metabolic & Bariatric Surgery. Obes (Silver Spring, Md) 2013;21(Suppl. 1):S1–27.
[104] Booth CM, Moore CE, Eddleston J, Sharman M, Atkinson D, Moore JA. Patient safety incidents associated with obesity: a review of reports to the National Patient Safety Agency and recommendations for hospital practice. Postgrad Med J 2011;87(1032):694–9.

[105] Gynaecologists RCoO. A practical guide to surgical positioning in Obstetrics & Gynaecology E-learning – RCOG Basic Practical Skills Course, <https://elearning.rcog.org.uk//sites/default/files/Generalprinciples/2014-08-13gynaepositioningpresentation_slides.pdf>; 2014.

[106] Ozdemir A, Gungorduk K, Ulker K, et al. Umbilical stalk elevation technique for safer Veress needle insertion in obese patients: a case-control study. Eur J Obstet Gynecol Reprod Biol 2014;180:168–71.

[107] Thepsuwan J, Huang K-G, Wilamarta M, Adlan A-S, Manvelyan V, Lee C-L. Principles of safe abdominal entry in laparoscopic gynecologic surgery. Gynecol Minim Invasive Ther 2013;2(4):105–9.

[108] Pickett SD, Rodewald KJ, Billow MR, Giannios NM, Hurd WW. Avoiding major vessel injury during laparoscopic instrument insertion. Obstet Gynecol Clin North Am 2010;37(3):387–97.

[109] Pasic R, Levine RL, Wolf Jr. WM. Laparoscopy in morbidly obese patients. J Am Assoc Gynecol Laparosc 1999;6(3):307–12.

[110] Bailey CR, Ahuja M, Bartholomew K, et al. Guidelines for day-case surgery 2019: guidelines from the Association of Anaesthetists and the British Association of Day Surgery. Anaesthesia 2019;74(6):778–92.

Chapter 27

Obesity and venous thromboembolism

Julia Czuprynska and Roopen Arya

King's College Hospital, London, United Kingdom

Introduction

The prevalence of obesity has dramatically increased in developed countries during the last decades and, consequently, is a major challenge to public health and healthcare systems [1]. Obesity is associated with a number of chronic conditions, including venous thromboembolism (VTE) [2], and given the increasing epidemic of obesity in the United States and worldwide, the association between obesity and thrombosis is particularly important [3]. VTE is a term encompassing deep vein thrombosis of the leg or pelvis and pulmonary embolism (PE). VTE is a multifactorial disease, involving interaction between acquired and inherited predispositions to thrombosis [4]. The link between obesity and VTE has been recognized for some time. In the 1970s postmortem studies reported that the risk of PE in obese subjects was one-half to two times that observed in the nonobese [5]. The role of obesity in the development of postoperative PE was described earlier in 1927 and confirmed in 1940 by Barker et al. who reported a twofold increase in PE in obese women after hysterectomy [6].

Obesity [body mass index (BMI) $\geq$ 30 kg/m^2] is an independent risk factor for both first VTE and VTE recurrence for both men and women [4]. The incidence of VTE among healthy women of reproductive age is 1–5 per 10,000 women years (WY) [7]. The baseline risk of VTE in obese women ranges from 6 to 11/10,000 WY, and the risk increases with age [8]. Women face a greater risk of thrombosis throughout their reproductive years due to transient risk factors, such as hormonal therapy and pregnancy [9], and this risk is further increased by the presence of obesity [10]. While obesity alone is of itself a moderate risk factor for VTE, it exerts a synergistic effect when combined with additional VTE risk factors.

A prospective study of 112,822 women who were followed for 16 years (which generated 1619,770 person years of follow-up) determined that, on multivariate analysis, obesity was an independent predictor of PE. Specifically, obese women (BMI $\geq$ 29 kg/m^2) had an almost threefold increased relative risk (RR) of primary (unprovoked) PE [multivariate RR = 2.9; 95% confidence interval (CI), 1.5–5.4] [11]. Similarly, in an analysis from the Multiple Environmental and Genetic Assessment of risk factors for venous thrombosis study (which examined 3834 patients with a first venous thrombosis and 4683 control subjects), relative to those with a normal BMI (<25), being overweight (BMI $\geq$ 25 and <30) increased the risk of venous thrombosis 1.7-fold [odds ratio (OR) adjusted for age and sex 1.7, (95% CI: 1.55–1.87)] and obesity (BMI $\geq$ 30) 2.4 fold [OR (adj) 2.44, 95% CI: 2.15–2.78]. Significantly, when the combined effect of obesity with oral contraceptive use and prothrombotic mutations on the risk of venous thrombosis was analyzed, the synergistic effect of combining risk factors was exemplified by the 24-fold higher thrombosis risk seen in obese women who used oral contraceptives [OR (adj) 23.78, 95% CI: 13.35–42.34]. The combined effect of factor V Leiden and obesity was associated with a 7.9-fold increased risk [OR (adj) 7.86, 95% CI: 4.70–13.15], and for obesity and the prothrombin gene mutation, the VTE risk was increased 6.6-fold [OR (adj) 6.58, 95% CI: 2.31–18.69] [10].

Furthermore, obesity features in a commonly used prediction score to help stratify risk VTE recurrence in women with unprovoked VTE. The HERDOO2 score (Hyperpigmentation, Edema, or Redness in either leg; raised D-dimer; obesity with BMI $\geq$ 30; or Older age, $\geq$ 65 years), can be used to select a group of women who are at lower risk of recurrence and can discontinue anticoagulation after short-term treatment [12].

The interplay between obesity and venous thromboembolism risk

The association between arterial thrombosis and obesity is explained partly by the strong relationship between central obesity and hypertension, type 2 diabetes mellitus, and dyslipidemia (all of which are major risk factors for arterial thrombosis). However, these are not established risk factors for VTE, and the precise mechanisms responsible for the increased risk of VTE seen with obesity are yet to be precisely determined [13].

Several mechanisms have been proposed to explain the increased risk of venous and arterial thromboembolism, including increased procoagulant activity, impaired fibrinolysis, increased inflammation, endothelial dysfunction, and altered lipid and glucose metabolism in metabolic syndrome [14]. Fat mass, independent of its distribution in the body, is positively associated with VTE and this is biologically plausible, because adipose tissue is more than an energy-storage organ; it is also metabolically active and secretes several biologically active substances [13]. Adipose tissue is known to produce several cytokines (known as adipokines), including leptin and adiponectin, tumor necrosis factor-α, and PAI-1 (plasminogen activator inhibitor-1) [14]. Enlarged fat cells produce a higher amount of these substances that normal-sized fat cells [13]. A number of these substances are associated with procoagulant activity or inhibition of fibrinolysis [13], and a decrease in blood fibrinolytic activity has been reported in obese patients [5]. The effect of inflammation may also be relevant, and IL-6 has also been implicated in mediating the link between abdominal obesity and VTE [15]. In addition, oxidative stress associated with adipose tissue leads to platelet activation, endothelial damage, and the shedding of thrombogenic endothelial cell-derived microparticles [16].

Estrogen produced by fat cells might also be a potential mediator of VTE risk, because plasma estrogen level is positively associated with obesity [13]. Lastly, obesity can be associated with venous stasis that promotes venous thrombosis [13]. Although poorly studied, larger sized deep veins with reduced flow velocities have been observed in obese subjects [17], and, hypothetically, there may be an additional effect from local compression on veins and valvular dysfunction [18]. Hence all three components of Virchow's triad can be present in obesity.

Hormonal contraception

Obese women are at increased risk of VTE at baseline, and while the risk of VTE will increase with hormonal contraceptive use, it should be remembered that these women are at significantly higher risk of pregnancy-related complications, and therefore safe and effective pregnancy prevention and planning are paramount [19]. Combined hormonal contraceptives do not appear to affect body weight (although this remains a significant concern for women) and provide additional noncontraceptive benefits that include the regulation of menses and a significant reduction in the risk of endometrial and ovarian cancer [8]. However, there are obvious concerns regarding safety of their use when considering the thromboembolic and cardiovascular risks, and some additional concern remains as to whether there is reduced efficacy due to the altered pharmacokinetics associated with obesity [8].

Hormonal contraception and venous thromboembolism risk

The increased risk of VTE associated with hormonal contraception was recognized in the 1960s in case reports published after the contraceptive pill was introduced [20] and supported by the findings of retrospective case–control studies conducted in the latter part of that decade. The mass of data that have accumulated since have concluded that estrogens increase the risk of developing VTE [5]. Overall, the use of combined hormonal contraceptives (CHC) confers a two- to fourfold increase in VTE risk, and greatest increase is within the first 3 months of use [21]. A Cochrane review published in 2014 established that the use of combined oral contraceptives (COCP) was associated with an increased VTE risk compared with nonuse (RR 3.5, 95% CI: 2.9–4.3), and that the risk of VTE increases with both the dose of estrogen and the type and dose of progestin [22]. A Medicines and Healthcare products Regulatory Agency (MHRA) drug safety update published in the same year underlined that the absolute risk of VTE with all low-dose CHC, that is, containing ethinylestradiol <50 μg, remains small, and products with the lowest risk of VTE are those containing the progestogens levonorgestrel (LNG), norethisterone, and norgestimate [23]. Table 27.1 summarizes the progestogen-specific estimates of VTE incidence replicated from the MHRA update.

VTE risk is not limited to oral preparations. A Danish cohort study assessed the relative and absolute risk of first-time VTE for users of oral contraceptives with different progestogens, different doses of estrogen, as well as in users of progesterone-only pills (POP) and hormone-releasing intrauterine devices (IUDs) [24]. The incident rate of VTE in nonusers of COCP was 3.7 per 10,000 WY. Women using COCP with drospirenone were at similar risk of VTE to those using COCP with desogestrel, gestodene, or cyproterone and higher that those using COCP with LNG. The risk of VTE was not reduced by using 20 μg ethinylestradiol instead of 30 μg ethinylestradiol in oral contraceptives with drospirenone. Importantly, progestogen-only products conferred no increased risk of VTE, whether taken as low-dose norethisterone pills, as desogestrel-only pills, or in the form of hormone-releasing IUDs [24]. Of note, two different estrogens are used in

TABLE 27.1 Risk of venous thromboembolism with combined hormonal contraceptives (CHC).

Progestogen in CHC (combined with ethinylestradiol, unless stated)	Relative risk versus levonorgestrel	Estimated incidence (per 10,000 women per year of use)
Nonpregnant nonuser	–	2
Levonorgestrel	Ref	5–7
Norgestimate, norethisterone	1.0	5–7
Gestodene, desogestrel, drospirenone	1.5–2.0	9–12
Etonogestrel, norelgestromin	1.0–2.0	6–12
Dienogest (combined with ethinylestradiol or estradiol)/nomegestrol acetate (estradiol)	To be confirmed[a]	To be confirmed[a]

[a] Further studies ongoing or planned to collect sufficient data to estimate risk for these products.

CHC: estradiol (E2), which is either micronized or valerate, and ethinyloestradiol (EE). For many years, and until 2009, EE was the only estrogen present in COCP. E2 had low bioavailability, but this has been enhanced by micronization and esterification [25]. Although the VTE risk of a particular combination of COCP containing E2V and dienogest appears to be lower than that associated with other COCPs, in women with VTE risk factors, such as obesity, it should not be viewed as a valid alternative to progestin-only contraceptives [26].

Focusing on nonoral hormonal contraception, a retrospective study of four national Danish registries, which included 1.9 million women, was conducted to assess the risk of VTE in current users of nonoral hormonal contraception. In nonusers, the incidence rate of confirmed VTE was 2.1 per 10,000 years. The RR of confirmed VTE associated with combined contraceptive transdermal patches and CHC vaginal rings (CVR) was significantly associated with VTE with a RR of 7.9 (95% CI: 3.5–17.7) and 6.5 (95% CI: 4.7–8.9), respectively, compared with nonusers [27]. The RR of VTE was modestly but not significantly increased in women who used subcutaneous implants (RR 1.4, CI: 0.6–3.4) but not in those who used the LNG IUD (RR 0.6, CI: 0.4–0.8) [27].

Despite the association with type and dose of progestin in combined contraceptive preparations, progestin-only oral preparations are generally considered low risk for VTE [9]. A previous meta-analysis assessing the risk of VTE in women taking progestin-only contraception concluded that there was no association between VTE risk and either progestin-only pills (RR 0.9, 95% CI: 0.57–1.45) or a progestin IUD (RR 0.61, 95% CI: 0.24–1.53), while the RR of VTE for users of an injectable progestin versus nonusers was 2.67 (1.29–5.53) [28].

Due to the increased of both VTE and arterial thromboembolic risks associated with COCP [8] and the reduced VTE risk of progestin-only methods of contraception, the latter are generally preferred for obese women and regarded as the safest option [29]. Progestin-only methods can be administered at various doses through a variety of delivery systems, including oral POP with norethisterone, LNG or desogestrel, injectable depo medroxyprogesterone acetate (DMPA), subcutaneous implant [releasing etonogestrel (ENG) or LNG], and LNG-releasing intrauterine systems (LNG-IUS) [8].

Obesity and contraceptive efficacy

There is ongoing concern that obese women may experience reduced efficacy and, therefore, contraceptive failure due to differences in pharmacokinetics. While the majority of observational and prospective studies do not indicate a decreased efficacy in CHC, a small increase in failure rate in women with BMI > 35 cannot be excluded [8]. There are limited data specifically examining obesity and data for women with BMI > 40 are sparse [19]. Intrauterine systems exert their effects locally, and an increased BMI would not be expected to affect effectiveness. The limited available evidence suggests that progesterone-only methods are effective in obesity [19]. Pharmacokinetic studies using the COCP and CVR have found reduced EE levels but maintained progestin levels in obese women together with maintained follicular suppression, which are reassuring from an efficacy viewpoint [8]. Similarly, there are concerns surrounding the use of ENG implants as to whether obese women receive a sufficient dose to suppress ovulation due the increased volume of distribution and altered plasma clearance in obese individuals. However, a cross-sectional study examining the relationship between ENG level and BMI in women using the contraceptive implant for more than 1 year found therapeutic ENG levels were maintained regardless of BMI, across a wide BMI range (21–56), supporting the use of ENG implants in obesity [29]. The limited evidence available also indicates that CHC methods remain effective in obesity [19], although the manufacturer's recommendations for the combined contraceptive patch (Evra) advised avoiding Evra in women who weigh ≥ 90 kg due to a concern over effectiveness [30].

Contraception and the fear of weight gain

Many women avoid using hormonal contraception due to a belief that contraceptive hormones promote weight gain [31] and perceived weight gain is one of the leading causes of discontinuation [19]. However, both adolescent and reproductive women tend to gain weight over time regardless of contraceptive use, which can make proving a causal association difficult unless accounted for in studies, and perceptions of weight gain while using contraception have been shown to be incongruent with actual weight [19].

The results of Cochrane reviews to evaluate the potential association between contraceptive use, and weight gain have examined combination contraceptive use and progesterone-only contraceptives. They concluded that most comparisons of different combination contraceptives showed no substantial difference in weight and that while available evidence was insufficient to determine the effect of combination contraceptives on weight, no large effect was evident [32]. For progesterone-only contraception (POC) the overall quality of evidence was moderate to low, and limited evidence of weight gain when using POC was found. Mean gain was less than 2 kg for most studies up to 12 months. Weight change for the POC group generally did not differ significantly from that of the comparison group using another contraceptive [33].

Although no clear causal relationship has been established for the majority of contraceptive methods, it does appear that the use of DMPA can be associated with weight gain [19]. In a recent retrospective study to determine the prevalence of excessive weight gain in association with DMPA use in young women (aged 10–24 years) in Thailand, out of 231 DMPA users, 28 women (12.1%, 95% CI: 7.8–16.3) had an excessive weight gain at 6 months. Age, baseline BMI, or race did not affect the likelihood of excessive weight gain. Mean baseline BMI was 21.00 (SD ± 3.49). Six of the thirteen (46.2%) patients, who had gained excessive weight at 6 months and who continued DMPA use, gained even more weight (>10% of their baseline weight) at 12 months. Therefore while the majority of adolescent girls using DMPA had no excessive weight gain in 6 months, those who had excessive weight at 6 months were at higher risk of gaining even more weight at 1 year [34].

A multicentre randomized controlled trial to evaluate weight changes in women randomized to either the ENG- or LNG-releasing contraceptive implants and to compare with users of the TCu380A IUD was conducted in Switzerland and included a total of 995, 997, and 971 users in the ENG implant, LNG implant, and IUD groups, respectively. At 36 months of use, ENG and LNG implant users had similar significant mean weight increase of 3.0 kg (95% CI: 2.5–3.5) and 2.9 kg (95% CI: 2.4–3.4), respectively ($P < .0001$), while IUD users had an increase of 1.1 kg (95% CI: 0.5–1.7) ($P = .0003$). On adding the group–time interaction term to the stratified baseline weight models, implant users gained 0.759 kg [standard error (SE) 0.11] and 0.787 kg (SE 0.22) more weight than their IUD user counterparts per year since placement if their baseline weight was in the category 51–69 kg ($P < .0001$) or ≥ 70 kg ($P = .0005$), respectively. In conclusion, ENG and LNG implant as well as IUD users had a small but significant weight increase with little clinical significance during the 3 years of follow-up, and it was slightly higher among implant than IUD users weighing >50 kg [35].

It is important to note that obese women are underrepresented in the majority of studies researching the impact of contraception on weight [19]. However, an older prospective study from 2006, which examined weight changes over 18 months in a large cohort of obese and nonobese adolescent girls initiating DMPA, OC, or no hormonal contraceptive method (control), found that baseline obesity was associated with weight gain. Adolescent girls who were obese at initiation of DMPA gained significantly more weight than did obese girls starting OC or control ($P < .001$ for both). At 18 months, mean weight gain was 9.4, 0.2, and 3.1 kg for obese girls receiving DMPA, receiving OC, and control, respectively. Weight gain in obese girls receiving DMPA was also greater than weight gain in all nonobese categories (4.0 kg, DMPA; 2.8 kg, OC; 3.5 kg, control; $P < .001$). A significant interaction ($P = 0.006$) between length of time receiving DMPA and weight gain was evident for obese subjects [36].

Weight-loss treatment and contraception

It is recommended that women with raised BMI lose weight; however, in addition to dietary and lifestyle modifications, weight-loss treatment may also comprise antiobesity medications, such as orlistat and laxatives. As such, women should be advised that it is possible that medications that induce diarrhea and/or vomiting could reduce the effectiveness of oral contraceptives [19].

Guideline recommendations

The European Society of Contraception Statement on Contraception in Obese Women recommends that CHC should only be used if no other acceptable contraceptive methods, such as progestin-only contraceptives or IUD, are available or acceptable and that obese women should be informed of their risk of thrombosis and counseled on the added risk of taking CHC [8]. Advising on progestin-only contraceptives, the consensus highly recommends desogestrel 75 μg as an important and safe option. ENG-releasing implants are also an option but caution is recommended as ENG levels decline over time, and earlier replacement of the plant may be considered in some obese women. DMPA can be used but has been associated with a negative effect on insulin resistance. Copper IUDs are not affected by BMI and do not affect metabolic parameters or VTE risk and are, therefore, highly recommended while the LNG-IUS is also considered an alternative, particularly in women with heavy menstrual bleeding [8].

The UK Medical Eligibility Criteria for Contraceptive Use (UKMEC) assigns different categories to guide healthcare providers and when applied in a clinical setting, a UKMEC Category 1 indicates that there is no restriction for use. A UKMEC Category 2 indicates that the method can generally be used, but more careful follow-up may be required. A contraceptive method with a UKMEC Category 3 can be used; however, it may require expert clinical judgment and/or referral to a specialist contraception provider since use is not usually recommended unless other methods are not available or acceptable. A UKMEC Category 4 indicates that use in that condition poses an unacceptable health risk and should not be used [37].

The Faculty of Sexual and Reproductive Healthcare (FSRH) guideline from January 2019 (Amended July 2019) recommends women with BMI < 35 kg/m^2 generally can use CHC (UKMEC 2). Women with BMI ≥ 35 generally should not use CHC (UKMEC 3), although CHC may be prescribed by a specialist provider. It is also recommended that BMI should be documented before starting CHC [38]. There is very limited safety information regarding the use of hormonal and nonhormonal contraception in obese women who have additional comorbidities and very little safety information for women using contraceptives with a BMI ≥ 40 [19].

The FSRH guidance, specifically focusing on increased BMI and contraception, was published in April 2019 entitled: Overweight, Obesity and Contraception. In summary, for obesity alone, that is, without coexistent medical conditions, all progestogen-only contraceptives and intrauterine contraceptives are categorized as UKMEC1, that is, no restriction of use. All estrogen-containing contraceptives (including COCP, patch and ring) are categorized as UKMEC 2 or 3 depending on BMI largely due to increased VTE risk. For women with raised BMI and additional risk factors for cardiovascular disease, the copper IUD is categorized as UKMEC 1, the LNG-IUS, contraceptive implants and POP are UKMEC 2, and DMPA, norethisterone enanthate and CHC are classed as UKMEC 3 [19]. Table 27.2 summarizes the UKMEC recommendations from this guidance.

Hormone replacement therapy

The risk of VTE increases two- to fourfold in postmenopausal women using hormone replacement therapy (HRT) [9], and obesity further increases this risk. Postmenopausal HRT typically contained lower doses of hormones than the COCP, and, compared with COCP use,

TABLE 27.2 UK Medical Eligibility Criteria for Contraceptive Use (UKMEC) recommendations for contraceptives in obese women [19].

Method	BMI	UKMEC category (BMI alone)	UKMEC category if additional cardiovascular disease risk factors present	History of bariatric surgery
COC (vaginal ring, patch)	≥ 30–34	2	3	2
	≥ 35	3		3
Progesterone-only pill	≥ 30–34	1	2	1[a]
	≥ 35			
Progesterone-only implant	≥ 30–34	1	2	1
	≥ 35			
Progesterone-only injectable (DMPA or NET-EN)	≥ 30–34	1	3	1
	≥ 35			
Copper intrauterine device	≥ 30–34	1	1	1
	≥ 35			
Levonorgestrel-releasing intrauterine system	≥ 30–34	1	2	1
	≥ 35			

BMI, Body mass index; COC, combined oral contraceptive; DMPA, depot medroxyprogesterone acetate; NET-EN, norethisterone enanthate.
[a]UKMEC categories for contraceptive use after bariatric surgery related to safety of use rather than effectiveness. Safety considerations after bariatric surgery relate to ongoing high BMI. Women should be advised that the effectiveness of oral contraception (OC), including emergency oral contraception, could be reduced by bariatric surgery and OC should be avoided in favor of nonoral methods.

concerns regarding the link with VTE were not raised until later. Indeed, early studies evaluating the possible relationship between HRT and VTE did not suggest an association [39]. The results from studies that were adequately powered to fully assess the relationship started to become available in the 1990s and consistently demonstrated an increased VTE risk [39], and later prospective randomized controlled studies removed any remaining doubt [40,41]. In the Women's Health Initiative trial the hazard ratio of VTE adjusted for age and treatment was 1.96 (95% CI: 1.33−2.88) for overweight and 3.09 (95% CI: 2.13−4.49) for obesity. The incidence rate of VTE was the highest among obese women taking a combination of oral estrogen and progestin, and these women had a nearly sixfold higher risk than normal-weight women who were taking placebo [40].

Obese women now make up an increasing proportion of peri- and postmenopausal women cared for by gynecologists and general practitioners, and while obesity rates are increasing, it is also noteworthy that the natural process of menopause itself may be obesogenic [42]. Various longitudinal studies have lent support to the increase in BMI, waist circumference, and fat, which is associated with the menopause and may not simply be due to the ageing process [42]. Intraabdominal fat accumulation rapidly rises in the 2 years before menopause that alters the hormonal milieu and can accelerate the development of metabolic syndrome [42]. Furthermore, obese women may experience more menopausal symptoms when compared to their leaner counterparts. In the SWAN study a community-based survey of over 16,000 women aged 40−55 years, hot flashes or night sweats, urine leakage, and stiffness or soreness were associated with a high BMI (OR 1.15−2.18 for women with a BMI ≥ 27 vs 19−26.9 kg/m^2) [43].

Importantly, the VTE risk varies according to the mode of administration of estrogen [4]. Data suggest that patients using transdermal preparations of HRT have a lower risk of VTE compared with those on an oral preparation [9]. An analysis to evaluate the impact of the route of estrogen administration on the association between BMI and VTE risk was performed in women included in the ESTHER study, a multicentre case−control study aimed at investigating the impact of the route of estrogen administration on VTE risk among postmenopausal women aged 45−70 years [44]. The OR for VTE was 2.5 (95% CI: 1.7−3.7) for overweight (BMI 25−30) and 3.9 (95% CI: 2.2−6.9) for obese (BMI > 30) women. Oral, not transdermal, estrogen was associated with an increased. Compared with nonusers with normal weight, the combination of oral oestrogen use and overweight or obesity further enhanced VTE risk: OR 10.2 (95% CI: 3.5− 30.2) and OR 20.6 (95% CI: 4.8−88.1) respectively.

However, transdermal users with increased BMI had a similar risk to nonusers with increased BMI: OR 2.9 (95% CI: 1.5−5.8) and OR 2.7 (95% CI: 1.7−4.5), respectively, for overweight; OR 5.4 (95% CI: 2.1−14.1) and OR 4.0 (95% CI: 2.1−7.8), respectively, for obesity [44]. A record linkage study using NHS hospital admission and death records that evaluated 1,058,259 postmenopausal UK women (as part of the Million Women Study) found that during 3.3 million years of follow-up, 2200 women were diagnosed with VTE. The risk varied by formulation and while the RR was significantly greater for oral estrogen-progestin than oral estrogen-only therapy [RR = 2.07 (95% CI: 1.86−2.31) vs 1.42 (1.21−1.66)], there was no increased risk with transdermal estrogen-only therapy [0.82 (0.64−1.06)] [45]. Two recent nested case−control studies performed to assess the association between risk of VTE and use of different types of HRT using UK general practice databases have confirmed the safety of using transdermal preparations. A number of 80 396 women aged 40−79 with a primary diagnosis of VTE between 1998 and 2017 were matched to 391,494 female controls. Overall, 5795 (7.2%) women who had VTE and 21,670 (5.5%) controls had been exposed to HRT within 90 days before the index date. Of these two groups, 4915 (85%) and 16 938 (78%) women used oral therapy, respectively, which was associated with a significantly increased risk of VTE compared with no exposure (adjusted OR 1.58, 95% CI: 1.52−1.64), for both estrogen-only preparations (1.40, 1.32−1.48) and combined preparations (1.73, 1.65−1.81). Transdermal preparations were not associated with risk of VTE that was consistent for different regimens (overall adjusted OR 0.93, 95% CI: 0.87−1.01) [46].

Thrombin generation has been demonstrated to be significantly increased in women who use HRT administered by the oral route compared with the transdermal route, and this may be explained by the avoidance of hepatic first-pass metabolism of estrone (the main metabolite of oral estradiol) when administered by the transdermal route [47]. When the effects of oral, transdermal, and no HRT (controls) on thrombin generation were examined in postmenopausal women, all parameters of thrombin generation were altered in women using oral HRT as compared with controls ($P < .001$ for all comparisons). No such differences were found in women using transdermal HRT. Estrone levels correlated with peak thrombin generation ($R = 0.451$, $P < .001$) in women using oral HRT, but there was no correlation in women using the transdermal route [47]. Similarly, a number of additional studies have demonstrated statistically significant increased levels of markers of coagulation activation (such as prothrombin fragment 1 + 2) in patients using oral HRT but not with transdermal HRT or placebo [39].

In summary, in obese women, the use of transdermal estrogen does not appear to confer an increased risk of VTE, and, at least from a VTE perspective, this would appear to be a safer option in obese menopausal women who require HRT.

Assisted conception

Assisted reproductive technologies (ART) are increasingly used in couples with fertility issues, and there is an association with an increased risk of VTE. VTE can occur during superovulation induction or in the following pregnancy, with a reported incidence after in vitro fertilization (IVF) of between 0.1% and 0.5% of treatment cycles [48]. Ovarian hyperstimulation syndrome (OHSS) is an iatrogenic and potentially fatal IVF complication, which is often associated with the administration of human chorionic gonadotropin (hCG) to induce ovulation after ovarian stimulation or endogenous hCG produced during pregnancy [48]. A fluid shift into the third space results in hemoconcentration and a hypercoagulable state [48], and it is conceivable that obesity would further increase this risk, although data examining the effect obesity on VTE in this particular group of patients are lacking. In an analysis of patients enrolled in the RIETE registry between March 2001 and October 2016, women of childbearing age who experienced VTE during ART procedures were identified and defined as the study population and the remaining women with VTE events that occurred outside ART representing the reference cohort [48]. Overall, 41 (0.6%) out of 6718 women of childbearing age with VTE experienced an ART-related event, but mean BMI was 26.4 (SD 6.4) [48].

A prospective Italian cohort study, which examined 684 ART cycles carried out by 234 women, reported a statistically significant increase in VTE in these women when compared with a reference population of pregnant women who had conceived naturally [49]. A population-based cohort study of all (964,532) inpatient deliveries in Sweden over a 10-year period (1999–2008) reported an antepartum incidence of VTE of 0.27% in women receiving IVF compared with 0.1% in the background population [OR 2.7 (95% CI: 2.1%–3.6%)], and the greatest risk was in the first trimester (0.17% compared with 0.02% in the general population; OR 9.8; 95% CI: 7.5%–14.3%) [50]. There was no statistically significant increase in VTE associated with IVF in the second or third trimester or in the postpartum period. Women conceiving with frozen embryos were not at increased risk of VTE, presumably due to less frequent or absent ovulation induction [50]. The VTEs in conjunction with IVF were diagnosed at a mean gestational age of 62 days; there was no increased risk of VTE related to frozen embryo replacement cycles or IVF after the first trimester [50]. The 6%–7% of IVF pregnancies that were complicated by OHSS showed a 100-fold increased risk of VTE, as opposed to the fivefold increased risk seen in the absence of OHSS [50]. Interestingly, in a similar national-based cohort Danish study, the overall VTE incidence rate ratio during IVF pregnancies compared with reference pregnancies was 3.0 (95% CI: 2.1–4.3), and ovarian hyperstimulation syndrome did increase the VTE [51].

The current UK guidelines for reducing the risk of VTE during pregnancy and the puerperium from the Royal College of Obstetricians and Gynecologists advise that while both obesity and IVF/ART are risk factors for antenatal VTE, thromboprophylaxis is not recommended in the absence of additional risk factors [52]. Female obesity is, however, negatively associated with pregnancy outcomes. D-Dimers have been shown to be increased in maternal obesity during pregnancy [53], and obesity may exaggerate the prothrombotic state of pregnancy, even in early pregnancy when IVF-related VTE occurs.

Gynecological surgery

While obesity and surgery are known risk factors for VTE, there is limited information about the independent effects of obesity on the incidence of postoperative VTE. Questionnaire data were linked from the aforementioned Million Women Study with hospital admission and death records to examine the risk of VTE in relation BMI both in the absence of surgery and in the first 12 weeks following an operation. Overall, 1,170,495 women (mean age, 56.1 years) recruited in 1996–2001 through the National Health Service Breast Screening Programme in England and Scotland were followed for an average of 6 years, during which time 6438 were admitted to hospital or died of VTE. The adjusted RRs of VTE increased progressively with increasing BMI and women with a BMI ≥ 35 were 3–4 times as likely to develop VTE as those with a BMI 22.5–24.9 [RR 3.45 (95% CI: 3.09–3.86)]. Overweight and obese women were more likely than lean women to be admitted for surgery and also to develop postoperative VTE. During a 12-week period without surgery, the incidence rates of VTE per 1000 women with a BMI < 25 and ≥ 25 were 0.10 (0.09–0.10) and 0.19 (0.18–0.20); the corresponding rates in the 12 weeks following day and inpatient surgery were, respectively, about 4 and 40 times higher [54]. VTE risk increases with increasing BMI, and this is particularly relevant after surgery.

Current UK NICE guidelines for reducing the risk of hospital-associated VTE (NG89) recommend that patients undergoing abdominal surgery (including gynecological) are offered VTE prophylaxis with mechanical (antiembolism stockings or intermittent pneumatic compression) until mobility returns to normal, and that pharmacological thromboprophylaxis should be added (where VTE risk outweighs bleeding risk) for a minimum of 7 days [55]. As regards extended thromboprophylaxis (to 28 days), this is usually reserved for major cancer surgery in the abdomen [55]. Pharmacological thromboprophylaxis in this context comprises low-molecular-weight heparin or fondaparinux, and the guidelines acknowledge that while higher doses are often used in obesity, there is continued uncertainty about the optimal dose in obese patients [55].

Gynecological cancer

The effect of obesity on VTE risk is also relevant for VTE in gynecological cancers. In a population of patients diagnosed and treated for ovarian cancer at a tertiary hospital database, the overall incidence of VTE was 9.7% (33) in 344 patients. The median (interquartile range) age of the patients was 57 (48−67) years and 21% of patients had a BMI $\geq$ 30. Significantly, on multivariate analysis, BMI $\geq$ 30 was an independent risk factor for VTE [OR = 3.1 (95% CI: 1.2−7.9)] [56].

Conclusion

The inexorable rise in obesity has significant and wide-reaching implications for all aspects of health care, and gynecology is no exception. From a thrombosis perspective, obesity is an important factor when hormonal treatment is being considered, whether for contraception or HRT, and the increased VTE risk is also relevant for ARTs (as well as the ensuing pregnancy). Obesity increases the risk of VTE following gynecological surgery and that associated with gynecological cancer. There are limited data for women with obesity, particularly in morbid and super-morbid obesity, and while public health measures aim to tackle the obesity epidemic, the increasing number of these patients offers the chance for further studies.

References

[1] Borch KH, Brækkan SK, Mathiesen EB, Njølstad I, Wilsgaard T, Størmer J, et al. Anthropometric measures of obesity and risk of venous thromboembolism. The Tromsø Study. Arterioscler Thromb Vasc Biol 2010;30:121−7.

[2] Baker C. House of commons briefing paper, obesity statistics, number 3336. <https://researchbriefings.files.parliament.uk/documents/SN03336/SN03336.pdf>; 2019.

[3] Practice Committee of the American Society for Reproductive Medicine.. Combined hormonal contraception and the risk of venous thromboembolism: a guideline. Fertil Steril 2017;107(1):43−51.

[4] Heit JA, Spencer FA, White RH. The epidemiology of venous thromboembolism. J Thromb Thrombolysis 2016;41:3−14.

[5] Coon WW. Epidemiology of venous thromboembolism. Ann Surg 1977;186:149−64.

[6] Barker NW, Nygaard KK, Walters W, Priestly JT. A statistical study of postoperative venous thromboembolism. II. predisposing factors. Mayo Clin Proc 1941;16:1.

[7] Ageno A, Auizzato A, Garcia D, Inerti D. Epidemiology and risk factors of venous thromboembolism. Semin Thormb Hemost 2006;32(7):651−8.

[8] Merki-Feld GS, Skouby S, Serfaty D, Lech M, Bitzer J, Crosignani PG, et al. European Society of Contraception Statement on Contraception in Obese Women. Eur J Contracept Reprod Health Care 2015;20:19−28.

[9] Speed V, Roberts LN, Patel JP, Arya R. Venous thromboembolism and women's health. Br J Haematol 2018;183:346−63.

[10] Pomp ER, le Cessie S, Rosendaal FR, Doggen CJ. Risk of venous thrombosis: obesity and its joint effect with oral contraceptive use and prothrombotic mutations. Br J Haematol 2007;139:289−96.

[11] Goldhaber SZ, Grodstein F, Stampfer MJ, Manson JE, Colditz GA, Speizer FE, et al. A prospective study of risk factors for pulmonary embolism in women. JAMA 1997;277:642−5.

[12] Kruger P, Eikelboom J. HERDOO2 identified women at low risk for recurrence after 5 to 12 months of anticoagulation for a first unprovoked VTE. Ann Intern Med 2017;167(6):JC33.

[13] Severinsen MT, Kristensen SR, Johnsen SP, Dethlefsen C, Tjønneland A, Overvad K. Anthropometry, body fat and venous thromboembolism: a Danish follow-up study. Circulation 2009;120(19):1850−7.

[14] Chitongo PB, Roberts LN, Yang L, Patel RK, Lyall R, Luxton R, et al. Visceral adiposity is an independent determinant of hypercoagulability as measured by thrombin generation in morbid obesity. TH Open 2017;1(2):e146−54 5.

[15] Matos MF, Lourenço DM, Orikaza CM, Gouveia CP, Morelli VM. Abdominal obesity and the risk of venous thromboembolism among women: a potential role of interleukin-6. Metab Syndr Relat Disord 2013;11(1):29−34.

[16] Reaven G, Scott EM, Grant PJ, Lowe GD, Rumley A, Wannamethee SG, et al. Hemostatic abnormalities associated with obesity and the metabolic syndrome. J Thromb Haemost 2005;3:1074−85.

[17] Engelberger RP, Indeermuhle A, Baumann F, et al. Diurnal changes of lower leg volume in obese and non-obese subjections. Int J Obes 2014;38:801−5.

[18] Hunt BJ. The effect of BMI on haemostasis: implications for thrombosis in women's health. Thromb Res. 2017;151(Suppl. 1)::53−1:55.

[19] FSRH guideline, overweight, obesity & contraception. <https://www.fsrh.org/standards-and-guidance/documents/fsrh-clinical-guideline-overweight-obesity-and-contraception/>; 2019.

[20] Jordan WM, Anand JK. Pulmonary embolism. Lancet 1961;278:1146−7.

[21] van Hycklama Vlieg A, Helmerhorst FM, Vandenbroucke JP, Doggen CJ, Rosendaal FR. The venous thrombotic risk of oral contraceptives, effects of oestrogen dose and progestogen type: results of the MEGA case−control study. BMJ 2009;339:b2921.

[22] de Bastos M, Stegeman BH, Rosendaal FR, Van Hylckama Vlieg A, Helmerhorst FM, Stijnen T, et al. Combined oral contraceptives: venous thrombosis. Cochrane Database Syst Rev 2014; (Issue 3):CD010813 Art. No.:.

[23] MHRA Drug Safety Update volume 7 issue 7, February 2014: A2.

[24] Lidegaard O, Nielsen LH, Skovlund CW, Skjeldestad FE, Løkkegaard E. Risk of venous thromboembolism from use of oral contraceptives containing different progestogens and oestrogen doses: Danish cohort study, 2001−2009. BMJ 2011;343:d6423.

[25] Fruzzetti F, Cagnacci A. Venous thrombosis and hormonal contraception: what's new with estradiol-based hormonal contraceptives? Open Access J Contracept 2018;9:75—9.

[26] Cushman M. Epidemiology and risk factors for venous thrombosis. Semin Hematol 2007;44(2):62—9.

[27] Lidegaard A, Nielsen LH, Skovlund CW, Skjeldestad FE, Løkkegaard E. Risk of venous thromboembolism from use of oral contraceptives containing different progestogens and oestrogen doses: Danish cohort study, 2001—2009. BMJ 2012;343:d6423.

[28] [P] Mantha S, Karp R, Raghavan V, Terrin N, Bauer KA, Zwicker JI. Assessing the risk of venous thromboembolic events in women taking progestin-only contraception: a meta-analysis. BMJ 2012;345:e4944.

[29] Morrell KM, Cremers S, Westhoff CL, Davis AR. Relationship between etonogestrel level and BMI in women using the contraceptive implant for more than 1 year. Contraception 2016;93:263—5.

[30] Summary of Product Characteristics for Evra, <https://www.medicines.org.uk/emc/medicine/12124>; 2019 [accessed 24.09.19].

[31] Rocha ALL, Campos RR, Miranda MMS, Raspante LBP, Carneiro MM, Vieira CS, et al. Safety of hormonal contraception for obese women. Expert Opin Drug Saf 2017;16(12):1387—93.

[32] Gallo MF, Lopez LM, Grimes DA, Carayon F, Schulz KF, Helmerhorst FM. Combination contraceptives: effects on weight. Cochrane Database Syst Rev 2014;29(1):CD003987.

[33] Lopez LM, Edelman A, Chen M, Otterness C, Trussell J, Helmerhorst FM. Progestin-only contraceptives: effects on weight. Cochrane Database Syst Rev 2013;2(7):CD008815.

[34] Jirakittidul P, Somyaprasert C, Angsuwathanaa S. Prevalence of documented excessive weight gain among adolescent girls and young women using depot medroxyprogesterone acetate. J Clin Med Res 2019;11(5):326—31.

[35] Bahamondes L, Brache V, Ali M, Habib N, WHO Study Group on Contraceptive Implants for Women. A multicenter randomized clinical trial of etonogestrel and levonorgestrel contraceptive implants with nonrandomized copper intrauterine device controls: effect on weight variations up to 3 years after placement. Contraception 2018;98(3):181—7.

[36] Bonny AE, Ziegler J, Harvey R, Debanne SM, Secic M, Cromer BA. Weight gain in obese and nonobese adolescent girls initiating depot medroxyprogesterone, oral contraceptive pills, or no hormonal contraceptive method. Arch Pediatr Adolesc Med 2006;160(1):40—5.

[37] UKMEC April 2016 (Amended December 2017), <https://www.fsrh.org/standards-and-guidance/external/ukmec-2016-digital-version/>.

[38] FSRH Guideline - Combined Hormonal Contraception 2019, <https://www.fsrh.org/documents/combined-hormonal-contraception/>.

[39] Eisenberger A, Westfoff C. Hormone replacement therapy and venous thromboembolism. J Steroid Biochem Mol Biol 2014;142:76—82.

[40] Cushman M, Kuller LH, Prentice R, Rodabough RJ, Psaty BM, Stafford RS, et al. Estrogen plus progestin and risk of venous thrombosis. JAMA. 2004;292(13):1573—80.

[41] Hulley S, Furberg C, Barrett-Connor E, Cauley J, Grady D, Haskell W, Knopp R, et al. Noncardiovascular disease outcomes during 6.8 years of hormone therapy: Heart and Estrogen/progestin Replacement Study follow-up (HERS II). JAMA 2002;288(1):58—66.

[42] Verhaeghe J. Menopause care for obese and diabetic women. Facts Views Vis Obgyn 2009;1(2):142—52.

[43] Gold EB, Sternfeld B, Kelsey JL, Brown C, Mouton C, Reame N, et al. Relation of demographic and lifestyle factors to symptoms in a multi-racial/ethnic population of women 40—55 years of age. Am J Epidemiol 2000;152(5):463—73.

[44] Canonico N, Oger E, Conard H, Meyer G, Le Vesque H, Trillot N, et al. Obesity and risk of venous thromboembolism among postmenopausal women: differential impact of hormone therapy by route of oestrogen administration. The ESTHER study. J Thromb Haemost 2006;4:1259—65.

[45] Sweetland S, Beral V, Balkwill A, Liu B, Benson VS, Canonico M, et al. Venous thromboembolism risk in relation to use of different types of postmenopausal hormone therapy in a large prospective study. Thromb Haemost 2012;10(11):2277—86.

[46] Vinogradova Y, Coupland C, Hippisley-Cox J. Use of hormone replacement therapy and risk of venous thromboembolism: nested case—control studies using the QResearch and CPRD databases. BMJ 2019;364:k4810.

[47] Bagot CN, Marsh MS, Whitehead M, Sherwood R, Roberts L, Patel RK, et al. The effect of estrone on thrombin generation may explain the different thrombotic risk between oral and transdermal hormone replacement therapy. J Thromb Haemost 2010;8(8):1736—44.

[48] Grandone E, Di Micco PP, Villani M, Colaizzo D, Fernández-Capitán C, Del Toro J, et al. Venous thromboembolism in women undergoing assisted reproductive technologies: data from the RIETE Registry. Thromb Haemost 2018;118(11):1962—8.

[49] Villani M, Dentali F, Colaizzo D, Tiscia GL, Vergura P, Petruccelli T, et al. Pregnancy-related venous thrombosis: comparison between spontaneous and ART conception in an Italian cohort. BMJ Open 2015;5:e008213.

[50] Rova K, Passmark H, Lindqvist PG. Venous thromboembolism in relation to in vitro fertilization: an approach to determining the incidence and increase in successful cycles. Fertil Steril 2012;979(1):95—100.

[51] Hansen AT, Kesmodel US, Juul S, Hvas AM. Increased venous thrombosis incidence in pregnancies after in vitro fertilization. Hum Reprod 2014;29(3):611—17.

[52] RCOG Green Top Guidelines Reducing the Risk of Venous Thromboembolism during Pregnancy and the Puerperium, Green-top Guideline No. 37a, April 2015

[53] Grossman KB, Arya R, Peixoto AB, Akolekar R, Staboulidou I, Nicolaides K. Maternal and pregnancy characteristics affect plasma fibrin monomer complexes and D-dimer reference ranges for venous thromboembolism in pregnancy. Am J Obstet Gynecol 2016;215(4):466.e1—8.

[54] Parkin L, Sweetland S, Balkwill A, Green J, Reeves G, Beral V. Body mass index, surgery, and risk of venous thromboembolism in middle-aged women: a cohort study. Circulation 2012;125(15):1897—904.

[55] National Institute for Health and Care Excellence (NICE) guideline [NG89], <https://www.nice.org.uk/guidance/ng89/resources/resource-impact-statement-4787147917>.

[56] Abu Saadeh F, Norris L, O'Toole S, Gleeson N. Venous thromboembolism in ovarian cancer: incidence, risk factors and impact on survival. Eur J Obstet Gynecol Reprod Biol 2013;170:214—18.

Chapter 28

Obesity and cardiovascular disease in reproductive health

Isioma Okolo[1] and Tahir A. Mahmood[2]

[1]Obstetrics & Gynaecology, NHS Lothian, Edinburgh, United Kingdom, [2]Department of Obstetrics and Gynaecology, Victoria Hospital, Kirkcaldy, United Kingdom

Introduction

Since 1975 the incidence of obesity has increased by threefold worldwide. In 2016 over 1.9 billion adults in the world were classed as overweight and 650 million were classed as obese [1,2]. In 2016, 41 million children under the age of 5 were classed as obese or overweight, while 340 million children and adolescents aged 5–19 years were classed as overweight or obese. Globally, obesity accounts for 17.9 million deaths annually. A significant proportion of these deaths are due to cardiovascular disease (CVD). In the United Kingdom, one in five women of reproductive age is now obese [3,4].

In the United Kingdom, 1 in 12 women dies from coronary heart disease (CHD). CHD kills more than twice as many women in the United Kingdom than breast cancer [5]. Overweight and obese women suffer excess mortality and morbidity from ischemic heart disease (HR 1.35, CI 1.28–143), stroke (HR 1.3, CI 1.19–1.42) compared to women with normal body mass index (BMI) [6].

Obesity prepregnancy and antenatally is a recognized risk factor for the metabolic syndrome in later life—diabetes, hypertension, and ischemic heart disease. CVD has been reported as the leading cause of maternal mortality in the United Kingdom since the 2000 confidential enquiry into maternal deaths. In the most recent 2019 report, one-third of mothers who died were classified as obese [7].

Around 29% of children in the United Kingdom are classified as overweight or obese. It is well recognized that the maternal in utero environment has an impact on fetal programing from neonatal health to long-term adult health [8]. The exact pathways that map this progression of maternal and fetal cardiovascular risk are still not well understood. Several animal studies have been used to explore these correlations.

Risk assessment in obese individuals

The most commonly used anthropometric tool used to assess relative weight and classify obesity is the BMI [9]. Additional risk factors for CVD include smoking, high-/low-density lipoproteins, cholesterols, hypertension, and dysfunction in glucose metabolism (Table 28.1).

It is not known to which extent these factors might act as independent contributors to increase morbidity. However, in obese individuals, these risk factors are often cumulative and may act as confounding factors. For example, overweight and obese individuals who smoke have increased levels of tumor necrosis factor alpha (TNF-α), a powerful cytokine that might decrease adiponectin levels and induce insulin resistance. Lower adiponectin levels found in chronic smokers confirm that they are insulin resistant. Endothelial dysfunction, an early marker of atherosclerosis, is present in obese individuals as well as chronic cigarette smokers. Therefore smoking among obese people will compound endothelial dysfunction and increase risk of CVD [11].

Obstructive sleep apnea (OSA) is more prevalent in morbidly obese individuals. OSA is also more common in smokers, women, in particular postmenopausal women. The male-to-female ratio is 2:1 but the gap narrows as women reach postmenopausal age [12]. OSA is associated with increased cardio metabolic diseases—systemic hypertension, ischemic heart disease or atherosclerosis, diastolic dysfunction, congestive heart failure, cardiac arrhythmias, stroke, and increased mortality or sudden death [12].

Being overweight or obese is linked with elevated oxidative stress and systemic inflammation. There is activation of the coagulation cascade, disturbance of the renin–angiotensin system, and most importantly enhanced lipid and protein

TABLE 28.1 Contribution of various factors in patients with cardiovascular disease.

↑ Risk of cardiovascular disease	Modifiable factors to reduce secondary risk for cardiovascular disease
Higher Smoking Hypertension Dyslipidemia Diabetes Abdominal obesity and metabolic syndrome Visceral obesity/ectopic fat Physical inactivity Poor cardiorespiratory fitness Poor nutritional quality ↑ Waist–hip ratio	↓ Weight loss ↓ BMI ↓ Waist/hip ratio ↑ Exercise to improve cardiorespiratory fitness ↓↓ Smoking ↑ Lean body mass

Source: Adapted from M. Bastien, P. Poirier, I. Lemieux and J.P. Despres, Overview of epidemiology and contributions of obesity to cardiovascular disease, Prog Cardiovasc Dis 56, 2014, 361–381[9].

oxidation resulting in the generation of oxidized low-density lipoproteins (LDLs). A decrease in oxidative stress after dietary restriction and weight loss has been reported in obese individuals [12].

The significance of adipose distribution

The BMI is a relatively simple anthropometric index of total adiposity which does not discriminate between muscle and fat mass. The latter is associated with metabolic abnormalities. For instance, markers of absolute and relative accumulation of abdominal fat such as elevated waist circumference and waist-to-hip ratio have been associated with an increased risk of myocardial infarction, hypertension, heart failure, and total mortality in patients with CVD [13,14]. An increase in both waist circumference and waist-to-hip ratio predicts an increased risk of CVD in men and women. The Nurses' health study found that a 1 cm increase in waist circumference and a 0.01-U increase in waist:hip ratio were, respectively, associated with a 2% and 5% increase in the risk of future CVD events [15].

Therefore in order to predict high risk, the BMI should be measured alongside specific indices of fat distribution, including waist circumference, waist-to-hip ratio, or weight-to-height ratio. Visceral adiposity can be measured accurately by computed tomography, magnetic resonance imaging, and with less precision by dual energy X-ray absorptiometry.

Based on experts' consensus, the WHO has proposed six specific cutoff values of waist circumference associated with increased CVD risk—94 cm in men and 80 cm in women for increased risk, 102 cm in men and 88 cm in women for substantially increased risk [15].

Pathogenesis of visceral obesity

Adipose tissue is now recognized as a key metabolically active endocrine organ. The amount and distribution of adipose tissue determines the level of metabolic risk. Visceral obesity is associated with chronic inflammation and insulin resistance.

This induces a variety of structural adaptations in cardiovascular structures and function, as well as in other vital organs and tissues such as the brain, liver, skeletal muscle, heart, and blood vessels themselves. These changes all contribute to the increased metabolic risks [15].

Surplus abdominal visceral adipose tissue is associated with adverse metabolic abnormalities such as insulin resistance, higher triglycerides (TGs) and lipoprotein B levels, low high-density lipoprotein (HDL) cholesterol, and an increased proportion of small LDLs and HDLs. These contribute to the increased risk of diabetes [16–18]. Adipocytokines secreted by adipose tissue are also involved in modulating processes promoting atherosclerosis such as endothelial vasomotor dysfunction, hypercoagulability, and dyslipidemia [19].

An important distinction should be made between the nonectopic fat (or subcutaneous fat) that appears to be less metabolically deleterious and the excess ectopic fat. Ectopic fat occurs when there is excess lipid accumulation in the visceral tissue and organs. When the adipose tissue has reached the maximum expansion capacity, a spillover of lipids from adipose sites occurs resulting in the increase of circulating free fatty acids (FFAs). Lipids then start to accumulate in ectopic sites (visceral adipose tissue, intrahepatic, intramuscular, renal sinus, pericardial, myocardial, and perivascular fat), leading to a phenomenon called lipotoxicity [20].

Visceral adipose tissue seems to have up to five times the number of plasminogen activator inhibitor 1 (PAI-1) producing stromal cells compared with subcutaneous adipose tissue. Plasma PAI-1 levels are more closely related to fat accumulation in the liver suggesting that in insulin-resistant individuals, the fatty liver is an important site of PAI-1 production [10]. In obese individuals, only PAI-1 levels were increased in those with metabolic syndrome. PAI-1 is expressed in visceral adipose tissue and is mainly expressed in stromal cells, including monocytes, smooth muscle cells, and preadipocytes [21].

Levels of many inflammatory mediators are altered in obesity. CRP and TNF production is increased in adipose tissue. But other mediators such as interleukin 6 and 1-beta and monocyte-chemoattractant protein-1 and hormones such as adiponectin and leptin are also known to potentially contribute to the inflammatory profile observed in obesity, particularly visceral obesity. Cytokines produced by adipose tissue have been classified as unhealthy adipokines (TNF, interleukin 6, PAI-1, adipocyte fatty acid−binding protein, lipocalin 2, chemerin, leptin, visfatin, vaspin, and resistin) that are upregulated in obesity [22,23]. In patients with chronic heart failure, adiponectin levels are increased, and such increased levels are associated with a worsened prognosis [24]. RBP4 is an adipocyte-secreted molecule that is elevated in the serum before the development of diabetes and seems to signal the presence of insulin resistance and associated cardiovascular risk factors [25].

Lipid metabolism in nonpregnant obese women

Obesity is characterized by increased altered energy storage with increased ectopic fat deposition (liver, pancreas, heart, and skeletal muscle) and VAT distribution (fat storage in intraperitoneal and retroperitoneal spaces). Due to an increased mass of metabolically active adipose tissue, there is excess lipolysis, increased plasma TG concentration, and circulatory FFAs [26]. This leads to increased oxidative stress, adipose tissue hypoxia, and lipotoxicity. This proinflammatory state can directly promote atherosclerosis and endothelial cell dysfunction and indirectly lead to cardiometabolic disease.

Basal lipolysis is stimulated by the lipolytic adipokine serum amyloid A (SAA). The autocrine feedback mechanism of lipolysis on increased production of SAA produced by enlarged adipocytes may contribute toward insulin resistance [8].

These changes are associated with an increased risk of diabetes, hypertension, atherosclerotic CVD, cardiac dysfunction/failure, and certain cancers such as breast and endometrial cancers [27,28].

Cardiac adaptation to obesity

The Framingham heart study showed that obese subjects were two times more at risk of developing heart failure than normal-weight individuals. This study suggested a direct link between excess body fat and cardiac dysfunction. An increased risk of 5% for men and 7% for women for every unit increase in BMI was observed after adjustment for established risk factors [29].

The cardiovascular system in obese patients adapts to this new challenge by increasing blood volume, cardiac output, and stroke volume, whilst reducing peripheral vascular resistance. The expanded blood volume contributes to increased heart rate and preload, shifting the Franck−Starling curves to the left. In the long term, such changes induce ventricular remodeling with enlargement of the cardiac cavities and increased wall tension. This may eventually lead to left ventricular hypertrophy (LVH), decreased diastolic chamber compliance, and an increase in left ventricular filling pressure. A possible long-term end result of this is left ventricular diastolic dysfunction. In addition to LVH, muscular degeneration, increased total blood volume, and diastolic and systolic dysfunction are the main precursors of heart failure in obesity. These cardiac adaptations to obesity are also modulated by the duration of the obesity [30−32].

A large retrospective cohort study of 111,847 patients with unstable angina and non-ST segment elevation myocardial infarction (NSTEMI) investigated the potential relationship between BMI classes and the incidence of NSTEMI. Obesity was the strongest factor associated with NSTEMI at a younger age followed by tobacco use [33]. Respectively for all BMI classes(overweight, obesity class 1, 2, and 3), the mean age of onset of NSTEMI was 3.5, 6.8, 9.4, and 12 years earlier compared to normal-weight individuals. In young adults, obesity through lifetime is positively related with atherosclerosis development as measured by carotid intimal medial thickness. This finding supports the notion of a potential cumulative cardiovascular effect of childhood obesity on adult cardiovascular outcomes [34] (Table 28.2).

Lipid metabolism in pregnant obese women

Pregnancy progresses from an anabolic state in first trimester to a catabolic state in the third trimester. Adaptations in the first trimester such as hyperphagia and relative low insulin resistance increase fat stores. In later gestation, insulin-resistance facilitates increased lipolysis and gluconeogenesis to allow for fetal growth and weight gain [35].

This mechanism is deranged in cases of maternal obesity or significant gestational weight gain. Hyperlipidemia is exaggerated in these women. There are higher circulating levels of serum TGs, cholesterol, and FFA. Excess FFA accumulate in ectopic sites such as the liver and skeletal muscle. There is reduced insulin-mediated uptake

TABLE 28.2 Abnormalities noted among overweight/obese individuals at ↑ risk of cardiovascular disease.

Myocardial fat + cardiac function	↑ Stroke risk, ↑ peripheral resistance ↓ cardiac output ↑ LV thickness, altered left ventricle function Impaired myocardial metabolism Reduced metabolic flexibility
Perivascular adipose tissue	Local inflammation Impaired vascular function ↑ Macrophages in atherosclerotic lesions
Visceral adipose tissue Liver Pancreas Renal Muscle Subcutaneous	↑ Glucose production ↓ insulin degradation ↑ VLDL production ↓ apolipoprotein-B Inflammation, apoptosis, ↓ β-cell function ↑ Blood pressure Insulin resistance/inflammation Postprandial uptake, protects against lipid spillover

Source: Adapted from M. Bastien, P. Poirier, I. Lemieux and J.P. Despres, Overview of epidemiology and contributions of obesity to cardiovascular disease, Prog Cardiovasc Dis 56, 2014, 361–381[9]

in skeletal muscle and hepatic extraction of insulin [35]. These pregnancies are characterized by increased central adiposity and visceral fat at these ectopic sites. Visceral fat is a known marker for insulin resistance and adverse obstetric outcomes such as miscarriages, gestational diabetes, pregnancy-induced hypertension, preeclampsia, and stillbirth [36].

A large cohort study by Yaniv-Salem et al. showed that obesity was an independent risk factor for CVD. Moreover, the average time from the index pregnancy to cardiovascular event is shorter. Thus obese mothers are at higher risk of premature cardiac morbidity and mortality. Among the 46,688 women followed up over a decade, the adjusted risk of hospitalization due to cardiac event was also higher (adjusted HR 2.6, 95% CI 2.0–3.4) [6].

Gynecology practice

Obesity in women is associated with an increase in leptin and a decrease in adiponectin levels in the circulation. Leptin has a stimulatory effect on the hypothalamo–pituitary axis but inhibits ovarian folliculogenesis as well as luteinizing hormone and insulin-mediated steroid production by granulose and theca cells. Obesity-related hyperinsulinemia causes hyperandrogenemia. Obesity encourages phenotypic expression of polycystic ovaries (PCOs) and causes deterioration of hormonal and metabolic parameters. A majority of women with PCO have central obesity and they tend to suffer from menstrual disorders and anovulatory infertility [37]. A series of systematic review of studies on the impact of obesity on natural and assisted reproduction have demonstrated that women with a BMI greater than 25 have lower pregnancy rates (OR: 0.71, 95% CI: 0.62–0.81) and a higher risk of miscarriage [38–43].

A cross-sectional study on 726 Australian women aged 26–36 years reported that higher odds of having irregular cycles in obese women (OR 2.61, 95% CI: 1.28–5.35) compared to those women with normal BMI. This further provides the link between visceral adiposity and menstrual irregularities. The risk of developing fibroids is tripled in women with a body weight of 70 kg or more in comparison with those who weigh less than 50 kg [44,45].

The prevalence of pelvic organ prolapse and urinary incontinence is highest in morbidly obese women (57%) compared with that of 44% in obese women with BMI above 30. WHI study has reported an incremental rise in the proportion of women with significant prolapses as the BMI rises. Surgically induced weight loss has a beneficial effect on symptoms of pelvic floor disorders in morbidly obese women [46]. The relationship between obesity and cancers of the female genital tract and breast has been addressed elsewhere within this book. Table 28.3 summarizes the risk of comorbidities associated with obesity [47].

In addition, in gynecological surgery, obese women are at increased risk of anesthetic, operative, and postoperative complications. Examples include difficulty siting regional anesthesia and/or peripheral vascular devices, failed intubation, aspiration, prolonged operating time, visceral injury, blood transfusion, pressure sores, venous thromboembolism, wound infection, wound dehiscence, and prolonged postanesthetic care in high dependency units [36,48,49].

TABLE 28.3 Risk of comorbidities associated with obesity and overweight.

Comorbidity	BMI 25–30 [RR (95% CI)]	BMI > 30 (95% CI)
Diabetes	3.92 (3.10–4.97)	12.41 (9.03–17.06)
Hypertension	1.65 (1.24–2.19)	2.42 (1.59–3.67)
Coronary artery disease	1.80 (1.64–1.98)	3.10 (2.81–3.43)
Osteoarthritis	1.80 (1.75–1.85)	1.96 (1.88–2.04)
Stroke	1.15 (1.00–1.32)	1.49 (1.27–1.74)
Asthma	1.25 (1.05–1.49)	1.78 (1.36–2.32)

BMI, Body mass index.
Source: Adapted with permission from Guh DP, Zhang W, Bansback N, Amarsi Z, Birmingham CL, Anis AH. The incidence of co-morbidities related to obesity and overweight; a systemic review and meta-analysis. BMC Public Health 2009;9(88).[47]

Collectively, all these risk factors contribute to the burden of premature mortality and excess morbidity in obese women.

Maternal obesity and in utero programing for cardiovascular disease

In 1997 Baker first described the effect on the in utero environment on short-, medium-, and long-term health outcomes in childhood and adulthood. Van De Maele et al. described the association between low birth weight and adult hypertension, diabetes, and obesity [37]. Animal studies reveal that an in utero environment of chronic inflammation leads to vasoconstriction and increased platelet aggregation and lipid storage in the placenta. This leads to poor placental function, hypoxia, and fetal growth. All these factors are linked with the development of atherosclerotic plaques in placental arterioles and an adaptive shift in fetal blood supply away from the kidneys and heart [49]. Consequently, reduced proliferation and maturation of cardiomyocytes lead to myocardial hypertrophy and fibrosis. Eventually, this manifests as premature cardiac dysfunction and cardiac death [31].

Human studies show that maternal obesity is associated with early CVD and premature death of offspring in adolescence and adulthood [26,49,50]. These studies do not prove a causal relationship and neither prove nor refute Baker's hypothesis.

A large cohort analysis by Lee et al. showed a positive correlation between maternal obesity and premature death from CVD in adult offspring. Of the 28,540 women and their 37,709 offspring followed up, obese mothers were at significant risk of major cardiovascular event and hospitalization for a cardiac event compared to mothers with a normal BMI [6].

Children exposed to maternal environments secondary to obesity-related metabolic syndrome are not only at increased risk of developing cardiovascular risk but are also at increased risk of transmitting intergenerational risk on to the future generations. The insulin-like growth factor receptor (IGFIR) expression is critical for insulin signaling and glucose transport when there is evidence of the fetal growth restriction, with increased placental IGFIR [51].

Various hypotheses have been proposed to explain intergenerational obesity. They include the following:

- The developmental overnutrition hypothesis which proposes that high maternal glucose and high FFA and amino acid plasma concentration result in permanent change in appetite control, neuroendocrine function, and energy metabolism in the developing fetus, leading to the risk of adiposity in later life and to a greater risk of insulin resistance and glucose intolerance [52].
- Two large studies (Australian birth cohort and the Evan longitudinal study of parents and children) have clearly demonstrated a plausible link between maternal overnutrition during intrauterine and during breastfeeding periods, and fetal programing [53,54].
- Fetal skeletal development is influenced by maternal obesity by shifting mesenchymal stem cells differentiation from myogenesis toward adipogenesis. This shift permanently impairs the physiological function of the offspring's skeletal muscle [54].
- Epigenetic modification secondary to exposure to an altered intrauterine milieu or metabolic perturbation may influence the phenotype much later in life; thus impaired glucose tolerance during pregnancy leads to adaptation in leptin gene DNA methylation [44,55].

Maternal obesity induced by diet prior to and throughout pregnancy and lactation results in exposure to very high levels of leptin to the offspring with a hyperphagic and obese phenotype in adulthood [45]. Raised levels of leptin are noted in neonates of those whose mothers had obesity during pregnancy [53]. There is strong evidence that type 2 diabetes mellitus is more prevalent among subjects that were exposed to maternal diabetes in utero. HAPO study has also reported an association between increased maternal BMI and fetal hyperinsulinemia as assessed by cord serum C peptide levels even after adjustment for maternal glycaemia [56,57].

Interventions to address obesity in reproductive health

Physical activity

Prevention of obesity and treatment strategies both at the individual and population levels has not been successful in the long term. Lifestyle and behavior interventions aimed at reducing calorie intake and increasing energy expenditure have limited effectiveness because complex and persistent hormonal, metabolic, and neurochemical adaptations challenge weight loss and promote weight regain. Reducing the obesity burden requires approaches that combine individual intervention with changes in the environment and society [1].

Both increased adiposity and reduced physical activity are strong and independent predictors of CHD and deaths. In general, for each unit of BMI increment, the risk of CHD increases by 8%. On the other hand, each 1-hour-met (metabolic equivalent) increase in activity score is associated with an 8% decrease in CHD risk. Physical activity improves glucose tolerance and sensitivity in improving noninsulin-dependent glucose intake; it improves the ratio between HDL and LDL cholesterol because it increases the activity of lipoprotein lipase. Physical activity decreases triglycerides, increases fibrinolysis, decreases platelet aggregation, improves oxygen uptake in the heart as well as in peripheral tissue, lowers the resting heart rate by increasing vagal tone, and lowers blood pressure. Physical activity also directly increases myocardial oxygen supply, thus improving myocardial contraction and electrical stability [58].

It has been suggested that individuals with type 2 diabetes mellitus who maintain weight loss after intensive lifestyle intervention have sustained improvements of cardio metabolic risk factors. The most recent analysis of data from "Look AHEAD (Action for Health in Diabetes)" suggests that particularly among participants who initially lost at least 10% of their body weight, maintaining the weight loss over the course of 4 years, led to significant improvements in cholesterol levels, blood glucose, and blood pressure among other parameters, compared with those who regained weight and even those who regain weight could still experience cardio metabolic benefits as long as they regain no more than 25% of the original weight loss [59].

Medical treatment

In insulin-resistant states the dyslipidemia as seen in obesity is characterized by a different composition and distribution of LDL particles, resulting in increased concentration of small, dense LDL. Hepatic overproduction of VLDL seems to be the primary and crucial defect in obesity. Insulin resistance in the liver, muscles, and adipose tissue leads to an inability to suppress hepatic glucose production, impaired glucose uptake, and oxidation and inability to suppress release of nonesterified fatty acids from adipose tissue. Visceral obesity limits the beneficial effects of the lipid-lowering strategies [60,61].

Approved antiobesity medications include the following:

Orlistat: which inhibits pancreatic and other lipases, thus inhibiting fat absorption.
Sibutramine: which inhibits reuptake of serotonin and norepinephrine and acts by increasing satiety and thermogenesis, primarily by modifying CNS neurotransmitters.
Rimonabant: which is a cannabinoid receptor antagonist that reduces the drive to eat. It reduces concentrations of HDL-cholesterol and TG.

To date, there is lack of published data on the long-term morbidity and mortality for all antiobesity drugs [62].

Surgical treatment

The National Institute for Health Care and Excellence (NICE) guidelines recommends that bariatric surgery should be considered in women with BMI over 40 or 35–40 with comorbidities, where other nonsurgical interventions have been unsuccessful in achieving significant sustained weight loss [48]. Bariatric surgery results in 15%–20% of body weight and a reduction in the prevalence of cardiometabolic syndrome in these women [63].

A Swedish study demonstrated an estimated 24% reduction in the adjusted overall mortality rate in those treated surgically compared with conventionally treated controls [64]. A systemic review of 136 studies involving 22,094 patients, 72.6% being women, reported that a substantial majority of patients with diabetes, hyperlipidemia, hypertension, and OSA experienced complete resolution or improvement of metabolic indices following bariatric surgery [65]. Karlsson et al. have provided long-term follow-up data of up to 7 years following gastric bypass surgery compared to BMI-matched subjects not receiving bariatric surgery reporting a 49% lower mortality from CVD and a 60% lower mortality due to cancer in patients receiving surgery [66]. Another study also reported significant improvement or resolution of the three components of nonalcoholic fatty liver disease [67].

Vest et al. conducted a systematic literature review to quantify the impact of bariatric surgery on cardiovascular risk factors, cardiac structure, and function. Postoperative resolution or improvement of hypertension occurred in 63% of subjects, of diabetes in 73%, and of hyperlipidemia in 65%. Echocardiographic studies revealed significant improvements in left ventricular mass, E/A ratio, and isovolumic relaxation time postoperatively [68]. A metaanalysis by Kwok et al. of 14 studies that included 29,208 patients who underwent bariatric surgery reported over 50% reduction in mortality, 58% reduced risk of composite cardiovascular adverse events, 79% reduction in myocardial infarction, and 59% reduction in the risk of stroke as compared to control group of nonsurgical treatment [68]. These findings were largely confirmed in a Swedish nationwide, matched, observational cohort study that also reported a 58% relative overall risk reduction, 49% risk reduction of fatal or nonfatal myocardial infarction, and 59% lower cardiovascular risk in the bariatric surgery group as compared to control group [69]. In women of reproductive age, menstrual irregularities and PCOs may be completely resolved following bariatric surgery [70].

Interventions to improve outcomes in pregnancy

Weight management

NICE advocates weight management before, during, and after pregnancy. Antenatal weight loss is not recommended, as this can adversely affect the health of the woman and her unborn child. Weight management should be individualized, focus on dietary intake and exercise as well as psychosocial factors that drive behavioral change [71].

Prepregnancy, women with a BMI of 30 and above should be encouraged to lose weight aiming for a healthy BMI of 18.5–24.9. Pregnant women who are overweight or obese should be risk assessed and managed with appropriate multidisciplinary team support. Focus should be placed on weight maintenance versus weight loss. Women should be supported in making healthy lifestyle choices around exercise and diet. It is important to dispel myths such as the need to "eat for two" or "drink full fat milk" early on. Women can be reassured that 30 minutes of daily moderate intensity activity is safe [63,72].

NICE has identified the period after childbirth and pregnancy as a time when women are likely to gain weight and may easily conceive again [36]. Therefore proactively managing a woman's weight during this time may reduce the chance of her being overweight or obese at the start of the next pregnancy. Postnatal opportunistic lifestyle advice should be offered when possible, for example, at the 6- to 8-week check [46]. Advice provided should take into consideration the mode of delivery, support network, and psychological state of the mother. Professionals should be mindful of challenging issues new mothers face such as depression, fatigue, lack of sleep, urinary and fecal incontinence, and backache. A sensitive and supportive approach should be used. Referral to weight-management services and/or a dietician is recommended [63].

Bariatric Surgery

The role of bariatric surgery prenatally is inconclusive. Two retrospective studies have reported conflicting data on the effects of bariatric surgery and the reduction in the risk of recurrent miscarriage. Women who had bariatric surgery have a significant reduction in pregnancy-induced hypertension but no difference in gestational diabetes mellitus. Furthermore, the risk of a large-for-dates-baby was 50% lower, whereas the risk for small-for-gestational-age infant was around 80% as compared to control group [57].

Conclusions

CVD is still the leading cause of premature deaths globally and maternal mortality in the United Kingdom. Maternal obesity is a recognized risk factor for CVD in mothers and their offspring in adulthood. Pregnancy should be considered an excellent window of opportunity to engage with obese women as a way of modifying not just the cardiovascular risks of their unborn children, but their very own cardiovascular risk in the short to long term.

In the political declaration of the high-level meeting of the UN General Assembly on the prevention and control of noncommunicable diseases of September 2011, the importance of reducing unhealthy diet and physical inactivity was recognized. The WHO acknowledges that healthy eating and increasing physical activity in the entire population should be promoted by policies and action implements in societies [57].

In summary, multifaceted proactive approaches to tackling this problem are required. Planning and implementing interventions must involve stakeholders—obese women, their families, health commissioners, government, and health-care workers.

References

[1] Bluher M. Obesity: global epidemiology and pathogenesis. Nature 2019;15:288–98.
[2] Fontaine KR, Redden DT, Wang C, Westfall AO, Allison DB. Years of life lost due to obesity. JAMA 2003;289:187–93.
[3] Prospective SC. Body-mass index and cause-specific mortality in 900,000 adults: collaborative analyses of 57 prospective studies. Lancet 2009;373(9669):1083–96.
[4] Berrington de Gonzalez A, et al. Body mass index and mortality among 1.46 million white adults. N Engl J Med 2010;363:2210–19.
[5] World Health Organisation. Obesity & overweight. Available from: <https://www.who.int/topics/obesity/en/>; 2019 [accessed 11.11.19].

[6] The Global BMI Mortality Collaboration. Body-mass index and all-cause mortality: individual-participant-data meta-analysis of 239 prospective studies in four continents. Lancet 2016;388 (10046):776–86.
[7] Knight M, Bunch K, Tuffnell D, Jayakody H, Shakespeare J, Kotnis R, et al., editors. Saving lives, improving mothers care—lessons learned to inform maternity care from the UK and Ireland Confidential Enquiries into Maternal Deaths and Morbidity 2015-17; 2019. Oxford: National Perinatal Epidemiology Unit, University of Oxford; 2019.
[8] Reynolds RM, Allan KM, Raja EA, Bhattacharya S, McNeill G, Hannaford PC, et al. Maternal obesity during pregnancy and premature mortality from cardiovascular event in adult offspring: follow-up of 1 323 275 person years. BMJ 2013;347:f4539.
[9] Bastien M, Poirier P, Lemieux I, Despres JP. Overview of epidemiology and contributions of obesity to cardiovascular disease. Prog Cardiovasc Dis 2014;56:361–81.
[10] Alessi MC, Juhan-Vague I. PAI-1 and the metabolic syndrome: links, causes, and consequences. Arterioscler Thromb Vasc Biol 2006;26:200–7.
[11] Blann AD, Steele C, McCollum CN. The influence of smoking on soluble adhesion molecules and endothelial cell markers. Thromb Res 1997;85:433–8.
[12] Holvoet P, et al. The metabolic syndrome, circulating oxidized LDL, and risk of myocardial infarction in well-functioning elderly people in the health, aging and body composition cohort. Diabetes 2004;53:1068–73.
[13] Dagenais GR, Yi Mann JF, Bosch J, Pogue J, Usuf S. Prognostic impact of body weight and abdominal obesity in women and men with cardiovascular disease. Am Heart J 2005;149(1):54–60.
[14] de Koning L, Merchant AT, Pogue J, Anand SS. Waist circumference and waist-to-hip ratio as predictors of cardiovascular event: meta-regression analysis of prospective studies. Eur Heart J 2007;28(7):850–6.
[15] Rexrode KM, et al. Abdominal adiposity and coronary heart disease in women. J Am Med Assoc 1998;280:1843–8.
[16] Despres JP. Body fat distribution and risk of cardiovascular disease: an update. Circulation 2012;126(10):1301–13.
[17] Cornier MA, Despres JP, Davis N, et al. Assessing adiposity: a scientific statement from the American Heart Association. Circulation 2011;124(18):1996–2019.
[18] Mertens I, Gaal LFV. Obesity, haemostasis and the fibrinolytic system. Obesity Reviews 2002;3:85–101. Available from: https://doi.org/10.1046/j.1467-789X.2002.00056.x.
[19] Van Gaal LF, Mertens IL, De Block CE. Mechanisms linking obesity with cardiovascular disease. Nature 2006;44(7121):875–80.
[20] Gray SL, Vidal-Puig AJ. Adipose tissue expandability in the maintenance of metabolic homeostasis. Nutr Rev 2007;65(6 pt 2):S7–12.
[21] Bastelica D, et al. Stromal cells are the main plasminogen activator inhibitor-1-producing cells in human fat: evidence of differences between visceral and subcutaneous deposits. Arterioscler Thromb Vasc Biol 2002;22:173–8.
[22] Dyck DJ, Heigenhauser GJ, Bruce CR. The role of adipokines as regulators of skeletal muscle fatty acid metabolism and insulin sensitivity. Acta Physiol 2006;186(1):5–16.
[23] Coullard C, Lamarche B, Mauriege P, et al. Leptinemia is not a risk factor for ischemic heart disease in men. Prospective results from the Quebec Cardiovascular Study. Diabetes Care 1998;21(5):782–6.
[24] Villarreal-Molina MT, Antuna-Puente B. Adiponectin anti-inflammatory and cardioprotective effects. Biochimie 2012;94 (10):2143–9.
[25] Graham TE, et al. Retinol-binding protein 4 and insulin resistance in lean, obese, and diabetic subjects. N Engl J Med 2006;354:2552–63.
[26] Thakali KM, Wahl EC, Wankhade UD, Zhong Y, Shankar K. Maternal obesity during pregnancy uniquely programs offspring aortic and perivascular adipose tissue transcriptomes. FASEB J 2017;31:1.
[27] Yaniv-Salem S, Shoham-Vardi I, Kessous R, Pariente G, Sergienko R, Sheiner E. Obesity in pregnancy: what's next? Long-term cardiovascular morbidity in a follow-up period of more than a decade. J Matern Fetal Neonatal Med 2016;29(4):619–23.
[28] Singla P, Bardoloi A. Metabolic effects of obesity: a review. World J Diabetes 2010;1(3):76–88.
[29] Mathieu P, Poirier P, Pibarot P, Lemieux I, Despres JP. Visceral obesity: the link among inflammation, hypertension and cardiovascular disease. Hypertension 2009;53(4):577–84.
[30] Neeland IJ, Gupta S, Ayers CR, et al. Relation of regional fat distribution to left ventrical structure and function. Circ Cardiovasc Imaging 2013;6(5):800–7.
[31] Shimabukuro J, Hirata Y, Tabata M, et al. Epicardial adipose tissue volume and adipocytokine imbalance are strongly linked to human coronary atherosclerosis. Arterioscler Throm Vasc Biol 2013;33(5):1077–84.
[32] Wilson PW, D'Agostino RB, Sullivan L, Parise H, Kannel WB. Overweight and obesity as determinants of cardiovascular risk: the Framingham experience. Arch Intern Med 2002;162(16):1867–72.
[33] Baker AR, Silva NF, Quinn DW, et al. Human epicardial adipose tissue expresses a pathogenic profile of adipocytokines in patients with cardiovascular disease. Cardiovasc Dabetol 2006;5:1.
[34] Freedman DS, Dietz WH, Tang R, et al. The relation of obesity throughout life to carotid intima-media thickness in adulthood: the Bogalusa Heart Study. Int J Obes Relat Metab Disord 2004;28 (1):159–66.
[35] Huda S, Nelson S. Pregnancy and metabolic syndrome of pregnancy. In: Mahmood T, Arulkumaran S, editors. Obesity. 1st ed Elsevier; 2013. p. 299–314.
[36] Denison FC, Aedla NR, Keag O, Hor K, Reynolds RM, Milne A, et al. Care of women with obesity in pregnancy. BJOG 2019;126(3): e62–e106.
[37] Van De Maele K, Devlieger R, Gies I. In utero programming and early detection of cardiovascular disease in the offspring of mothers with obesity. Atherosclerosis 2018;275:182–95.
[38] Metwally M, Li TC, Ledger WL. The impact of obesity on female reproductive function. Obes Rev 2007;8:515–23.
[39] Lord J, Thomas R, Fox B, Acharya U, Wilkin T. The central issue? Visceral fat mass is a good marker of insulin resistance and metabolic disturbance in women with polycystic ovary syndrome. BJOG 2006;113:1203–9.
[40] Metwally M, Ong KJ, Ledger WL, Li TC. Does high body mass index increase the risk of miscarriage after spontaneous and assisted conception? A meta-analysis of the evidence. Fertil Steril 2008;90:714–26.
[41] Wattanakumtornkul S, Damario MA, Steens Hall SA, Thornhill AR, Tummon IS. Body mass index and uterine receptivity in the oocyte donation model. Fertil Sertil 2003;80:336–40.
[42] Bellver J, Melo MA, Bosch E, Serra V, Remohi J, Pellicer A. Obesity and poor reproductive outcome: the potential role of the endometrium. Fertil Sertil 2007;88:446–51.

[43] Bellver J, Ayllon Y, Ferrando M, et al. Female obesity impairs in vitro fertilization outcome without affecting embryo quality. Fertil Sertil 2009;93.

[44] Vickers MH. Developmental programming of the metabolic syndrome—critical windows for intervention. World J Diabetes 2011;2(9):137–48.

[45] Li M, Sloboda DM, Vickers MH. Maternal obesity and developmental programming of metabolic disorders in offspring: evidence from animal models. Exp Diabetes Res 2011;592408.

[46] Kudish NI, Iglesia CB, Sokol RJ, et al. Effect of weight change on natural history of pelvic organ prolapse. Obstet Gynecol 2004;103:674–80.

[47] Guh DP, Zhang W, Bansback N, Amarsi Z, Birmingham CL, Anis AH. The incidence of co-morbidities related to obesity and overweight; a systemic review and meta-analysis. BMC Public Health 2009;9(88).

[48] Sholtz S, Balen AH, le Roux CW. The role of bariatric surgery in improving reproductive health. Scientific impact paper no. 17; 2010.

[49] Nyrnes SA, Garnæs KK, Salvesen A, Timilsina AS, Moholdt T, Ingul CB. Cardiac function in newborns of obese women and the effect of exercise during pregnancy. A randomized controlled trial. PLoS One 2018;13(6):e0197334.

[50] Andraweera PH, Dekker GA, Leemaqz S, McCowan L, Roberts CT, on behalf of the, SCOPE Consortium. The obesity associated FTO gene variant and the risk of adverse pregnancy outcomes: evidence from the SCOPE study. Obesity (Silver Spring) 2016;24(12):2600–7.

[51] Shankar K, et al. Maternal obesity at conception programs obesity in the offspring. Am J Physiol Regul Integr Comp Physiol 2008;294(2):R528–38.

[52] Lawlor DA. Exploring the developmental overnutrition hypothesis using parental-offspring associations and FTO as an instrumental variable. PLoS Med 2008;5(3):e33.

[53] Dabelea D, et al. Intrauterine exposure to diabetes conveys risks for type 2 diabetes and obesity: a study of discordant sibships. Diabetes 2009;49(12):2208–11.

[54] Aguiari P, et al. High glucose induces adipogenic differentiation of muscle-derived stem cells. Proc Natl Acad Sci USA 2008;105(4):1226–31.

[55] Egger G, et al. Epigenetics in human disease and prospects for epigenetic therapy. Nature 2004;429(6990):457–63.

[56] HAPO Study Cooperative, Research Group. Hyperglycaemia and Adverse Pregnancy Outcome (HAPO) study: associations with maternal body mass index. BJOG 2010;117(5):575–84.

[57] Mahmood T, Thanoon O. The role of bariatric surgery on female reproductive health. Obstet Gynecol Rep Med 2016;20(5):155–7.

[58] Li TY, et al. Obesity as compared with physical activity in predicting risk of coronary heart disease in women. Circulation 2006;113:499–506.

[59] Berger SE, Huggins GS, McCafferty JM, Jacques PF, Lichtenstein AH. Changes in cardiometabolic risk factors associated with magnitude of weight regain 3 years after a 1 year intensive life style intervention in type 2 Diabetes Mellitus. The Look AHEAD Trial. J Am Heart Ass 2019;8(20):1–15.

[60] Howard BV, et al. LDL cholesterol as a strong predictor of coronary heart disease in diabetic individuals with insulin resistance and low LDL: The Strong Heart Study. Arterioscler. Thromb. Vasc. Biol. 2000;20:830–5.

[61] Chan DC, Watts GF, Redgrave TG, Mori TA, Barrett PHR. Apolipoprotein B-100 kinetics in visceral obesity: associations with plasma apolipoprotein C-III concentration. Metabolism 2002;51:1041–6.

[62] Padwal RS, Majumdar SR. Drug treatments for obesity: orlistat, sibutramine and rimonabant. Lancet 2007;369(9555):71–7.

[63] National Institute for Health Care and Excellence. Weight management before, during and after pregnancy. 2010; Available at: https://www.nice.org.uk/guidance/ph27.

[64] Sjostrom L, Narbro K, Sjostrom CD, et al. Effects of bariatric surgery on mortality in Swedish obese subjects. N Engl J Med 2007;357:741–52.

[65] Buchwald H, Avidor Y, Braunwald E, et al. Bariatric surgery: a systematic review and meta-analysis. J Am Med Assoc 2004;292:1724–37.

[66] Karlsson J, Taft C, Ryden A, et al. Ten years trends in health-related quality of life after surgical and conventional treatment for severe obesity: The SOS intervention study. Int J Obes 2007;31(8):1248–61.

[67] Mummadi RR, Kasturi KS, Chennareddygari S, Sood JK. Effect of geriatric surgery on non-alcoholic fatty liver disease: systematic review and meta-analysis. Clin Gastroenterol Hepatol 2008;6(12):1396–402.

[68] Vest A, Heneghan HM, Agarwal S, et al. Bariatric surgery and cardiovascular outcomes: a systematic review. Heart 2012;98:1763–77.

[69] Kwok CS, Pradhan A, Khan MA, et al. Bariatric surgery and its impact on cardiovascular disease and mortality: A systematic review and meta-analysis. Int J Cardiol 2014;173:20–8.

[70] Eliasson B, Liakopoulos V, Franzen S, et al. Cardiovascular disease and mortality in patients with type 2diabetes after bariatric surgery in Sweden: a nationwide, matched, observational cohort study. The Lancet Diabetes & Endocrinology 2015;3(11):828–9.

[71] Escobar-Morreale HF, Botella-Carretero JI, Alvarez-Blasco F, et al. The polycystic ovary syndrome associated with morbid obesity may resolve after weight loss induced by bariatric surgery. J Clin Endocrinol Metab 2005;90:6364–9.

[72] National Institute for Health and Care Excellence. QS98: Maternal and child nutrition. 2015; Available at: https://www.nice.org.uk/guidance/qs98. Accessed QS98.

Chapter 29

Female obesity and osteoporosis

Rashda Bano[1] and Tahir A. Mahmood[2]

[1]Obstetrics and Gynaecology, Royal Infirmary of Edinburgh, Edinburgh, United Kingdom, [2]Department of Obstetrics and Gynaecology, Victoria Hospital, Kirkcaldy, United Kingdom

Introduction

Obesity and osteoporosis are two important global health problems with a high impact on both mortality and morbidity. The development of obesity is due to an imbalance when energy intake exceeds energy expenditure over a prolonged period. In healthy adults, body weight is tightly regulated by several environmental, nutritional, and hormonal factors. For instance, postmenopausal women often show increased body weight, likely due to a decrease in basal metabolism, alteration of hormonal levels, and reduced physical activity. Moreover, obese postmenopausal women are often affected by hypertension, dyslipidemia, diabetes mellitus (DM), and cardiovascular disease and also have an increased risk of developing some cancers. Interestingly, these women have always been considered protected against osteoporosis [1–10].

Osteoporosis is a metabolic bone disease characterized by excessive skeletal fragility (due to a reduction in both bone quantity and quality) leading to an increased risk of developing spontaneous and traumatic bone fractures [7] and even death [11,12]. Osteoporosis is typically defined in an individual when a bone mineral density (BMD) T-score is 2.5 or less than standard deviations below normal (T score ≤ -2.5) [13]. The rate of bone loss in adults reflects the interaction between genetic and environmental factors, which also influences the extent of bone acquisition during growth, known as peak bone mass [14].

A common pathophysiological linkage has been proposed between obesity and osteoporosis [15]. Both diseases are affected by genetic and environmental factors, or the interaction between them, and there is some overlap between the genetic and environmental factors influencing both diseases. Normal aging is associated with a high incidence of osteoporosis and bone marrow adiposity [15]. Bone remodeling and adiposity are both regulated through the hypothalamus and sympathetic nervous system [15]. Adipocytes (the cell for storing energy) and osteoblasts (the bone from a common progenitor)—the mesenchymal stem cell (Fig. 29.1).

Normal bone metabolism

Osteoblasts, osteoclasts, and osteocytes are the main cells of the bone. Osteoblasts are nonproliferative bone-building cells that originate from osteoblast progenitor cells and aid in the formation of the bone matrix by secreting osteoid, a substance responsible for bone mineralization. A mature osteoblast is known as osteocyte. Osteocytes are unable to divide and no longer secrete matrix components. Osteoclasts originate from macrophage monocyte cell lineage and participate in bone resorption, ultimately leading to decreased bone mass.

Among the three cell types of the bone, osteoblasts play the most important role in bone formation. Preosteoblasts express receptors for different types of growth factors, proinflammatory cytokines and hormones, including bone morphogenic proteins, Wnt, transforming growth factor-beta, parathyroid hormone (PTH), interleukin-6 (IL-6), 5-HT, insulin/insulin-like growth factor, and tumour necrosis factor (TNF). Binding of these ligands with their receptors induces the activation of different types of transcription factors responsible for osteoblast differentiation, maturation, and survival.

Parathormone stimulates osteoclasts and releases calcium and phosphate in the blood. Calcitonin inhibits osteoclasts and increases deposition of the calcium in the bone. Calcitriol stimulates the absorption of calcium and phosphate from the small intestine and ensures availability in the bone. It also activates the osteoblasts to synthesize collagen. Estrogen inhibits bone resorption. Growth hormone stimulates bone formation. Insulin increases synthetic activity of osteoblasts while glucocorticoids inhibit osteoblasts.

Obesity and Gynecology. DOI: https://doi.org/10.1016/B978-0-12-817919-2.00029-2
© 2020 Elsevier Inc. All rights reserved.

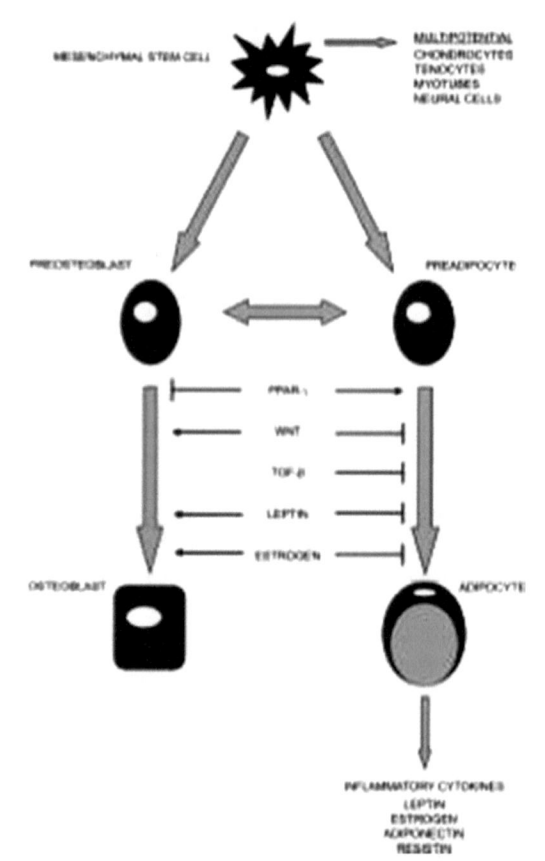

FIGURE 29.1 Common factors shared in osteoblast and adipocyte differentiation. Osteoblasts and adipocytes originate from common progenitor—mesenchymal stem cells. The balance of their differentiation is determined by several common factors, such as PPAR-γ, Wnt, TGF-β, leptin, and estrogen. Adipocytes express and secrete a variety of bioactive peptides, such as estrogen, resistin, leptin, adiponectin, and inflammatory cytokines. Some of these peptides affect human energy homeostasis and may be involved in bone metabolism. *Adapted from Rosen CJ, Bouxsein ML. Mechanisms of disease: is osteoporosis the obesity of bone? Nat Clin Pract Rheumatol 2006;2:35—43.*

Epidemiology of osteoporosis

Among the US population, there are 10 million aged more than 50 years, who have osteoporosis [16]. Osteoporosis has become a significant health problem as approximately 200 million people worldwide are estimated to have osteoporosis [17]. In a Brazilian population-based study the prevalence of fragility fracture in women and men aged higher than 40 years was 15.1% and 12.8%, respectively [18]. The WHO's World Health Statistics report in 2015 shows that in the European region the overall obesity rate among adults is 21.5% in males and 24.5% in females. It has been further projected that 60% of the world's population, that is, 3.3 billion people, could be overweight (2.2 billion) or obese (1.1 billion) by 2030 if recent trends continue.

Age and female gender increases the risk of developing both obesity and osteoporosis [3,5,7]. Age-related changes in body composition, metabolic factors, and hormonal levels after menopause, accompanied by a decline in physical activity, may all provide mechanisms for the propensity to gain weight and, in particular, for an increase in fat mass often characterized by replacement of lean mass (LM) by adipose tissue [3,4].

Soon after menopause, the process of bone loss begins due to increased bone resorption by osteoclasts that exceed bone formation by osteoblasts [14]. Moreover, osteoblast function declines with aging, determining the imbalance between bone resorption and bone formation [19]. Traditionally, osteoporosis has been regarded as a disorder associated only with fracture and skeletal disability in old age, but recent studies demonstrate that BMD appears to be a better long-term predictor of death than blood pressure or cholesterol [12]. Total body fat and LM are correlated with BMD, with obesity apparently conferring protection against bone loss after menopause [5,6]. A recent study demonstrated that premenopausal women with increased central adiposity had poorer bone quality and stiffness and markedly lower bone formation.

Relationship between fat and bone: epidemiologic and clinical observations

It has been well accepted that the most powerful and measurable determinant of fracture risk is the amount of bone in the skeleton, as defined by either BMD or BMC [20,21]. Extensive data have shown that high body weight or BMI is correlated with high BMD or BMC and a decrease in body weight leads to bone loss [22–24]. These correlations are seen in both men and women, across the entire adult age range, and within the skeleton [6,22,25]. This relationship is also found in children and adolescence, although its significance is less clear because of intensive bone acquisition in this period. There is also evidence to support the view that fat mass, a component of total body weight and one of the most important indices of obesity, has a similar beneficial effect on increasing bone mass, thereby reducing the risk of osteoporosis. In normal pre- and postmenopausal women, total body fat was positively related to BMD throughout the skeleton, and this effect was found in both white [25,26] and Japanese subjects [26,27]. The EPIC study also reported that "rapid" bone losers had significantly lower fat mass than the "slow" bone losers [28]. Finally, Lau et al. showed that men with severe vertebral deformity had much lower fat mass and BMD than controls [29].

However, in contrast to the abovementioned reported results, other independent groups have shown that excessive fat mass may not protect against decreases in bone mass [30,31]. In a large-scale study of Chinese and white subjects, a positive correlation was found between fat mass and bone mass when results had not been corrected for the mechanical loading effect caused by total body weight. When the mechanical loading effect of total body weight was statistically removed, then fat mass was negatively correlated with bone mass thus suggesting that fat mass actually has a detrimental effect on bone [32]. A study conducted on evaluation of BMD in individuals with high body mass index has shown that obese patients have significant reduction in bone mineral mass for age and BMI. It also showed the evidence that morbid obesity may not be considered a protective factor against osteoporosis in both female and male population [33].

Evidence from environmental factors and medical interventions also support an inverse correlation between fat mass and bone mass. For instance, physical exercise increases bone mass while reducing fat mass [34]. Consumption of milk and tea is believed to be beneficial for the prevention of both osteoporosis and obesity [35]. Milk is a good source of highly absorbable calcium. Increased milk intake has been shown to increase peak bone mass at puberty, slow bone loss, and reduce the incidence of osteoporotic fractures in the elderly [36]. Menopause has also been associated with increased bone loss, increased fat mass, and decreased LM. Hormone replacement therapy is an effective means of attenuating loss of LM and bone [37] and reversing menopause-related obesity [38] in postmenopausal women. Osteoporosis and obesity are side effects of treatment with gonadotropin-releasing hormone agonists. The clinical use of glucocorticoids has been shown to cause decreased bone mass and an increase in central obesity [39]. The finding that all of these interventions have opposite effects on fat versus bone mass supports the concept that there is an inverse correlation between fat and bone mass and that fat does not have a protective effect on the bone.

Another study has suggested that LM is the strongest predictor of BMD at all sites. It is important that LM should also be the target for improvement when considering prevention and/or management of osteoporosis [40].

Diabetes mellitus treatment and osteoporosis

Thiazolidinediones (TZD) were introduced as powerful glucose lowering agents. The first of these drugs, troglitazone, was withdrawn from the market because of hepatotoxicity. ADOPT study, a 4-year investigation designed to compare rosiglitazone, metformin, and glyburide for the maintenance of glycemic control, showed that women, not men, on rosiglitazone had a higher incidence of appendicular fractures than women on metformin or glyburide. Increased limb fractures were also observed in patients treated with this TZD, and once again this finding was only in female diabetic patients.

It is recommended that clinical fracture risk in patients be determined before initiation of TZD therapy. Screening for BMD may be useful in postmenopausal women with DM, particularly those on TZD therapy. Osteoporosis therapy should be initiated in those women whose bone density and other risk factors place them at an increased risk for fracture.

Beta-cell hormones (*pancreatic hormones*)

Insulin is a potential regulator of bone growth, since osteoblasts have insulin receptors [41] as well as IGF-1 receptor. Hyperinsulinemic patients develop a cluster of abnormalities, including androgen and estrogen overproduction in the ovary and reduced production of sex hormone—binding globulins in the liver. As a result, there is increased free concentration of sex hormones resulting in reduced osteoclasts activity and possibly increased osteoblasts activity, leading to increased bone mass. Insulin resistance is highly correlated with obesity, and several studies support the concept that insulin is a potential regulator of bone metabolism. Fasting insulin levels were significantly and positively associated with BMD of the radius and spine in middle-aged women [42]. The complex effects of insulin on the bone are similar to the complicated relationship between fat and bone. Insulin is cosecreted with Amylin that directly stimulates osteoblasts proliferation in vitro and in vivo.

Adipocyte hormones

Studies of adipocyte function have revealed that adipose tissue is not just an inert organ for energy storage. It expresses and secretes a variety of biologically active molecules, such as estrogen, resistin, leptin, adiponectin, and IL-6. These molecules affect human energy homeostasis and may be involved in bone metabolism that may contribute to the complex relationship between fat mass and bone (Fig. 29.2).

The enzyme aromatase is found in the gonadal tissue and adipocytes and plays an important role in the synthesis of estrogen. The adipocyte has been recognized as an estrogen-producing cell particularly in postmenopausal women. Estrogen inhibits bone turnover by reducing osteoclast-mediated bone resorption and by stimulating osteoblasts-mediated bone formation. In postmenopausal women as ovaries no longer secrete estrogens, extragonadal estrogen synthesis in fat tissue becomes the dominant estrogen source. Hence the role of adipocyte as estrogen

FIGURE 29.2 Possible mechanisms by which fat mass may influence bone cell function, and thus bone mass, independent of the effect of feeding. *SHBG*, Sex hormone–binding globulin. *Copyright IR Reid, used with permission.*

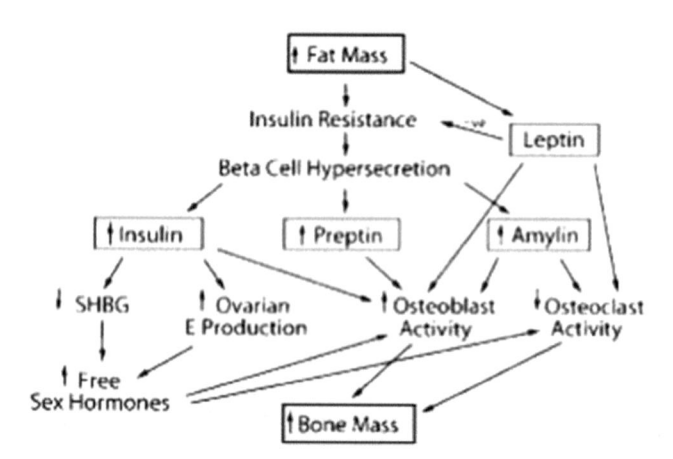

producers may become more important for the bone metabolism in postmenopausal women. Thus early postmenopausal women who lose bone rapidly have lower levels of both estrone and estradiol than slow losers, and this may be accounted for by their lower fat mass.

Leptin: Leptin is the most widely recognized adipocyte-derived hormone. It is mainly known for its function of suppressing appetite and increasing energy expenditure. The effect of leptin on obesity is mediated by a series of integrated neuronal pathways, including the catabolic pathway represented by proopiomelanocortin (POMC) neurons and the anabolic pathway represented by neuropeptide Y (NPY). Leptin stimulates POMC neurons, and it results in reduced food intake and increased energy expenditure. Leptin controls bone formation through a hypothalamic relay, thereby suggesting central mechanism also to be involved in its action on the bone. NPY is a hypothalamus-derived peptide, essential for the regulation of food consumption, energy homeostasis, and bone remodeling [43]. NPY expression stimulated food intake, inhibited energy expenditure, can lead to the development of obesity and its related phenotypes. Leptin inhibits NPY gene expression in the hypothalamus. In addition to regulating the appetite for food consumption, leptin is also a major regulator of bone remodeling.

Adiponectin: Adiponectin is another adipocyte-derived hormone that regulates energy homeostasis and has antiinflammatory and antiatherogenic effects [44,45]. Adiponectin increases insulin sensitivity and its circulating levels are reduced in obesity and diabetes. Adiponectin levels may increase with moderate weight loss. Adiponectin and corresponding receptors are expressed in primary human osteoblasts, suggesting a link between adiponectin and bone. Generally, an inverse relationship between serum adiponectin level and BMD has been reported [46]

Resistin: It, also known as adipocyte-secreted factor, was discovered recently while screening for substances that are downregulated in response to insulin-sensitizing antidiabetic drugs [47]. Thommesen et al. [47] showed that resistin may play a role in bone remodeling. Their study indicates that resistin is expressed in mesenchymal bone marrow stem cells, osteoblasts, and osteoclasts and increases osteoblasts proliferation and cytokine release, as well as osteoclast differentiation [47].

IL-6: IL-6, a pluripotent inflammatory cytokine, is released from adipocytes, the adipose tissue matrix, and elsewhere [48]. Adipose tissue accounts for one-third of the circulating levels of IL-6. Just like leptin, overweight and obese children and adults generally have elevated serum levels of IL-6 [49].

Proinflammatory cytokines, including TNF-a, IL-1, and IL-6, are key mediators in the process of osteoclast differentiation and bone resorption and IL-6 antagonizes osteoblasts differentiation. Chronic inflammation and increased proinflammatory cytokines induce bone resorption and bone loss in patients with periodontitis, pancreatitis, inflammatory bowel disease, and rheumatoid arthritis. The accelerated bone loss at menopause is linked to increased production of proinflammatory cytokines, including TNF-a, IL-1, and IL-6 [50]. The significant increase in the development of osteoarthritis in obese human subjects provides evidence that chronic inflammation influences bone metabolism.

Adipsin: Adipsin similarly has been shown to negatively affect osteoporosis, particularly DM induces osteoporosis. Increased expression is seen among DM and obese patients, and it has been shown to decrease bone formation [51].

Beside these factors some other environmental important factors have been proposed in the genesis of osteoporosis. Among obese women, vitamin D insufficiency and decreased bioavailability of vitamin D3 have been reported as cutaneous body fat has been reported to have long-term negative influence. This factor becomes very important as there is already a high prevalence of vitamin D deficiency reported in obese women [52].

Obesity of the bone

Bone marrow mesenchymal stromal cells are the common precursors for both osteoblasts and adipocytes. Aging may shift composition of bone marrow by increasing adipocytes, osteoclast activity, and decreasing osteoblasts activity, resulting into osteoporosis [53].

Clinical and diagnostic implication of the concept-obesity of bone

Errors in BMD determinations commonly seen in markedly obese individuals are because of fat deposition in bone marrow. Dual-energy X-ray absorptiometry measurements may be falsely elevated by increased body fat, whereas measurements of trabecular BMD by quantitative computed tomography may be decreased by greater marrow fat [15]. However, all the secondary causes of osteoporosis such as Cushing's syndrome, DM, glucocorticoids levels, and immobility are associated with obesity. It was difficult in the past to explain that obesity was not protective for osteoporosis development but as a cause of osteoporosis in such cases. More published data have helped in our understanding of the basis of associated increase in bone marrow adiposity seen in such cases [15]. Similarly at extremes of life, puberty, and old age, fat infiltration in the bone marrow might be detrimental for skeletal strength and may negatively affect the optimal function of the bone remodeling unit [15].

Treatment implications of the concept-obesity of the bone

The present treatment options for osteoporosis primarily known are either antiosteoclastogenesis or proosteoblastogenesis in nature. However, recently vitamin D has been shown to act by inhibiting bone marrow adipogenesis as an additional mechanism beside its known actions on bone [54]. Furthermore, there is evidence that vitamin D affects body fat mass by inhibiting adipogenic transcription factors and lipid accumulation during adipocyte differentiation. Some recent studies have also demonstrated that vitamin D metabolites also influence adipokine production and the inflammatory response in adipose tissue [55].

High-fat diet has a pivotal role in bone formation because it markedly reduces the rate of Ca^{2+} absorption by the intestine and thereby decreases the availability of Ca^{2+} required for osteogenesis. Decreased levels of vitamin D are a hallmark of osteoporosis and bone fractures. Vitamin D deficiency in the serum prevents intestinal uptake of Ca^{2+} from the diet and hereby signals the parathyroid gland to secret increased levels of PTH. Increased secretion of PTH induces osteolysis and prevents osteogenesis by supplying adequate levels of calcium and phosphorus in the blood necessary for metabolic processes and neuromuscular function. Although a normal level of PTH is beneficial to bone health, elevated secretion of PTH has been identified as a negative regulator of osteoblastogenesis.

Alendronate is a widely used bisphosphonate and recently has been reported to stimulate osteoblastic differentiation while inhibiting adipogenesis in vitro, thereby suggesting anabolic effect on bone through the differentiation of mesenchymal stem cells [56].

PTH also has been shown in the past to induce osteoblasts differentiation, inhibits adipogenesis, and suppresses osteoclasts apoptosis [57]. Similarly, PTH-related protein has been shown to induce a mild osteogenic effect and inhibit adipocytic effect in human mesenchymal stem cell, thereby helping in osteoporosis beside its known action to modulate bone formation through promoting osteoblasts differentiation [58].

Strontium ranelate has both antiresorptive and anabolic effects on bone. However, it has been recently shown that adipogenesis is negatively affected in the presence of strontium ranelate with a concomitant dose-dependent decrease in the expression of adipogenic markers and changes in adipokine profile, thereby generating a favorable osteogenic effect within the bone marrow milieu [59].

Dietary relevant mixtures of isoflavones and their metabolites, lignans and their metabolites, coumestrol, and a mixture containing all of them have been shown to inhibit adipocyte differentiation as their additional mechanism of action in preventing osteoporosis independent of their concentration [60].

Bariatric surgery and bone health

The need to develop therapeutic options, the efficacy in which is less dependent on the patient's determination, and the challenging effort to change habits has favored the use of gastric reduction surgery as a desirable alternative. Weight loss evolves rapidly and currently bariatric surgery has increasingly become recognized as a highly effective alternative for achieving major weight reduction for obese patients. A large prospective cohort study (Swedish Obese Subjects study) found that weight loss was still apparent 10 years following surgery, whereas patients receiving conventional treatment had gained weight [61].

Bariatric surgery adversely affects bone health. The skeletal effects of bariatric surgery are presumably multifactorial, and mechanisms may involve nutritional factors, mechanical unloading, hormonal factors, and changes in body composition and bone marrow fat.

The Roux-en-Y gastric bypass (RYGB) procedure combines restriction and malabsorption techniques and involves creating both a small gastric pouch and a deviation of a segment of the small intestine.

Presently, RYGB is the most frequent and efficient surgery option to treat severe obesity. Metabolic bone disease is a well-documented long-term complication of obesity surgery. Abnormalities in calcium and vitamin D metabolism begin shortly after gastrointestinal bypass operations; however, clinical and biochemical evidence of metabolic bone disease may not be detected until many years later. RYGB can potentially be associated with two metabolic bone disorders: osteomalacia and osteoporosis.

Vitamin D has an important role in calcium absorption. Estrogen levels, dietary intake, age, and body weight also influence calcium absorption. RYGB has significant impact on calcium and vitamin D metabolism. Diet restriction reduces the exogenous load of calcium and vitamin D and decreases intake of macronutrients that positively affect their absorption. In RYGB the proximal jejunum is bypassed, excluding an important site of calcium absorption, which contributes to the decreased calcium load. In addition, the reduction in food intake leads to increased release of cortisol and decrease in IGF-I serum levels, both adaptations potentially impair calcium absorption [62].

Management after Roux-en-Y gastric bypass

Follow-up includes careful examination to detect subclinical fracture. Patient stature should be measured before and at regular intervals after surgery.

Patients undergoing RYBG should be screened for osteoporosis with bone density measurement. Laboratory evaluation includes calcium, albumin, magnesium, PTH, and 25(OH) D. Operated patients should be encouraged to perform regular weight-bearing physical exercise. Recent data suggest increased fall risk in postoperative bariatric surgery patients [63]. Thus mechanical risks for falls should be addressed, and physical activity is part of the strategy to reduce fracture risk. Calcium and vitamin D supplementation should be prescribed in all bariatric patients. Patients should be advised to take slightly higher daily doses of vitamin D and Calcium than RDI recommendations: 1500 mg calcium and 2000 IU vitamin D. 25(OH) D serum levels should be checked every 2 months to ensure adequate levels. One should consider pharmacological treatment in patients if BMD is below −1.5.

Pragmatic approach for obese women

What a physician should advise to an obese woman?

Physicians have a responsibility to recognize obesity as a gateway disease and help patients with appropriate prevention and treatment pathways for obesity and its comorbidities, including osteoporosis.

Treatment should be based on good clinical care and evidence-based interventions, and it should be individualized, multidisciplinary, and focused on realistic goals of prevention of weight regain and weight maintenance. Advice, treatment, care, and the information given to the patients should be nondiscriminatory and culturally appropriate. It should also be accessible to people with additional needs such as physical, sensory, or learning disabilities and to people who do not speak or read English [64].

The main requirement of a dietary approach to weight loss and osteoporosis is that total energy intake should be less than energy expenditure, and they should eat plenty of fiber-rich foods, including five portions of a variety of fruit and vegetables each day [64].

Interventions to increase physical activity should focus on activities that fit easily into people's everyday life such as walking and should be tailored to people's individual preferences and circumstances. Attention should be paid to women who are at risk of developing obesity just like pregnant and menopausal women. Women should be encouraged to increase their physical activity to lose weight, as evidence suggests that physical activity can reduce risk of type 2 diabetes, cardiovascular disease, sudden death, cancer especially cancer of the endometrium, depression due to body image, and osteoporosis. Adults should be encouraged to do at least 30 minutes of at least moderate-intensity physical activity on 5 or more days a week [64]. Obese women should take higher dose folic acid and vitamin D along with additional calcium.

Pharmacological and surgical treatment should be initiated based on clinical assessment [64].

Conclusion

The relationship between fat mass and bone is confounded by complex genetic backgrounds and by interactions between metabolic factors and regulatory pathways influencing both obesity and osteoporosis. The previous concept that obesity is protective for osteoporosis may not stand to careful scrutiny as new concept of bone marrow fat deposition seen in obesity has emerged supporting the detrimental effect of obesity for bone health. Thus obesity especially central obesity may not be considered protective for osteoporosis. Considering that obesity can be associated with fracture and that obesity treatment also can damage skeleton, it is reasonable to conclude that the primary target should be obesity prevention.

Conflict of interest

None.

References

[1] Kado DM, Huang MH, Karlamangla AS, Barrett-Connor E, Greendale GA. Hyperkyphotic posture predicts mortality in older community-dwelling men and women: a prospective study. J Am Geriatr Soc 2004;52:1662−7.

[2] Rossner S. Obesity: the disease of the twenty-first century. Int J Obes Relat Metab Disord 2002;26(Suppl. 4):S2−4.

[3] Hu FB. Overweight and obesity in women: health risks and consequences. J Women Health (Larchmt) 2003;12(2):163−72.

[4] World Health Organization. Obesity: preventing and managing the global epidemic. Report of a WHO consultation. World Health Organization Technical Report Series 894; 2000. pp. 1−253.

[5] Albala C, Yanez M, Devoto E, Sostin C, Zeballos L, Santos JL. Obesity as a protective factor for postmenopausal osteoporosis. Int J Obes Relat Metab Disord 1996;20:1027−32.

[6] Reid IR. Relationships among body mass, its components, and bone. Bone 2002;31:547−55.

[7] NIH. Consensus development panel on osteoporosis. JAMA 2001;285:785−95.

[8] Cagnacci A, Zanin R, Cannoletta M, Generali M, Caretto S, Volpe A. Menopause, estrogens, progestin, or their combination on body weight and anthropometric measurements. Fertil Steril 2007;88(6):1603−8.

[9] Lebovitz HE. The relationship of obesity to the metabolic syndrome. Int J Clin Pract Suppl 2003;134:18−27.

[10] Sowers JR. Obesity as a cardiovascular risk factor. Am J Med 2003;8:37S−41S.

[11] Zhao LJ, Jiang H, Papasian CJ, et al. Correlation of obesity and osteoporosis, effect of fat mass on the determination of osteoporosis. J Bone Miner Res 2008;23:17−29.

[12] Johansson C, Black D, Johnell O, Oden A, Mellstrom D. Bone mineral density is a predictor of survival. Calcif Tissue Int 1998;63:190−6.

[13] Kanis JA, Melton III LJ, Christiansen C, Johnston CC, Khaltaev N. The diagnosis of osteoporosis. J Bone Miner Res 1994;9:1137−41 [PubMed: 7976495].

[14] Brown S, Rosen CJ. Osteoporosis. Med Clin North Am 2003;87:1039−63.

[15] Rosen CJ, Bouxsein ML. Mechanisms of disease: is osteoporosis the obesity of bone? Nat Clin Pract Rheumatol 2006;2:35−43.

[16] U.S. Department of Health and Human Services. Bone health and osteoporosis: a report of the surgeon general. Rockville, MD: U.S. Department of Health and Human Services; 2004.

[17] Roy B. Biomolecular basis of the role of diabetes mellitus in osteoporosis and bone fractures. World J Diabetes 2013;4:101−13. Available from: https://doi.org/10.4239/wjd.v4.i4.101.

[18] Pinheiro MM, Ciconelli RM, Martini LA, Ferraz MB. Clinical risk factors for osteoporotic fractures in Brazilian women and men: the Brazilian Osteoporosis Study (BRAZOS). Osteoporos Int 2009;20(3):399−408.

[19] Kveiborg M, Flyvbjerg A, Rattan SI, Kassem M. Changes in the insulin-like growth factor-system may contribute to in vitro age-related impaired osteoblast functions. Exp Gerontol. 2000, 35 (8):1061-1074.

[20] Cummings SR, Black DM, Nevitt MC, Browner W, Cauley J, Ensrud K, et al. Bone density at various sites for prediction of hip fractures. The Study of Osteoporotic Fractures Research Group. Lancet 1993;341:72−5.

[21] Melton III LJ, Atkinson EJ, O'Fallon WM, Wahner HW, Riggs BL. Long-term fracture prediction by bone mineral assessed at different skeletal sites. J Bone Miner Res 1993;8:1227−33.

[22] Felson DT, Zhang Y, Hannan MT, Anderson JJ. Effects of weight and body mass index on bone mineral density in men and women: The Framingham study. J Bone Miner Res 1993;8:567−73.

[23] Marcus R, Greendale G, Blunt BA, Bush TL, Sherman S, Sherwin R, et al. Correlates of bone mineral density in the postmenopausal estrogen/progestin interventions trial. J Bone Miner Res 1994;9:1467−76.

[24] Mazess RB, Barden HS, Ettinger M, Johnston C, Dawson-Hughes B, Baran D, et al. Spine and femur density using dual-photon absorptiometry in US white women. Bone Miner 1987;2:211−19.

[25] Reid IR, Plank LD, Evans MC. Fat mass is an important determinant of whole body bone density in premenopausal women but not in men. J Clin Endocrinol Metab 1992;75:779−82.

[26] Douchi T, Oki T, Nakamura S, Ijuin H, Yamamoto S, Nagata Y. The effect of body composition on bone density in preand postmenopausal women. Maturitas 1997;27:55−60.

[27] Douchi T, Yamamoto S, Oki T, Maruta K, Kuwahata R, Nagatan Y. Relationship between body fat distribution and bone mineral density in premenopausal Japanese women. Obstet Gynecol 2000;95:722−5.

[28] Riis BJ, Rodbro P, Christiansen C. The role of serum concentrations of sex steroids and bone turnover in the development and occurrence of postmenopausal osteoporosis. Calcif Tissue Int 1986;38:318−22.

[29] Lau EM, Chan YH, Chan M, Woo J, Griffith J, Chan HH, et al. Vertebral deformity in chinese men: Prevalence, risk factors, bone mineral density, and body composition measurements. Calcif Tissue Int 2000;66:47−52.

[30] De Laet C, Kanis JA, Oden A, Johanson H, Johnell O, Delmas P, et al. Body mass index as a predictor of fracture risk: a meta-analysis. Osteoporos Int 2005;16:1330−8.

[31] Janicka A, Wren TA, Sanchez MM, Dorey F, Kim PS, Mittelman SD, et al. Fat mass is not beneficial to bone in adolescents and young adults. J Clin Endocrinol Metab 2007;92:143−7.

[32] Zhao LJ, Liu YJ, Liu PY, Hamilton J, Recker RR, Deng HW. Relationship of obesity with osteoporosis. J Clin Endocrinol Metab 2007;92:1640−6.

[33] Greco EA, Fornari R, Rossi F, Santiemma V. Is obesity protective for osteoporosis. 2010.

[34] Reid IR, Legge M, Stapleton JP, Evans MC, Grey AB. Regular exercise dissociates fat mass and bone density in premenopausal women. J Clin Endocrinol Metab 1995;80:1764–8.

[35] Stonge MP. Dietary fats, teas, dairy, and nuts: potential functional foods for weight control? Am J Clin Nutr 2005;81:7–15.

[36] Reid IR. Therapy of osteoporosis: calcium, vitamin D, and exercise. Am J Med Sci 1996;312:278–86.

[37] Manson JE, Martin KA. Postmenopausal hormonereplacement therapy. N Engl J Med 2001;345:34–40.

[38] Sorensen MB, Rosenfalck AM, Hojgaard L, Ottesen B. Obesity and sarcopenia after menopause are reversed by sex hormone replacement therapy. Obes Res 2001;9:622–6.

[39] de Gregorio LH, Lacativa PG, Melazzi AC, Russo LA. Glucocorticoid-induced osteoporosis. Arq Bras Endocrinol Metab 2006;50:793–801.

[40] Bolaji Lilian Ilesanmi-Oyelere, Jane Coad, Nicole Roy, and Marlena Cathorina Kruger. Lean Body Mass in the Prediction of Bone Mineral Density in Postmenopausal Women. BioResearch Open Access Volume 7.1, 2018 DOI: 10.1089/biores.2018.0025

[41] Pun KK, Lau P, Ho PW. The characterization, regulation and function of insulin receptors on osteoblast like clonal osteosarcoma cell line. J Bone Miner Res 1989;4.853–62.

[42] Barrett-Connor E, Kritz-Silverstein D. Does hyperinsulinemia preserve bone? Diabetes Care 1996;19:1388–92.

[43] Herzog H. Neuropeptide Y and energy homeostasis: Insights from Y receptor knockout models. Eur J Pharmacol 2003;480:21–9.

[44] Combs TP, Berg AH, Obici S, Scherer PE, Rossetti L. Endogenous glucose production is inhibited by the adiposederived protein Acrp30. J Clin Invest 2001;108:1875–81.

[45] Berg AH, Combs TP, Du X, Brownlee M, Scherer PE. The adipocyte-secreted protein Acrp30 enhances hepatic insulin action. Nat Med 2001;7:947–95.

[46] Lenchik L, Register TC, Hsu FC, Lohman K, Nicklas BJ, Freedman BI, et al. Adiponectin as a novel determinant of bone mineral density and visceral fat. Bone 2003;33:646–51.

[47] Thommesen L, Stunes AK, Monjo M, Grosvik K, Tamburstuen MV, Kjobli E, et al. Expression and regulation of resistin in osteoblasts and osteoclasts indicate a role in bone metabolism. J Cell Biochem 2006;.

[48] Fain JN, Madan AK, Hiler ML, Cheema P, Bahouth SW. Comparison of the release of adipokines by adipose tissue, adipose tissue matrix, and adipocytes from visceral and subcutaneous abdominal adipose tissues of obese humans. Endocrinology 2004;145 (5):2273–82.

[49] Fernandez-Real JM, Ricart W. Insulin resistance and chronic cardiovascular inflammatory syndrome. Endocr Rev 2003;24:278–301.

[50] Mundy GR. Osteoporosis and inflammation. Nutr Rev 2007;65(12 Pt 2):S147–51.

[51] Botolin S, Faugere MC, Malluche H, Orth M, Meyer R, McCabe LR. Increased bone adiposity and peroxisomal proliferator-activated receptor-gamma2 expression in type I diabetic mice. Endocrinology 2005;146:3622–31 [PMCID: PMC1242186] [PubMed: 15905321].

[52] Wortsman J, Matsuoka LY, Chen TC, Lu Z, Holick MF. Decreased bioavailability of vitamin D in obesity. Am J Clin Nutr 2000;72:690–3 [PubMed: 10966885].

[53] Horowitz MC, Lorenzo JA. The origin of osteoclasts. Curr Opin Rheumatol 2004;16:464–8 [PubMed: 15201612].

[54] Duque G, Macoritto M, Kremer R. Vitamin D treatment of senescence accelerated mice (SAM-P/6) induces several regulators of stromal cell plasticity. Biogerontology 2004;5:421–9 [PubMed: 15609106].

[55] Ding C, Gao D, Wilding J, Trayhurn P, Bing C. Vitamin D signalling in adipose tissue. Br J Nutr 2012;108:1915–23 [PubMed: 23046765].

[56] Duque G, Rivas D. Alendronate has an anabolic effect on bone through the differentiation of mesenchymal stem cells. J Bone Miner Res 2007;22:1603–11 [PubMed: 17605634].

[57] Chan GK, Miao D, Deckelbaum R, Bolivar I, Karaplis A, Goltzman D. Parathyroid hormone-related peptide interacts with bone morphogenetic protein 2 to increase osteoblastogenesis and decrease adipogenesis in pluripotent C3H10T 1/2 mesenchymal cells. Endocrinology 2003;144:5511–20 [PubMed: 12960089].

[58] Casado-Diaz A, Santiago-Mora R, Quesada JM. The N- and C-terminal domains of parathyroid hormone-related protein affect differently the osteogenic and adipogenic potential of human mesenchymal stem cells. Exp Mol Med 2010;42:87–98 [PMCID: PMC2827833] [PubMed: 19946180].

[59] Vidal C, Gunaratnam K, Tong J, Duque G. Biochemical changes induced by strontium ranelate in differentiating adipocytes. Biochimie 2013;95:793–8 [PubMed: 23186800].

[60] Taxvig C, Specht IO, Boberg J, Vinggaard AM, Nellemann C. Dietary relevant mixtures of phytoestrogens inhibit adipocyte differentiation *in vitro*. Food Chem Toxicol 2013;55:265–71 [PubMed: 23348407].

[61] Sjostrom L. Bariatric surgery and reduction in morbidity and mortality: experiences from the SOS study. Int J Obes 2008;32(Suppl. 7):S93–7.

[62] Shapses SA, Riedt CS. Bone, body weight and weight reduction what are the concern. J Nutr 2006;136(6):1453–6.

[63] Berarducci A, Murr MM, Haines K. Risk and incidence of falls and skeletal fragility following Roux-en-Y gastric bypass surgery for morbid obesity. Osteoporos Int 2007;18(S1):201.

[64] National institute of Clinical Excellence (NICE) Osteoporosis: assessing the risk of fragility fracture. Clinical guideline [CG146] 07 February 2017. (www.nice.org.UK).

Future research

Exploring the connection between bone and fat at a molecular and cellular level is likely to lead to a better understanding of several risk factors and pathogenesis basis and the development of drugs with dual mechanism for both osteoporosis and obesity in future.

Chapter 30

Obesity, menopause, and hormone replacement therapy

Marta Caretto, Andrea Giannini, Tommaso Simoncini and Andrea R. Genazzani

Division of Obstetrics and Gynecology, Department of Clinical and Experimental Medicine, University of Pisa, Pisa, Italy

Introduction

The menopausal transition marks a period of physiologic changes as women approach reproductive senescence. Evidence supports the clinical importance of the transition for many women as period of temporal changes in health and quality of life [vasomotor symptoms (VMS), sleep disturbance, and depression] and longer term changes in several health outcomes (urogenital symptoms, bone, and lipids) that may influence women's quality of life and the likelihood of healthy aging [1].

Loss of sex hormones during aging contributes to changes in body mass, musculoskeletal integrity, sexual dysfunction, and long-term risks of health and disease.

The metabolic syndrome increases in prevalence after menopause and consists of insulin resistance, abdominal obesity, dyslipidemia, elevated blood pressure, and proinflammatory and prothrombotic states. This syndrome, also known as insulin resistance syndrome, usually precedes the development of diabetes mellitus (DM) and carries a twofold increased risk for cardiovascular events [2]. Obesity is a very common public health problem, especially in the Western hemisphere. The World Health Organization estimates that in 2016 more than 1.9 billion adults aged 18 years and older were overweight, of these over 650 million adults were obese; in 2016 39% of adults aged 18 years and over (39% of men and 40% of women) were overweight, and about 13% of the world's adult population (11% of men and 15% of women) was obese in 2016 [3]. The prevalence of obesity [body mass index (BMI) > 30 kg/m^2] is higher in postmenopausal women than in premenopausal women. This is a consequence of a multifactorial process that involves reduced energy expenditure due to physical inactivity, which is sometimes compounded by depression, as well as due to muscle atrophy and a lower basal metabolic rate. Whereas menopause per se is not associated with weight gain, it leads to an increase of total body fat and a redistribution of body fat from the periphery to the trunk, which results in visceral adiposity. Increased BMI and upper body fat distribution (indicated by waist-to-hip ratio) and menopausal estrogen decline are associated with adverse metabolic changes such as insulin resistance, a propensity to develop type 2 DM and dyslipidemia characterized by high triglyceride levels, low high-density lipoprotein cholesterol levels, and an increased frequency of small, dense low-density lipoprotein particles. Cross-sectional and longitudinal studies using waist circumference or the waist-to-hip ratio report no effect of menopause on body fat distribution. By contrast, studies using dual-energy X-ray absorptiometry (DXA) showed increased trunk fat in postmenopausal women. Moreover, studies using computed tomography and magnetic resonance imaging show that postmenopausal women have greater amounts of intraabdominal fat compared to premenopausal women. Collectively, these studies confirm that the menopause transition is associated with an accumulation of central fat and, in particular, intraabdominal fat. Postmenopausal women had 36% more trunk fat, 49% greater intraabdominal fat area, and 22% greater subcutaneous abdominal fat area than premenopausal women. The menopause-related difference in intraabdominal fat persisted after statistical adjustment for age and fat mass, whereas no differences were noted in trunk or abdominal subcutaneous fat [4].

The menopausal obesity: role of estrogens

In vivo and in vitro studies indicate that the estrogen receptors (ERs) are mechanistically implicated in endocrine-related diseases. Recent studies with ER knockout mice have helped to unravel the role of the ERs in brain degeneration, osteoporosis, cardiovascular diseases (CVDs), and obesity [5].

In humans the hormones help integrate metabolic interaction among major organs that are essential for metabolically intensive activities such as reproduction and metabolic function. Sex steroids are required to regulate adipocytes' metabolism and also influence the sex-specific remodeling of particular adipose depots. Concentrations of sex hormones partially control fat distribution: men have less total body fat but more central/intraabdominal adipose tissue, whereas women tend to have more total fat in gluteal/femoral and subcutaneous depots. Weight and fat abdominal distribution differ among women of reproductive age and menopausal women.

The function of estrogens is mediated by nuclear receptors that are transcription factors that belong to the superfamily of nuclear receptors. Two types of ERs have been identified: the alpha (ERα) and beta (ERβ). Human subcutaneous and visceral adipose tissues express both ERα and ERβ, whereas only ERα mRNA has been identified in brown adipose tissue. ERα plays a major role in the activity of adipocytes and sexual dimorphism of fat distribution. Polymorphism of ERα in humans has been associated with risk factors for CVDs. Lipolysis in humans is controlled primarily by the action of β-adrenergic receptors (lipolytic) and α2-adrenergic receptors (antipolytic) [6].

Genazzani and Gambacciani [7] evaluated the effects of climacteric modifications on body weight and fat distribution: he selected 2175 untreated normal healthy women attending a menopause clinic. He divided them into three groups: premenopausal, perimenopausal, and postmenopausal and compared them with 354 postmenopausal women receiving different forms of hormone replacement therapy (HRT). The total body fat tissue mass and distribution were analyzed using DXA. Body weight and BMI were significantly higher in perimenopausal and postmenopausal than in premenopausal women. Fat tissue and regional fat tissue as a percentage of total fat tissue were higher in the trunk and arms in perimenopausal and postmenopausal than in premenopausal women. Instead, in age-matched HRT-treated postmenopausal women, the fat tissue was similar to that in the premenopausal group. Perimenopausal and postmenopausal women show a shift to a central, android fat distribution that can be counteracted by HRT [7].

A number of studies looked at the pattern of hormonal changes during the menopausal transition between obese and nonobese women. In two studies, "Study of Women's Health Across the Nation (SWAN)" and "Penn Ovarian Aging Study (POAS)," obese women had lower estradiol (E2) and follicular-stimulating hormone (FSH) levels than nonobese women, and in the POAS, lower luteinizing hormone and inhibin B levels as well. More rigorous analysis of hormonal changes before and after the final menstrual period (FMP) between obese and nonobese women has found that the patterns of change in FSH and E2 in relation to the FMP were not statistically different when comparing obese to nonobese women, although significant differences in the mean FSH and E2 levels were observed. The E2 change was less pronounced in obese women when compared with nonobese women, because obese women had lower premenopausal mean E2 levels but higher postmenopausal mean E2 levels. The rate of E2-blunted decline observed among obese women is physiologically corroborated by a similarly blunted FSH rise surrounding the FMP in obese versus nonobese women. Ultrasound data have shown no difference in antral follicle count between obese and nonobese women in their late reproductive age (40−52 years). This lack of difference does not support low ovarian reserve as the mechanism underlying lowers E2 levels in obese women premenopausally, and this mechanism is currently unclear. In POAS, AMH was found to be lower in obese women compared to nonobese women in the late reproductive years, demonstrating the complex relationship between obesity and reproductive hormones in women approaching menopause. Follicular dysfunction and alterations in central nervous system regulation of hormonal levels among obese women may be factors, but additional research in this area is needed. The blunted magnitude of change in reproductive hormones in obese women during menopausal transition may be related to the change in the primary source of circulating E2 as the menopause transition progresses; the primary source of circulating E2 premenopausally is the ovary, whereas in postmenopause the primary source of circulating E2 is the aromatization of androgens within the adipose tissue. This change in E2 source provides postmenopausal obese women with a nonovarian reservoir of estrogen that normal-weight women do not have, which may blunt the gonadotropin rises and mitigate ovarian estrogen loss with menopause. These hormonal alterations may also blunt menopause-associated adverse health effects [8].

Obesity, lifestyle intervention, and hormone replacement therapy

Obese postmenopausal women differ from the general postmenopausal population. Hot flushes and menopausal symptoms, in general, are more frequent in obese women compared to women with normal BMI. In the SWAN study the odds ratio for hot flushes was 1.27 for each standard deviation increase in percentage of body fat. Women who gain weight during the menopausal transition are more prone to have menopausal symptoms [9]. Obese postmenopausal women are at increased risk of developing coronary heart disease (CHD). According to the Nurses' Health Study, 5 kg/m² increase in BMI is associated with a 30% increase in the incidence of CHD in women, independently of other CHD risk factors, such as age, smoking, physical activity, alcohol intake, or family history of CHD [10]. Stroke risk increases linearly with increasing BMI independently of sex and race. Data from the Nurses' Health Study show that women with BMI > 32 kg/m² have a relative risk of 2.37 of developing ischemic stroke. Furthermore, women who gain 10–20 kg during their adult life have a 69% increase in the risk of ischemic stroke [11]. Obesity is associated with the increased risk of venous thromboembolism (VTE). VTE is rare in premenopausal and young postmenopausal women and its incidence increases with age, BMI, and the presence of prothrombotic mutations (factor V-Leiden and prothrombin G20210A). Obese women in the placebo arm of the Women's Health Initiative (WHI) trial had 2.9 times increased risk of developing VTE compared to women with normal BMI [12]. Obese postmenopausal women are at increased risk of developing breast cancer. Obesity is associated with a relative risk of breast cancer ranging between 1.26 and 2.52. According to a metaanalysis on 2.5 million women, a 5 kg/m² increase in BMI is associated with 12% increase in the incidence of breast cancer. Possible explanations are the higher endogenous estrogens produced by the aromatization of precursor adrenal and ovarian androgens in adipose tissue and mitogenic IGF-1 activity associated with insulin resistance. Apart from absolute body weight, the weight gained after 30th–40th year of age and especially perimenopausally appears to constitute an extra risk of breast cancer [13].

Data from the Women's Healthy Lifestyle Project provide clear evidence that weight gain and increased waist circumference, along with elevations in lipid levels and other CVD risk factors, are preventable through the use of lifestyle intervention in healthy menopausal-aged women. In fact, although these changes are inevitable with age and menopause, physical activity may attenuate the impact of both events. Thus weight gain prevention should be recognized as an important health goal for women before they approach menopause, and women should make regular physical activity [14].

All women at midlife should be encouraged to maintain or achieve a normal body weight, be physically active, adopt a healthy diet, limit alcohol consumption, and not smoke. Some women find that avoidance of spicy food, hot drinks, and alcohol lessens their VMS. Obesity is associated with a greater likelihood of VMS, although women who are overweight (BMI from 25 to <30 kg/m²), as opposed to obese (BMI ≥ 30 kg/m²), are more likely to have severe symptoms [15]. For obese women, weight loss may lessen VMS as well as reduce the risks of CVD, diabetes, urinary incontinence, breast, pancreatic and endometrial cancers, and dementia.

Estrogens seem to influence glucose homeostasis through increased glucose transport into the cells, whereas lack of estrogens has been associated with a progressive decrease in glucose-stimulated insulin secretion and insulin sensitivity as well as with insulin resistance. These may explain why HRT administration to postmenopausal women is associated with a significant decrease in the incidence of type 2 diabetes.

Estrogen deficiency is the principal pathophysiological mechanism that underlies menopausal symptoms and various estrogen formulations are prescribed as menopausal hormone therapy, which remains the most effective therapeutic option available. The addition of progesterone aims to protect against the consequences of systemic therapy with estrogen only in women with intact uteri [16], namely, endometrial pathologies, including hyperplasia and cancer. The risk–benefit ratios of all treatment options must be considered, taking into account the nature and severity of symptoms, and individual treatment-related risks.

In the systemic circulation, E2 and estrone are partly bound to sex hormone–binding globulin (SHBG), as well as to albumin, as is testosterone. Increasing or decreasing SHBG levels will affect the amount of unbound estrogen and testosterone in the circulation [17].

Obesity is a biologically plausible risk factor for (VTE, but the mechanisms underlying the relation of obesity with VTE are not totally understood. A strong positive correlation between plasminogen activator inhibitor-1 (PAI-1) level and BMI has been reported. PAI-1 is the main fibrinolytic inhibitor, and reduced plasma fibrinolytic potential may be a risk factor for venous thrombosis. Decreased fibrinolysis because of a high level of PAI-1 could explain in part the association of VTE with overweight and obesity. Moreover, other studies suggested that an increased BMI was associated with higher levels of prothrombotic factors such as fibrinogen and factor (F) VII. Thus both oral estrogen and obesity may have synergistic effects on the unbalance between procoagulant factors and antithrombotic mechanisms. By contrast, transdermal estrogen appears to have little or no effect on hemostasis. Alternatively, increased C-reactive protein levels have been reported in obese individuals with a history of VTE, and low-grade inflammation could explain in part our findings. In addition to the effects on hemostasis and inflammation, obesity may also have direct mechanic effects on the venous area. An increased BMI may result in a higher VTE risk through an increased intraabdominal pressure and a decreased venous return. These effects may result in venous hypertension, varicose veins, and venous stasis, which promote the development of VTE [18].

For those who require pharmacological therapies, average-dose HRT is the most effective treatment for

VMS [19] with reductions in both frequency and severity in the order of 75% [20], and HRT may improve quality of life in symptomatic women [21].

In 2010 the European Menopause and Andropause Society formulated a position statement to provide evidence-based advice on the management of obese postmenopausal women [22]. Lambrinoudaki et al. concluded that obese postmenopausal women requiring HT should be thoroughly evaluated at baseline and the severity of symptoms and risk of fracture should be weighed against individual risks of breast cancer, CVD, and VTE. Although, there is a lack of specific data in obese patients, once the decision is made to commence HT, there is a rationale to use the lowest effective dose [oral conjugated equine estrogens (CEEs) 0.300–0.400 mg or estradiol 0.5–1 mg orally or 25–50 μg transdermally], and it may be preferable using the transdermal route [22]. HRT should be avoided in those with unexplained vaginal bleeding, active liver disease, previous breast cancer, CHD, stroke, personal history of thromboembolic disease, or known high inherited risk. CVD risk factors do not automatically preclude HRT but should be taken into account. Upregulation of the hepatic synthesis of procoagulants is another known effect of oral estrogens. Transdermal estradiol does not seem to increase the risk of venous thromboembolic events. Evidence shows that transdermal estrogen (≤50 μg) is associated with a lower risk of deep vein thrombosis, stroke, and myocardial infarction compared to oral therapy [23] and may be the preferred mode of treatment in women with an increased thrombosis risk, such as obese women and smokers. In addition, unlike oral estrogen, transdermal estradiol does not increase the risk of gallbladder disease [24,25].

Genitourinary syndrome (GSM) is a relatively new terminology describing vulvovaginal changes at menopause, as well as urinary symptoms of frequency, urgency, nocturia, dysuria, and recurrent urinary tract infections. Vaginal dryness is common after menopause and unlike VMS usually persists and may worsen with time [26]. Urogenital symptoms are effectively treated with either local (vaginal) or systemic estrogen therapy [27]. Pelvic floor dysfunction is more common in the overweight and obese women. Risk factors for developing pelvic organ prolapse can be divided into obstetric, lifestyle, comorbidity, aging, social, pelvic floor factors, and surgical factors. The most important lifestyle factor is a higher BMI. Obesity may impair pelvic floor function increasing intraabdominal pressure that damages pelvic musculature and nerve, this is linked to conduction abnormalities and obesity-related comorbidities, including diabetic neuropathy and intervertebral disc herniation.

Estrogen therapy restores normal vaginal flora, lowers the pH, and thickens and revascularizes the vaginal lining. The number of superficial epithelial cells is increased, and symptoms of atrophy are alleviated. Importantly, low-dose vaginal estrogen improves vaginal atrophy without causing proliferation of the endometrium. Given the documented efficacy and proven safety, vaginal estrogen is the first-line approach to treat the symptoms of vaginal atrophy in the majority of women: vaginal estrogen is effective and while systemic absorption does occur, it does not induce endometrial hyperplasia. Concerns about systemic absorption mean that vaginal estrogens may be avoided in breast cancer patients taking aromatase inhibitors. The relationship between HRT and urinary incontinence depends on the delivery route. Systemic HRT worsens urinary incontinence, but vaginal treatment may improve urge incontinence and prevent recurrent urinary tract infections. Using very low doses for the first few weeks is helpful if irritation occurs, and indeed lower doses of vaginal estrogens, with less frequent administration, often yield satisfactory results [28].

In conclusion, initiation of HRT is usually contraindicated in women with a personal history of breast cancer or VTE, or those with a high risk for breast cancer, thrombosis, or stroke. Transdermal estrogen therapy may be considered and preferred when highly symptomatic women with type 2 DM or obesity, or those at high risk of CVD, do not respond to nonhormonal therapies. In general, commencement of hormone therapy is not recommended for women who are aged >60 years [29].

In order to avoid undue chronic stimulatory effects on the endometrium, control menstrual bleeding, avoid abnormal bleeding and cancer development, the combination of the estrogen with a progestogen is needed. Endometrial cancer is the most common gynecologic cancer: it is estimated that risk of endometrial cancer increases about 59% for every 5 U increase in BMI (kg/m^2), and overweight and obesity are responsible for 57% of all cases of endometrial cancer in the United States. Obesity increases exposure to estrogen unopposed by progesterone in pre- and postmenopausal women. The inclusion of progesterone appears to increase breast cancer risk, but progestogens are still indicated to prevent endometrial hyperplasia and cancer risk [28].

Progesterone are naturally produced in the ovaries (particularly the corpus luteum), in the placenta, and, to a certain extent, in the adrenals, and there are a variety of synthetic progestogens. One of these progestogens, dydrogesterone, is a retroprogesterone and, another, drospirenone (DRSP), is spironolactone derivative. The "newer" progestogens belong to different classes based on their structure. For each of them the progestogenic, as well as the antiestrogenic action, is common. The antiandrogenic effect is relevant for dienogest and DRSP and minor for nomegestrol acetate. None of them have a glucocorticoid effect. DRSP is different due to its strong antimineralocorticoid action and has a favorable effect on blood pressure. In addition, these progestogens do not interfere with the positive effect of estrogens on lipid and carbohydrate metabolism, do not augment hemostasis processes as monotherapy, and avoid induction of abnormal proliferation of the endometrium in doses clinically tested. Therefore all three progestogens appear to be suitable for the treatment of menopausal women [30].

The most recent Position Statement of the North American Menopause Society on HRT, published in 2017, provides easy recommendations for menopause management and suggests that HRT may help attenuate abdominal adipose accumulation and the weight gains that are often associated with the menopause transition, and it significantly reduces the diagnosis of new-onset type 2 DM, but the US government does not approve the HRT use for this purpose [31].

Nonetheless, considering that HRT could create important health risks, it is highly desirable to discover new alternatives in the menopause-related symptoms management, with minor side effects. Over the past 15 years, hormone preparations of dehydroepiandrosterone (DHEA) have been available over the counter and have been sold as the "fountain of youth" [32]. DHEA serves as a precursor for estrogens and androgens from fetal life to postmenopause, and many people believe that DHEA is merely an inactive precursor pool for the formation of bioactive steroid hormones. DHEA-sulfate (DHEAS) represents the most abundant sex steroid in plasma in humans (more than 1000 times higher than estradiol and testosterone levels), but its serum concentration goes down to 10%−20% of its maximum level by around the age of 70 years. The large difference between low and high serum DHEA levels has a major clinical impact. Among postmenopausal women with coronary risk factors, lower DHEA levels were linked with higher mortality from CVD and all-cause mortality [33]. Several studies had previously demonstrated that 1-year treatment [34,35], using administration of 10 mg DHEA daily in symptomatic postmenopausal women with lower (5th percentile) baseline DHEAS levels, improved climacteric and sexual symptoms and directly reversed some age-related changes in adrenal enzymatic pathways, including adrenal DHEA and progesterone synthesis.

Emerging menopausal therapies

In the past years, two new pharmaceutical preparations were approved in the United States and Europe for the treatment of menopausal symptoms. An oral selective ER modulator (SERM), ospemifene, has been approved for the treatment of moderate-to-severe pain during intercourse associated with vulvovaginal atrophy [36], and a tissue-specific SERM—estrogen complex [a combination of oral CEE and bazedoxifene (BZA) (a SERM)] has been approved for the management of moderate-to-severe VMS in women with an intact uterus [37]. Tissue selectivity is achieved through the concurrent use of estrogen and a SERM, which replaces a progestogen and selectively blocks the undesirable actions of estrogen. In the case of CEE− BZA, the proliferative effects of estrogen are blocked in the uterus and possibly also the breast, whereas the bone-sparing actions of estrogen are preserved.

The role of testosterone for the treatment of postmenopausal desire or arousal disorders and the long-term implications of such a therapy in postmenopausal women are unclear. The rationale for combining estrogens with a SERM [*T-SEC, combination of CEE and BZA (SERM)*] is to retain beneficial effects of estrogens on VMS, VVA, and bone while incorporating the antiestrogenic effects of the SERM on the breast and endometrium to improve the overall safety profile [38]. The tissue-selective estrogen complex [combination of 0.45 mg of oral CEE and 20 mg BZA (a SERM)] has been approved for the management of moderate-to-severe VMS in the United States and Europe [39].

Tibolone is a synthetic steroid that is rapidly converted to two metabolites with estrogenic activity and to a third metabolite characterized by a mixed progestogenic/androgenic activity. Tibolone controls hot flushes, sweating, and mood symptoms and is effective in improving libido, due to its androgenic component. Randomized, controlled studies show that tibolone increases bone mineral density and reduces fracture risk. These beneficial effects are seen over long-term treatments [40] (over 10 years) and both in early and late postmenopausal women as well as in women with established osteoporosis. The combined analysis of randomized clinical studies on tibolone indicates no increase in risk of breast cancer development compared with placebo. Tibolone treatment is associated with a reduction of proliferation and a stimulation of apoptosis in normal breast cells that is possibly attributable to the impact of this compound on the activity of estrogen-metabolizing breast enzymes [41]. The metabolization of tibolone is tissue selective, and the conversion to the progestogenic metabolite is particularly active in the endometrium. Investigation of endometrial histology in women treated with tibolone shows no hyperplasia and a high level of atrophic endometrium, indicating no proliferative effect of this molecule.

A number of *nonhormonal therapies* are efficacious against menopausal VMS and should be considered for women who do not wish to take estrogen or those with contraindications.

For VMS, many drugs have demonstrated efficacy in several studies: paroxetine, fluoxetine, and citalopram (which are selective serotonin reuptake inhibitors); venlafaxine and desvenlafaxine (selective noradrenaline reuptake inhibitors); clonidine (α2-adrenergic receptor agonist); and anticonvulsants (gabapentin and pregabalin). Paroxetine and fluoxetine are potent cytochrome P450 2D6 (CYP2D6) inhibitors and as they decrease the metabolism of tamoxifen (a SERM used in the treatment of breast cancer)—which may reduce its anticancer effects—these drugs should be avoided in tamoxifen users. However, consistency of treatment response and efficacy of the various alternative options remain questionable.

Conclusion

Estrogen therapy may partly prevent menopause-related change in body composition and the associated metabolic sequelae. The decision to start HRT in a woman transitioning toward menopause requires a personalized discussion on the unique balance of risks and benefits in that particular individual. Thorough counseling on the relevance of improving lifestyle, dietary habits and implementing physical activity should be provided. Menopause physicians should appropriately stress how these behavioral changes are far more important to prevent cardiovascular risk than any pharmacologic or hormonal intervention.

Within this frame the available evidence shows that the balance of benefits and risks for HRT is most favorable within the first 10 years of menopause. Expert agreement is that HRT should still be primarily initiated for the management of climacteric symptoms to improve quality of life. Long-term preventive strategies, particularly cardiovascular protection, should not be a primary goal even if HRT has long-term protective effects on the cardiovascular system, reflected in an approximately 40%–50% reduction in cardiovascular events in most clinical studies, when started during this "window of opportunity."

New and emerging menopausal therapies have the potential to fill an unmet need in the post-WHI era for effective relief of menopausal symptoms with improved safety profiles. Based on the WHI, the greatest risk appears to be associated with combined estrogen--progestin therapy; therefore recent strategies have focused on eliminating the need for progestins either through the use of topical estrogens without a progestin for VVA or by combining estrogen(s) or DHEA with potentially safer options (e.g., micronized progesterone, SERMs) to reduce endometrial stimulation.

Menopausal hormone therapy remains the most effective treatment of VMS and is also indicated for GSM (previously called *vulvovaginal atrophy*) and bone protection. With no fixed duration of treatment the guidelines now state that HRT should be individualized to account for each patient's unique risk–benefit profile. The ultimate goal is to get closer to the profile of the ideal menopausal therapy—that is, to relieve bothersome menopausal symptoms and reduce the risk of osteoporosis and CVD, without increasing the risk of endometrial or breast cancer.

Conflict of interest

None.

References

[1] Roberts H, Hicke M. Managing the menopause: An update. Maturitas 2016;86:53–8.
[2] Salpeter SR, Walsh JME, Ormiston TM, Greyber E, Buckley NS. Meta-analysis: effect of hormone-replacement therapy on components of the metabolic syndrome in postmenopausal women. Diabetes Obes Metab 2006;8:538–54.
[3] World Health Organization. [2019, Online]. Obesity and overweight. Fact sheet: World Health Organisation. https://www.who.int/news-room/fact-sheets/detail/obesity-and-overweight.
[4] Toth MJ, Tchernof A, Sites CK, Poehlman ET. Menopause-related changes in body fat distribution. Ann N Y Acad Sci 2000;904:502–6.
[5] Mueller SO, Korach KS. Estrogen receptor and endocrine disease: lessons from estrogen receptor knockout mice. Curr Opin Pharmacol 2001;1:613–19.
[6] Lizcano F, Guzmán G. Estrogen deficiency and the origin of obesity during menopause. Biomed Res Int 2014;2014:757461.
[7] Genazzani AR, Gambacciani M. Effect of climacteric transition and hormone replacement therapy on body weight and body fat distribution. Gynecol Endocrinol 2006;22(3):145–50.
[8] Al-Safi ZA, Polotsky AJ. Obesity and menopause. Best Pract Res Clin Obstet Gynaecol 2015;29:548–53.
[9] Thurston RC, Sowers MR, Sternfeld B, et al. Gains in body fat and vasomotor symptom reporting over the menopausal transition: the study of women's health across the nation. Am J Epidemiol 2009;170(6):766–74.
[10] Schenck-Gustafsson K. Risk factors for cardiovascular disease in women. Maturitas 2009;63(3):186–90.
[11] Rexrode KM, Hennekens CH, Willett WC, et al. A prospective study of body mass index, weight change, and risk of stroke in women. JAMA 1997;277(19):1539–45.
[12] Cushman M, Kuller LH, Prentice R, et al. Women's health initiative investigators. Estrogen plus progestin and risk of venous thrombosis. JAMA 2004;292(13):1573–80.
[13] Renehan AG, Tyson M, Egger M, Heller RF, Zwahlen M. Body-mass index and incidence of cancer: a systematic review and meta-analysis of prospective observational studies. Lancet 2008;371(9612):569–78.
[14] Marioribanks J, Farquhar C, Roberts H, Lethaby A. Long term hormone therapy for perimenopausal and postmenopausal women. Cochrane Database Syst Rev 2012;7.
[15] Davis SR, Castelo-Branco C, Chedraui P, Lumsden MA, Nappi RE, et al. Understanding weight gain at menopause. Climacteric 2012;15:419–29.
[16] Woods DC, White YA, Tilly JL. Purification of oogonial stem cells from adult mouse and human ovaries: an assessment of the literature and a view toward the future. Reprod Sci 2013;20:7–15.

[17] Dunn JF, Nisula BC, Rodboard D. Transport of steroid hormones. Binding of 21 endogenous steroids to both testosterone-binding globulin and corticosteroid-steroid binding globulin in human plasma. J Clin Endocrinol Metab 1981;53:58–68.

[18] Canonico M, Oger E, Conard J, Meyer G, Le vesque H, et al. Obesity and risk of venous thromboembolism among postmenopausal women: differential impact of hormone therapy by route of estrogen administration. The ESTHER Study. J Thromb Haemost 2006;4:1259–1265.

[19] Gartoulla P, Worsley R, Bell RJ, Davis SR. Moderate to severe vasomotor and sexual symptoms remain problematic for women aged 60 to 65 years. Menopause 2018;25(11):1331–8.

[20] MacLennan AH, Broadbent JL, Lester S, Moore V. Oestrogen and combined oestrogen/progestogen therapy versus placebo for hot flushes. Cochrane Database Syst Rev 2004;4.

[21] Welton AJ, Vickers MR, Kim J, Ford D, Lawton B, et al. Health related quality of life after combined hormone replacement therapy: randomised controlled trial. BMJ 2008;21.

[22] Lambrinoudaki I, Brincat M, Erel CT, Gambacciani M, Moen MH, Schenck-Gustafsson K, et al. EMAS position statement: managing obese postmenopausal women. Maturitas. 2010;66(3):323–6.

[23] The North American Menopause Society. The 2012 hormone therapy position statement of the North American Menopause Society. Menopause 2012;19:257–71.

[24] Olie V, Plu-Bureau G, Conard J, Horellou MH, Canonico M, Scarabin PY. Hormone therapy and recurrence of venous thromboembolism among postmenopausal women. Menopause 2011;18:488–93.

[25] Hoibraaten E, Qvigstad E, Arnesen H, Larsen S, Wickstrøm E, Sandset PM. Increased risk of recurrent venous thromboembolism during hormone replacement therapy—results of the randomized, double-blind, placebo-controlled estrogen in venous thromboembolism trial (EVTET). Thromb Haemost 2000;84:961–7.

[26] Palacios S, Castelo-Branco C, Currie H, Mijatovic V, Nappi RE, Simon J, et al. Update on management of genitourinary syndrome of menopause: a practical guide. Maturitas 2015;82:308–13.

[27] Santen RJ. Vaginal administration of estradiol: effects of dose, preparation and timing on plasma estradiol levels. Climacteric 2014;17:1–14.

[28] Roberts H, Hickey M, Lethaby A. Hormone therapy in postmenopausal women and risk of endometrial hyperplasia Cochrane review summary. Maturitas 2014;77:4–6.

[29] Davis SR, Lambrinoudaki I, Lumsden M, Mishra GD, Pal L, et al. Menopause. PRIMER 2015;I.

[30] Schindler AE. The "newer" progestogens and postmenopausal hormone therapy (HRT). J Steroid Biochem Mol Biol 2014;142:48–51.

[31] The NAMS 2017 Hormone Therapy Position Statement Advisory Panel. The 2017 hormone therapy position statement of The North American Menopause Society. Menopause 2017;24(7):728–75.

[32] Pluchino N, Carmignani A, Cubeddu A, Santoro A, Cela V, Errasti T. Androgen therapy in women: for whom and when. Arch Gynecol Obstet 2013;288:731–7.

[33] Shufelt C, Bretsky P, Almeida CM, Johnson BD, Shaw LJ, Azziz R, et al. DHEA-S levels and cardiovascular disease mortality in postmenopausal women: results from the National Institutes of Health – National Heart, Lung, and Blood Institute (NHLBI)-sponsored Women's Ischemia Syndrome Evaluation (WISE). J Clin Endocrinol Metab 2010;95:4985–92.

[34] Genazzani AR, Stomati M, Valentino V, Pluchino N, Pot E, Casarosa E, et al. Effect of 1-year, low-dose DHEA therapy on climacteric symptoms and female sexuality. Climacteric 2011;14:661–8.

[35] Pluchino N, Ninni F, Stomati M, et al. One-year therapy with 10 mg/day DHEA alone or in combination with HRT in postmenopausal women: effects on hormonal milieu. Maturitas 2008;59:293–303.

[36] Constantine G, Graham S, Portman DJ, Rosen RC, Kingsberg SA. Female sexual function improved with ospemifene in postmenopausal women with vulvar and vaginal atrophy: results of a randomized, placebo-controlled trial. Climacteric 2015;18:226–32.

[37] Pinkerton JV, Abraham L, Bushmakin AG, Cappelleri JC, Racketa J, Shi H, et al. Evaluation of the efficacy and safety of bazedoxifene/conjugated estrogens for secondary outcomes including vasomotor symptoms in postmenopausal women by years since menopause in the Selective Estrogens, Menopause and Response to Therapy (SMART) trials. J Womens Health (Larchmt) 2014;23:18–28.

[38] Pickar JH, Yeh IT, Bachmann G, Speroff L. Endometrial effects of a tissue selective estrogen complex containing bazedoxifene/conjugated estrogens as a menopausal therapy. Fertil Steril 2009;92:1018–24.

[39] Genazzani AR, Komm BS, Pickar JH. Emerging hormonal treatments for menopausal symptoms. Expert Opin Emerg Drugs 2015;20(1):31–46.

[40] Rymer J, Robinson J, Fogelman I. Ten years of treatment with tibolone 2.5 mg daily: 283 effects on bone loss in postmenopausal women. Climacteric 2002;5:390–8.

[41] Valdivia I, Campodonico I, Tapia A, Capetillo M, Espinoza A, Lavin P. Effects of 287 tibolone and continuous combined hormone therapy on mammographic breast density and 288 breast histochemical markers in postmenopausal women. Fertil Steril 2004;81:617–23.

Chapter 31

Obesity and chronic pelvic pain

I-Ferne Tan[1] and Andrew W. Horne[2]

[1]Department of Obstetrics and Gynaecology, Nepean Hospital, Sydney, Australia, [2]MRC Centre for Reproductive Health, University of Edinburgh, Edinburgh, United Kingdom

Introduction

Chronic pelvic pain in women is defined as intermittent or constant pain in the pelvic or lower abdominal area that persists beyond 6 months duration [1]. Beyond this, the characteristics of the pain vary between countries, governing bodies, and gynecological societies due to the lack of consensus in the international community regarding its definition. The Royal College of Obstetricians and Gynaecologists (RCOG) exclude pain that occurs exclusively during menstruation, intercourse, or associated with pregnancy [2]. The Royal Australian and New Zealand College of Obstetricians and Gynaecologists (RANZCOG) consider it a diagnosis of exclusion when all other causes of pelvic pain have been excluded [3]. The American College of Obstetrics and Gynecology (ACOG) include constant, intermittent, cyclical, and pain related to menstrual cycles in their definition [4]. The Society of Obstetricians and Gynaecologists of Canada (SOGC) suggest that an element of physical or psychosocial dysfunction is present in conjunction with the chronic pain [5]. Regardless of the definition used, the prevalence of chronic pelvic pain is estimated to be 4% in the general female population and 15% in reproductive-aged women [5–8].

In reality, chronic pelvic pain is not a diagnosis but a symptom. It may be associated with various gynecological and nongynecological conditions (Table 31.1). However, even if a diagnosis is made and a treatment is implemented, the pain may still continue. This implies that there may be ongoing stimulation of pain, either through alternate mechanisms or concomitant pathology [9]. In this chapter, we will focus on the common potential gynecological causes of chronic pelvic pain.

Obesity is a condition of excess adipose tissue. In 2016 the World Health Organization (WHO) estimated that 40% of adult women were overweight and 15% of adult women were obese. These figures have tripled since 1975 to 1.9 billion and 650 million women, respectively [10]. There are many different measurements used to categorize obesity, with the most commonly used being a body mass index (BMI) of greater than 30 kg/m^2 regardless of gender. Unlike the BMI the alternate measures of obesity have different cutoff ranges for men and women. The female-specific obesity ranges are a waist circumference greater than 88 cm (35 in.), a waist-to-hip ratio greater than 0.85, and a body fat percentage greater than 30% [11–13]. Unfortunately, the most appropriate tool to assess and identify obesity has not been agreed upon. The use of BMI to diagnose obesity is losing favor in lieu of more accurate measures of adiposity [12].

The relationship between chronic pelvic pain and obesity is poorly researched, with no identifiable studies specifically addressing these two entities. They are usually only mentioned in passing as a secondary outcome, though the association is usually one of increased severity of pain associated with increased BMI [14]. Therefore most of the information regarding obesity and chronic pelvic pain are extrapolated from studies of other chronic pain syndromes.

Obesity and pain physiology

Pain is felt when nociceptors are stimulated, transmitting impulses via the spinal cord, thalamus, and limbic system, and to the cerebral cortex of the brain where it is perceived [15]. Along the entire pathway, from the offending area to the cerebral cortex where pain is interpreted, various chemical mediators are released that can interact with and manipulate the pain signal. Adipose tissue is in fact a source of many inflammatory proteins including tumor necrosis factor alpha (TNF-α), interleukin-6 (IL-6), and IL-1beta (IL-1β) [16].

Appetite and hunger are also under signaling peptide control and some of the most well-known regulators of body habitus homeostasis demonstrate an involvement with inflammation and pain signal modulation. Leptin is

TABLE 31.1 Conditions associated with chronic pelvic pain (CPP) in women and the effect of obesity.

Causes of CPP	Association with obesity	Impact of obesity on assessment	Impact of obesity on management	Recommendations
Endometriosis	Controversial; Commonly associated with low BMI. Increased aromatase activity in adipose tissue increases estrogen levels. Higher leptin levels and lower ghrelin levels in peritoneal fluid	Abdominopelvic assessment is limited by a thick subcutaneous layer, difficult positioning, inadequate instruments. Ultrasonography requires a greater depth of insonation and an increase in signal amplitude. CT and MRI machines have weight and diameter limitations	The OCP increases the risk of VTE and myocardial infarct. Analgesic doses may need adjusting, e.g., sedating analgesics in breathing disorders associated with obesity such as sleep apnea. Excess surface and visceral adiposity impedes laparoscopy. Respiratory and gut changes during general anesthesia complicated by obesity, steep Trendelenburg, and high intraperitoneal pressures	Transvaginal ultrasound provides better resolution than transabdominal ultrasound and ergonomics are better. MRI has greater sensitivity for endometriosis and greater specificity for adenomyosis. Only progestogen hormones should be the first-line treatment. Intraperitoneal pressure and angle of recline during surgery should be negotiated to optimize surgery while avoiding respiratory and gastrointestinal complications. Consider comorbidities with analgesic selection. Use sedating analgesics cautiously in patients with preexisting or at risk of breathing disorders
Adenomyosis	Higher incidence of adenomyosis in obese women. Increased aromatase activity in adipose tissue results in higher estrogen levels			
Abdominal myofascial syndrome	Associated with pelvic organ prolapse and obesity	Palpation abdominal and pelvic floor muscles are obscured by subcutaneous fat. Pelvic assessment is impeded by subpar positioning and instruments	Analgesia doses may need adjusting, e.g., the sedating analgesics on breathing disorders associated with obesity such as sleep apnea	Consider comorbidities with analgesic selection. Antidepressants such as amitriptyline may be appropriate in patients with concomitant depression. Use sedating analgesics cautiously in patients with preexisting or at risk of breathing disorders
Interstitial cystitis	Does not appear to be associated with obesity	Respiratory and gastrointestinal changes during general anesthesia are complicated by obesity		Consider awake cystoscopy in high-risk patients
Irritable bowel syndrome	Controversial, suggested increased incidence in obese population	No direct impact on assessment or management		
Lower back pain	Strong dose–response relationship with obesity		Analgesia doses may need adjusting, e.g., the sedating analgesics on breathing disorders associated with obesity such as sleep apnea	Consider comorbidities with analgesic selection. Use sedating analgesics cautiously in patients with preexisting or at risk of breathing disorders
Psychiatric somatization	Associated with obesity, particularly victims of sexual, psychological, or physical abuse	No direct impact on assessment or management		

BMI, Body mass index; *CPP*, chronic pelvic pain; *CT*, computed tomography; *MRI*, magnetic resonance imaging; *OCP*, oral contraceptive pill; *VTE*, venous thromboembolism.

produced by adipose tissue to inhibit hunger and can also cause a rise in inflammatory markers such as C-reactive protein (CRP) with a positive correlation between leptin levels and pain perception [17]. Ghrelin is a neuropeptide hormone secreted from the stomach to increase hunger and is involved in glucose metabolism. Plasma levels of ghrelin are decreased in obesity. Further stimulation of appetite occurs through the activation of neuropeptide Y (NPY). Ghrelin reduces inflammation by inhibiting the expression of IL-6, IL-1β, and TNF-α [18]. It also increases the production of nitric oxide synthase which in turn modulates μ-opioid receptors to produce an antinociceptive effect [19]. The result is an increased level of inflammation and heightened susceptibility to pain in obesity due to a reduction of the protective antinociceptive and antiinflammatory effects of ghrelin.

NPY stimulates the appetite and reduces energy expenditure [20]. It stimulates glucocorticosteroid production, gluconeogenesis, and glycogen storage. It further contributes toward obesity through its interactions with leptin, ghrelin, and insulin-like growth factor. In fact, ghrelin can directly activate NPY to enhance its effect [21]. The nociceptive effect of NPY is somewhat controversial, with evidence to support both increased and decreased analgesia. The answer appears to lie in the site of the NPY receptor. Central nervous system (CNS) NPY receptors stimulate analgesia, while peripherally located postsynaptic receptors trigger hyperalgesia [22–24]. Obese women are known to have higher circulating serum levels of NPY [25].

The orexinergic system consists of neuropeptides, orexin A, and orexin B that stimulate the appetite and can reduce the perception of pain [26]. In times of stress, such as acute or chronic pain, the orexin system is stimulated. This reaction is part of the fight or flight response to inhibit the transmission of pain signals and improve physical performance [27]. Via the orexinergic system, chronic pain can stimulate the appetite and contribute to weight gain.

As a result of these associations, obesity is increasingly being considered a proinflammatory state [17].

Obesity also appears to contribute to the risk of developing neuropathic pain disorders. In both diabetic and nondiabetic population, obese individuals have a higher incidence of developing peripheral neuropathy [28–30].

However, the relationship between obesity and pain is not quite clear. While some studies demonstrate a positive correlation between pain sensitivity and BMI [31–33], several studies contrarily demonstrate a higher pain threshold for this group [34,35]. Recently, sleep disorders such as obstructive sleep apnea (OSA) have been linked to chronic pain and obesity, with studies demonstrating lower pain thresholds in obese patients with OSA and higher pain tolerance in OSA patients treated with high-capacity continuous positive airway pressure [36].

Finally, the prevalence of obesity is demonstrably higher among chronic pain sufferers [37]. The experience of pain can limit physical activity. This can lead to weight gain and deconditioning, which further hinders physical activity [38]. This vicious cycle of decreased activity, weight gain, and decondition can be difficult to break. Alternately, excessive weight may cause structural changes in the body that can result in an increased risk of developing chronic pain [39]. Quality-of-life surveys show that weight loss can significantly improve bodily pain scores [40,41].

The genetics of obesity and chronic pain

It is generally well established that genetics play a contributing role in the obesity epidemic [42]. Genetic predisposition is also likely to play a role in the development of chronic pain syndromes through a heightened level of pain perception and sensitivity [43–45]. Sensitization of the CNS may also be triggered by neuroplasticity resulting in allodynia, hyperalgesia, and other abnormal experiences of pain [5,46]. Twin studies from the United Kingdom and Australia agree that there is likely a hereditable component to chronic pelvic pain [47,48].

No studies have been conducted to investigate any genetic associations between obesity and chronic pelvic pain; however, there is a growing body of research demonstrating a genetic link between obesity and pain. A systematic review of twin studies and lower back pain demonstrates a strong genetic association with obesity [49]. Heterozygous mutations in the melanocortin receptor 4 (MC4R) protein can cause dysregulation of appetite and hyperphagia [50]. MC4R deficiency has an incidence of 0.1% in the general population compared with 2%–3% of obese children and 5% of young adults with severe early-onset obesity [50,51]. Melanocortin receptors, including MC4R, also play an important role in the regulation of pain and are therefore implicated in various chronic pain syndromes [52].

Another suggested shared etiological pathway for the development of both chronic pain and obesity may be due to glucocorticoid receptor gene polymorphisms [53]. Glucocorticoid production is controlled by the hypothalamic–pituitary–adrenal (HPA) axis, and its receptors are found throughout the body including the CNS and in adipose tissue. Impaired receptor function can lead to glucocorticoid resistance. The resulting negative feedback cycle can cause HPA axis dysregulation and an incongruently excessive production of glucocorticoid. The discordance between hyperactive HPA axis production of glucocorticoids and end target tissue resistance to glucocorticoids can result in stimulation of inflammatory cytokines such as CRP and accumulation of visceral fat [53,54]. Many of the comorbidities of chronic pain or obesity are also implicated including type 2 diabetes, metabolic syndrome, anxiety, depression, hypertension, atherosclerosis, cardiovascular disease, and inflammatory autoimmune disorders [53].

The psychological impact of obesity and chronic pelvic pain

There is at least a moderate link between psychiatric disorders, mainly anxiety and depression, and both obesity and chronic pelvic pain. This association is likely to be bidirectional [55–58]. The presence of either can reduce one's quality of life and the presence of both is likely to have an additive effect [59,60].

One explanation may be "body image"—the way a person perceives their own physical appearance. Obesity is at least modestly associated with body image distortion and body image dissatisfaction [55]. Chronic pain sufferers also demonstrate an element of body image distortion [61,62]. Nevertheless, there are currently very few studies assessing the psychological effect of having both chronic pelvic pain and obesity. A small study by Gurian et al. suggests an increase in depressive symptoms, but no change in anxiety symptoms, and in obese women with chronic pelvic pain [35]. Further insights into the connection between obesity, chronic pelvic pain, and psychopathology may be provided by reviewing the greater medical literature on chronic pain syndromes. Data extracted from health-related quality-of-life surveys show that the presence of both chronic pain and obesity in an individual has a cumulative negative effect on psychological morbidity [63–65]. Anxiety and depression can also stimulate an increase in circulating IL-6 and CRP levels in the obese individual [66,67]. Weight loss treatment enhances the quality of life and improves pain management in chronic pain sufferers [40,41,68]. Interestingly, the effect of pain treatment on weight reduction and quality of life has not been well studied.

There is a negative relationship between obesity and socioeconomic status for women in highly developed countries, meaning that women from low socioeconomic groups are more likely to be obese [69,70]. Interestingly, this association is not seen in men. Chronic pelvic pain also displays a negative relationship with socioeconomic status [71,72]. Prolonged economical hardship is known to have an adverse effect on mental health, which is compounded by both obesity and chronic pain. It is likely that the presence of both conditions in low socioeconomic groups will have the greatest impact on mental health, but this is yet to be definitively proven [73].

The impact of obesity on the assessment of chronic pelvic pain

The assessment of chronic pelvic pain usually requires a physical examination of the abdomen, bimanual palpation to assess the pelvic organs and the location of any tenderness, and speculum inspection of the vaginal walls and cervix. Rectal examination, pelvic floor assessment, and mobility assessments can also be performed as necessary. A thorough pelvic examination is essential in distinguishing between different etiologies and will also include an assessment for central sensitization which frequently accompanies chronic pelvic pain syndromes [74]. However, the presence of extra adiposity around the abdomen, pelvis, buttocks, and thighs will increase the difficulty and limit the effectiveness of pelvic examinations. Achieving a comfortable position for the patient, the ability to identify pelvic organs, and the visibility of vaginal structures are all increasingly hampered by increasing thickness of subcutaneous tissue. A study of physician attitudes toward the assessment of obese patients highlights these issues. 77% had difficulty performing a bimanual examination, 54% complained of inability to adequately separate the thighs, and 49% had difficulties with speculum examination [75]. The lack of appropriate equipment, such as larger instruments and bariatric examination couches, can contribute to the problem.

Women with chronic pelvic pain and obesity and of low socioeconomic status all appear to have similar pitfalls and barriers to overcome. Access to healthcare services, the relationship with their primary care provider and gynecologist, trust in healthcare professionals, communication with healthcare professionals, frequent comorbidities, and fear of negative labels are prevalent in all three groups [71,76]. It stands to reason that these factors may compound and present even higher barriers to women in overlapping groups.

The reliance on imaging in the assessment of chronic pelvic pain in the obese population is increasing due to the difficulties with clinical examination. Ultrasound is the most widely used imaging modality as it is readily available and relatively inexpensive. However, a greater depth of insonation and adjusting for the reduced signal amplitude produced by the attenuation of the ultrasound beam by fat is required for obese individuals. This helps to mitigate the effects of obesity on the quality of the image produced [77]. This can be done by reducing mean array emission frequency to improve penetration or by using filters such as harmonic or compound imaging to increase the signal-to-background noise ratio [78]. The increased body habitus can provide ergonomic challenges for the sonographer. Care must be taken to avoid awkward positions which can result in muscular strain injuries [77].

Given the user-dependent nature of this imaging modality, access to sonographers with experience in optimizing the images of obese patients may present a challenge. Transvaginal ultrasound provides better resolution than transabdominal ultrasound, particularly in obese patients, and is the modality of choice [79].

The results of computed tomography (CT) and magnetic resonance imaging (MRI) present much better interobserver agreement, particularly with obesity [80]. However, there are still restrictions on their use. Both machines have an upper limit of capacity for weight and gantry diameter. The supine position adopted can potentially cause hypoxia and hypotension secondary to aortocaval compression by a large pannus [77]. CT use presents additional dilemmas in calculating intravenous contrast dosage, which is based on lean body weight, and the increased radiation exposure, which is required for optimization of image quality in obesity [81].

The impact of obesity on the treatment of women with chronic pelvic pain

Regardless of whether a cause for the pain is diagnosed, the long-term treatment of chronic pelvic pain in women often involves ovarian hormone suppression, analgesia, or a combination of both. The oral contraceptive pill (OCP) is often used as empirical treatment of chronic pelvic pain as estrogen is a pain modulator and therefore can also influence pain perception of nongynecological causes of chronic pelvic pain. Hormonal therapy is consistently the first-line treatment for common potential gynecological causes of chronic pelvic pain, such as endometriosis and adenomyosis [5,82]. While obesity may reduce the efficacy of the OCP in preventing pregnancy, its effect on the treatment of chronic pelvic pain has not been adequately assessed [83,84].

Obesity and OCP use are both known to increase the risk of venous thromboembolism (VTE) by two and three times, respectively [85]. This risk increases with increasing BMI and the effect is likely to be cumulative in obese women using the OCP. Compared to nonusers with a normal BMI, the risk of VTE is 10 times greater [85,86]. There is also some concern regarding an increased risk of acute myocardial infarction in obese women using the OCP [85]. Other cardiovascular events with a known association with estrogen, such as embolic stroke, do not appear to be affected by OCP use in obese women. Neither the risk of pregnancy nor cardiovascular disease is increased with progesterone-only subdermal implants and intrauterine devices, and therefore these methods of hormone suppression are preferred in an obese cohort [84].

The dosing of common analgesic medication in obese patients can present a challenge since dosing recommendations are based on total body weight deduced from studies that often exclude obese individuals [87]. Obesity is associated with greater fat mass, greater muscle mass, increased blood and plasma volume, greater cardiac output, decreased pulmonary function, increased glomerular filtration rate, increased free fatty acids, and derangement of enzymes required for drug metabolism. The magnitude of these changes in clinical efficacy and drug metabolism are thought to be related to the lipophilicity of the drug [88]. However, in practice, this is not always the case as demonstrated in a trial of a fixed dose of morphine administered to subjects in varying BMI categories [89]. The use of opioids and other sedating analgesics also needs careful consideration as they can result in or exacerbate abnormal breathing patterns such as central sleep apnea, OSA, and ataxic breathing [90]. The risk of hypoxemia is therefore compounded in these situations. Another point for consideration is the effect of any proposed medical treatments on obesity. Neuropathic pain medications, such as pregabalin, and hormonal suppressants may cause weight gain and therefore be unacceptable to a woman already struggling to control her weight [91]. Increasing BMI will also worsen any obesity-related conditions, including chronic pain.

Finally, consideration should be given to the management of the psychological component of chronic pelvic pain in obese individuals since studies suggest that this combination reduces the success of pain management strategies [65,92]. Cognitive—behavioral therapy, goal-setting techniques, and development of coping strategies have demonstrated effectiveness in obese patients, as well as chronic pain sufferers [93].

The impact of obesity on the surgical management of women with chronic pelvic pain

The investigative work-up and initial management of chronic pelvic pain may require surgical intervention such as a diagnostic laparoscopy. The majority of the difficulty with obesity in gynecological surgery is due to the need for steep Trendelenburg positioning during laparoscopy to access and navigate the pelvis. This is made even more challenging in the obese patient who often demonstrates a visceral adipose tissue thickness to match their exterior, obscuring visceral organs and increasing the risk of injury [94]. Distortion or obscuring of anatomical landmarks commonly used to guide entry and trocar placement in laparoscopic surgery by a large panniculus or a thick subcutaneous layer in the anterior abdominal wall will further increase the risk of visceral injury.

Obese patients have larger gastric volumes, more acidic gastric juice, and prolonged gastric emptying times. All of these increase the risk of aspiration of gastric contents during surgery [94]. Obesity is associated with detrimental respiratory changes including increased oxygen consumption, decreased chest wall and lung compliance, and decreased functional residual capacity [94]. This is compounded by an additional 20% with the use of general anesthesia, increasing the risk of hypoxemia and a higher degree of intrapulmonary shunting [95]. These alternations in physiology are further exacerbated by the pneumoperitoneum pressures required for laparoscopic surgery [96,97]. Increasing the depth of incline and insufflation pressures is directly related to the increasing impact on respiratory physiology, therefore they are subject to frequent change as a compromise with the anesthetist since the resulting trade-off in patient positioning often contributes to surgeon fatigue.

If surgery is indicated, laparoscopy is preferred over laparotomy. It has a lower rate of complications including postoperative ileus, fever, and wound infections [98,99]. In addition, the availability of robot-assisted laparoscopy is increasing and growing in popularity, particularly for the obese patient, and there are studies that demonstrate that the use of robots in gynecological endoscopy may reduce operating time, blood loss, and length of stay in hospital [100]. However, it should be acknowledged that there is little evidence yet to support robot use outside of oncological procedures, and there is no available literature specifically on their use for chronic pelvic pain.

Obesity and endometriosis

Endometriosis is an estrogen-dependent inflammatory condition, defined by the presence of endometrial-like tissue (lesions) outside the uterine cavity and is the most common gynecological disease associated with chronic pelvic pain in reproductive-aged women [101]. Endocrinological changes thought to contribute to the disease include increased aromatase enzyme activity resulting in increased circulating estrogen, increased conversion of the less-active estrone to the more-active estradiol by the β-hydroxyl steroid dehydrogenase type 1 enzyme, increased numbers of estrogen receptors, and decreased number of progesterone receptors [102]. Higher levels of leptin and lower levels of ghrelin in the peritoneal fluid of women with endometriosis directly correlate with the increase in inflammatory markers such as CRP and IL-6 [103]. The increased numbers of adipose cells in obese women result in increased aromatase expression and ultimately higher circulating levels of estrogen. This is likely the underlying reason for the association between endometriosis and estrogen-related cancers of the ovary, breast, and endometrium and is exacerbated by obesity [104]. However, whether there is direct causality or merely an association due to their similar risk profile has yet to be proven.

Contradictorily, the presence of endometriosis is generally associated with women being at the lower end of the BMI scale [105,106]. However, the correlation between BMI and the severity of endometriosis observed is controversial [105–107]. At the very least the increased levels of estradiol associated with obesity can result in neuromodulation since estradiol is known to influence the sensitivity of peripheral nerves and modulate CNS activity [101].

A causal association between endometriosis and severe dysmenorrhea probably exists, but the conventionally accepted symptoms of dysmenorrhea, dyspareunia, and dyschezia are a poor predictor of finding endometriosis at surgery [108,109]. There also does not appear to be an association between the extent of disease and the severity of pain since debilitating pelvic pain can also be present in women with minimal or mild endometriosis [109]. For this reason a thorough surgical exploration of the abdominopelvic cavity is important, though inherently more difficult in obese patients. Surgical management of endometriosis poses extra challenges in the obese population as discussed previously, including difficulties with adequate Trendelenburg positioning and insufflation pressures for laparoscopy.

Minimal-to-mild disease cannot be identified with imaging modalities, but transvaginal ultrasound and MRI are frequently used to identify or exclude severe disease (deep infiltrating endometriosis) and for subsequent surgical planning. The limitations of both are discussed earlier. The specificity of both modalities is greater than 90% regardless of the organ in question, but the sensitivity can vary quite markedly by location [110–112]. MRI is more likely to identify endometriosis than ultrasound [111].

Long-term management of endometriosis requires hormonal suppression. As previously discussed, the use of the OCP increases the risk of VTE and is of particular concern in obese patients. Therefore progestogen-only forms of hormonal suppression are recommended over combined estrogen and progestogen formulations. As a last resort, oophorectomy will result in surgical menopause and is used for the treatment of refractory endometriosis in women who have completed their family. Despite this, up to 40% of women will experience a recurrence of symptoms within 5 years [113,114]. In obese women, this is possibly due to an ongoing source of estrogen, such as high levels of aromatase in adipose tissue. Menopause is associated with an increased risk of coronary heart disease, depression, anxiety, and all-cause mortality. These conditions are all exacerbated by obesity [55–58,115–117].

Obesity and adenomyosis

Adenomyosis is the presence of ectopic endometrial-like tissue within the myometrium. Whether this condition represents a separate entity or is merely a clinical manifestation of endometriosis is debatable. Unlike endometriosis, there appears to be a higher incidence of adenomyosis in obese women [118,119].

The classic symptoms of adenomyosis, dysmenorrhea, and menorrhagia can be managed with hormonal suppression or hysterectomy. The challenges with hormonal suppression and surgical management are the same as for endometriosis and are described in detail earlier.

For suspected adenomyosis, MRI has a greater specificity and positive predictive value compared to transvaginal ultrasound and a greater ability to distinguish between adenomyomas and leiomyomas [120,121]. Given the added benefits of MRI over ultrasound in an obese population, it is the imaging modality of choice for identifying adenomyosis.

Obesity and abdominal myofascial pain syndrome

An underappreciated cause of chronic pelvic pain is abdominal myofascial pain syndrome, which can affect up to 93% of women attending pain clinics and 30% of women attending primary care centers [122]. It causes inflammation and intense pain in the pelvis and is activated by trigger points in the muscle fascia. These can be palpated as taut bands or spasm of the rectus abdominus and pelvic floor muscles [123].

Pelvic organ prolapse, for which obesity is a well-documented risk, may also be a source of pain [2,124,125]. Pelvic floor examination of the obese patient to identify a prolapse or elicit signs of myofascial pain syndrome of the levator muscles can be more challenging due to difficulties in positioning the patient and the need for appropriately sized instruments. Identification of taut bands or abdominal rectus muscle spasm is also increasingly challenging with increasing thickness of the subcutaneous layer.

Obesity and nongynecological causes of chronic pelvic pain

Nongynecological sources of chronic pelvic pain are usually urological, gastrointestinal, musculoskeletal, or psychological in nature.

Interstitial cystitis is commonly associated with chronic pelvic pain in women but does not appear to be related to obesity. However, it is interesting to note its frequent coexistence with endometriosis and myofascial pain syndrome, where it can be seen in two-thirds and three-fourths of patients, respectively [126,127].

Irritable bowel syndrome is also associated with endometriosis [128,129]. Its association with obesity is contentious, with some studies reporting an increase in incidence or severity in obese population [130]. This is likely due to the difficulty in diagnosis and a poorly understood etiology.

Lower back pain is a well-recognized cause of chronic pain which can be referred to the pelvic region. The association between obesity and chronic back pain is well documented in the literature and demonstrates a dose—response relationship, directly correlating increasing BMI with increasing prevalence and pain scores [131,132].

Somatization of psychological disorders is another well-recognized cause of chronic pain, including chronic pelvic pain. This is particularly evident in victims of abuse, who also demonstrate an increased incidence of obesity, regardless of whether the abuse is of a sexual, verbal, or physical nature [133].

Conclusion

Though chronic pelvic pain is a well-recognized symptom in women, the inconsistency in its definition presents a problem in coalescing ideas. The association between chronic pelvic pain and obesity is poorly researched, but there are several pathophysiological and genetic connections that predict a probable link between these two entities. Although many gynecological and nongynecological conditions are associated with chronic pelvic pain, there appears to be a significant overlap in the presence of these disorders. It is likely that true chronic pelvic pain is multifactorial in origin and therefore all coexisting pathologies need to be identified and addressed for successful treatment. Multidisciplinary management of chronic pain also has beneficial effects on reducing pain, reducing somatization, and improving mood symptoms [134—136]. There is a further economic benefit with a reduction in the use of healthcare resources and a faster return to work [134].

Obesity presents significant challenges to the assessment and management of an already challenging "chronic pain syndrome." Adequate equipment is vital for the comfort of both patient and physician. MRI is the imaging modality of choice in obese women with chronic pelvic pain, though the cost may be prohibitive. Progestogen-only formulations, such as the levonorgestrel intrauterine device or etonogestrel subdermal implant, are the preferred methods of hormonal suppression to avoid the complications associated with estrogen use in obese patients. To complicate matters, establishing a successful and respectful patient—physician relationship is more difficult in a cohort that is often suspicious and poorly understood, but it is essential for a favorable outcome.

TABLE 31.2 Take-home messages.

Take-home points

Common etiological pathways: The association between chronic pelvic pain and obesity is poorly researched, but there are several pathophysiological and genetic connections that predict a probable link between these two entities:
- Adipose tissue is a source of many inflammatory proteins including IL-6, IL-1β, and TNF-α, resulting in a proinflammatory state
- Appetite and nociception share common signaling peptide pathways including the orexinergic system, leptin, ghrelin, and neuropeptide Y
- There are genetic links between chronic pain syndromes and obesity such as the melanocortin receptor 4 protein which is associated with hyperphagia and pain dysregulation
- Glucocorticoid receptor dysfunction can cause excessive adrenal production of glucocorticoids despite glucocorticoid resistance resulting in increased inflammatory cytokines and accumulation of visceral fat

Similar psychological aspects:
- Both chronic pain and obesity have a cumulative negative effect on psychological morbidity and quality of life
- Mistrust of healthcare professionals and fear of negative labels are barriers for both obese patients and those with chronic pelvic pain

Challenges in assessment:
- Physical examination of chronic pain patients is limited by the presence of excess adiposity due to difficulties with adequate positioning and lack of appropriate equipment
- Imaging for investigation of chronic pelvic pain is impeded by obesity. Ultrasonography requires a greater depth of insonation and an increase in signal amplitude. CT and MRI machines have weight and diameter limitations

Management issues:
- The oral contraceptive pill is commonly used as first-line or empirical treatment of chronic pelvic pain. Obesity increases the risk of venous thromboembolism and myocardial infarcts. Progestogen-only hormonal treatment, such as levonorgestrel intrauterine devices or etonogestrel subdermal implants, is the preferred treatment option to minimize these risks
- The dose of common analgesia has not been tested in the obese population. Sedating analgesics can worsen disorders with abnormal breathing patterns commonly associated with obesity such as sleep apnea.
- Surgical assessment or treatment of chronic pelvic pain is more challenging in obese patients due to distortion of anatomical landmarks and obscuring of visceral organs by visceral adiposity. Laparoscopy is preferred to laparotomy due to the increased risk of ileus, fever, and wound infections with the latter. The complications associated with general anesthesia are more common in obese patients due to changes in the respiratory and digestive systems. High intra-abdominal pressures required for pneumoperitoneum and steep Trendelenburg positioning can further exacerbate these complications

Condition-specific points:
- Endometriosis is a common, estrogen-dependent cause of chronic pelvic pain. Increased aromatase activity due to greater amounts of adipose tissue results in higher circulating levels of estrogen. However, the presence of endometriosis is more commonly associated with a lower BMI
- Adenomyosis is an estrogen-dependent cause of chronic pelvic pain that has a higher incidence in obese women. For suspected cases, MRI is the imaging modality of choice for obese women. It has a higher specificity and positive predictive value and a greater ability to distinguish adenomyomas from leiomyomas
- Abdominal myofascial pain syndrome is an underappreciated cause of chronic pelvic pain that may be associated with obesity
- Lower back pain and somatization in psychological disorders are common nongynecological causes of chronic pelvic pain that are associated with and exacerbated by obesity
- Interstitial cystitis and irritable bowel syndrome are common nongynecological causes of chronic pelvic pain that do not have an established association with obesity. However, they are commonly associated with endometriosis and abdominal myofascial pain syndrome

BMI, Body mass index; *IL-1β*, interleukin-1beta; *IL-6*, interleukin-6; *MRI*, magnetic resonance imaging; *TNF-α*, tumor necrosis factor alpha.

Patient education, respectful communication, legitimization of their pain, and multidisciplinary holistic care will help overcome many barriers in managing chronic pelvic pain in the obese woman (Table 31.2).

References

[1] Howard FM. Chronic pelvic pain. Obstet Gynecol 2003;101(3):594–611.
[2] Kennedy S, Moore J. The initial management of chronic pelvic pain. Green top guidelines. London: Royal College of Obstetricians and Gynaecologists; 2005.
[3] RANZCOG. Chronic pelvic pain. 2017.
[4] American College of Obstetricians and Gynecologists. Frequently asked questions: gynecologic problems, FAQ099; 2011.
[5] Jarrell JF, et al. No. 164-consensus guidelines for the management of chronic pelvic pain. J Obstet Gynaecol Can 2018;40(11):e747–87.
[6] Mathias SD, et al. Chronic pelvic pain: prevalence, health-related quality of life, and economic correlates. Obstet Gynecol 1996;87(3):321–7.

[7] Zondervan KT, et al. Prevalence and incidence of chronic pelvic pain in primary care: evidence from a national general practice database. BJOG 1999;106(11):1149—55.

[8] Zondervan K, Barlow DH. Epidemiology of chronic pelvic pain. Best Pract Res Clin Obstet Gynaecol 2000;14(3):403—14.

[9] Baranowski AP. Chronic pelvic pain. Best Pract Res Clin Gastroenterol 2009;23(4):593—610.

[10] World Health Organization. Obesity and overweight; 2018.

[11] World Health Organization. Physical status: the use of and interpretation of anthropometry. In: Report of a WHO Expert Committee; 1995.

[12] Okorodudu D, et al. Diagnostic performance of body mass index to identify obesity as defined by body adiposity: a systematic review and meta-analysis. Int J Obes 2010;34(5):791.

[13] World Health Organization. Waist circumference and waist-hip ratio: report of a WHO expert consultation, Geneva: World Health Organization; 2011.

[14] Yosef A, et al. Multifactorial contributors to the severity of chronic pelvic pain in women. Am J Obstet Gynecol 2016;215(6):760.e1—760.e14.

[15] Steeds CE. The anatomy and physiology of pain. Surgery (Oxford) 2009;27(12):507—11.

[16] Cesari M, et al. Sarcopenia, obesity, and inflammation—results from the Trial of Angiotensin Converting Enzyme Inhibition and Novel Cardiovascular Risk Factors study. Am J Clin Nutr 2005;82(2):428—34.

[17] Younger J, et al. Association of leptin with body pain in women. J Womens Health 2016;25(7):752—60.

[18] Dixit VD, et al. Ghrelin inhibits leptin-and activation-induced proinflammatory cytokine expression by human monocytes and T cells. J Clin Investig 2004;114(1):57—66.

[19] Gaskin FS, et al. Ghrelin-induced feeding is dependent on nitric oxide. Peptides 2003;24(6):913—18.

[20] Loh K, Herzog H, Shi Y-C. Regulation of energy homeostasis by the NPY system. Trends Endocrinol Metab 2015;26(3):125—35.

[21] Guneli E, Gumustekin M, Ates M. Possible involvement of ghrelin on pain threshold in obesity. Med Hypotheses 2010;74(3):452—4.

[22] Tracey DJ, Romm MA, Yao NN. Peripheral hyperalgesia in experimental neuropathy: exacerbation by neuropeptide Y. Brain Res 1995;669(2):245—54.

[23] Lin Q, et al. Involvement of peripheral neuropeptide Y receptors in sympathetic modulation of acute cutaneous flare induced by intradermal capsaicin. Neuroscience 2004;123(2):337—47.

[24] Brumovsky P, et al. Neuropeptide tyrosine and pain. Trends Pharmacol Sci 2007;28(2):93—102.

[25] Minor RK, Chang JW, De Cabo R. Hungry for life: how the arcuate nucleus and neuropeptide Y may play a critical role in mediating the benefits of calorie restriction. Mol Cell Endocrinol 2009;299(1):79—88.

[26] Razavi BM, Hosseinzadeh H. A review of the role of orexin system in pain modulation. Biomed Pharmacother 2017;90:187—93.

[27] Xie X, et al. Hypocretin/orexin and nociceptin/orphanin FQ coordinately regulate analgesia in a mouse model of stress-induced analgesia. J Clin Investig 2008;118(7):2471—81.

[28] Lean ME. Pathophysiology of obesity. Proc Nutr Soc 2000;59(3):331—6.

[29] Miscio G, et al. Obesity and peripheral neuropathy risk: a dangerous liaison. J Peripher Nerv Syst 2005;10(4):354—8.

[30] Tesfaye S, et al. Vascular risk factors and diabetic neuropathy. N Engl J Med 2005;352(4):341—50.

[31] Pradalier A, et al. Relationship between pain and obesity: an electrophysiological study. Physiol Behav 1981;27(6):961—4.

[32] McKendall MJ, Haier RJ. Pain sensitivity and obesity. Psychiatry Res 1983;8(2):119—25.

[33] Yancy Jr. WS, et al. Relationship between obesity and health-related quality of life in men. Obes Res 2002;10(10):1057—64.

[34] Zahorska-Markiewicz B, Kucio C, Pyszkowska J. Obesity and pain. Hum Nutr Clin Nutr 1983;37(4):307—10.

[35] Gurian MB, et al. Measurement of pain and anthropometric parameters in women with chronic pelvic pain. J Eval Clin Pract 2015;21(1):21—7.

[36] Okifuji A, Hare BD. The association between chronic pain and obesity. J Pain Res 2015;8:399.

[37] Loevinger BL, et al. Metabolic syndrome in women with chronic pain. Metabolism 2007;56(1):87—93.

[38] Janke EA, Collins A, Kozak AT. Overview of the relationship between pain and obesity: what do we know? Where do we go next? J Rehabil Res Dev 2007;44(2):245—62.

[39] Fransen M, et al. Risk factors associated with the transition from acute to chronic occupational back pain. Spine 2002;27(1):92—8.

[40] Dixon JB, Dixon ME, O'Brien PE. Quality of life after lap-band placement: influence of time, weight loss, and comorbidities. Obes Res 2001;9(11):713—21.

[41] Kaukua J, et al. Health-related quality of life in obese outpatients losing weight with very-low-energy diet and behaviour modification—a 2-y follow-up study. Int J Obes Relat Metab Disord 2003;27(10):1233—41.

[42] Farooqi SI. Genetic, molecular and physiological mechanisms involved in human obesity: Society for Endocrinology Medal Lecture 2012. Clin Endocrinol 2015;82(1):23—8.

[43] Mogil JS. The genetic mediation of individual differences in sensitivity to pain and its inhibition. Proc Natl Acad Sci USA 1999;96(14):7744—51.

[44] Diatchenko L, et al. Genetic basis for individual variations in pain perception and the development of a chronic pain condition. Hum Mol Genet 2004;14(1):135—43.

[45] Edwards RR. Genetic predictors of acute and chronic pain. Curr Rheumatol Rep 2006;8(6):411—17.

[46] Brawn J, et al. Central changes associated with chronic pelvic pain and endometriosis. Hum Reprod Update 2014;20(5):737—47.

[47] Zondervan KT, et al. Multivariate genetic analysis of chronic pelvic pain and associated phenotypes. Behav Genet 2005;35(2):177—88.

[48] Vehof J, et al. Shared genetic factors underlie chronic pain syndromes. Pain 2014;155(8):1562—8.

[49] Ferreira PH, et al. Nature or nurture in low back pain? Results of a systematic review of studies based on twin samples. Eur J Pain 2013;17(7):957—71.

[50] Farooqi IS, et al. Clinical spectrum of obesity and mutations in the melanocortin 4 receptor gene. N Engl J Med 2003;348(12):1085—95.

[51] Farooqi IS, et al. Dominant and recessive inheritance of morbid obesity associated with melanocortin 4 receptor deficiency. J Clin Investig 2000;106(2):271—9.

[52] Tao Y-X. The melanocortin-4 receptor: physiology, pharmacology, and pathophysiology. Endocr Rev 2010;31(4):506—43.

[53] Chrousos GP, Kino T. Glucocorticoid action networks and complex psychiatric and/or somatic disorders. Stress 2007;10(2):213–19.

[54] Silverman MN, Sternberg EM. Glucocorticoid regulation of inflammation and its functional correlates: from HPA axis to glucocorticoid receptor dysfunction. Ann NY Acad Sci 2012;1261(1):55–63.

[55] Friedman MA, Brownell KD. Psychological correlates of obesity: moving to the next research generation. Psychol Bull 1995;117(1):3.

[56] Atlantis E, Baker M. Obesity effects on depression: systematic review of epidemiological studies. Int J Obes 2008;32(6):881.

[57] De Wit L, et al. Depression and obesity: a meta-analysis of community-based studies. Psychiatry Res 2010;178(2):230–5.

[58] Luppino FS, et al. Overweight, obesity, and depression: a systematic review and meta-analysis of longitudinal studies. Arch Gen Psychiatry 2010;67(3):220–9.

[59] Ayorinde AA, et al. Chronic pelvic pain in women: an epidemiological perspective. Womens Health (Lond) 2015;11(6):851–64.

[60] Gambadauro P, Carli V, Hadlaczky G. Depressive symptoms among women with endometriosis: a systematic review and meta-analysis. Am J Obstet Gynecol 2019;220(3):230–41.

[61] Osumi M, et al. Negative body image associated with changes in the visual body appearance increases pain perception. PLoS One 2014;9(9):e107376.

[62] Senkowski D, Heinz A. Chronic pain and distorted body image: implications for multisensory feedback interventions. Neurosci Biobehav Rev 2016;69:252–9.

[63] Barofsky I, Fontaine KR, Cheskin LJ. Pain in the obese: impact on health-related quality-of-life. Ann Behav Med 1997;19(4):408–10.

[64] Heo M, et al. Obesity and quality of life: mediating effects of pain and comorbidities. Obes Res 2003;11(2):209–16.

[65] Marcus DA. Obesity and the impact of chronic pain. Clin J Pain 2004;20(3):186–91.

[66] Capuron L, et al. Relationship between adiposity, emotional status and eating behaviour in obese women: role of inflammation. Psychol Med 2011;41(7):1517–28.

[67] Daly M. The relationship of C-reactive protein to obesity-related depressive symptoms: a longitudinal study. Obesity 2013;21(2):248–50.

[68] Shapiro JR, Anderson DA, Danoff-Burg S. A pilot study of the effects of behavioral weight loss treatment on fibromyalgia symptoms. J Psychosom Res 2005;59(5):275–82.

[69] Sobal J, Stunkard AJ. Socioeconomic status and obesity: a review of the literature. Psychol Bull 1989;105(2):260.

[70] McLaren L. Socioeconomic status and obesity. Epidemiol Rev 2007;29(1):29–48.

[71] Grace VM. Problems of communication, diagnosis, and treatment experienced by women using the New Zealand health services for chronic pelvic pain: a quantitative analysis. Health Care Women Int 1995;16(6):521–35.

[72] Roth RS, Punch MR, Bachman JE. Educational achievement and pain disability among women with chronic pelvic pain. J Psychosom Res 2001;51(4):563–9.

[73] Everson SA, et al. Epidemiologic evidence for the relation between socioeconomic status and depression, obesity, and diabetes. J Psychosom Res 2002;53(4):891–5.

[74] Stratton P, et al. Association of chronic pelvic pain and endometriosis with signs of sensitization and myofascial pain. Obstet Gynecol 2015;125(3):719–28.

[75] Ferrante JM, et al. Family physicians' practices and attitudes regarding care of extremely obese patients. Obesity (Silver Spring) 2009;17(9):1710–16.

[76] Forman-Hoffman V, Little A, Wahls T. Barriers to obesity management: a pilot study of primary care clinicians. BMC Family Pract 2006;7(1):35.

[77] Glanc P, et al. Challenges of pelvic imaging in obese women. Radiographics 2012;32(6):1839–62.

[78] Paladini D. Sonography in obese and overweight pregnant women: clinical, medicolegal and technical issues. Ultrasound Obstet Gynecol 2009;33(6):720–9.

[79] Uppot RN, et al. Effect of obesity on image quality: fifteen-year longitudinal study for evaluation of dictated radiology reports. Radiology 2006;240(2):435–9.

[80] van Randen A, et al. A comparison of the accuracy of ultrasound and computed tomography in common diagnoses causing acute abdominal pain. Eur Radiol 2011;21(7):1535–45.

[81] Bae KT. Intravenous contrast medium administration and scan timing at CT: considerations and approaches. Radiology 2010;256(1):32–61.

[82] Gambone JC, et al. Consensus statement for the management of chronic pelvic pain and endometriosis: proceedings of an expert-panel consensus process. Fertil Steril 2002;78(5):961–72.

[83] Yamazaki M, et al. Effect of obesity on the effectiveness of hormonal contraceptives: an individual participant data meta-analysis. Contraception 2015;92(5):445–52.

[84] Lopez LM, et al. Hormonal contraceptives for contraception in overweight or obese women. Cochrane Database Syst Rev 2016;18(8).

[85] Horton LG, Simmons KB, Curtis KM. Combined hormonal contraceptive use among obese women and risk for cardiovascular events: a systematic review. Contraception 2016;94(6):590–604.

[86] Simmons KB, Edelman AB. Hormonal contraception and obesity. Fertil Steril 2016;106(6):1282–8.

[87] Ingrande J, Lemmens HJ. Dose adjustment of anaesthetics in the morbidly obese. Br J Anaesth 2010;105(Suppl. 1):i16–23.

[88] Lloret-Linares C, et al. Challenges in the optimisation of postoperative pain management with opioids in obese patients: a literature review. Obes Surg 2013;23(9):1458–75.

[89] Patanwala AE, Holmes KL, Erstad BL. Analgesic response to morphine in obese and morbidly obese patients in the emergency department. Emerg Med J 2014;31(2):139–42.

[90] Yue HJ, Guilleminault C. Opioid medication and sleep-disordered breathing. Med Clin North Am 2010;94(3):435–46.

[91] Cabrera J, et al. Characterizing and understanding body weight patterns in patients treated with pregabalin. Curr Med Res Opin 2012;28(6):1027–37.

[92] Janke AE, Kozak AT. "The more pain I have, the more I want to eat": obesity in the context of chronic pain. Obesity 2012;20(10):2027–34.

[93] Narouze S, Souzdalnitski D. Obesity and chronic pain: systematic review of prevalence and implications for pain practice. Reg Anesth Pain Med 2015;40(2):91–111.

[94] Lamvu G, et al. Obesity: physiologic changes and challenges during laparoscopy. Am J Obstet Gynecol 2004;191(2):669–74.

[95] Söderberg M, Thomson D, White T. Respiration, circulation and anaesthetic management in obesity. Investigation before and after jejunoileal bypass. Acta Anaesthesiol Scand 1977;21(1):55–61.

[96] Fahy BG, et al. The effects of increased abdominal pressure on lung and chest wall mechanics during laparoscopic surgery. Anesth Analg 1995;81(4):744–50.

[97] Fahy BG, et al. Effects of Trendelenburg and reverse Trendelenburg postures on lung and chest wall mechanics. J Clin Anesth 1996;8(3):236–44.

[98] Brezina PR, Beste TM, Nelson KH. Does route of hysterectomy affect outcome in obese and nonobese women? JSLS: J Soc Laparoendosc Surg 2009;13(3):358.

[99] Siedhoff MT, et al. Effect of extreme obesity on outcomes in laparoscopic hysterectomy. J Minim Invasive Gynecol 2012;19(6):701–7.

[100] Gehrig PA, et al. What is the optimal minimally invasive surgical procedure for endometrial cancer staging in the obese and morbidly obese woman? Gynecol Oncol 2008;111(1):41–5.

[101] Stratton P, Berkley KJ. Chronic pelvic pain and endometriosis: translational evidence of the relationship and implications. Hum Reprod Update 2010;17(3):327–46.

[102] Dassen H, et al. Estrogen metabolizing enzymes in endometrium and endometriosis. Hum Reprod 2007;22(12):3148–58.

[103] Rathore N, et al. Distinct peritoneal fluid ghrelin and leptin in infertile women with endometriosis and their correlation with interleukin-6 and vascular endothelial growth factor. Gynecol Endocrinol 2014;30(9):671–5.

[104] Zanetta GM, et al. Hyperestrogenism: a relevant risk factor for the development of cancer from endometriosis. Gynecol Oncol 2000;79(1):18–22.

[105] Ferrero S, et al. Body mass index in endometriosis. Eur J Obstet Gynecol Reprod Biol 2005;121(1):94–8.

[106] Holdsworth-Carson SJ, et al. The association of body mass index with endometriosis and disease severity in women with pain. J Endometriosis Pelvic Pain Disord 2018;10(2):79–87.

[107] Lafay Pillet M-C, et al. Deep infiltrating endometriosis is associated with markedly lower body mass index: a 476 case–control study. Hum Reprod 2011;27(1):265–72.

[108] Eskenazi B, et al. Validation study of nonsurgical diagnosis of endometriosis. Fertil Steril 2001;76(5):929–35.

[109] Fauconnier A, Chapron C. Endometriosis and pelvic pain: epidemiological evidence of the relationship and implications. Hum Reprod Update 2005;11(6):595–606.

[110] Moore J, et al. A systematic review of the accuracy of ultrasound in the diagnosis of endometriosis. Ultrasound Obstet Gynecol 2002;20(6):630–4.

[111] Abrao MS, et al. Comparison between clinical examination, transvaginal sonography and magnetic resonance imaging for the diagnosis of deep endometriosis. Hum Reprod 2007;22(12):3092–7.

[112] Guerriero S, et al. Accuracy of transvaginal ultrasound for diagnosis of deep endometriosis in uterosacral ligaments, rectovaginal septum, vagina and bladder: systematic review and meta-analysis. Ultrasound in Obstet Gynecol 2015;46(5):534–45.

[113] Namnoum AB, et al. Incidence of symptom recurrence after hysterectomy for endometriosis. Fertil Steril 1995;64(5):898–902.

[114] Shakiba K, et al. Surgical treatment of endometriosis: a 7-year follow-up on the requirement for further surgery. Obstet Gynecol 2008;111(6):1285–92.

[115] Manson JE, et al. A prospective study of obesity and risk of coronary heart disease in women. N Engl J Med 1990;322(13):882–9.

[116] Adams KF, et al. Overweight, obesity, and mortality in a large prospective cohort of persons 50 to 71 years old. N Engl J Med 2006;355(8):763–78.

[117] Flegal KM, et al. Association of all-cause mortality with overweight and obesity using standard body mass index categories: a systematic review and meta-analysis. JAMA 2013;309(1):71–82.

[118] Templeman C, et al. Adenomyosis and endometriosis in the California Teachers Study. Fertil Steril 2008;90(2):415–24.

[119] Trabert B, et al. A case-control investigation of adenomyosis: impact of control group selection on risk factor strength. Womens Health Issues 2011;21(2):160–4.

[120] Reinhold C, et al. Diffuse adenomyosis: comparison of endovaginal US and MR imaging with histopathologic correlation. Radiology 1996;199(1):151–8.

[121] Dueholm M, et al. Magnetic resonance imaging and transvaginal ultrasonography for the diagnosis of adenomyosis. Fertil Steril 2001;76(3):588–94.

[122] Montenegro ML, et al. Abdominal myofascial pain syndrome must be considered in the differential diagnosis of chronic pelvic pain. Eur J Obstet Gynecol Reprod Biol 2009;147(1):21–4.

[123] Yap E-C. Myofascial pain—an overview. Ann Acad Med Singapore 2007;36(1):43.

[124] Olsen AL, et al. Epidemiology of surgically managed pelvic organ prolapse and urinary incontinence. Obstet Gynecol 1997;89(4):501–6.

[125] Hendrix SL, et al. Pelvic organ prolapse in the Women's Health Initiative: gravity and gravidity. Am J Obstet Gynecol 2002;186(6):1160–6.

[126] Chung MK, Chung RP, Gordon D. Interstitial cystitis and endometriosis in patients with chronic pelvic pain: the "evil twins" syndrome. JSLS: J Soc Laparoendosc Surg 2005;9(1):25.

[127] Bassaly R, et al. Myofascial pain and pelvic floor dysfunction in patients with interstitial cystitis. Int Urogynecol J 2011;22(4):413–18.

[128] Ballard K, et al. Can symptomatology help in the diagnosis of endometriosis? Findings from a national case–control study—Part 1. BJOG 2008;115(11):1382–91.

[129] Seaman H, et al. Endometriosis and its coexistence with irritable bowel syndrome and pelvic inflammatory disease: findings from a national case–control study—Part 2. BJOG 2008;115(11):1392–6.

[130] Pickett-Blakely O. Obesity and irritable bowel syndrome: a comprehensive review. Gastroenterol Hepatol 2014;10(7):411.

[131] Shiri R, et al. The association between obesity and low back pain: a meta-analysis. Am J Epidemiol 2010;171(2):135–54.

[132] Vismara L, et al. Effect of obesity and low back pain on spinal mobility: a cross sectional study in women. J Neuroeng Rehabil 2010;7(1):3.

[133] Williamson DF, et al. Body weight and obesity in adults and self-reported abuse in childhood. Int J Obes 2002;26(8):1075.

[134] Flor H, Fydrich T, Turk DC. Efficacy of multidisciplinary pain treatment centers: a meta-analytic review. Pain 1992;49(2):221–30.

[135] Milburn A, Reiter RC, Rhomberg AT. Multidisciplinary approach to chronic pelvic pain. Obstet Gynecol Clin N Am 1993;20(4):643–61.

[136] Stones R, Mountfield J. Interventions for treating chronic pelvic pain in women (Cochrane Review). Cochrane Lib 2003;4.

Chapter 32

Obesity and clinical psychosomatic women's health

Mira Lal[1,2] and Abhilash H.L. Sarhadi[3]
[1]*St James's University Hospital, Leeds, United Kingdom,* [2]*The Dudley Group NHS Foundation Trust, Dudley, United Kingdom,* [3]*Independent Scholar, Stourbridge, United Kingdom*

Introduction

The term "Psychosomatic" stems from the Greek words *psūkhe*, referring to "mind," and *sōmatikos*, concerning *sōma*—"the body." The concept "*clinical psychosomatic*" brings together and emphasizes the connection between mind and body as being relevant to clinical medicine when evaluating diseases that affect both physical and mental health concomitantly [1]. Likewise, it describes an approach to medicine that epitomizes the connection between the higher centers of the brain, including the limbic system, and the rest of the body, as mediated by the neural networks and messenger hormones that comprise the hypothalamic−pituitary−adrenal (HPA) axis and its target organs. Above all, the concept aims to account for the fact that the body and mind are not disparate entities. Rather, they are anatomically and physiologically linked via the neuroendocrine system, with their interplay influencing the maintenance of overall health and the generation of clinical psychosomatic disease conditions [2].

While obesity is considered a serious health condition in the United Kingdom [3], it is viewed as a disease in its own right in North America [4,5], with nearly 40% of the US adult population (and 18.5% of youth) being obese [6]. However, the continuing upward trend in the United Kingdom, currently approaching 30% adult prevalence [7], has prompted recent debate as to whether it should also be reclassified as a disease there too [8]. Obesity is often associated with diseases due to clinical psychosomatic interactions that can affect women's reproductive health, such as menstrual problems, metabolic disorders, infertility, gender-related violence, and cancer. Both physical and mental illnesses in such despondent patients can lead to overeating and obesity.

When the patient's body weight crosses the threshold of 29.9 kg/m^2, she is considered to be obese, with the upper limit in this category reaching 39.9 kg/m^2. Increasing weight variations above this range of obesity can reach up to 50 kg/m^2 or higher; this is then designated as being "superobese." The preceding weight range, below 30 kg/m^2 and above the normal weight of 24.9 kg/m^2, is regarded as being overweight. As over 1.9 billion adults and 18% of children and adolescents worldwide are now classified as being overweight or obese according to the World Health Organization (WHO) [9], and with over 21% of women estimated to be obese by 2025 [10], it is imperative that greater attention be paid to these preventable health conditions. Moreover, since being overweight precedes obesity, early recognition and appropriate attention to relentless weight gain, often due to burgeoning psychosomatic issues, would likely prevent many cases of obesity. This discourse will mainly focus on obesity in gynecology, which should be of interest to both practicing clinicians and gynecological trainees. We will begin by considering the theoretical justification as to why a psychosomatic approach can often prove fruitful in the clinical management of patients with obesity.

Why are mind−body approaches likely to be of benefit in the management of obese patients?— an analysis

As described previously, obesity is considered to be present when an individual bears a greater amount of weight than might be considered healthy for their body frame, typically defined on a physiologically informed epidemiological basis, such as BMI ranges. In clinically relevant obesity (as opposed to the technical but misleading kind associated with extensively muscular bodybuilders or athletes), the excess tissue of note is fat. Given the requirement for sufficient accumulation to proceed beyond "mere" overweight status, the process of becoming obese necessarily involves an extended period where caloric intake through food exceeds energy expended, often referred to as *positive energy balance* [11].

Nutritional energy balance rests upon three aspects—intake, expenditure, and storage. Thus, conceptually, one can reason that the proximate causes of obesity can fall into one of three categories: a decrease in expenditure, a boosted caloric intake, or an increase in storage. Naturally, the first two of these are closely intertwined, since an excessive intake of food is best defined relative to a comparatively low metabolic expenditure. Indeed, intake can be considered primary, for without intake being excessive in the first place, storage does not come into consideration.

Hence, new-onset obesity may often be expected to arise from changes in social environment that affect eating choices or activity patterns: for example, a shift in job role; a different, more sedentary hobby; a new patisserie or coffee shop opening; recreation center closures; fluctuations in disposable income—even an appealing advertising campaign. The practicing clinician may come across women affected by any or all of these. For such individuals, their obesity can be conceived of as having a strong environmental component, and adjustments focusing on issues relating to body weight or adiposity may therefore be sufficient to guide management.

Nevertheless, one must consider the fact that, broadly speaking, obesity is generally linked with poorer health outcomes, particularly at higher grades [12–15]. As a result, it is unlikely that a person who cares about their future wellbeing would wish to remain in such a state for long, provided they have sufficient knowledge of the dangers involved. Thus, leaving aside those who still lack awareness of the risks, from a management perspective, it is important to divide those with persistent obesity into two groups—those who consume food excessively as part of a deliberate lifestyle choice and those who would like to lose weight but are unable to do so.

Considering the former group—whether epicure gourmands, strongly present-focused, or rebels against those recommending moderation—although they can hardly be seen as being healthy, they are personally unlikely to regard themselves as having a problem. From their perspective, all is as intended. It is with the latter group, however, where clinical sense must be honed to achieve the best outcomes. This group can itself be further subdivided by assessing the specific reason that is preventing a shift away from positive energy balance.

Where the root cause is a lack of knowledge, treatment should ideally begin with education or guidance. If the problem is due to a lack of awareness concerning the importance of regular physical activity, examples can be given of exercises or hobbies that might help foster change. On the other hand, if the gap lies in essential dietary knowledge, such as of food groups and basic nutritional science, beneficial interventions may include information campaigns, a consultation with a dietitian to create a personalized food plan, keeping a food diary to follow up with an app or website, or a well-considered decision to join a weight loss organization, which can help by providing guidance on portion sizes as well as moral support. In such situations, it is also still essential to take social environment into account, as this remains a key influence on the availability, acceptability, and ultimate efficacy of any given solution [16–18].

A further useful development in terms of better understanding food intake could be a shift from purely numeric labeling of nutritional information, as with Reference Intakes/Guideline Daily Amounts, toward a more interpretive front-of-pack label, such as the "traffic light" system pioneered by the United Kingdom in the early 21st century [19–21], or the closely related French Nutri-Score, which is based on similar principles but forgoes numeric display in favor of a summary health grade [22]. While the detail of the Multiple Traffic Light system and the simplicity of the Nutri-Score each have their own merits, the bright colors both employ offer a useful aid to comprehension, which appears to be recognized and appreciated internationally [23]. Indeed, having a more intuitive indication of the potential unhealthiness of any given food item may be enough to tip the scales of health back in the desired direction [24], particularly where combined with educational interventions to raise awareness [11,25–28].

In theory, a traffic light system also sidesteps some of the risks associated with pure numeric labeling; for example, some individuals will have a disinclination to involve themselves with anything seemingly mathematical, while others with a keen financial sense (albeit less nutritionally well-informed) may feel they are getting good value by purchasing large quantities of discounted sugar-sweetened beverages or ultraprocessed snacks, failing to take into account the potential costs of future ill-health. This touches on the notion that, for some, the personal situation preventing a shift away from positive energy balance is predominantly financial and/or time-related (two sides of the same coin). For such people, it is important that dietary guidance should take into account the price and availability of local foodstuffs for best results; targeted financial incentives may prove effective when cost or time poses a potential barrier [26]. Cooking lessons may also be helpful in expanding the variety of foods considered acceptable for everyday use, and for increasing efficiency, as could addressing a disinclination for trying unfamiliar foods (sometimes characterized as "food neophobia" [29,30]) if present.

Having addressed those who recognize the dangers of being obese, but originally had difficulty with the more technical and practical aspects of how to go about losing weight, this leaves those who have a more serious issue: namely, that they are unable to reduce their caloric intake, despite being aware of the consequences. Where knowledge and economics do not pose barriers, engaging in a behavior that one wishes to stop, but just cannot seem to, has the hallmarks of an addiction. This dual nature of food is reflected in the efficacy of the front-of-pack "warning label," an alternative/complement to the more colorful label types described earlier. Boldly implemented in Chile [31], though showing promise across the Americas, particularly among those with lower nutritional literacy [32–35], these stark black-and-white warnings calling out undesirably elevated levels of certain nutrients echo those found on products containing alcohol and nicotine in many other nations; the legal imposition of stringent marketing restrictions [36] being a further parallel.

As with other "legal drugs," there is an important distinction to be made between true addiction and excessive consumption for hedonic reasons [37,38]. Nevertheless, there are compelling similarities between how the neuroendocrine reward system responds to excessive food intake in obese individuals (with considerable evidence for sugar in particular [39–41]) and how the brains of addicts respond to their drug of choice [42–44]. Indeed, there is reason to believe that certain aspects of reward system activation by food are partly independent of taste [44–46], implying that it would be shortsighted to characterize all instances of persistent overeating as epicurean sorties into the realms of gustatory gratification.

Furthermore, even with hedonic overeating, the reason for continuously seeking to drown in the pleasures of gastronomic excess may well merit attention—one would expect there to be a spectrum running from the passionate yet picky gourmet, through the habitual comfort eater [47], and onward to something potentially far more serious. To paraphrase the health maxim (oft incorrectly ascribed to Hippocrates [48]), if a given individual is in fact endeavoring to "let food be their medicine," they may well be "self-medicating" (to the extent such a concept is valid). Viewed through this lens, when faced with an obese individual, the clinician ought to consider the possibility that there may be an underlying psychological or social issue, in the same way that they might when managing an intoxicated patient. If obesity can be regarded as a potential sign of self-medicating, there may well be concomitant psychological comorbidity [49], as with the parallel in alcohol or drugs [50]; with such individuals, treating the obesity as a disease in its own right can mask the true underlying cause, leading to suboptimal clinical outcomes.

Where obesity is associated with excessive eating as a counterpoint to significant stresses of past or present, the contrast with the gourmand bon vivant could not be sharper. In the former, their behavior may represent a legacy of chronic stress with subsequent neurophysiological alterations [51–53], in some cases reflecting a history of difficulties in coping with challenging early life experiences [54,55], which may have been further amplified by environmental and social factors [56]. Not only do such early stresses increase the likelihood of developing a concomitant bona fide food addiction [57], but they also cause long-lasting alterations in the HPA axis, which can result in lifelong mental health vulnerabilities [58]. What is perhaps less well appreciated is that the same sorts of alterations can influence a broad range of "physical" health conditions, which should more properly be conceived of as having a psychosomatic dimension [2], with individual presentations determining the extent to which this warrants attention.

A final consideration is that there is increasing evidence that the gut–brain axis [59] also plays a key role in the interplay between obesity, psychological factors, and the psychosomatic comorbidities that may arise; a reminder that, although the human body may be conceptually divided up into a range of physiological systems as an aid to understanding, it should never be forgotten that the integrated whole remains ever present and interconnected—a system in its own right, with all the potential complexity that engenders.

On a related note, women's health conditions that concern clinical psychosomatic wellbeing or ill-health associated with obesity can occasionally transgress the boundaries of gynecology. For example, an obese woman can also have bulimia during pregnancy [60]. However, the inclusion of obstetric presentations in this chapter would distract from the intended focus on the gynecological aspect of women's diseases, and their associations with obesity, and so will not be addressed further here.

In addition, in this chapter the theoretical aspect of selected disease conditions will be illustrated with anonymized clinical vignettes from the United Kingdom. These convey patients' stories, including early life experiences that encouraged obesity-promoting lifestyles, illustrating how childhood and adolescent experiences can shape patient attitudes toward overeating. Familial/workplace/social issues that promote being overweight, and then facilitate progression to obesity, will be analyzed. Neglecting a healthy lifestyle because of concurrent conflicting biopsychosociocultural issues would fall into this category. Any clinical psychosomatic health issues that result from such interactions could be crucial in generating obesity in many patients. As shown in the vignettes, these initiating clinical psychosomatic factors that potentiate obesity may have arisen when the woman was an adolescent or even earlier [61], rather than when she first attended the gynecology clinic in adulthood. Therefore a consideration of traditional beliefs and behavioral factors [60] (as occasionally prompted by the attention-seeking behavior exhibited by some individuals) should be addressed earlier by health professionals and family/friends. Timely management of potential psychosomatic issues could curb the progression toward obesity.

The remainder of the chapter will now move on to a discussion of certain common clinical psychosomatic disease conditions related to obesity that also affect women's gynecological health, preceded by a brief reminder of the pathophysiology of clinical presentations where relevant.

Section 1: menstrual problems and obesity

Pathophysiology of psychosomatic menstrual issues

With the onset of menarche around 12 years of age (range 10–16.5 years), and gradual transition into regular ovulatory menstrual cycles, the physiological alterations to one's body remain unique to each teenager. Changes occur from the fifth day of the menstrual cycle in response to the anterior pituitary–secreted follicle-stimulating hormone. This initiates release of estrogen from the dominant ovarian follicle, along with maturation of a few follicles. The estrogen level peaks around ovulation (14th day of cycle) and then falls. Concurrently, progesterone is released at ovulation after stimulation of the ovarian corpus luteal cells by luteinizing hormone released from the anterior pituitary. Subsequently, progesterone peaks in the luteal phase and then falls around the 28th day of the cycle along with estrogen. Only one maturing egg reaches full maturation as the others become atretic. The released mature egg at ovulation travels to and through a fallopian tube facilitated by the physiological effects of progesterone; in the absence of fertilization, menstruation begins [62]. Normally, menstrual bleeding lasts for about 5 days, accompanied by cramping abdominal pains that radiate to the thighs, hips, and lower back (sometimes associated with clotting), though pains may start before menstruation begins. These features are considered as normal by many who obtain symptomatic relief by rest, distractions, local heat, or analgesics (NSAIDs). Irregular menstrual cycles may occur at menarche and when the pattern changes to shorter premenopausal cycles; at such times, regular ovulation is uncommon (oligoovulation). These phases can be associated with painful/heavy menstrual bleeding that affects psychosomatic welfare, more so in the obese.

Psychosomatic insights into menstrual presentations in the obese

In concordance with the abovementioned characteristics, the obese woman can present with complaints of dysmenorrhea (menstrual pain), menorrhagia (heavy bleeding), menometrorrhagia (heavy, prolonged, irregular bleeding), premenstrual syndrome (PMS), oligomenorrhea (scanty menstrual loss), amenorrhea (temporary cessation of menstruation), or menopause (complete cessation); all of these are usually considered benign conditions. Nonetheless, they may cause dysphoria (anxiety and/or depression) in those experiencing these menstrual deviations, which can compel some women to seek relief by comfort eating, even if they might become overweight. Likewise, nonmalignant growths, such as fibroids, ovarian cysts, and endometriosis, may be associated with heavy blood loss and dysmenorrhea, which can promote overeating with an obesogenic body habitus. In addition, endometrial cancer or other malignancies of the female reproductive organs could present with menstrual problems, and these conditions occur more frequently in the obese. Latterly, these cancers are affecting obese women at a younger age [61] so are of increasing concern. These cancers may be reduced by concerted efforts to discourage lifestyle developments that could lead to obesity.

The clinical psychosomatic approach and the patients' perspective

Menstrual problems may relate to perceived ill-health, even if these problems are considered as normal by many, thereby encouraging some to gormandize in order to "feel good." Such problems can be of significance in the obese teenager undergoing menarche, especially if discussions with parents/caregivers medicalize reasonable lifestyle restrictions as so-called menstrual abnormalities. Moreover, a medical referral that sometimes involves a gynecological assessment could ensue. Verily, the health professional who receives the referral may not be comfortable in assessing associated dysphoria that could lead to obesity. Many young patients may need gentle handling during the history-taking, examination, and investigations, even if these methods are noninvasive. There could be other personal factors that caused dysphoria with concomitant overeating. Any relevant initiating/aggravating biopsychosocial factors deserve appropriate consideration when revealed at a medical consultation for menstrual irregularities; a clinician with psychosomatic expertise can often be successful in bringing about symptom relief with less invasive methods.

If a constitutional delay at menarche is diagnosed without any pathological contributing factors, reassurance and lifestyle alterations could enable the patient to reach her normal weight for age. This would limit the risk of overeating because of anxiety/depression [63], which could lead to becoming overweight/obese. However, if endocrinological or chromosomal anomalies are confirmed, and complex hormonal/surgical treatments considered necessary to alleviate the menstrual problem [64], the patient may experience feelings of shock, grief, denial, or guilt, due to perceived loss of femininity. Many of these patients, who habitually overeat without paying heed to consequences that hamper their overall fitness, would benefit from a clinical psychosomatic approach with patient-centered management. Clear and empathetic communication can be enhanced by taking a psychosomatic viewpoint during discussions regarding the management of these health issues [60], even more so in the overweight/obese. A greater appreciation for the psychosomatic management perspective could also benefit the gynecologist facing such scenarios, by fostering better doctor—patient rapport, thereby making patients more likely to adhere to recommended lifestyle interventions with the goal of preventing obesity.

Clinical psychosomatic gynecological referrals for menstrual irregularities

Many gynecological encounters arise when obese patients are troubled with various types of menstrual irregularities that may affect their physical and mental health with varied clinical implications [65]. Such patients may have an urge to overeat even if overweight or obese. This would preferably warrant an evaluation by a gynecologist with a clinical psychosomatic-oriented approach to deal with presenting symptoms potentially involving mind—body interactions. Even so, facilities for such evaluations are sparse, and there are limited numbers of trained staff who can deliver the necessary patient-centered assessments. Adequate funding for such tailored care is lacking for many patients in the United States, and also in the United Kingdom [66].

Clinical psychosomatic angles to menstrual irregularities, premenstrual syndrome, and polycystic ovarian syndrome

Menstrual irregularities are common gynecological disorders, which seem to have a greater penchant for the obese. Although management strategies for these varied physical/mental presentations exist, including medication/surgery, patients may consider certain approaches as being too invasive. This is more evident for the anxious patient who wants to start a family. The clinician should consider any aggravating biopsychosocial factors that may affect the patient's physical and mental wellbeing to generate overeating. These factors could influence the patient's perception of her symptomatology, and disregard for her weight gain despite becoming obese. Sometimes treatment may be sought even if the objectively assessed menstrual loss (about 80 mL) is considered normal by the attending clinician, conceivably suggesting veiled clinical psychosomatic issues. Again, the menstrual loss may be preceded by dysmenorrhea, bloating, dysphoria, irritability, headaches, bowel symptoms, and breast tenderness, which are collectively referred to as premenstrual molimina. Premenstrual molimina is perceived as normal and dealt with as trivial discomfort by many. However, if these symptoms do distress patients, they could likely benefit from advice given by a psychosomatically aware gynecologist.

Despite their acceptance by many women as being normal, in recent times there has been a trend among many gynecologists to classify such menstrual symptomatology as PMS. Not infrequently, following patient dissatisfaction with the results of pharmacotherapy, invasive surgical managements are offered as a potential cure. However, there is some recent consensus that noninvasive options targeting physical and mental health issues that promote obesity in select patients with PMS could also result in symptom relief. A more severe form of premenstrual symptomatology is associated with tearfulness, sleeplessness, or a preference for being bed-bound, overeating or having food cravings—all factors that increase the risk of obesity, besides amplifying the psychological impact of menstrual molimina. Such a variable symptom complex is classified as premenstrual dysphoric disorder (PMDD). It has been reported in 3%—8% of patients with manifestation soon after ovulation, continuing until menstruation. Again, the gynecologist unfamiliar with psychosomatic issues usually prescribes hormonal medication, failing which surgical treatment is advised. Yet, PMS may persist even after major surgery, such as hysterectomy [60]; this remains no less of a challenge in the obese. Specific invasive treatments are advised by some gynecologists but these may be considered unsatisfactory by many patients who feel that the negative effects of menstruation are ill-understood by medical professionals [67].

Notwithstanding, PMS and the more severe form PMDD have been considered as psychosomatic disorders related to endocrinological and autonomic alterations associated with changes in the woman's body and certain brain centers during the menstrual cycle [2]. These conditions can also promote weight gain from being sedentary and overeating, accordingly promoting obesity. Making etiology-tailored management decisions for these psychosomatic disorders remains clinically challenging, with several options on offer to alleviate associated symptoms of physical discomfort, anxiety, depressed mood, tiredness, hopelessness, irritability, and reduced quality of life (QOL). Suicidal tendencies can arise with severe menstrual symptoms. Cognitive behavioral self-help therapy/counseling could aid affected women who are dissatisfied with their ongoing medical treatments [68], and despair of ever getting relief from their severe symptoms. In certain social groups, comfort eating is encouraged for women with PMS/PMDD as a coping strategy, often with little regard for potential weight increase. Severe clinical cases who may have suffered for years, leading to habitual comfort eating and subsequent obesity, would benefit from individualized treatment for altering their lifestyles. Consultation with a psychosomatically oriented clinician may persuade the patient to accept a trial of noninvasive management with risk of fewer potential side effects, including lifestyle modifications that could also remedy the ill effects of obesity.

Polycystic ovarian syndrome and obesity—an increasing clinical psychosomatic conundrum

Polycystic ovarian syndrome (PCOS) is said to occur in 15%–20% of young adults and seems more common in the United States than in Europe [69]. It can present with irregular menstrual cycles or amenorrhea. PCOS is frequently associated with obesity. Moreover, PCOS with obesity can be linked to bulimia or binge-eating behavior (where an individual eats within 2 hours of having eaten a quantity of food that others may consider as being excessive within the same period of time). Along with the rising proportion of the overweight in the West, which seems to particularly affect women [70], the prevalence of PCOS has risen. Having PCOS along with being overweight can progress on to the woman becoming obese, along with associated clinical psychosomatic problems such as depression, insomnia [71], hyperlipidemia, hyperandrogenism, and overeating [72]; these symptoms are linked to obesity [73]. The diagnosis of PCOS is made when there are two of three of the following criteria: the level of androgens (male sex hormones) is high, periods are missed or irregular, and there are small cysts (fluid-filled sacs) in the ovaries. PCOS is classified according to the Rotterdam criteria, which include four types relating to prognosis [74], or is otherwise categorized into three types according to WHO criteria that guide management [75]. Nevertheless, these criteria have been questioned as not being representative for PCOS manifestations that are prevalent in certain populations, thereby increasing the risk of a referral bias [76].

There can be a familial association for PCOS with obesity, insulin resistance, and heart disease [77]. Besides, others [78] have reported that PCOS in some mothers may have negative effects on growth, cardiometabolic health, reproductive health, and neurodevelopment in the offspring, but further research is warranted. A systematic review and meta-analysis of 28 studies has concluded that women with PCOS are at a heightened risk of emotional distress [79,80]. There may be accompanying clinical psychosomatic presentations with symptoms of dysphoria, anxiety, bipolar disorder, and obsessive–compulsive disorder [81], along with glucose intolerance or Type-2 diabetes, obesity, cardiovascular disease, infertility, and malignancies. In order to deal comprehensively with PCOS, the menstrual problems and associated obesity, as well as other physical/mental health conditions present that can increase suicide ideation [82], attention to relevant biopsychosocial factors affecting sufferers, along with their siblings who may also exhibit such symptoms, gain high import. These factors reduce the QOL in the obese [83] so should be duly considered. Besides biennial follow-ups with a lipid profile [84], documentation of associated clinical features has been endorsed. Any unhelpful mind–body interactions that initiated the overeating should also be incorporated into the management plan, along with pertinent clinical psychosomatic health care as required.

To illustrate, the clinical conundrum of an obese adult with PCOS and PMS who sought surgical intervention—but later favored clinical psychosomatic care—is presented in Table 32.1.

TABLE 32.1 Ms. N. Lee, a gynecological outpatient attender (British Caucasian) [Clinical Vignette (01)].

Presentation and management	• Ms. Lee, a 22-year-old woman, resided in her university's accommodation; she had recently broken off with a boyfriend • She had a family history of PCOS and diabetes • At her first gynecological visit 4 years back, PMS was confirmed and she had wanted a hysterectomy for cure as she mentioned no relief from routinely prescribed medications; she was persuaded to change her mind by a rereferral to an Ob/Gyn with psychosomatic specialization (Ob/Gyn–Psyc); the Ob/Gyn–Psyc added the psychosomatic aspect to her management • Currently, Ms. Lee was on an outpatient visit about her resurfaced menstrual problems along with her overeating with periodic vomiting • She feared persistence of her symptoms although previously she had significant relief from her PMS after consulting an Ob/Gyn–Psyc specialist

(Continued)

TABLE 32.1 (Continued)

	• She had dysmenorrhea and menorrhagia 5 days back, besides bloating, breast tenderness, muscle cramps, irritability, headaches, and dysphoria • Her last medication of a gonadotrophin analog with add—back therapy was stopped because of her headaches and a low mood • She had a long history of these premenstrual symptoms, which subsided at the onset of menstrual flow but now persisted during her menstrual loss • The intervals between her periods were getting longer (35—36 days); she had developed acne, hirsutism, and had gained considerable weight • At this visit during her premenstrual phase, she felt low • On examination, general observations were normal, her BMI was 32 and her abdomen soft with tenderness in both iliac fossae on deep palpation • She consented for pelvic examination with pipelle biopsy, which revealed a normal uterus with cervix, but tender ovaries and fleshy curettings • A pelvic ultrasound on the third and eighth day of her cycle and at ovulation revealed polycystic ovaries (14 mL each), with follicles sized between 2 and 9 mm; each ovary had 8—10 cysts, but other pelvic features were normal • Hormonal tests were sent off and a GTT arranged • Analgesia (NSAID) was prescribed and a review planned for after 24 days • At review: 17-hydroxyprogesterone and anti-Müllerian hormone levels were raised, the GTT was impaired; the endometrial histology was normal • The final diagnosis was of PCOS (Rotterdam criteria—nonclassic no. 3, or WHO 2) with PMS, dysmenorrhea, and impaired GTT in an obese nullipara • A fourth-generation OCP and NSAIDs were prescribed; concurrently CBT for dysphoria and lifestyle/dietary advice was arranged • After 5 months, when attending with her mother, she mentioned clearance of her acne, and pain relief, though her dysphoria had persisted • She had lost 2 kg in weight yet had menorrhagia and prolonged cycles • She declined further OCP or antiandrogens, for fear of their side effects • She also refused treatment for her dysphoria, which was getting better • She had started a new supportive relationship and was back to university • She agreed to continue with metformin tablets (2 g/day) and analgesics • After 6 months, she had lost 5 kg (now overweight), her periods were regular monthly, her GTT was normal but the polycystic ovaries persisted • A pregnancy was planned on graduating; meanwhile she sought GP care • Ms. Lee was discharged with plans for annual physical and screening tests for her PCOS; an Ob/Gyn—Psyc review would be undertaken, if requested
Psychological initiating, maintaining, and alleviating factors	• Ms. Lee's perception of her menstrual symptoms was influenced by her anxiety/mood and lifestyle issues • Her mother, a community nurse, was domineering but well-meaning, yet their contact seemed to exacerbate the severity of Ms. Lee's symptoms • Her mother's great concern that something was grossly wrong with her daughter had amplified Ms. Lee's perception of the severity of symptoms • Her sudden break-off with her first boyfriend made her feel low initially • The referral to an Ob/Gyn—Psyc specialist stopped her from demanding a hysterectomy for symptom relief; she was eager to start a family later
Impact on regional health-care system	• Ms. Lee's long history of PMS and her recent PCOS, along with an unassuming personality and maternal overprotection, made her feel low • She continued with her university and personal tasks after consulting the Ob/Gyn—Psyc specialist regarding her clinical psychosomatic issues

GP, General practitioner; *GTT*, glucose tolerance test; *OCP*, oral contraceptive pill; *PCOS*, polycystic ovarian syndrome; *PMS*, premenstrual syndrome; *WHO*, World Health Organization.

Learning points: Help from family/friends before a clinical psychosomatic condition gets entrenched in adulthood can eliminate dysphoric symptoms in a young adult who has menstrual problems with PCOS and is obese due to overeating and neglect of exercising. Modifications toward a healthier lifestyle were not possible for Ms. Lee until she had received evaluation with advice from an Ob/Gyn—Psyc specialist in addition to dietary/lifestyle guidance. This form of management brought her weight down from the obese category, which reduced menstrual irregularities and added to her self-esteem. She remained motivated to lose weight through lifestyle changes and attending follow-up appointments.

Section 2: infertility/subfertility and clinical psychosomatic aspects

Successes/failures in conceiving, distress, and obesogenic behavior

Infertility or subfertility is the inability to conceive despite regular unprotected intercourse for a year. However, as a woman ages, and fecundity becomes lower, a lesser period of inability to conceive despite trying is also accepted [85]. Infertility results in psychosomatic consequences that are often associated with obesity in women [86]. Since time immemorial, the woman has been blamed for failures to conceive [87], which has persisted in many cultures even if investigations disclose that she is fertile. Obesity can impact on fertility by impairing the development of ovarian follicles, causing defective oocyte maturation, and disrupting meiosis that causes abnormal embryo preimplantation; all factors that can cause poorer outcomes [88]. Disappointing pregnancy outcomes can result in stress with guilt and self-blame because of unforgiving sociocultural attitudes. Thus infertile women facing failed attempts at conceiving, notably if obese, can acquire depression, phobic anxiety, and paranoid ideation.

Moreover, psychosocial distress can promote overeating and obesity, which further potentiates infertility due to ovulatory or unexplained causes in >50% of couples [89]. In addition, it promotes an endocrinological milieu that causes early miscarriages or pregnancy complications that result in the noncontinuation of pregnancy [85], thereby encouraging binge-eating with lowered self-esteem in these women with compromised fertility [90]. It has been recognized that the loss of a much-wanted pregnancy not only causes physical pain but also dysphoria with emotional pain that stems in part from the impact on the patient's self-identity and social relationships [91]. Persisting dysphoria can in turn lead to overeating and obesity, potentially bringing about reproductive disorders [92], and complicating a planned pregnancy [85]. It would be prudent for the obese infertile woman to accept clinical psychosomatic health care after pregnancy loss and discuss about the possibility of losing weight to improve her chances of conceiving (whether naturally or by assisted conception). Nonetheless, satisfactory pregnancy outcomes may not be guaranteed [93]. Besides, an obese habitus can lead to pregnancy failures even after assisted conception, which can then give rise to depression [94], as will be elaborated upon shortly.

While a successful natural conception is the ideal outcome for any couple, persisting infertility (particularly if unexplained [95]) or continuing pregnancy losses would suggest that assisted conception may be a better option. Nevertheless, undergoing in vitro fertilization (IVF), as opposed to expectant management, clomiphene citrate−stimulated ovulation, intrauterine insemination (with or without ovarian stimulation), or gamete intrafallopian transfer, is said to bring about lower live birth rates for couples with unexplained infertility [96], along with a higher risk of complications in the obese. Also, infertility is three times higher in obese women, with an increase in failures of assisted reproduction techniques even after they have undergone pharmacotherapy or weight-reducing surgery [97]. All the same, despite reduced live birth rates, assisted conception, in particular IVF, remains on offer for the obese infertile woman [98], thereby leading to additional biopsychosocial challenges for these couples.

Couples can be steeped in dysphoria following repeated failed attempts to conceive and may welcome an obesogenic diet that is comforting [99]. Again, the obese can have medical comorbidities so are more prone to complications such as ovarian hyperstimulation or thromboembolism while undergoing assisted conception procedures [100]. Knowing the inherent risks with assisted conception in obese women, some clinicians do not offer assisted conception to them until they have lost weight, while others consider individualizing management for the overweight or obese [101]. When providing assisted conception, in addition to the other complications of obesity, pregnancy-related complications [102] such as hypertensive disorders may occur; rarely fatalities can occur [103]. Stress-related dropout from assisted conception may also arise due to uncertainty from undergoing procedures without guaranteed success [103].

Moreover, if couples opt for assisted techniques such as IVF and intracytoplasmic sperm injection, recent reports on overall outcomes describe long-term deleterious effects that can cause early malignancies in the male partner [104], and also their offspring. These are worrying and would further increase biopsychosocial stress for couples, particularly if they are cognizant of such outcomes. Therefore, despite great advances that permit an offer of preimplantation genetic diagnosis to help couples with genetic disorders have healthy offspring, or techniques that promote fertility by providing ovarian tissue transplantation to assist cancer survivors, the physical, psychosocial, and ethical issues associated with various fertility treatments persist [105]. Hence, the clinical psychosomatic repercussions of managing infertility, particularly in the obese [106], remain germane.

Stress can affect the woman's endocrinological milieu, resulting in menstrual problems such as longer cycles or amenorrhea that may engender infertility, with possible knock-on effects on the patient's lifestyle and dietary intake. Whether the psychological factor is the cause or the effect of infertility is controversial, but obesity remains an important contributory factor affecting successful conception. Publications repeatedly confirm that women undergoing these assisted conception methods have high levels of stress [102]. Stress associated with infertility may also cause sexual or marital conflicts, which have long-lasting effects [107], and comorbid dysphoria could manifest in couples. This in turn can impact negatively on any offspring exposed to such parental feelings [108], to the detriment of on their wellbeing. In addition, pregnancy rates may increase in obese infertile women after they undergo behavioral treatment for stress management and weight loss. A gynecologist with clinical psychosomatic skills could help resolve intense situations in the infertile couple by assessing specific needs, and appropriately counseling the obese who want to start a family.

Table 32.2 presents the clinical conundrum of an obese adult (BMI 43) with abdominal pain and atypical endometrial hyperplasia who suffered from anxiety/depression upon being informed of her condition (after Table 8.2 of Lal's Clinical Psychosomatic Obstetrics and Gynaecology [61]). The dysphoria continued after surgical management. Subsequently, she favored clinical psychosomatic care.

TABLE 32.2 Mrs. R. Mead, a gynecological outpatient attender (British Caucasian) [Clinical Vignette (02)].

Presentation and management	
	• Mrs. Mead, a 38-year-old married volunteer with two teenage children, was on a gynecological follow-up visit
	• At this visit, she complained of intermittent abdominal pain for 5 days; oral analgesia (tramadol, codeine) was ineffective; her GP prescribed a tranquilizer for insomnia; the pain was on her right flank of a "stabbing type" preventing her from sitting or standing up without her partner's support; she had been "sick" twice at the start of her painful episodes but later she was only "queasy"
	• She complained of being feverish and had a vaginal discharge with pain "below" so avoided sex; episodes of pain persisted for 2 months without relief from analgesics; she feared that she had cancer and had also developed IBS
	• Hematological investigations were negative for infection and tumor markers; a pelvic ultrasound scan showed a small (2.5 × 3.0 cm) cyst in the right ovary
	• [Past H/o: Irregular, painful, heavy menses; ultrasound confirmed a "normal uterus, cervix and adnexa, the endometrium is proliferative and consistent with dates, myometrium is normal, the left ovary shows a cystic corpus luteum with numerous follicular cysts, no malignancy or other abnormality seen"; endometrial sampling revealed atypical hyperplasia; hysterectomy and salpingo-oophorectomy were advised; the healthy right ovary would be conserved; she had completed her family
	• Following the consultation, Mrs. Mead was overwhelmed; at a second visit with her husband when the risk of malignancy from atypia was explained again, she accepted surgery; a presurgical CT scan reported: "Axial contrast enhanced section from the diaphragm down to the ischial tuberosities, showed no abnormal soft tissue mass nor fluid collection in the abdomen; a small 2.5 × 3.0 cm diameter heterogeneous part cystic/part solid mass is seen deep in the pelvis?? endometriosis, no intravesicular lesion, no pelvic or retroperitoneal lymphadenopathy, no sign of appendicitis, normal sized kidneys with no hydronephrosis, no focal hepatic lesion, the gall bladder distended normally, normal caliber bile ducts, no bile stones seen, no pancreatic mass, normal sized spleen, no signs of intestinal obstruction or pleural infusion, lung bases clear"; a hysterectomy and left salpingo-oophorectomy were carried out with technical difficulty as her BMI was 44; atypical hyperplasia was confirmed with no sign of malignancy in the uterus, cervix, or left tube and ovary
	• Postoperatively she was slow to mobilize; routine care with anticoagulants was given; she was discharged on the fifth postoperative day but readmitted on the ninth day for fever and leaking of serosanguinous fluid from the wound; the wound was dressed and parenteral antibiotics given, she was discharged on oral antibiotics; she attended again for a "grumbling" pyrexia with wound infection; a second course of antibiotics/dressings was added for 14 days; follow-up was for 18 months]
	• At this visit, she was afebrile, BMI was 43, her BP 142/84 mmHg, and her pulse 88
	• Examination confirmed an obese abdomen, with tenderness and rebound, but no guarding, the liver was impalpable, and no masses felt; per speculum exam revealed a healthy vaginal vault, well hitched up with serous vaginal discharge; tenderness of the vault was elicited on the right side; a high vaginal swab for culture was taken

(Continued)

TABLE 32.2 (Continued)

	• She consented for a diagnostic laparoscopy and wanted her right ovary conserved • The obese abdomen made laparoscopy a challenge; dense adhesions of the intestine to the abdominal wall made visualization of the right ovary impossible; on proceeding to a laparotomy with careful adhesolysis, visualization of the right ovary was unclear; a ruptured follicular cyst was noted with no active bleeding; a normal saline pelvic wash was given and a pelvic drain left in; the abdomen was closed • The drain was removed the next day and Mrs. Mead mobilized early; she felt reassured after the operation and made a remarkable postoperative recovery; the cultures and cytology were negative; she was discharged on the fourth day on oral analgesia; Appointments with an Ob/Gyn with psychosomatic specialization (Ob/Gyn–Psyc) and a psychotherapist were set up to assist her with weight reduction
Psychosocial initiating, maintaining, and alleviating factors	• Mrs. Mead was a caring individual who tried helping others; being the third child in a family of six she somehow felt neglected, and binged on sweets/savory food • She had felt dysphoric after the treatment of her "precancer" and later developed IBS—a psychosomatic condition • She became a volunteer at a cancer hospice and sometimes felt low when patients passed away despite treatment; her husband detected her surreptitious overeating • The Ob/Gyn–Psyc and a psychotherapist had to deal with her dysphoria and bingeing that increased her weight after she began supporting cancer patients again
Impact on regional health-care system	• Mrs. Mead had her first operation due to a premalignant condition and the second to investigate her pain with cancer phobia—both were related to her becoming obese • She was affected by clinical psychosomatic issues when volunteering for the support of cancer patients and needed help from an Ob/Gyn–Psyc specialist • The Ob/Gyn–Psyc and therapist would help reduce her dysphoria and overeating

IBS, Irritable bowel syndrome.

Learning points: The clinical vignette of Mrs. Mead confirms that a lack of empathy rather than help from one's family can give rise to a clinical psychosomatic health condition. Such a familial situation can promote overeating of sweets/savory food without exercising, and a relentless progression toward dysphoria and obesity. An Ob/Gyn–Psyc and a psychotherapist can team up to enact behavioral change toward a healthier lifestyle and weight reduction.

Section 3: physical, mental, and sexual violence with obesity in migrants

The clinical psychosomatic aspect of gender-related violence

Population shift and sociocultural strife can contribute to the generation of interpersonal violence (IPV) with clinical psychosomatic sequelae. The breakdown of personal and social support networks in mobile populations can initiate and maintain physical and mental illnesses that have both individual and collective health implications. Resettling can aggravate any biopsychosocial health problems by affecting customary lifestyles along with creating rifts in one's emotional support structure, which limits the ability to cope with stress, notably when adjusting to new environments [109]. Migrants may become part of an urban underclass, undermining their overall health and wellbeing. Under such circumstances, psychosomatic illnesses that negatively affect diet can accrue. Therefore medical/social services need to be proactive in managing the migrant's problems of increased weight gain from high-caloric diets, or emaciation from low caloric intake.

Under uncertain conditions when settling into a new environment, stressful circumstances that promote comfort eating and unchecked weight gain are not uncommon. Moreover, gender-related IPV may plague these populations and pave the way for obesogenic cravings. It is common knowledge that domestic and sexual violence are prevalent in many populations that face uncertainty, and their health repercussions on psychosomatic wellbeing demand recognition. Yet the familial/social hierarchy may be loath to prevent gender-related violence, even when there are existing legal deterrents to prevent such behavior. Data from the United States confirms that IPV can lead to obesity [110]. Although underreported for fear of reprisal from the male partner, IPV causes 4 million injuries annually, with 36% of women reporting rape, physical violence, or stalking, and 48% complaining of psychological aggression [111]. These can result in clinical psychosomatic issues and obesity in many affected women. Again, women who experience both physical and nonphysical IPV are at an increased risk of experiencing numerous adverse health problems, such as chronic pain, gastrointestinal disorders, depression, anxiety, posttraumatic stress disorder, and sleep disorders [112], along with their overeating habits. Besides, women may experience nonphysical forms of IPV, such as control through humiliation, verbal abuse, or threats of abuse to her or someone she loves, so that she lives in fear of her partner, thereby making her resort to comfort eating as a panacea for her dysphoria.

This concern is not merely country-specific, but relevant to the practice of gynecology worldwide; these women ignore weight gain even if it leads to obesity [111]. A large proportion of obesity in many populations would be prevented if IPV and linked overeating could be stopped. The vital importance of ending violence against women to facilitate improved psychosomatic health internationally has been recognized by key international organizations such as the WHO and United Nations (UN) for some time [109], having been incorporated into the Millennium Development Goals [113], and later the Sustainable Development Goals [114]. Indeed, in support of this, the UN has recently launched the "Spotlight Initiative" in conjunction with the European Union, which has committed €500 million of seed funding, with the noble aim of ending "all forms of violence against women or girls" worldwide [115]. The important role that clinicians can play in this has also long been recognized by FIGO, with the stated priority to "address the barriers of clinicians to respond to violence against women through the use of advocacy, training and services" [116]. In fact, the shared goal with Project Spotlight was articulated around 15 years earlier [117] and reaffirmed more recently in the FIGO and WHO Global Declaration on Violence Against Women, which includes an important section underscoring the role that clinicians can and must play due to their exclusive and privileged positions on the front lines, where it would be both immoral and unethical to close one's eyes [118].

Progress in creating awareness and developing support systems for affected women has been prioritized in the United Kingdom as in the United States, besides many other developed countries. Nonetheless, further clinical psychosomatic issues continue to evolve, particularly with population shift and changes in the economy. These generate recurrent, diverse health conundrums that need to be addressed concurrently. The United Kingdom's National Institute for Health and Clinical Excellence, after a public consultation, developed relevant guidance to prevent/stop IPV [119]. Nevertheless, more efforts should go into helping affected individuals and their partners. Educating health professionals to deal effectively with women's clinical psychosomatic illnesses due to gender-related violence remains vital.

The next two vignettes, in Table 32.3 (after Table 12.2 in Lal's Clinical Psychosomatic Obstetrics and Gynaecology [109]), illustrate the predicaments of women who seek hospitalization for covert IPV.

Characteristics of hospital catchment area where patients resided

Gynecological vignettes 03 and 04 were selected from a teaching (tertiary) hospital in the United Kingdom with a catchment population that included lower middle−class women from Eastern Europe and Asia, besides local British Caucasian citizens. Miss B. Oates and Mrs. J. Čierna, who needed patient-centered care in hospital, exemplify a category of dysphoric women who attended between 9:00 p.m. and 1:00 a.m. daily while seeking gynecological attention. They often overwhelmed the emergency overnight health-care arrangements with symptoms such as headaches, abdominal pain, vague aching of limbs, and bodily injuries; often symptoms did not correlate with signs. Many of these women, who appeared oblivious to being overweight/obese (probably due to "comfort eating"), were reluctant to go home. The clinical psychosomatic behavior of these attendees was suggestive of IPV (to say nothing of the physical signs), but, despite assurances about confidentiality, they were reticent, as they feared repercussions from partners (IPV would worsen if the partner suspected disclosure to others, thus hastening a descent into overeating and obesity).

Learning points—Miss Oates had faced assault as a child and developed into a timorous personality who was vulnerable to future assaults; this also led to comfort eating. A mother can protect a child from assault at home, but, when circumstances do not allow this, she could be compelled to collude in such acts. Miss Oates experienced this as a child and decided to leave home while still yearning for parental affection. This is what perhaps led her to select a partner who appeared to be supportive but had a personality similar to her violent father. Miss Oates began perceiving symptoms created by her emotional triggers, but no diagnosis was confirmed after medical examination and investigations. Her discussions with the clinical psychosomatic physician persuaded her to seek support for moving to a "safe haven." Mrs. Čierna had faced sexual violence as a teenager with clinical psychosomatic sequelae and had developed apareunia when trying for steady relationships. Reticence due to social constraints prevented help-seeking. The clinical psychosomatic consultation convinced her to seek help to reduce her apprehensions.

TABLE 32.3 Clinical vignettes (03 and 04) exemplifying how physical/mental and sexual violence can generate clinical psychosomatic manifestations, leading to gynecological hospitalizations.

	03 Gynecology: an acute emergency admission	04 Gynecology: a nonemergency gynecological referral
Presentation and management	<ul><li>Miss Oates was a 19-year-old British Caucasian with a history of repeat hospitalizations</li><li>She was admitted for pelvic pain via the accident/emergency route</li><li>Her general observations were stable; she was nonpregnant so transferred to the Gyn ward</li><li>Miss Oates had a BMI of 30</li><li>On examination, her abdomen was soft with suprapubic tenderness</li><li>She refused a pelvic examination</li><li>She also refused analgesia</li><li>She was discharged home the next day</li><li>She was readmitted with pain as an acute abdomen</li><li>She was laparoscoped and found to have a normal pelvis/abdomen</li></ul>	<ul><li>Mrs. Čierna was a 37-year-old East-European Caucasian, teacher's assistant, married for the third time 2 years back</li><li>She was referred by her GP because of infrequent intercourse (four times in the last 2 years) and dyspareunia, despite having a husband who showed sexual interest</li><li>Mrs. Čierna had a BMI of 34</li><li>She had avoided intercourse "unless inebriated"</li><li>She had refused routine cervical smears after the first one at 25 years of age</li><li>She consented to an examination after a detailed explanation</li><li>She was given a pelvic examination with a lubricated nulliparous speculum</li><li>She had rejected her husband's subtle sexual advances; he remained supportive</li></ul>
Psychosocial initiating, maintaining, and alleviating factors	<ul><li>Miss Oates left home at 16 and went abroad for 3 years as she felt her parents "put her down," and "whacked her" for "no reason"</li><li>She worked as an administrative assistant but left the job at her current partner's request to be home when he returned</li><li>She suffered repeated physical and sexual assaults by her partner when he returned from work</li><li>She was vulnerable because of low self-esteem, loneliness, and economic dependence on her abusive partner</li><li>She appeared as very anxious and frightened and was low in mood</li></ul>	<ul><li>At consultation, Mrs. Čierna disclosed for the first time about her teenage experience of sexual assaults by a religious leader</li><li>Her parents had migrated from Europe; she was timid and respected elders</li><li>She did not confide in anyone as she was afraid that the perpetrator would hurt her or her family—he lived in the same street</li><li>She had nightmares about it</li><li>Her lack of trust in other males prevented steady relationships even when married</li><li>Her complex thinking had generated a clinical psychosomatic health condition that made her abhor sex, which led to dyspareunia and then apareunia</li></ul>
Impact on regional health-care systems	<ul><li>Miss Oates accepted intervention by a social worker who arranged transfer to a "shelter home"</li><li>She moved to an undisclosed address for attention from a clinical psychosomatic specialist and a psychotherapist; she felt that she was "living" again</li><li>Additional costs for health and social care provision were added to her costs for getting better</li></ul>	<ul><li>Counseling for rape was started and Mrs. Čierna kept up with wifely activities in order to be "fair" to her husband</li><li>Psychosexual support to improve her sexual relationship would be started after rape counseling by a clinical psychosomatic specialist so that her fear/anxiety about having sex could be reduced</li><li>Such provision would have added to costs</li></ul>

Other forms of presentations and behavior after sexual assault

- Overt presentations of physical abuse, menstrual problems, vaginal discharge, unwanted pregnancies, anxiety, depression, posttraumatic stress disorder, and symptoms of sexually transmitted infections such as hepatitis, syphilis, and/or HIV
- Symptoms of miscarriage, sexual problems, and infertility
- May hide facts or refuse social intervention
- Unusual problems such as a vaginal fistula following sexual assault involving foreign objects
- Inpatient hospital admissions for long periods with vague complaints
- Could go unrecognized, as the woman is loath to speak for fear of repercussions
- Partner may appear concerned, overprotective or seem unconcerned

GP, General practitioner.

Section 4: obesity and severe pelvic/perineal dysfunction

The clinical psychosomatic viewpoint

Pelvic floor/perineal dysfunction [120], also referred to as pelvic/perineal dysfunction, relates to symptoms that bring about physical and mental ill-health [121]. The pelvic/perineum diaphragm comprises myofascial structures that support the pelvic organs and facilitate normal urogenital function by preserving their anatomical integrity and innervation from segments L3−5, S2−5 of the spinal cord. Derangement of the nerve supply or injury to the muscles and ligaments of the pelvic floor or perineum can lead to the symptoms of pelvic/perineal dysfunction. Verily, bladder and bowel continence is a voluntarily acquired, socially appropriate behavior learnt through a process of conditioning during childhood, so the loss of continence, especially in obese women, can impair physical and mental health. It is usually apparent that severe biopsychosocial repercussions occur in many who suffer from pelvic floor symptoms (which comprise urinary incontinence, anal incontinence, dyspareunia, prolapse, and hemorrhoids). These symptoms are worrying to the patient who experiences them and subsequently acquires related physical and mental ill-health, yet studies of the impact on the relevant clinical psychosomatic health issues that are generated remain scarce [122].

While the agony of excruciating symptoms of incontinence when measured by perineal protection for urinary loss was reported by a few, the perceived psychosomatic misery borne by those with severe symptoms was not given enough attention [85]. Hence, it remains clinically important to evaluate the relationship of pelvic/perineal symptoms with physical and mental wellbeing that is compromised by urinary/fecal incontinence, too often in the obese. It has been observed that incontinence causes anxiety/depression/phobias, which impact on overall welfare and health-seeking behavior, thereby underscoring the importance of the psychosomatic approach toward symptom relief [123,124].

Despite various methods being used to investigate pelvic floor dysfunction, studies that include all symptoms and also advocate help for the sufferer's psychosomatic issues are scarce [85,125], particularly in obese/overweight patient samples [126]. With recent attention on questionable operative choices resulting in lifelong physical/mental ill-health following botched operations for pelvic floor disorders, it is incumbent on specialists treating these disorders to select the most appropriate management strategy for each patient, including factors particular to the obese. Further research on obesity and the psychosomatic aspects of pelvic floor/perineal dysfunction in the obese would be clinically useful for improving patient-centered care as novel management insights accrue.

Section 5: psychosomatic impact of gynecological tumors in the obese

The obese are prone to developing both benign and malignant gynecological tumors, to the joint detriment of physical and mental health. This discussion will concentrate on gynecological malignancies that frequently affect the obese female population, as the psychosomatic aspects of benign tumors have been given due attention elsewhere [85]. The main focus will be on Type-1 endometrial cancer, as it has been increasingly affecting obese females at a younger age [61]. This suggests that preventing obesity would likely also reduce the risk of Type-1 endometrial cancer. In addition, fetal metabolic programming in the obese gravida can promote obese offspring [127], who would in turn be expected to be at increased risk of cancer. Thus reduction of obesity would positively influence prevention of this cancer and other malignancies [128], thereby averting considerable clinical psychosomatic health burden.

The global disease burden of endometrial cancer/malignancy

It is known that cancer or malignancy occurs when cells of a specific part of the body multiply uncontrollably to form a tumor that has the potential to spread to other parts of the body. Endometrial cancer may be silent initially, like many malignancies (such as ovarian), but can also present as abnormal bleeding, a lump, a change in bowel function, or otherwise inexplicable weight loss or fever. Screening tests can detect certain cancers, but clinicians unfamiliar with the changing epidemiological trends may not think to apply these in the younger age groups. Endometrial cancer is most common in countries with a "High" or "Very High" Human Development Index [129], with the United Kingdom reporting an annual frequency of 8475 new cases in 2011 [130], making it the fourth most common women's cancer there. The incidence of endometrial cancer has risen over the last three decades [131] to affect many at a younger age, mirroring the trend in obesity; both health conditions mandate concurrent attention.

The etiopathogenesis of endometrial cancer

Obese women have inflamed hypertrophied adipocytes that release leptin, adiponectin, and tumor necrosis factor. These molecules modify cellular adhesion and disrupt normal tissue architecture, along with increasing angiogenesis and carcinogenesis. Carcinogenesis (or tumorigenesis) is promoted by these alterations in the endometrium at the expense of apoptosis [132], with obesity playing a major role in the genesis of endometrial cancer. While the mechanisms underlying this are not yet fully understood, they include promotion of a proinflammatory environment, which affects carcinogenesis by upsetting obese patients' immune systems and disturbing tissue homeostasis, together with increasing oxidative stress [133]. Moreover, an American case−control study suggested that the incidence of endometrial cancer in women who have undergone gastric bypass surgery is substantially reduced compared to nonoperated severely obese controls [134], implying a strong association. Indeed, as the rates of obesity increase, the incidence of this cancer is increasing in concert, with a predicted US incidence of 42.13 cases per 100,000 women by 2030 [135]. Thus, by averting obesity, along with associated precancerous changes, one would expect a corresponding decrease in cancer risk [61].

Of the two types recognized traditionally [136], Type-1 cancer is the most common, even more so in obese women [137]. A rise in BMI by 5 kg/m^2 is associated with a 1.6-fold increased risk of incident endometrial cancer [138]. Besides BMI, several other anthropometric measures show a positive association with endometrial cancer risk, including waist circumference, waist-to-hip ratio, hip circumference, height, and adult weight gain [139]. Further, dose−response analysis shows a significant nonlinearity whereby the risk of endometrial cancer at a BMI of 40 kg/m^2 is approximately 10-fold higher than that for women with a BMI within the normal range [130]. In addition, endometrial cancer has a worse prognosis in the obese [140].

Preventing/treating obesity to prevent endometrial cancer

Given the risk factors, it seems sensible to try and prevent endometrial cancer in those with a raised BMI by advising weight loss [139], ideally beginning with individualized lifestyle interventions. This would be of particular relevance to the clinical psychosomatic health of younger women who wish to start a family, as cancer treatment may permanently preclude natural childbearing. If prevention is unsuccessful, however, sensible eating habits remain valuable, for there is an inverse relationship between obesity and QOL as appraised in cancer survivors treated for early-stage endometrial cancer [141], with physical, mental, and social wellbeing inversely related to BMI. A relatively longer cancer-free survival period would also improve QOL in these women.

Some young women treated for gynecological cancer may mistakenly believe that the hysterectomy and salpingo-oophorectomy have cured them, so reducing obesity is inconsequential. Patients should accordingly be counseled on the importance of maintaining a healthy weight following successful surgery, and the risks that obesity poses to long-term health, being associated with the development of cardiovascular disease, cerebrovascular disease, diabetes, and cancer (both recurrence and novel primary malignancies) [142]. Furthermore, the importance of regular follow-up should be emphasized since late detection of recurrence not only reduces the chances of successful treatment but can also be devastating to QOL, particularly in patients who sincerely believed they had been permanently cured.

While primary and secondary preventative strategies for obesity and its complications seem sensible, such prudence may seem unacceptable to those who overeat due to familial/social pressures. Many patients who reach the obese habitus require tertiary management measures such as weight-reducing or bariatric surgery. Bariatric surgery under the United Kingdom's National Health Service can be considered if a structured program for reducing obesity is unable to maintain weight loss consistently [143], but further clarification of the selection criteria for operation/reoperation is required. Suitability for bariatric surgery should be evaluated according to each patient's personal circumstances. Although a Cochrane review found that bariatric surgery appeared to be better than nonsurgical methods at achieving sustainable weight loss in obese adults, postoperative complications and reoperation rates were generally not well reported [144]. This is unfortunate, given that such operations can have serious complications, even requiring emergency readmissions [145]—under such circumstances long-term psychosomatic repercussions detrimental to QOL can manifest [61]. Similarly, a CADTH technology review concluded that limited information precluded assessing the generalizability of costs and outcomes relating to bariatric surgery [146]. Evidently, further research is warranted. Even when the surgery itself is successful, it is important to remember that the causes of obesity are multifactorial [18]—addressing contributory biopsychosociocultural influences could be instrumental in advancing the patient's postsurgical recovery and optimizing long-term outcomes.

Finally, it should be noted that the American Society for Clinical Oncology has been particularly active in promoting awareness among professionals and public alike of the links between obesity and cancer, including preparing a guide to aid physicians in advising weight reduction in obese women [147,148]. Likewise, the importance of weight loss has also been highlighted by a review aiming to reduce the incidence of gynecological cancers [149]. Implementing such guidance would aid significantly in preventing obesity-related gynecological cancer, thereby averting a range of adverse clinical psychosomatic outcomes.

Conclusions

Clinical psychosomatic health issues linked with obesity are not uncommon in today's world and deserve greater recognition. The trend for coping with menstrual irregularities using a calorie-rich diet seems unnecessary but is nevertheless followed by certain groups. Other treatments for menstrual problems, including PMS and PMDD, have ranged from pharmacotherapy to surgery, but these maladies could merit less-invasive management in many. Latterly, the incidence of polycystic ovaries with menstrual irregularities, obesity, and associated psychosomatic presentations has risen worldwide, thereby garnering more attention. A clinical psychosomatic approach aiming to reduce the urge to overeat seems reasonable for many facing these health issues.

Infertility/subfertility, with its psychosomatic implications, continues to be of great significance for many women who desire an offspring but are obese. Assisted conception remains in demand for achieving this. In many who undergo these procedures, which include IVF, success is not guaranteed; procedural/operative complications, besides thromboembolism, depression, pregnancy-related problems, and even fatalities, may occur, especially in overweight/obese women. Many women feel disillusioned by repeated unsuccessful attempts, which cause significant dropout rates with ensuing dysphoria. Natural childbearing has proponents who report of successful conception after psychotherapy sessions. Similarly, psychosomatic approaches that promote a healthy lifestyle could enable natural conception to childbirth after desired weight reduction in the obese.

Obesity can be promoted by inescapable IPV. This behavior is associated with major clinical psychosomatic implications that promote overeating and has recently gained wider recognition, with increases in both internal and international migration by women intending to escape such violence, yet experiencing it again in novel environs, whether at home, while traveling, or in temporary accommodation as refugees. Although the UN, WHO, and FIGO have declared the intent to eliminate all forms of violence against women, it deserves further attention from caregivers, health professionals, volunteer organizations, and NGOs, to help deliver women from these situations and facilitate safe relocation, where they may have better opportunities to pursue a healthy lifestyle/diet, and achieve an appropriate weight.

Bowel and bladder incontinence, along with other pelvic floor disorders, also have a penchant for the obese. Such disorders have an associated clinical psychosomatic aspect and are aggravated by the overweight habitus. Symptom relief may be obtained by methods nonsurgical or surgical, but the pros and cons of each merit evaluation; success is encouraged by weight loss in the obese. Careful, patient-centered, decision-making concerning operative methods would reduce the risk of postoperative complications, which have gained prominence more recently when affected patients sought legal redress. Applying the psychosomatic perspective to health needs would improve patient outcomes, particularly in the overweight/obese.

Obesity increases the risk of younger age Type-1 endometrial cancer occurrence (unlike for Type-2). Metabolic programming in the female fetus of the obese gravida can promote the birth of obese offspring, who would in turn be expected to be similarly prone to such cancers. Some obese patients seek bariatric surgery for weight reduction but this can be expensive and has potential complications that impact both physical and mental health. Primary and secondary prevention of cancer and obesity could be a sound clinical psychosomatic approach but such health-care provision may not be available to many women globally, so tertiary prevention is offered. Prevention of obesity would reduce the disease burden of such cancers.

Going forward, health providers ought to consider the biological, psychological, social, and cultural factors that influence clinical psychosomatic interactions and promote obesogenic behavior. In light of the range of obesity-related health issues that have been discussed here, nutritionally informed lifestyle guidance should begin in childhood and aim to promote futuristic psychosomatic health with appropriate weight for age.

References

[1] Margetts EL. The early history of the word "psychosomatic". Can Med Assoc J 1950;63(4):402–4.

[2] Lal M. Chapter 1. Clinically significant mind-body interactions: evolutionary history of the scientific basis. In: Lal M, editor. Clinical psychosomatic obstetrics and gynaecology: a patient-centred biopsychosocial practice. Oxford: Oxford University Press; 2017. p. 1–38.

[3] National Institute for Health and Care Excellence. Surveillance report 2018 — Obesity: identification, assessment and management (2014) NICE guideline CG189 and BMI: preventing ill health and premature death in black, Asian and other minority ethnic groups (2013) NICE guideline PH46. [Online]. Available from: <https://www.nice.org.uk/guidance/cg189/resources/surveillance-report-2018-obesity-identification-assessment-and-management-2014-nice-guideline-cg189-and-bmi-preventing-ill-health-and-premature-death-in-black-asian-and-other-minority-ethnic-groups-2-4847559661>; 2018 [accessed 10.11.19].

[4] Kyle TK, Dhurandhar EJ, Allison DB. Regarding obesity as a disease: evolving policies and their implications. Endocrinol Metab Clin North Am 2016;45(3):511–20.

[5] Gonzalez-Campoy JM. Obesity in America: a growing concern. [Online]. Available from: <https://www.endocrineweb.com/conditions/obesity/obesity-america-growing-concern>; 2019 [accessed 10.11.19].

[6] Hales CM, Carroll MD, Fryar CD, Ogden CL. Prevalence of obesity among adults and youth: United States, 2015–2016, NCHS data brief, no 288. Hyattsville, MD: National Center for Health Statistics; 2017.

[7] Baker C. Obesity statistics. Briefing paper number 3336. London: House of Commons Library; 2019.

[8] Wilding JPH, Mooney V, Pile R. Should obesity be recognised as a disease? BMJ 2019;366:l4258.

[9] World Health Organization. Obesity and overweight. [Online]. Available from: <https://www.who.int/news-room/fact-sheets/detail/obesity-and-overweight>; 2018 [accessed 26.01.20].

[10] FIGO Committee on Pregnancy and Non-communicable Diseases, Committee on pregnancy and NCDs statement: PONI. [Online]. Available from: <https://www.figo.org/news/committee-pregnancy-and-ncds-statement-poni-0016284>; 2019 [accessed 29.01.20].

[11] Hill JO, Wyatt HR, Peters JC. Energy balance and obesity. Circulation 2012;126(1):126–32.

[12] Flegal KM, Kit BK, Orpana H, Graubard BI. Association of all-cause mortality with overweight and obesity using standard body mass index categories: a systematic review and meta-analysis. JAMA 2013;309(1):71–82.

[13] Bhaskaran K, dos-Santos-Silva I, Leon DA, Douglas IJ, Smeeth L. Association of BMI with overall and cause-specific mortality: a population-based cohort study of 3·6 million adults in the UK. Lancet Diabetes Endocrinol 2018;6(12):944–53.

[14] Izumida T, Nakamura Y, Ishikawa S. Impact of body mass index and metabolically unhealthy status on mortality in the Japanese general population: the JMS cohort study. PLoS One 2019;14(11): e0224802.

[15] Kuk JL, Ardern CI. Are metabolically normal but obese individuals at lower risk for all-cause mortality? Diabetes Care 2009;32(12):2297–9.

[16] Green MA, Subramanian SV, Strong M, Cooper CL, Loban A, Bissell P. 'Fish out of water': a cross-sectional study on the interaction between social and neighbourhood effects on weight management behaviours. Int J Obes 2015;39(3):535–41.

[17] Anekwe TD, Rahkovsky I. Economic costs and benefits of healthy eating. Curr Obes Rep 2013;2(3):225–34.

[18] Lee BY, Bartsch SM, Mui Y, Haidari LA, Spiker ML, Gittelsohn J. A systems approach to obesity. Nutr Rev 2017;75(Suppl. 1):94–106.

[19] Food Standards Agency. Eat well, be well – traffic light labeling. [Online]. Available from: <https://web.archive.org/web/20110201180054/http://www.eatwell.gov.uk/foodlabels/trafficlights/>; 2006 [accessed 01.02.11].

[20] Department of Health, Food Standards Agency, UK Government. Guide to creating a front of pack (FoP) nutrition label for prepacked products sold through retail outlets. [Online]. Available from: <https://www.food.gov.uk/sites/default/files/media/document/fop-guidance_0.pdf>; 2016 [accessed 05.01.20].

[21] Food Standards Agency. Supporters of FSA's approach to signpost labeling. [Online]. Available from: <https://web.archive.org/web/20091203192639/http:/www.food.gov.uk/foodlabelling/signposting/supportfsasignp>; 2008 [accessed 03.12.09].

[22] Julia C, Hercberg S. Development of a new front-of-pack nutrition label in France: the five-colour Nutri-Score. Public Health Panorama 2017;3(4):712–25.

[23] Talati Z, Egnell M, Hercberg S, Julia C, Pettigrew S. Consumers' perceptions of five front-of-package nutrition labels: an experimental study across 12 countries. Nutrients 2019;11(8):1934.

[24] Labonté M-E, Emrich TE, Scarborough P, Rayner M, L'Abbé MR. Traffic light labelling could prevent mortality from noncommunicable diseases in Canada: a scenario modelling study. PLoS One 2019;14(12):e0226975.

[25] Finkelstein EA, Ang FJL, Doble B, Wong WHM, van Dam RM. A randomized controlled trial evaluating the relative effectiveness of the multiple traffic light and Nutri-Score front of package nutrition labels. Nutrients 2019;11(9):2236.

[26] Franckle RL, Levy DE, Macias-Navarro L, Rimm EB, Thorndike AN. Traffic-light labels and financial incentives to reduce sugar-sweetened beverage purchases by low-income Latino families: a randomized controlled trial. Public Health Nutr 2018;21 (8):1426–34.

[27] Moore SG, Donnelly JK, Jones S, Cade JE. Effect of educational interventions on understanding and use of nutrition labels: a systematic review. Nutrients 2018;10(10):1432.

[28] da Costa Souza SMF, Lima KC, Costa Feitosa Alves MdS. Promoting public health through nutrition labeling – a study in Brazil. Arch Public Health 2016;74:48.

[29] Pliner P, Salvy S-J. Food neophobia in humans. In: Shepherd R, Raats M, editors. The psychology of food choice. Wallingford: CABI; 2006. p. 75–92.

[30] Guzek D, Pęska J, Głąbska D. Role of food neophobia and allergen content in food choices for a polish cohort of young women. Nutrients 2019;11(11):2622.

[31] Reyes M, Garmendia ML, Olivares S, Aqueveque C, Zacarías I, Corvalán C. Development of the Chilean front-of-package food warning label. BMC Public Health 2019;19:906.

[32] Vargas-Meza J, Jáuregui A, Pacheco-Miranda S, Contreras-Manzano A, Barquera S. Front-of-pack nutritional labels: understanding by low- and middle-income Mexican consumers. PLoS One 2019;14(11):e0225268.

[33] Khandpur N, de Morais Sato P, Mais LA, Bortoletto Martins AP, Spinillo CG, Garcia MT, et al. Are front-of-package warning labels more effective at communicating nutrition information than traffic-light labels? A randomized controlled experiment in a Brazilian sample. Nutrients 2018;10(6):688.

[34] Arrúa A, Curutchet MR, Rey N, Barreto P, Golovchenko N, Sellanes A, et al. Impact of front-of-pack nutrition information and label design on children's choice of two snack foods: comparison of warnings and the traffic-light system. Appetite 2017;116:139–46.

[35] Nieto C, Jáuregui A, Contreras-Manzano A, Arillo-Santillan E, Barquera S, White CM, et al. Understanding and use of food labeling systems among Whites and Latinos in the United States and among Mexicans: results from the International Food Policy Study, 2017. Int J Behav Nutr Phys Act 2019;16:87.

[36] Sobre composición nutricional de los alimentos y su publicidad. (2012. Ley 20.606 (CL).

[37] Finlayson G. Food addiction and obesity: unnecessary medicalization of hedonic overeating. Nat Rev Endocrinol 2017;13:493–8.

[38] Fletcher PC, Kenny PJ. Food addiction: a valid concept? Neuropsychopharmacology 2018;43:2506–13.

[39] Wiss DA, Avena N, Rada P. Sugar addiction: from evolution to revolution. Front Psychiatry 2018;9:545.

[40] Freeman CR, Zehra A, Ramirez V, Wiers CE, Volkow ND, Wang G-J. Impact of sugar on the body, brain, and behavior. Front Biosci (Landmark Ed) 2018;23:2255–66.

[41] Lennerz B, Lennerz JK. Food addiction, high-glycemic-index carbohydrates, and obesity. Clin Chem 2018;64(1):64–71.

[42] Volkow ND, Wang G-J, Fowler JS, Telang F. Overlapping neuronal circuits in addiction and obesity: evidence of systems pathology. Philos Trans R Soc B: Biol Sci 2008;363(1507):3191–200.

[43] Michaud A, Vainik U, Garcia-Garcia I, Dagher A. Overlapping neural endophenotypes in addiction and obesity. Front Endocrinol 2017;8:127.

[44] Berthoud H-R, Lenard NR, Shin AC. Food reward, hyperphagia, and obesity. Am J Physiol Regul Integr Comp Physiol 2011;300(6):R1266–77.

[45] Tzieropoulos H, Rytz A, Hudry J, le Coutre J. Dietary fat induces sustained reward response in the human brain without primary taste cortex discrimination. Front Hum Neurosci 2013;7:36.

[46] Chambers ES, Bridge MW, Jones DA. Carbohydrate sensing in the human mouth: effects on exercise performance and brain activity. J Physiol 2009;587(8):1779–94.

[47] Singh M. Mood, food, and obesity. Front Psychol 2014;5:925.

[48] Dalby M. Hippocratic misquotations: let thy quotations not be by Hippocrates. [Online]. Available from: <https://honey-guide.com/2017/10/31/hippocratic-misquotations-let-thy-quotations-not-be-by-hippocrates/>; 2017 [accessed 15.01.20].

[49] Olvera RL, Williamson DE, Fisher-Hoch SP, Vatcheva KP, McCormick JB. Depression, obesity, and metabolic syndrome: prevalence and risks of comorbidity in a population-based representative sample of Mexican Americans. J Clin Psychiatry 2015;76(10):e1300–5.

[50] Turner S, Mota N, Bolton J, Sareen J. Self-medication with alcohol or drugs for mood and anxiety disorders: a narrative review of the epidemiological literature. Depress Anxiety 2018;35(9):851–60.

[51] Osadchiy V, Mayer EA, Bhatt R, Labus JS, Gao L, Kilpatrick LA, et al. History of early life adversity is associated with increased food addiction and sex-specific alterations in reward network connectivity in obesity. Obes Sci Pract 2019;5(5):416–36.

[52] Adam TC, Epel ES. Stress, eating and the reward system. Physiol Behav 2007;91(4):449–58.

[53] Drewnowski A, Almiron-Roig E. Chapter 11. Human perceptions and preferences for fat-rich foods. In: Montmayeur JP, le Coutre J, editors. Fat detection: taste, texture, and post ingestive effects. Boca Raton, FL: CRC Press; 2010. p. 265–92.

[54] Faber A, Dubé L. Parental attachment insecurity predicts child and adult high-caloric food consumption. J Health Psychol 2015;20(5):511–24.

[55] Yang X, Casement M, Yokum S, Stice E. Negative affect amplifies the relation between appetitive-food-related neural responses and weight gain over three-year follow-up among adolescents. Neuroimage Clin 2019;24:102067.

[56] Schvey NA, Marwitz SE, Mi SJ, Galescu OA, Broadney MM, Young-Hyman D, et al. Weight-based teasing is associated with gain in BMI and fat mass among children and adolescents at-risk for obesity: a longitudinal study. Pediatr Obes 2019;14(10):e12538.

[57] Wei NL, Quan ZF, Zhao T, Yu XD, Xie Q, Zeng J, et al. Chronic stress increases susceptibility to food addiction by increasing the levels of DR2 and MOR in the nucleus accumbens. Neuropsychiatr Dis Treat 2019;15:1211–29.

[58] Cowan CSM, Callaghan BL, Kan JM, Richardson R. The lasting impact of early-life adversity on individuals and their descendants: potential mechanisms and hope for intervention. Genes Brain Behav 2016;15(1):155–68.

[59] Niccolai E, Boem F, Russo E, Amedei A. The gut–brain axis in the neuropsychological disease model of obesity: a classical movie revised by the emerging director "microbiome". Nutrients 2019;11(1):156.

[60] Lal M. Psychosomatic approaches to obstetrics, gynaecology and andrology. J Obstet Gynaecol 2009;29(1):1–12.

[61] Lal M. Chapter 8. Women's psychosomatic health promotion and the biopsychosociocultural nexus. In: Lal M, editor. Clinical psychosomatic obstetrics and gynaecology: a patient-centred biopsychosocial practice. Oxford: Oxford University Press; 2017. p. 199–236.

[62] Lumsden MA, Hickey M. Complete women's health. London: Thorsons; 2000.

[63] van Dammen L, Wekker V, de Rooij SR, Groen H, Hoek A, Roseboom TJ. A systematic review and meta-analysis of lifestyle interventions in women of reproductive age with overweight or obesity: the effects on symptoms of depression and anxiety. Obes Rev 2018;19(12):1679–87.

[64] Garvey WT, Mechanick JI, Brett EM, Garber AJ, Hurley DL, Jastreboff AM, et al. American Association of Clinical Endocrinologists and American College of Endocrinology Comprehensive Clinical Practice Guidelines for medical care of patients with obesity. Endocr Pract 2016;22(Suppl. 3):1–203.

[65] Kyrou I, Randeva HS, Tsigos C, Kaltsas G, Weickert MO. Clinical problems caused by obesity. In: Feingold KR, Anawalt B, Boyce A, Chrousos G, Dungan K, Grossman A, et al., editors. Endotext. [Online]. South Dartmouth, MA: MDText.com, Inc; 2000. Available from: <https://www.ncbi.nlm.nih.gov/books/NBK278973/> [accessed 11.01.18].

[66] Lainez NM, Cross D. Obesity, neuroinflammation, and reproductive function. Endocrinology 2019;160(11):2719–36.

[67] Chrisler JC, Johnston-Robledo I. Reproduction and mental health. In: Lundberg-Love PK, Nadal KL, Paludi MA, editors. Women and mental disorders. Santa Barbara, CA: Praeger; 2012. p. 122–7.

[68] Kues JN, Janda C, Kleinstäuber M, Weise C. Internet-based behavioural self-help for premenstrual syndrome: study protocol for a randomised controlled trial. Trials 2014;15:472.

[69] Wolf WM, Wattick RA, Kinkade ON, Olfert MD. Geographical prevalence of polycystic ovary syndrome as determined by region and race/ethnicity. Int J Environ Res Public Health 2018;15(11):2589.

[70] Flegal KM, Kruszon-Moran D, Carroll MD, Fryar CD, Ogden CL. Trends in obesity among adults in the United States, 2005 to 2014. JAMA 2016;315(21):2284–91.

[71] Himelein MJ, Thatcher SS. Polycystic ovary syndrome and mental health: a review. Obstet Gynecol Surv 2006;61(11):723–32.

[72] Farrell K, Antoni M. Insulin resistance, obesity, inflammation, and depression in polycystic ovary syndrome: biobehavioral mechanisms and interventions. Fertil Steril 2010;94(5):1565–74.

[73] Moran LJ, March WA, Whitrow MJ, Giles LC, Davies MJ, Moore VM. Sleep disturbances in a community-based sample of women with polycystic ovary syndrome. Hum Reprod 2015;30(2):466–72.

[74] Azziz R, Woods KS, Reyna R, Key TJ, Knochenhauer ES, Yildiz BO. The prevalence and features of the polycystic ovary

[74] syndrome in an unselected population. J Clin Endocrinol Metab 2004;89(6):2745−9.
[75] Hayek SE, Bitar L, Hamdar LH, Mirza FG, Daoud G. Poly cystic ovarian syndrome: an updated overview. Front Physiol 2016;7:124.
[76] Lizneva D, Suturina L, Walker W, Brakta S, Gavrilova-Jordan L, Azziz R. Criteria, prevalence, and phenotypes of polycystic ovary syndrome. Fertil Steril 2016;106(1):6−15.
[77] Mahalingaiah S, Diamanti-Kandarakis E. Targets to treat metabolic syndrome in polycystic ovary syndrome. Expert Opin Ther Targets 2015;19(11):1561−74.
[78] Vanky E, Engen Hanem LG, Abbott DH. Children born to women with polycystic ovary syndrome-short- and long-term impacts on health and development. Fertil Steril 2019;111 (6):1065−75.
[79] Teede HJ, Misso ML, Costello MF, Dokras A, Laven J, Moran L, et al. Recommendations from the international evidence-based guideline for the assessment and management of polycystic ovary syndrome. Fertil Steril 2018;110(3):364−79.
[80] Veltman-Verhulst SM, Boivin J, Eijkemans MJ, Fauser BJ. Emotional distress is a common risk in women with polycystic ovary syndrome: a systematic review and meta-analysis of 28 studies. Hum Reprod Update 2012;18(6):638−51.
[81] Brutocao C, Zaiem F, Alsawas M, Morrow AS, Murad MH, Javed A. Psychiatric disorders in women with polycystic ovary syndrome: a systematic review and meta-analysis. Endocrine 2018;62(2):318−25.
[82] Cesta CE, Månsson M, Palm C, Lichtenstein P, Iliadou AN, Landén M. Polycystic ovary syndrome and psychiatric disorders: co-morbidity and heritability in a nationwide Swedish cohort. Psychoneuroendocrinology 2016;73:196−203.
[83] Hahn S, Janssen OE, Tan S, Pleger K, Mann K, Schedlowski M, et al. Clinical and psychological correlates of quality-of-life in polycystic ovary syndrome. Eur J Endocrinol 2005;153 (6):853−60.
[84] Misso M, Boyle J, Norman R, Teede H. Development of evidenced-based guidelines for PCOS and implications for community health. Semin Reprod Med 2014;32(3):230−40.
[85] Lal M, Bitzer J. Chapter 6. Disease severity, pain and patient perception, themes in clinical practice and research. In: Lal M, editor. Clinical psychosomatic obstetrics and gynaecology: a patient-centred biopsychosocial practice. Oxford: Oxford University Press; 2017. p. 133−74.
[86] van der Steeg JW, Steures P, Eijkemans MJ, Habbema JD, Hompes PG, Burggraaff JM, et al. Obesity affects spontaneous pregnancy chances in subfertile, ovulatory women. Hum Reprod 2008;23(2):324−8.
[87] Davis G, Loughran T, editors. The Palgrave handbook of infertility in history: approaches, contexts and perspectives. London: Palgrave Macmillan; 2017.
[88] Silvestris E, Pergola G de, Rosania R, Loverro G. Obesity as disruptor of the female fertility. Reprod Biol Endocrinol 2018;16 (1):22.
[89] Talmor A, Dunphy B. Female obesity and infertility. Best Pract Res Clin Obstet Gynaecol 2015;29(4):498−506.
[90] Rodino IS, Byrne S, Sanders KA. Obesity and psychological wellbeing in patients undergoing fertility treatment. Reprod Biomed Online 2016;32(1):104−12.

[91] Leventhal H, Everhart D. Emotion, pain, and physical illness. In: Izard CE, editor. Emotions in personality and psychopathology. New York: Plenum Press; 1979. p. 261−99.
[92] Norman RJ, Noakes M, Wu R, Davies MJ, Moran L, Wang JX. Improving reproductive performance in overweight/obese women with effective weight management. Hum Reprod Update 2004;10 (3):267−80.
[93] Jungheim ES, Travieso JL, Hopeman MM. Weighing the impact of obesity on female reproductive function and fertility. Nutr Rev 2013;71(1):S3−8.
[94] Volgsten H, Skoog Svanberg A, Ekselius L, Lundkvist O, Sundström Poromaa I. Risk factors for psychiatric disorders in infertile women and men undergoing in vitro fertilization treatment. Fertil Steril 2010;93(4):1088−96.
[95] Hornstein MD, Gibbons WE. Unexplained infertility. Waltham, MA: UpToDate; 2019. . Available from: <https://www.uptodate.com/contents/unexplained-infertility> [accessed 11.01.20].
[96] Pandian Z, Gibreel A, Bhattacharya S. In vitro fertilisation for unexplained subfertility. Cochrane Database Syst Rev 2015;(11): CD003357.
[97] Sharma A, Bahadursingh S, Ramsewak S, Teelucksingh S. Medical and surgical interventions to improve outcomes in obese women planning for pregnancy. Best Pract Res Clin Obstet Gynaecol 2015;29(4):565−76.
[98] Sermondade N, Huberlant S, Bourhis-Lefebvre V, Arbo E, Gallot V, Colombani M, et al. Female obesity is negatively associated with live birth rate following IVF: a systematic review and meta-analysis. Hum Reprod Update 2019;25 (4):439−51.
[99] Andrews FM, Abbey A, Halman LJ. Stress from infertility, marriage factors, and subjective well-being of wives and husbands. J Health Soc Behav 1991;32(3):238−53.
[100] Selter J, Wen T, Palmerola KL, Friedman AM, Williams Z, Forman EJ. Life-threatening complications among women with severe ovarian hyperstimulation syndrome. Am J Obstet Gynecol 2019;220(6):575.e1−575.e11.
[101] Koning A, Mol BW, Dondorp W. It is not justified to reject fertility treatment based on obesity. Hum Reprod Open 2017;2017 (2):hox009.
[102] Hawkins LK, Rossi BV, Correia KF, Lipskind ST, Hornstein MD, Missmer SA. Perceptions among infertile couples of lifestyle behaviors and in vitro fertilization (IVF) success. J Assist Reprod Genet 2014;31(3):255−60.
[103] Quant HS, Zapantis A, Nihsen M, Bevilacqua K, Jindal S, Pal L. Reproductive implications of psychological distress for couples undergoing IVF. J Assist Reprod Genet 2013;30(11):1451−8.
[104] Al-Jebari Y, Elenkov A, Wirestrand E, Schütz I, Giwercman A, Giwercman YL. Risk of prostate cancer for men fathering through assisted reproduction: nationwide population based register study. BMJ 2019;366:l5214.
[105] Levine JM, Kelvin JF, Quinn GP, Gracia CR. Infertility in reproductive-age female cancer survivors. Cancer 2015;121 (10):1532−9.
[106] Crujeiras AB, Casanueva FF. Obesity and the reproductive system disorders: epigenetics as a potential bridge. Hum Reprod Update 2015;21(2):249−61.
[107] Greil AL, McQuillan J, Lowry M, Shreffler KM. Infertility treatment and fertility-specific distress: a longitudinal analysis of a

population-based sample of U.S. women. Soc Sci Med 2012;73 (1):87–94.
[108] Ludwig M, Diedrich K. Follow-up of children born after assisted reproductive technologies. Reprod Biomed Online 2002;5 (3):317–22.
[109] Lal M. Chapter 12. Migration, gender and cultural issues in healthcare: psychosomatic implications. In: Lal M, editor. Clinical psychosomatic obstetrics and gynaecology: a patient-centred biopsychosocial practice. Oxford: Oxford University Press; 2017. p. 293–322.
[110] Davies R, Lehman E, Perry A, McCall-Hosenfeld JS. Association of intimate partner violence (IPV) and healthcare provider-identified obesity. Women Health 2016;56(5):561–75.
[111] Black MC, Basile KC, Breiding MJ, Smith SG, Walters ML, Merrick MT, et al. The National Intimate Partner and Sexual Violence Survey (NISVS): 2010 summary report. Atlanta, GA: National Center for Injury Prevention and Control, Centers for Disease Control and Prevention; 2011.
[112] Flegal KM, Carroll MD, Kit BK, Ogden CL. Prevalence of obesity and trends in the distribution of body mass index among US adults: 1999–2010. JAMA 2012;307(5):491–7.
[113] World Health Organization. Addressing violence against women and achieving the Millennium Development Goals. Geneva: WHO Department of Gender, Women and Health; 2005.
[114] Ellsberg M. Violence against women and the Millennium Development Goals: facilitating women's access to support. Int J Gynecol Obstet 2006;94(3):325–32.
[115] Spotlight Initiative. What we do. [Online]. Available from: <https://spotlightinitiative.org/what-we-do>; 2019 [accessed 27.01.20].
[116] Benagiano G. The role of FIGO in addressing violence against women. Int J Gynecol Obstet 2002;78(Suppl. 1):S125–7.
[117] Benagiano G, Schei B. A FIGO initiative for the 21st century: eliminate all forms of violence against women worldwide 2004;86:328–34.
[118] International Federation of Gynecology and Obstetrics. FIGO shouts 'No' to violence against women. [Online]. Available from: <https://www.figo.org/Declaration-VAW>; 2018 [accessed 07.01.20].
[119] National Institute for Health and Care Excellence. Domestic violence and abuse – NICE CKS. [Online]. Available from: <https://cks.nice.org.uk/domestic-violence-and-abuse>; 2018 [accessed 14.01.20].
[120] Grant JCB, Basmajian JV, Slonecker CE. Pelvis and perineum. In grant's method of anatomy. 11th ed. Baltimore, MD: Lippincott Williams & Wilkins; 1989. p. 227–38.
[121] Lal M. Pelvic/perineal dysfunction & biopsychosocial morbidity: biological predictors and psychosocial associations in postcaesarean and vaginally delivered primiparae [Ph.D. thesis]. Birmingham: University of Birmingham; 2012. Available from: <https://etheses.bham.ac.uk//id/eprint/3729/>.
[122] Jundt K, Peschers U, Kentenich H. The investigation and treatment of female pelvic floor dysfunction. Dtsch Arztebl Int 2015;112(33–34):564–74.
[123] Vigod SN, Steward DE. Major depression in female urinary incontinence. Psychosomatics 2006;47(2):147–51.
[124] Debus G, Kästner R. Psychosomatic aspects of urinary incontinence in women. Geburtshilfe Frauenheilkd 2015;75(2):165–9.

[125] MacLennan AH, Taylor AW, Wilson DH, Wilson D. The prevalence of pelvic floor disorders and their relationship to gender, age, parity and mode of delivery. BJOG 2000;107(12):1460–70.
[126] Hunskaar S. A systematic review of overweight and obesity as risk factors and targets for clinical intervention for urinary incontinence in women. Neurourol Urodyn 2008;27(8):749–57.
[127] Zhu Z, Cao F, Li X. Epigenetic programming and fetal metabolic programming. Front Endocrinol 2019;10:764.
[128] Pischon T, Nimptsch K. Obesity and risk of cancer: an introductory overview. In: Pischon T, Nimptsch K, editors. Obesity and cancer. Cham: Springer; 2016. p. 1–15.
[129] The Global Cancer Observatory. Corpus uteri. [Online]. Available from: <http://gco.iarc.fr/today/data/factsheets/cancers/24-Corpus-uteri-fact-sheet.pdf>; 2019.
[130] Renehan AG, MacKintosh ML, Crosbie EJ. Obesity and endometrial cancer: unanswered epidemiological questions. BJOG 2016;123(2):175–8.
[131] Arnold M, Karim-Kos HE, Coebergh JW, Byrnes G, Antilla A, Ferlay J, et al. Recent trends in incidence of five common cancers in 26 European countries since 1988: analysis of the European Cancer Observatory. Eur J Cancer 2015;51(9):1164–87.
[132] Hickman JA. Apoptosis and tumourigenesis. Curr Opin Genet Dev 2002;12(1):67–72.
[133] Ramos-Nino ME. The role of chronic inflammation in obesity-associated cancers. ISRN Oncol 2013;2013:697521.
[134] Adams TD, Stroup AM, Gress RE, Adams KF, Calle EE, Smith SC, et al. Cancer incidence and mortality after gastric bypass surgery. Obesity 2009;17(4):796–802.
[135] Sheikh MA, Althouse AD, Freese KE, Soisson S, Edwards RP, Welburn S, et al. USA Endometrial Cancer Projections to 2030: should we be concerned? Future Oncol 2014;10(16):2561–8.
[136] Bokhman JV. Two pathogenetic types of endometrial carcinoma. Gynecol Oncol 1983;15(1):10–17.
[137] Talhouk A, McAlpine JN. New classification of endometrial cancers: the development and potential applications of genomic-based classification in research and clinical care. Gynecol Oncol Res Pract 2016;3:14.
[138] Crosbie EJ, Zwahlen M, Kitchener HC, Egger M, Renehan AG. Body mass index, hormone replacement therapy and endometrial cancer risk: a meta-analysis. Cancer Epidemiol Biomarkers Prev 2010;19(12):3119–30.
[139] Aune D, Navarro Rosenblatt DA, Chan DSM, Vingeliene S, Abar L, Vieira AR, et al. Anthropometric factors and endometrial cancer risk: a systematic review and dose–response meta-analysis of prospective studies. Ann Oncol 2015;26(8):1635–48.
[140] Onstad MA, Schmandt RE, Lu KH. Addressing the role of obesity in endometrial cancer risk, prevention, and treatment. J Clin Oncol 2016;34(35):4225–30.
[141] Smits A, Lopes A, Bekkers R, Galaal K. Body mass index and the quality of life of endometrial cancer survivors—a systematic review and meta-analysis. Gynecol Oncol 2015;137(1):180–7.
[142] Ligibel JA, Alfano CM, Courneya KS, Demark-Wahnefried W, Burger RA, Chlebowski RT, et al. American Society of Clinical Oncology position statement on obesity and cancer. J Clin Oncol 2014;32(31):3568–74.
[143] NHS Commissioning Board Clinical Reference Group for Severe and Complex Obesity. Clinical commissioning policy: complex and specialised obesity surgery. [Online]. Available from:

<https://www.england.nhs.uk/wp-content/uploads/2016/05/appndx-6-policy-sev-comp-obesity-pdf.pdf>; 2013.

[144] Colquitt JL, Pickett K, Loveman E, Frampton GK. Surgery for weight loss in adults. Cochrane Database Syst Rev 2014;(8): CD003641.

[145] Kassir R, Debs T, Blanc P, Gugenheim J, Ben Amor I, Boutet C, et al. Complications of bariatric surgery: presentation and emergency management. Int J Surg 2016;27:77–81.

[146] Canadian Agency for Drugs and Technologies in Health. Bariatric surgical procedures for obese and morbidly obese patients: a review of comparative clinical and cost-effectiveness, and guidelines. Ottawa, ON: Canadian Agency for Drugs and Technologies in Health; 2014.

[147] Ligibel JA, Alfano CM, Hershman DL, Merrill JK, Basen-Engquist K, Bloomgarden ZT, et al. American society of clinical oncology summit on addressing obesity through multidisciplinary provider collaboration: key findings and recommendations for action. Obesity 2017;25(Suppl. 2):S34–9.

[148] American Society of Clinical Oncology. Obesity and cancer: a guide for oncology providers. [Online]. Available from: <https://www.asco.org/sites/new-www.asco.org/files/content-files/blog-release/documents/obesity-provider-guide.pdf>; 2014.

[149] Aune D, Navarro Rosenblatt DA, Chan DSM, Vingeliene S, Abar L, Vieira AR, et al. Anthropometric factors and ovarian cancer risk: a systematic review and nonlinear dose-response meta-analysis of prospective studies. Int J Cancer 2015;136(8):1888–98.

Chapter 33

Obesity and psychosexual disorders

Ernesto González-Mesa
Obstetrics and Gynecology, Malaga University School of Medicine, Malaga, Spain

Sexual function is related to complex neurophysiological and psychological processes. In fact, for years, the relationship between obesity and sexual function has been considered one of the unsolved mysteries in psychosomatic medicine [1]. However, despite the many unresolved questions, it seems clear that obese individuals engage in less sexual activity than nonobese individuals, or even than themselves in periods of normal weight. Physical limitations, reduced self-esteem, and issues related to the perception of the own body as being less attractive by the individual or by others may be directly linked to reduced sexual activity. In general, sexual function disorders are more common in obese women than in overweight women, although most studies agree that there is a negative correlation between BMI and sexual function.

Biological and psychological mechanism

A large number of biological and psychological mechanisms could in theory relate obesity and sexual dysfunctions, although both aspects do not need to be altered simultaneously. Disorders of sexual desire or satisfaction may occur without physical or physiological impairments and vice versa, there are cases in which desire remains intact but there is severe vascular involvement [2]. In any case, we will always have to consider the great heterogeneity of sexual behavior among people, men and women, obese and nonobese. However, while there may be some differences in sexual function in cases of obesity, this is common for both men and women.

Disorders of sexual function are not isolated phenomena in the life of individuals, there are predisposing factors that increase the vulnerability of some women, precipitating factors that trigger the disorder, mechanisms that perpetuate the situation, and a series of contextual aspects that condition the interaction of all the above [3].

Different comorbidities associated with obesity have been reported to alter women's sexual function, such as diabetes mellitus, hypertension, dyslipidemia, or depression, making it difficult to establish the role of obesity in the dysfunction [4]. Also, studies that have controlled other variables show how the course of time leads to a reduced level of sexual activity at certain stages of life, such as menopause, which makes the process more difficult. It would, therefore, be appropriate to consider a certain physiological reduction in sexual function when reaching menopause, a stage when a tendency toward overweight and obesity is observed [5]. Multiple classifications of sexual function disorders have been drawn up. Tables 33.1 and 33.2 show the most commonly used nomenclature, based on DSM V [7] and the International Society for the Study of Women's Sexual Health (ISSWSH) [7]. The ISSWSH nomenclature includes two sexual dysfunctions not specifically included in the DSM V— orgasmic illness syndrome and persistent sexual arousal disorder.

Normal sexual functio

Proper sexual function involves a complex interaction of biological, sociocultural, and psychological factors. Biological disorders associated with obesity, which may have an impact on sexual function, include cardiovascular, metabolic, respiratory, and endocrine disorders. Other factors such as age, menopause or education, and income levels associated with lifestyles have been related to the integrity of female sexual function. On the other hand, in the emotional sphere, mood alterations or eating disorders can be observed in a significant number of obese female patients, affecting their sexual functioning.

Any situation associated with obesity affecting the physiological mechanisms that control clitoral erection will result in either vascular or neuropathic orgasmic dysfunction. It is worth noting that diabetic, hypertensive, or hyperlipidemic patients, who are frequently obese, could suffer some degree of clitoral hypoperfusion, leading to

TABLE 33.1 Classification of sexual dysfunctions *MAO*, monoamine oxidase; *SSRI*, selective serotonin reuptake inhibitors.

Dysfunction	Criteria	Risk factors
Female orgasmic disorder	• Delay in, infrequency of, or absence of orgasm, or a reduced intensity of orgasmic sensations during 75% of sexual activity encounters • Persistent for 6 months • It causes significant discomfort	Problems in couple relationships Multiple sclerosis Radical hysterectomy SSRI Menopause
Female sexual interest/arousal disorder	Reduced/absent interest in sexual activity, reduced/absent sexual thoughts/fantasies, reduced/absent sexual excitement or pleasure in response to partners (internal or external) invitation, reduced/absent genital or nongenital sensations during sexual activity events or reduced/absent initiative or receptivity of messages for the initiation of sexual activity Persistent for 6 months It causes significant discomfort	Problems in couple relationships Diabetes mellitus Hypothyroidism
Genito-pelvic pain/penetration disorder	• Difficult penetration, pain during sexual activity or penetration attempts, fear to feel vulvovaginal or pelvic pain before, during or after penetration, tensing or contraction of pelvic floor muscles during penetration attempts • Persistent for 6 months • It causes significant discomfort	History of sexual abuse History of physical abuse Genital infections Endometriosis
Drug-induced sexual dysfunction	• Symptoms occur shortly after poisoning or withdrawal from the substance or after exposure to a drug	Alcohol and nicotine Opioids, amphetamines, cocaine Antipsychotic and antidepressant drugs, including MAO inhibitors, SSRI Some gastrointestinal, hormonal, and cardiovascular drugs (antihypertensive drugs)

Source: Adapted from DSM-V. Manual Diagnóstico e Estatístico de Transtornos Mentais. American Psychiatric Association. 2014 [6].

fibrosis in the long term and some degree of atrophy of the corpora cavernosa.

Obesity is associated with the *metabolic syndrome* to a higher degree than any other condition, and fatty tissue behaves like an endocrine organ that produces proinflammatory hormones and proteins that promote endothelial damage leading to orgasmic dysfunction. Similarly, androgen aromatization occurs in the fatty tissue, and fatty tissue disorder leads to a certain increase in the production of estrone and a certain dysregulation of the androgenic metabolism, a fact related to the decrease in sexual desire. The underlying hyperinsulinemia in obese women results, among other effects, in a reduced biological availability of sex hormone binding globulin (SHBG)

TABLE 33.2 Adapted from International Society for the Study of Women's Sexual Health sexual disorders nomenclature [7].

Hypoactive sexual desire disorder	Lack of motivation for sexual activity, decreased or absent spontaneous desire, or decreased or absent responsive desire to erotic cues
Female genital arousal disorder	Inability to develop or maintain adequate genital response
Persistent genital arousal disorder	Unwanted or intrusive, distressing feelings of genital arousal, or being on the verge of orgasm (genital dysesthesia), not associated with concomitant sexual interest or fantasies
Female orgasm disorders	Persistent or recurrent distressing compromise of orgasm frequency, intensity, timing, and/or pleasure
Female orgasmic illness syndrome	Aversive symptomatology before, during, or after orgasm

and, therefore, an increased amount of circulating free androgens. Other substances, such as leptin, neuropeptide Y, or galanin, may directly interfere with steroidogenesis [8]. Levels of androgens such as testosterone are known to influence sexual function at the cognitive level, boosting the frequency of sexual fantasies and increasing sexual drive. However, the reasons why excessively high or low levels of this hormone are related to sexual dysfunction are not well known, thus, although testosterone is used as a treatment to increase sexual desire in women, sexual dysfunction is not always accompanied by decreased levels of circulating testosterone [9].

Effect of cardiovascular disease on sexual function

Several studies have associated *coronary artery disease*, of which obesity is an independent risk factor, with sexual dissatisfaction and a significant reduction in vaginal lubrication and orgasms [10]. In these women, it has been found that their fear or their partner's fear of a new episode of myocardial ischemia will affect sexual function, and that psychological support and specific educational strategies can contribute to maintaining adequate sexual function [9].

Despite the existence of sexual function problems in patients with *cardiovascular disorders*, often associated with overweight and obesity, the body mass index has been pointed out by some authors [11] as a protective factor of sexual function, especially in postmenopausal cardiac patients, in which some overweight with BMI below 30 seemed to favor higher levels of sexual satisfaction, as it happens with other conditioning factors such as not being married or being content with life in general.

The combination of alcohol, smoking, diabetes, and the need for medical treatment largely explains the impaired sexual activity found in patients with high blood pressure or heart failure, both present in a significant number of obese women, when compared to that seen in groups of healthy women [12]. In fact, the type of medical treatment for high blood pressure is important for future sexual activity [13]. The negative effect of drugs, such as central adrenergic inhibitors (methyldopa), beta-blockers, including beta-1-selective blockers, or thiazide diuretics, is known.

Mechanism and effect of obstructive sleep apnea

Obesity is a major risk factor for the development of *obstructive sleep apnea* (OSA), with 58% of OSA cases thought to be caused by obesity. Furthermore, there is increasing evidence that OSA contributes to weight gain, consequently the link to obesity is two-way [14]. More than 90% of patients scheduled for bariatric surgery present OSA criteria [11]. Most studies relate OSA to decreased sexual desire, decreased arousal, and orgasmic dysfunction, with a frequency of 30%−70% in women with OSA. On the other hand, treatment with continuous positive airway pressure techniques has improved sexual function in these female patients [15]. Since sexual function is related to complex neurophysiological and psychological processes, the pathophysiology of sexual dysfunction in women with OSA is multifactorial. In fact, factors such as testosterone levels, degree of peripheral neuropathy, mood alterations, or impairment of quality of life have been associated with poor sexual function in women with this disorder. Female genital tract is mainly innervated by the pudendal nerve and its integrity is a key element for normal female sexual function. Peripheral neuropathy has been reported to develop in OSA, related to the severity of chronic intermittent nocturnal hypoxia [13]. Although oxygen saturation levels are associated with orgasmic dysfunction, most studies found no evidence that BMI was an important factor in cases of sexual dysfunction, but it was negatively associated with couple relationship quality [16].

Polycystic Ovarian syndrome

Polycystic ovary syndrome (PCOS) is a condition that affects 5%−7% of women of childbearing age. The link between the metabolic syndrome and the polycystic ovary syndrome is established by the fact that around 60% of female carriers suffer from insulin resistance and compensatory hyperinsulinemia, the pathogenic basis of the metabolic syndrome. Up to 50%−60% of women with PCOS are estimated to be obese. Obesity itself is a factor of poor prognosis in all aspects of PCOS; in fact, menstrual disorders are more frequent and the symptoms associated with hyperandrogenism are more obvious. When the body has an android fat distribution, with a waist circumference >88 cm (>35 in.) or a waist-to-hip ratio >0.85, the prevalence of metabolic syndrome and cardiovascular disease is 3−5 times greater than that of the nonaffected population [17]. Although in PCOS androgen levels are high, when associated with obesity, the increase in its peripheral aromatization and its conversion to low-potent estrogens result in abnormal sexual desire and arousal.

However, while levels of testosterone and other androgens are important for proper sexual function, the observed decline in the sexuality of obese women with PCOS also depends on other factors, many of which are unknown [18]. In fact, obese people are 55% more likely to develop depression and 25% more likely to suffer from anxiety; obesity is therefore very difficult to treat and has a high recurrence rate [19].

TABLE 33.3 Contributing factors for impair sexual function.

Predisposing factors	• Congenital anatomical anomalies • Hormonal anomalies • Prior life experiences: dysfunctional attachments, restrictive education, sexual abuse, physical violence
Precipitating factors	• Contextual issues
Maintaining factors	• Psychological self-reaction: feelings of guilt, anxiety, anger, impaired self-image, impaired self-esteem, avoidance behaviors • Psychological partner's reaction: loss of sexual confidence, lack of communication, avoidance behaviors • Systemic behaviors (couple): avoidance
Contextual factors	• Laboral stress, economic problems, fatigue, lack of time

Source: Adapted from Paarlberg KM, van de Wiel HBM. Bio-psycho-social obstetrics and gynecology: a competency-oriented approach. 2017.

Impaired body image

Body image is the mental representation of one's own body and it constitutes a multidimensional construct that groups together attitudes regarding health, appearance, functional ability, social relations, or sexuality, including positive and negative evaluations. The representation of the body image includes the complex interaction of conscious and unconscious perceptual phenomena and leads to the consideration of three different registers: the first, most evident to the senses, corresponds to the form or figure; the second is associated to the content and development of physiological processes and involuntary interoceptive evaluations; and the third is related to their meaning. The level of form or figure refers to the perception of body size and limits. The content level includes all proprioceptive, preconscious phenomena related to the satisfaction of physiological needs such as hunger, cold, and pain. Finally, the meaning includes the ability to both communicate and symbolize. Although obesity affects body image, the stereotypes of current beauty and the so-called culture of thinness can exert a greater influence affecting the meaning of self-image. Today's society creates prejudices around the obese woman and considers them as individuals without the will to control behaviors and in general less attractive, which leads many obese women to develop a negative image of themselves. Obese women often feel dissatisfied with their body image and make negative evaluations of their physical appearance focused on loss of femininity and attractiveness, especially when associated with PCOS [20], predisposing them to depression and anxiety. Negative body image evaluations result in significant loss of self-esteem, shyness, embarrassment, difficulties establishing relationships, and isolation. In fact, obese women have been found to use the Internet to seek sexual partners more often than nonobese women [21] and try to avoid direct social contact. These women are at high risk of sexual dysfunction, primarily decreased sexual desire, lower levels of arousal, worse lubrication, and pain. Although overweight and moderate obesity have sometimes been related to outgoing personalities [22] and positive moods, many studies associate obesity with depression or altered moods. Epidemiological studies have shown that depression in adolescence may precede adult obesity, as well as other mood disorders such as late luteal phase dysphoric disorder, chronic anxiety states with binge eating disorders, and posttraumatic stress disorder, which are associated with significant weight gains that can occur even years after the triggering event [23]. However, the relationship between obesity and depression is two-way. The use of psychoactive drugs in women diagnosed with depression may predispose to overweight and obesity. However, hypoactivity and sometimes the accompanying eating disorder may also predispose to overweight and obesity in the absence of medical treatments. On the other hand, loss of self-esteem and negative body image evaluations will contribute to a depressed mood (Table 33.3).

References

[1] Kinzl JF, Trefalt E, Fiala M, Hotter A, Biebl W, Aigner F. Partnership, sexuality, and sexual disorders in morbidly obese women: consequences of weight loss after gastric banding. Obes Surg 2011;11(4):455—8.

[2] Larsen SH, Wagner G, Heitmann BL. Sexual function and obesity. Int J Obes 2007;31:1189—98.

[3] Paarlberg KM, van de Wiel HBM. Bio-psycho-social obstetrics and gynecology: a competency-oriented approach. 2017.

[4] Steinke EE. Sexual dysfunction in women with cardiovascular disease: what do we know? J Cardiovasc Nurs 2010;25:151—8.

[5] Simoncig Netjasov A, Tančić-Gajić M, Ivović M, Marina L, Arizanović Z, Vujović S. Influence of obesity and hormone disturbances on sexuality of women in the menopause. Gynecol Endocrinol 2016;32:762—6.

[6] DSM-V. Manual Diagnóstico e Estatístico de Transtornos Mentais. American Psychiatric Association; 2014.

[7] Parish SJ, Goldstein AT, Goldstein SW, Goldstein I, Pfaus J, Clayton AH, et al. Toward a more evidence-based nosology and nomenclature for female sexual dysfunctions—Part II. J Sex Med 2016;13(12):1888–906.

[8] Diamanti-Kandarakis E, Bergiele A. The influence of obesity on hyperandrogenism and infertility in the female. Obes Rev 2001;2(4):231–8.

[9] Eyada M, Atwa M. Sexual function in female patients with unstable angina or non-ST-elevation myocardial infarction. J Sex Med 2007;4:1373–80.

[10] Træen B, Olsen S. Sexual dysfunction and sexual well-being in people with heart disease. Sex Relatsh Ther 2007;25(2):151–8.

[11] Lopes Neto JM, Brandão LO, Loli A, Leite CVdeS, Weber SAT. Evaluation of obstructive sleep apnea in obese patients scheduled for bariatric surgery. Acta Cir Bras 2013;28(4):317–22.

[12] Steinke EE, Mosack V, Wright DW, Chung ML, Moser DK. Risk factors as predictors of sexual activity in heart failure. Dimens Crit Care Nurs 2009;28(3):123–9.

[13] Yilmaz Z, Sirinocak PB, Voyvoda B, Ozcan L. Sexual dysfunction in premenopausal women with obstructive sleep apnea. Urol J 2017;14(6):1352–7.

[14] Hamilton GS, Joosten SA. Obstructive sleep apnoea and obesity. Aust Fam Physician 2017;46(7):460–3.

[15] Petersen M, Kristensen E, Berg S, Midgren B. Long-term effects of continuous positive airway pressure treatment on sexuality in female patients with obstructive sleep apnea. Sex Med 2013;1(2):62–8.

[16] Steinke E, Palm Johansen P, Fridlund B, Broström A. Determinants of sexual dysfunction and interventions for patients with obstructive sleep apnoea: a systematic review. Int J Clin Pract 2016;70(1):5–19.

[17] Messinis IE, Messini CI, Anifandis G, Dafopoulos K. Polycystic ovaries and obesity. Best Pract Res Clin Obstet Gynaecol 2015;29(4):479–88.

[18] Pastoor H, Timman R, de Klerk C, Bramer WM, Laan ETM, Laven SEJ. Sexual function in women with polycystic ovary syndrome. A systemic review and meta analysis. Reprod Biomed. online 2018;37(6):750–60.

[19] Porter Starr K, Fischer JG, Johnson MA. Eating behaviors, mental health, and food intake are associated with obesity in older congregate meal participants. J Nutr Gerontol Geriatr 2014;33(4):340–56.

[20] Himelein MJ, Thatcher SS. Depression and body image among women with polycystic ovary syndrome. J Health Psychol 2006;11(4):613–25.

[21] Bajos N, Wellings K, Laborde C, Moreau C. Sexuality and obesity, a gender perspective: results from French national random probability survey of sexual behaviours. BMJ 2010;340:c2573. https://doi.org/10.1136/bmj.c2573.

[22] Armon G, Melamed S, Shirom A, Shapira I, Berliner S. Personality traits and body weight measures: concurrent and across-time associations. Eur J Pers 2013;27:398–408

[23] Wurtman J, Wurtman R. The trajectory from mood to obesity. Curr Obes Rep 2018;(7):1–8.

Chapter 34

Professionally responsible clinical management of obese patients before and during pregnancy

Frank A. Chervenak and Laurence B. McCullough

Department of Obstetrics and Gynecology, Zucker School of Medicine at Hofstra/Northwell, Lenox Hill Hospital, New York, NY, United States

Introduction

Professionally responsible clinical management of obese patients before and during pregnancy should be guided by professional ethics in obstetrics [1]. In this chapter, we provide an introduction to professional ethics in obstetrics, especially the ethical principles of beneficence and respect for autonomy and the professional virtues of compassion and self-effacement. On the basis of these ethical principles and professional virtues, we address professionally responsible clinical practice with obese patients before and during pregnancy. We describe the distinctive roles of directive counseling, or making recommendations, and nondirective counseling, or shared decision-making. We emphasize a preventive ethics approach to preconception counseling. We describe the ethically justified use of "nudging" patients to change their eating habits [2]. We close by identifying the tools that professional ethics provides for managing the strong responses that obese patients can evoke.

Professional ethics in obstetrics

Professional ethics in obstetrics is based on professional ethics in medicine generally. Two British 18th-century physician—ethicists, John Gregory (1724—73) of Scotland and Thomas Percival (1740—1804) of England, invented the ethical concept of medicine as a profession [3]. They did so in response to a crisis of distrust of physicians. Patients did not have confidence that physicians knew what they were talking about or that physicians were acting primarily for the clinical benefit of their patients and not out of self-interest in money, reputation, and power.

The ethical concept of medicine as a profession

Gregory and Percival proposed a two-part ethical concept of medicine as the antidote to this corrosive distrust [1,3]. Physicians should become scientifically and clinically competent, a radical proposal at a time in the history of medicine when very few physicians possessed such competence. Physicians should use their scientific and clinical competence for the clinical benefit of the patient and keep self-interest systematically secondary. Gregory and Percival called for physicians to make these two commitments and over the subsequent decades, enough physicians did so that the ethical concept of medicine as a profession became a clinical reality.

The commitments that define the ethical concept of medicine as a profession are put into clinical practice by identifying and following the implications of ethical principles and professional virtues for patient care. Ethical principles focus on the behaviors that implement the commitments, while professional virtues focus on traits of character that implement the commitments. Professionally responsible clinical management of obesity is guided by two ethical principles—beneficence and respect for autonomy—and two professional virtues—compassion and self-effacement.

The ethical principle of beneficence

Percival was perhaps the first in the global history of medical ethics to use "beneficence," by which he meant that physicians should identify and offer clinical management that had an evidence base for clinical benefit [4]. This became the basis of the ethical principle of beneficence in professional medical ethics: Physicians have the ethical obligation to provide forms of clinical management that have evidence to support a reliable prediction of net clinical benefit for the patient [1].

The clinical ethical concept of medically reasonable clinical management

The ethical principle of beneficence supports the clinical ethical concept of medically reasonable clinical management: A form of clinical management that is clinically feasible and in beneficence-based clinical judgment is reliably predicted to result in net clinical benefit for the patient. We emphasize that both conditions must be satisfied for a form of clinical management to be considered medically reasonable. In particular, clinical feasibility, by itself, does not establish that a form of clinical management is medically reasonable. When there are two or more forms of medically reasonable clinical management for a patient's condition, they are referred to as the medically reasonable alternatives for the clinical management of the patient's condition [1].

The ethical principle of respect for autonomy

Gregory was perhaps the first in the global history of medical ethics to express the concept of a patient's rights: "Every man has a right to speak where his own life is concerned, or that of a friend." ([5], p. 33). This expression of patients' rights later became the basis for the ethical principle of respect for autonomy in professional ethical in medicine: The physician has the ethical obligation to offer to the patient the medically reasonable alternatives for the management of her condition and the clinical benefits and risks of each medically reasonable alternative. The goal of doing so is to empower the patient to make informed decisions with her physician about the clinical management of her condition [1].

The professional virtue of compassion

The professional virtue of compassion is the habit of recognizing when one's patient is in pain, distress, or suffering and seeking to relieve them. Compassion is also the habit of recognizing when one's patient is at risk for pain, distress, or suffering and seeking to prevent them [1]. Obesity makes the patient's life difficult in many dimensions, from the biological to the psychological and social [6]. Obesity can cause pain, for example, hip, knee, and ankle pain from bearing extra weight, pain that pregnancy can exacerbate. Obesity can cause distress, the disruption of one's behavioral repertoire, for example, finding a comfortable position for driving, especially over long distances. Obesity can cause suffering, the experience of one's aims and hopes for the future being blocked or diminished, for example, dealing with everything from subtle insults to outright invidious discrimination. The professional virtue of compassion creates the physician's first response to obesity during pregnancy, the sustained commitment to mitigating the biopsychosocial effects of obesity and working diligently and patiently to prevent unacceptable outcomes.

The professional virtue of self-effacement

The professional virtue of self-effacement is the habit of a conscious effort not to be influenced by clinically irrelevant differences between oneself and one's patient [1]. Obesity during pregnancy can evoke a strong emotional response that, if not identified and mitigated, can bias clinical judgment. For example, one of us (FAC) received a credible report that a physician had pictures of pigs in the chart of an obese pregnant patient. This is incompatible with self-effacement and should therefore never occur.

Compassion and self-effacement considered together

Cultivating the professional virtues of compassion and self-effacement creates the basis for managing the sense of foreboding that physicians naturally experience when the risk of clinically unacceptable outcomes is high and difficult to mitigate. This foreboding can sometimes transmute into dread to the point that the physician despairs of the well-being of both the pregnant and fetal patients. Compassion and self-effacement should be drawn in tandem to keep the focus on the patient and not one's understandable response to the high risk of clinically unacceptable outcomes [1].

Professional ethics in clinical practice with obese patients before and during pregnancy

The ethical principles of beneficence and respect for autonomy as well as the professional virtues of compassion and self-effacement provide practical guidance to professionally responsible clinical practice with obese patients before and during pregnancy. These principles and virtues are often complementary and thus strengthen each other, as we will show.

The role of directive counseling or making recommendations

In the informed consent process the physician should make recommendations when it is ethically justified to do so [1]. It is ethically justified to make a recommendation when the evidence is clear, for example, cesarean delivery for intrapartum well documented, complete placenta previa. Cesarean delivery for this intrapartum complication is the only way to prevent the deaths of both the pregnant and fetal patients. The ethical principle of beneficence becomes the guiding consideration for making such recommendations. The ethical principle of respect for patient autonomy becomes complements beneficence and thereby supports making recommendations, in order to empower the patient to make decisions based on her incorporating recommendations into her reflection and judgment. Thus empowering the patient may help her to better cope with the biopsychosocial dimensions of obesity during pregnancy, an often unappreciated dimension of respect for the patient's autonomy.

The role of shared decision-making

Making recommendations is not ethically justified when the evidence is less clear, that is, when there are two more forms of clinical management of the patient's condition supported in beneficence-based clinical judgment [1]. For example, beneficence-based clinical judgment supports both planned cesarean delivery and vaginal delivery in appropriately staffed and equipped hospital for intrapartum management after a low transverse incision in a previous pregnancy [7]. The ethical principle of beneficence becomes the guiding consideration in presenting both alternatives with an explanation of the clinical benefits and risks of each. The ethical principle of respect for patient autonomy complements beneficence and thereby calls for supporting the patient as she assesses each alternative and makes a decision based on her values and beliefs. This is shared decision-making, precisely understood. It follows that shared decision-making is not a universal model for decision-making about the clinical management of obesity and pregnancy.

Compassion and self-effacement working together keep the physician's focus on the pregnant and fetal patients, as explained previously. These two professional virtues thus work in synergy with shared decision-making, an often unappreciated dimension of shared decision-making.

The role of preventive ethics in preconception counseling

Obesity is a chronic clinical condition with myriad complications [6]. Pregnancy increases the risks of these complications and adds its own distinctive complications. Like clinical management of all chronic conditions, preventing clinically unacceptable outcomes is good clinical care. A preventive ethics approach to decision-making with patients should therefore be in the physician's clinical toolbox [1].

Preventive ethics uses the decision-making process with patients to anticipate and address ethical challenges in the clinical management of chronic conditions. The pregnant woman should be informed about the complications of obesity and pregnancy that are described in the other chapters. To help her assess these complications and become motivated to mitigate them, the physician should emphasize a shared value: the best possible clinical outcome for her pregnancy. The patient can then use these values to assess the complications and work with her physician on a plan to mitigate them. The professional virtue of compassion creates the ethical obligation to provide sustained psychosocial support to the patient as she undertakes this important self-care.

In preconception counseling the patient should be informed that a significant complication of obesity for pregnancy is an increased risk of cesarean delivery [8,9]. She should also be informed that cesarean delivery for an obese woman carries additional risks. She should be informed that it is therefore likely that the physician may need to recommend cesarean delivery. The physician should point out that it is therefore prudent for the patient to reduce her weight as much as possible before initiation of pregnancy. The patient should be offered psychosocial support for doing so.

The role of ethically justified "nudging" patients toward more healthy eating habits

Discontinuing eating habits that contribute to obesity is difficult for many patients. A tool for dealing with this challenge has recently been developed in behavioral economics. For many decades, economics embraced the model of the rational decision-maker who has all pertinent information and therefore can make unbiased decisions. This model has been challenged recently as lacking an evidence base in behavior psychology. Behavioral psychology has shown that we often make decisions nonrationally, for example, from habit or under the influence of various, subtle biases [2]. For example, the anchoring bias occurs when one's judgment strongly influenced by vivid memory, for example, the patient who refuses treatment for cancer because a friend died from cancer 30 years ago. The patient therefore does not have confidence that the recommended treatment will help because he does not appreciate that significant advances have been made since his friend died. In such a case, information is the antidote.

Information alone is often not sufficient to change eating habits. The insight of behavioral psychology is that biases can be subtly influenced and thus align an individual's food choices with the patient's health-related interests. The means of doing so include a hospital cafeteria ceasing to offer beverages with sugar or placing healthy foods at the beginning of the serving area or in the easiest-to-reach position. Inasmuch, as these subtle influences on our decision-making about what to eat aim to improve health, implementing them is supported by the ethical principle of beneficence. These subtle influences may be introduced without informing those who eat in the cafeteria, to increase the efficacy of the influences. This approach limits the autonomy of individuals for their benefit, which is a form of paternalism. Such paternalism is questionable from the perspective of the ethical principle of respect for autonomy. It has been argued that such influences on decision-making limit but do not take away individual autonomy and are therefore ethically permissible [2]. It is therefore consistent with good patient care for physicians to advocate for the use of "nudging" in how food is presented to patients in their organizations.

Responsibly managing the strong responses that obese patients can evoke

The increased risks of pregnancy complicated by obesity can elicit strong responses, especially the sense of foreboding that, despite sustained effort to prevent problems, problems, even disasters, may nonetheless occur. This sense of foreboding should not be ignored, to prevent it from biasing patient care, for example, giving up on counseling about weight reduction when a patient makes no effort to lose weight. The ethical principle of beneficence creates an ethical obligation to continue to provide high-quality patient care even when the patient's decisions and behaviors reduce the efficacy of such care. The professional virtue of compassion creates the ethical obligation not to give up on any patient, no matter how refractory her eating habits may be.

Conclusion

The obese patient presents ethically significant clinical challenges before and during pregnancy. This chapter has provided the reader with tools to responsibly address these challenges that are based in professional ethics in obstetrics.

References

[1] McCullough LB, Coverdale JH, Chervenak FA. Professional ethics in obstetrics and gynecology. Cambridge, UK: Cambridge University Press; 2020.
[2] Blumenthal-Barby JS, Burroughs H. Seeking better health care outcomes: the ethics of using the "nudge". Am J Bioeth 2012;12:1–10.
[3] McCullough LB. John Gregory and the invention of professional medical ethics and the profession of medicine. Dordrecht, The Netherlands: Kluwer Academic Publishers (Now Springer); 1998.
[4] Percival T. Medical ethics; or, a code of institutes and precepts, adapted to the professional conduct of physicians and surgeons. London: Johnson & Bickerstaff; 1803.
[5] Gregory J. Lectures on the duties and qualifications of a physician. London: W. Strahan and T. Cadell, 1772. In: McCullough LB, editor. John Gregory's writings on medical ethics and the philosophy of medicine. Dordrecht, The Netherlands: Kluwer Academic Publishers (Now Springer); 1998. p. 161–248.
[6] Dutton HP, Borengasser SJ, Gaudet LM, Barbour LA, Keely EJ. Obesity in pregnancy – optimizing outcomes for mom and baby. Med Clin North Am 2018;102:87–106.
[7] Bangdiwala SI, Brown SS, Cunningham FG, Dean TM, Frederiksen M, Hogue CJ, et al. NIH consensus development conference draft statement on vaginal birth after cesarean: new insights. NIH Consens State Sci Statements 2010;27(3):1–42. Available from: <http://consensus.nih.gov.ezproxyhost.library.tmc.edu/2010/vbacstatement.htm> [accessed 06.05.19]..
[8] Rogers AJG, Harper LM, Mari G. A conceptual framework for the impact of obesity on cesarean delivery. Am J Obstet Gynecol 2018;219:356–63.
[9] Smid MC, Vladutiu CJ, Dotters-Katz SK, Boggess KA, Manuck TA, Stamilio DM. Maternal obesity and major intraoperative complications during cesarean delivery. Am J Obstet Gynecol 2017;216:614.e1–7.

Index

Note: Page numbers followed by "*f*" and "*t*" refer to figures and tables, respectively.

A

AACE. *See* American Association of Clinical Endocrinologists (AACE)
Abdominal myofascial pain syndrome, obesity and, 287
Abnormal uterine bleeding (AUB), causes, 171
ABS. *See* American Bariatric Society (ABS)
Acanthosis nigricans, 77
Acetylcholine (Ach), 109
Acidosis, 227
Acne, 77
ACOG. *See* American College of Obstetricians and Gynecologists (ACOG)
ACS-NSQIP. *See* American College of Surgeons National Surgery Quality Improvement Program (ACS-NSQIP)
Adenomyosis, obesity and, 287
Adipocyte hormones, 267–269
Adipokines, 24, 84, 84*t*, 203
Adiponectin, 84, 203, 268
  receptors, 25
    AdipoR1, 25
    AdipoR2, 25
Adipose distribution, significance of, 256
Adipose tissue, 84
  functions, 195
Adiposity, 3, 9
  childhood, 17
  truncal, 226
Adipsin, 268
Adjustable gastric band, 57
Adjusted relative risk (aRR), 69
ADNEX model. *See* Assessment of Different NEoplasias in adneXa model (ADNEX model)
Adolescence
  obesity in, 3, 15, 171
    on adult health, 19
    adverse outcomes, 16–17
    clinical manifestation, 9
    counseling, 9–11
    etiology, 4–7
    factors affecting, 16
    health consequences, 9*t*
    incidence, 3–4
    management, 11
    obstetric outcomes in obese adolescents, 18–19
    PCOS in, 17–18, 20–21

Adrenal androgens in obese women with hirsutism, 78
AES. *See* Androgen Excess and PCOS Society (AES)
AI. *See* Anal incontinence (AI)
Alendronate, 269
Altered endometrial receptivity, 128
Amenorrhea, 296
American Association of Clinical Endocrinologists (AACE), 223–224
American Bariatric Society (ABS), 152
American College of Surgeons National Surgery Quality Improvement Program (ACS-NSQIP), 204
American College of Obstetricians and Gynecologists (ACOG), 153–154, 281
American Society for Metabolic and Bariatric Surgery (ASMBS), 153–154, 233
American Society for Reproductive Medicine (ASRM), 153–154
AMH. *See* Anti-Müllerian hormone (AMH)
5'AMP-activated protein kinase (AMPK), 145–146
Anagen, 79
Anal incontinence (AI), 189–190. *See also* Urinary incontinence (UI)
  incontinence symptoms following weight loss, 192
  outcomes of incontinence procedures in obese women, 191–192
  pathophysiology of incontinence in obese population, 191
Anastrozole, 206
Androgen antagonists, 80–81
Androgen Excess and PCOS Society (AES), 23
Androgen production, obesity and, 77
Androstenedione, 78
Anovulation, 23–24, 83–84, 154, 172
Anti-Müllerian hormone (AMH), 99, 154
  inhibition, 23–24
Anxiety, 38
Apelin, 26
Area under the curve (AUC), 67
Aromatase, 196
Aromatase inhibitors, 28
aRR. *See* Adjusted relative risk (aRR)
ART. *See* Assisted reproduction treatment (ART); Assisted reproductive technologies (ART)

ASMBS. *See* American Society for Metabolic and Bariatric Surgery (ASMBS)
ASRM. *See* American Society for Reproductive Medicine (ASRM)
Assessment of Different NEoplasias in adneXa model (ADNEX model), 166
Assisted Conception Treatment (ART). *See* Assisted reproduction treatment (ART)
Assisted reproduction treatment (ART), 97, 251
  after bariatric surgery, 154–155
  clinical procedures, 99–100
  ethical issues relevant to access to services, 102
  evidence of reduced fertility in obese, 97–98
  cycle effects, 97–98
  effect of obesity, 100–101
  practical management of obese women, 99
    monitoring of stimulation, 99
    patient selection, 99
    stimulation regimes, 99
  prevalence of obesity in, 97
  rationale for use of, 98–99
  safety issues for mothers and offspring, 101–102
  specific issues, 98
    effects on embryos, 98
    effects on endometrium, 98
    effects on oocyte, 98
Assisted Reproductive Technology (ARTech). *See* Assisted reproduction treatment (ART)
AUB. *See* Abnormal uterine bleeding (AUB)
AUC. *See* Area under the curve (AUC)
Autonomy, ethical principle of respect for, 320
Avanafil, 115

B

Bariatric Outcomes Longitudinal Database (BOLD database), 239
Bariatric surgery (BS), 26–27, 57, 88, 135–136, 181–182, 261
  assisted reproduction after, 154–155
  impact of bariatric surgery on fertility, 153–154
  biliopancreatic diversion, 136
  and bone health, 269–270
  contraceptive issues after, 50–51

323

Bariatric surgery (BS) (*Continued*)
  emergency contraception, 63
  endoluminal procedures, 136
  gastric banding, 136
  gastric bypass, 136
  long-acting reversible contraception, 58–61
  and polycystic ovarian syndrome, 154
  potential of bariatric surgery for negative impact on fertility, 154
  practical considerations, 139–140
  pregnancy after, 154
  sleeve gastrectomy, 135–136
  types, 152–153
  as weight loss measure, 152
Barrier method, 63
Basal lipolysis, 257
BC. *See* Breast cancer (BC)
Beneficence, ethical principle of, 319–320
Beta-cell hormones, 267
BF. *See* Body fat (BF)
Biliopancreatic diversion (BPD), 57, 136
Biomarkers, 26
BMD. *See* Bone mineral density (BMD)
BMI. *See* Body mass index (BMI)
Body fat (BF), 223
Body image, 316
Body mass index (BMI), 3, 23, 37, 43, 57, 83, 91, 97, 105, 119, 135, 143, 171, 179, 189, 201, 209–210, 217, 245, 255, 273, 281
  classification of obesity based on, 43
  percentile curves for girls, 4*f*
BOLD database. *See* Bariatric Outcomes Longitudinal Database (BOLD database)
Bone, obesity of, 269
Bone mineral density (BMD), 51, 265
BPD. *See* Biliopancreatic diversion (BPD)
Breast cancer (BC), 196, 201
  diagnosis, 204
  epidemiology, 201–202
  obesity and, 210
  pathogenetic mechanisms, 202–204
    adipokines, 203
    chronic inflammation, 203–204
    hyperinsulinemia, 202–203
    microbiome, 204
    sex hormones, 202
  prognosis, 206
  therapy, 204–206
    chemotherapy, 205–206
    endocrine, 206
    radiotherapy, 205
    surgery, 204–205
BS. *See* Bariatric surgery (BS)

C

C-reactive protein (CRP), 86
CAD. *See* Coronary artery disease (CAD)
CADTH technology, 306
Cancer
  clinical implications for prevention and treatment of cancer, 197–198
  contraception and, 50

epidemiologic evidence for links between obesity and, 195–196
  potential mechanisms for oncogenesis, 196–197
  unique to or more common in women, 196
Carcinogenesis, 306
Cardiac adaptation to obesity, 257
  lipid metabolism in pregnant obese women, 257–258
Cardiopulmonary exercise test (CPET), 233
Cardiovascular assessment, 233
Cardiovascular disease (CVD), 218, 255, 273
  adipose distribution, significance of, 256
  cardiac adaptation to obesity, 257
  contraception and, 49
  gynecology practice, 258–259
  interventions to address obesity in reproductive health, 260–261
    medical treatment, 260
    physical activity, 260
    surgical treatment, 260–261
  interventions to improve outcomes in pregnancy, 261
    bariatric surgery, 261
    weight management, 261
  lipid metabolism in nonpregnant obese women, 257
  maternal obesity and in utero programing for, 259–260
  pathogenesis of visceral obesity, 256–257
  risk assessment in obese individuals, 255–256
Cardiovascular hypertrophy, 226
Cardiovascular risk, 111, 259
Caverject, 115–116
CCK. *See* Cholecystokinin (CCK)
Center for Disease Control and Prevention (CDC), 3, 67
Cervical cancer, 168
  obesity and, 210–211
CFU. *See* Colony-forming unit (CFU)
cGMP, 108–109
CHC. *See* Combined hormonal contraceptives (CHC)
CHD. *See* Coronary heart disease (CHD)
Chemerin, 85
Chemotherapy, 205–206
Childhood
  adiposity, 17
  obesity in, 3, 171
    on adult health, 19
    adverse outcomes, 17*t*
    clinical manifestation, 9
    comorbidities and complications, 10*t*
    counseling, 9–11
    etiology, 4–7
    factors affecting, 16
    incidence, 3–4
    management, 11
    prevalence of, 15–16
Cholecystokinin (CCK), 78
Cholesterol, 154
Chronic inflammation, 203–204
Chronic pelvic pain, 281

genetics of obesity and chronic pain, 283–284
obesity
  and abdominal myofascial pain syndrome, 287
  and adenomyosis, 287
  and endometriosis, 286
  impact on assessment, 282*t*, 284–285
  impact on surgical management of women with, 285–286
  impact on treatment of women with, 285
  and nongynecological causes of chronic pelvic pain, 287
  and pain physiology, 281–283
  psychological impact of obesity and chronic pelvic pain, 284
CI. *See* Confidence interval (CI)
Cialis. *See* Tadalafil
Class III obese patients, alternatives for, 230–231
Clinical psychosomatic women's health, 293
  infertility/subfertility and clinical psychosomatic aspects, 300–302
    successes/failures in conceiving, distress, and obesogenic behavior, 300–302
  menstrual problems and obesity
    clinical psychosomatic approach, 296–297
    clinical psychosomatic gynecological referrals, 297
    menstrual irregularities, 297–298
    pathophysiology of psychosomatic menstrual issues, 296
    patients' perspective, 296–297
    PCOS (polycystic ovarian syndrome), 297–298
    PCOS and obesity, 298–300
    premenstrual syndrome (PMS), 297–298
    psychosomatic insights into menstrual presentations, 296
  mind–body approaches, 293–296
  obesity and severe pelvic/perineal dysfunction, 305
    clinical psychosomatic viewpoint, 305
  physical, mental, and sexual violence with obesity in migrants, 302–305
  psychosomatic impact of gynecological tumors in obese, 305–307
Clomiphene citrate, 27
COCPs. *See* Combined oral contraceptive pills (COCPs)
COCs. *See* Combined oral contraceptives (COCs)
Colon cancer, 195
Colony-forming unit (CFU), 93
Combined hormonal contraceptives (CHC), 45, 62–63, 67, 246, 247*t*
Combined oral contraceptive pills (COCPs), 20, 80
Combined oral contraceptives (COCs), 46, 62
Combined semen parameters, 120–121
  sperm DNA damage, 120–121
Comorbidities
  of childhood obesity, 10*t*
  obesity-related, 110–113

Compassion, professional virtue of, 320
Compliance, 39, 88, 148
Complications of childhood obesity, 10t
Computed tomography (CT), 285
Confidence interval (CI), 135, 146, 201
Continuous positive airway pressure (CPAP), 233
Contraception, obesity and
 and cancer, 50
 and cardiovascular disease, 49
 classification of obesity based on BMI, 43
 contraceptive issues after bariatric surgery, 50–51
 effects of obesity on pregnancy outcomes, 44t
 efficacy, 44–45
  evidence, 45–46
  mechanisms, 44–45
 intrauterine contraceptive devices in obese women, 51
 potential concerns, 43–44
 risks of obesity in pregnancy, 43
 safety of hormonal contraceptives in obese women, 46–49
 sterilization procedures in obese women, 51–52
 and venous thromboembolism, 49–50
Contraceptive efficacy, obesity and, 247–248
Contraceptive failure, 39, 43
Contraceptive patch and ring, 63
Copeptin, 26
Copper intrauterine device (Cu-IUD), 58–59, 68–69
 additional health benefits, 68
 contraindications, 68
 efficacy, 68
 health risks, 68
 limitations of use, 68
 practical issues in obese women, 69
 side effects, 68
Coronary artery disease (CAD), 111, 218, 315
Coronary heart disease (CHD), 255, 274–275
Cost-effectiveness, 231
CPAP. *See* Continuous positive airway pressure (CPAP)
CPET. *See* Cardiopulmonary exercise test (CPET)
CRP. *See* C-reactive protein (CRP)
CT. *See* Computed tomography (CT)
Cu-IUD. *See* Copper intrauterine device (Cu-IUD)
Cushing's syndrome, 79
CVD. *See* Cardiovascular disease (CVD)
CYP19, 109
Cyproterone acetate, 20, 80
Cytochrome P450 2D6 (CYP2D6), 277–278
Cytokines, 196–197, 257

D

Decidualization defects, 86–87
Dehydroepiandrosterone (DHEA), 78–79
Depo-Provera, 61
Depot medroxyprogesterone acetate (DMPA), 61
Depression, 38
Detrusor overactivity (DO), 181
DHEA. *See* Dehydroepiandrosterone (DHEA)
DHEA-sulfate (DHEAS), 277
DHT. *See* Dihydrotestosterone (DHT)
Diabetes mellitus (DM), 105, 111–113, 265, 273
 evaluation, 111–113
  general considerations, 111–112
  sexual function, 112–113
 treatment and osteoporosis, 267
Dianette, 80
Diet, 26–27, 144
 fertility, 88
Diet-induced obesity mouse model (DIO mouse model), 83
Dietary interventions, 144
Dihydrotestosterone (DHT), 77
DIO mouse model. *See* Diet-induced obesity mouse model (DIO mouse model)
Directive counseling or making recommendations, 320–321
DM. *See* Diabetes mellitus (DM)
DMPA. *See* Depot medroxyprogesterone acetate (DMPA)
DO. *See* Detrusor overactivity (DO)
Drospirenone, 20
Drugs, 19–20
 pharmacology, 227
DSG. *See* 3-Keto-desogestrel (DSG)
Dual-energy X-ray absorptiometry (DXA), 269, 273
Dysbiosis. *See* Microbial alterations
Dysmenorrhea, 296
Dysphoria, 296

E

EAES. *See* European Association for Endoscopic Surgery (EAES)
Early puberty, 7–8
EC. *See* Emergency contraception (EC)
ECG. *See* Electrocardiogram (ECG)
ED. *See* Erectile dysfunction (ED)
EDV. *See* End diastolic velocity (EDV)
EE. *See* Ethinyl estradiol (EE)
Efficacy, 44–45, 68–71
 contraceptive, 247–248
Eflornithine, 81
Egg collection, 99–100
Ejaculation, 106–107
Electrocardiogram (ECG), 233
Embryo, effects on, 86
Embryo development, 128
Embryo transfer (ET), 100
Embryology/American Society for Reproductive Medicine (ESHRE/ASRM), 23
Emergency contraception (EC), 63
eMSC. *See* Endometrial mesenchymal stem cell (eMSC)
End diastolic velocity (EDV), 113
Endocrine theory, 121–122
Endocrine therapy, 206
Endoluminal procedures, 136

Endometrial cancer, 165–166, 305–306
 etiopathogenesis of, 306
 global disease burden of, 305–306
 obesity and, 209–210
 preventing/treating obesity to prevent, 306–307
Endometrial changes in obesity, 93
Endometrial hyperplasia, 165–166
Endometrial mesenchymal stem cell (eMSC), 93
Endometrial polyp, 171–172
Endometriosis, obesity and, 286
Endometrium, 173–174
 effects, 86–87
Endoplasmic reticulum (ER), 86
Endothelial nitric oxide synthase (eNOS), 108
Endothelin-1 (ET-1), 108
Energy-restricted diet, 26
ENG. *See* Etonogestrel (ENG)
eNOS. *See* Endothelial nitric oxide synthase (eNOS)
ER. *See* Endoplasmic reticulum (ER)
ERalpha (ERα), 274
ERbeta (ERβ), 274
Erectile dysfunction (ED), 38, 105, 139
ERK1. *See* Extracellular-signal-regulated kinase 1 (ERK1)
ERs. *See* Estrogen receptors (ERs)
ERV. *See* Expiratory reserve volume (ERV)
ERα. *See* ERalpha (ERα)
ERβ. *See* ERbeta (ERβ)
ESHRE/ASRM. *See* Embryology/American Society for Reproductive Medicine (ESHRE/ASRM)
Estradiol (E2), 8, 109
Estrogen receptors (ERs), 273
Estrogens, 273–274
 deficiency, 275
ET. *See* Embryo transfer (ET)
ET-1. *See* Endothelin-1 (ET-1)
Ethically justified "nudging" patients, 321–322
Ethics
 preventive ethics in preconception counseling, 321
 professional ethics in obstetrics, 319–320
Ethinyl estradiol (EE), 49–50
Etiological theories, 121–123
 endocrine theory, 121–122
 genetic theory, 122
 reactive oxygen species theory, 123
 sexual dysfunction theory, 122–123
 testicular hyperthermia theory, 123
Etonogestrel (ENG), 60, 70
Etonogestrel-releasing implant, 70–72
 contraindications, 72
 efficacy, 70–71
 health benefits, 71–72
 health risks, 71
 practical issues, 72
 progestogen implants, 70
 safe prescribing, 72
 side effects, 71
European Association for Endoscopic Surgery (EAES), 227–228

Evaluation of sexual dysfunction, general considerations, 111–112
Evaluation of sexual function, 112–113
Evidence-based assisted reproduction in obese women
    altered endometrial receptivity, 128
    impaired ovarian folliculogenesis, 127–128
    obesity and FET, 129
    obesity and in vitro fertilization, 128–129
    obesity and intrauterine insemination, 129
Exercise, 144–145
Expiratory reserve volume (ERV), 226
Extracellular-signal-regulated kinase 1 (ERK1), 203

F

FAI. *See* Free androgen index (FAI)
Fat mass, 246, 266
Fecal incontinence (FI), 182, 189
    obesity and, 182
Female infertility
    challenges of managing obese women, 87
    clinical effects of obesity on, 85–87
    epidemiology, 83
    pathophysiological basis of infertility in obese women, 83–85
    treatment options, 87–89
Female malignancies
    mechanisms relating obesity to risks, 210–211
    obesity
        and breast cancer, 210
        and cervical cancer, 210–211
        effect on management, 212–213
        and endometrial cancer, 209–210
        and ovarian cancer, 210
Female sex function index scores (FSFI scores), 39
Female sexual function, 313
Ferriman–Gallwey scoring system, 79
Fertility
    impact of bariatric surgery, 153–154
    diet, 88
    high body mass index, 143
    obesity and, 151
    outcomes, 145
    potential of bariatric surgery for negative impact, 154
FET. *See* Frozen embryo transfer (FET)
FFAs. *See* Free fatty acids (FFAs)
FI. *See* Fecal incontinence (FI); Fluorescence imaging (FI)
Finasteride, 81
Flatus incontinence, 182, 189
Fluorescence imaging (FI), 228
Fluoxetine, 277–278
Flutamide, 81
Foley lap lift technique, 235–236
Follicle-stimulating hormone (FSH), 23–24, 28–29, 84, 119–120, 136, 274
    follicle-stimulating hormone and luteinizing hormone ratio (FSH/LH ratio), 20
Food neophobia, 294
Fractures, 15

FRC. *See* Functional residual capacity (FRC)
Free androgen index (FAI), 79, 92
Free fatty acids (FFAs), 256
French Nutri-Score, 294
Frozen embryo transfer (FET), 127, 129
FSFI scores. *See* Female sex function index scores (FSFI scores)
FSH. *See* Follicle-stimulating hormone (FSH)
FSRH. *See* UK Faculty for Sexual and Reproductive Health (FSRH)
Functional residual capacity (FRC), 218, 226

G

Gastric banding, 136
Gastric bypass, 136
Gastric restriction, 57
    clinical psychosomatic aspect of, 302–303
Genetic theory, 122
Genitourinary syndrome (GSM), 276
GH. *See* Growth hormone (GH)
Ghrelin, 26, 85
Glucagon-like peptide-1 (GLP-1), 127, 145, 147–148
Glycodelin A, 59
GnRH. *See* Gonadotropin-releasing hormone (GnRH)
Gonadotropin-releasing hormone (GnRH), 7–8, 24, 127, 136
Growth hormone (GH), 26
    deficiency, 6
GSM. *See* Genitourinary syndrome (GSM)
Gynecological cancer, 252
Gynecological surgery, 251–252
    anesthetic challenges, 219
    equipment and general considerations, 219
    indications for surgery, 217
    intraoperative challenges, 219–220
    laparoscopic surgery, 219–220
    medicolegal implication, 221
    open abdominal surgery, 220
    physiological changes in obese patients, 218
    postoperative issues, 220–221
    preoperative evaluation, 218–219
    risk of obese women undergoing surgery, 218
    thromboprophylaxis, 219
Gynecological tumors, psychosomatic impact of, 305–307
    etiopathogenesis of endometrial cancer, 306
    global disease burden of endometrial cancer/malignancy, 305–306
Gynecology ultrasound, 165
    obesity and
        clinical applications, 164–168
        ergonomic considerations, 163–164
        pelvic ultrasound, 159–163
        ultrasound settings, 163

H

Hair cycle, 79
Hair growth, 79
HAIR-AN syndrome, 79
Hazard ratio (HR), 201

hCG. *See* Human chorionic gonadotrophin (hCG)
HD. *See* Heart disease (HD)
HDL. *See* High-density lipoprotein (HDL)
Health and Safety Executive (HSE), 219
Health Survey for England (HSFE), 223
Heart disease (HD), 255
Heavy menstrual bleeding (HMB), 173
HFD. *See* High-fat diet (HFD)
High body mass index on fertility and pregnancy, 143
High-density lipoprotein (HDL), 256
High-fat diet (HFD), 109, 120
Hirsutism, 77–78
    adrenal androgens in obese women with, 78
    clinical assessment, 79
    management associated with obesity, 78–79
    obesity and androgen production, 77
    obesity and ovarian function, 77
    obesity and polycystic ovarian syndrome, 78
    treatment, 79–81
HMB. *See* Heavy menstrual bleeding (HMB)
Hormonal contraception, 11, 246–248
    contraception and fear of weight gain, 248
    obesity and contraceptive efficacy, 247–248
    and VTE risk, 246–247
    weight-loss treatment and contraception, 248
Hormonal implants, 60
Hormone replacement therapy (HRT), 210, 249–251, 274–277
Hormone therapy (HT), 210
Hospital catchment area characteristics, 303–305
HPA axis.
    See Hypothalamic–pituitary–adrenal axis (HPA axis)
HPO axis.
    See Hypothalamic–pituitary–ovarian axis (HPO axis)
HR. *See* Hazard ratio (HR)
HRT. *See* Hormone replacement therapy (HRT)
HSE. *See* Health and Safety Executive (HSE)
HSFE. *See* Health Survey for England (HSFE)
HT. *See* Hormone therapy (HT)
HTN. *See* Hypertension (HTN)
Human chorionic gonadotrophin (hCG), 92–93, 129
    trigger, 100
Hyperandrogenemia, 92
Hyperandrogenism, 23–24
Hypercapnia, 227
Hypercholesterolemia, 110
Hyperglycemia, 109, 203
Hyperglycemic memory, 203
Hyperinsulinemia, 8, 24, 77, 85, 92, 202–203, 267
Hyperlipidemia, 257–258
Hyperplasia, 172
Hypertension (HTN), 105, 111
Hypertrichosis, 77–78
Hypothalamic–pituitary–adrenal axis (HPA axis), 283–284, 293
Hypothalamic–pituitary–gonadal axis, 136–138

Hypothalamic–pituitary–ovarian axis (HPO axis), 83
  effect, 85–86
Hypothyroidism, 79
Hysterectomy, 173

I

IAP. *See* Intraabdominal pressure (IAP)
IBW. *See* Ideal body weight (IBW)
IC arteries. *See* Intracavernosal arteries (IC arteries)
ICG. *See* Indocyanine green (ICG)
ICSI. *See* Intracytoplasmic sperm injection (ICSI)
Ideal body weight (IBW), 99
IGF-1. *See* Insulin-like growth factor 1 (IGF-1)
IGFBP-1. *See* Insulin-like growth factor-binding protein-1 (IGFBP-1)
IGFIR. *See* Insulin-like growth factor receptor (IGFIR)
IL-6. *See* Interleukin-6 (IL-6)
Immune system, 227
Immunological factors, 93
Impaired body image, 316
Impaired ovarian folliculogenesis, 127–128
Implanon, 60, 70
Implants
  hormonal, 60
  progestogen, 70
In utero programing for cardiovascular disease, 259–260
In vitro fertilization (IVF), 25, 29–30, 92, 119, 128–129, 143, 300
In vitro fertilization/intracytoplasmic sperm injection (IVF/ICSI), 127
Incontinence
  continence and prolapse surgery in obese woman, 184
  detrusor overactivity, 181
  incidence and prevalence, 179
  normal bladder function and causes in women, 179–180
  obesity
    and FI, 182
    obesity and prolapse, 182–183
    and urinary incontinence, 181–182
  pathophysiology in obese population
    anal incontinence, 191
    urinary incontinence, 190–191
  symptoms following weight loss, 192
  urodynamic stress incontinence, 180–181
Indocyanine green (ICG), 228
Infertility, 23–26. *See also* Fertility
  and clinical psychosomatic aspects, 300–302
  diet, lifestyle changes, and bariatric surgery, 26–27
  hyperandrogenism, 23–24
  hyperinsulinemia, 24
  hypersecretion of luteinizing hormone, 24
  treatment, impact of obesity on, 26
Inhibin B, 139
Insulin, 85. *See also* Diabetes mellitus (DM)
Insulin resistance (IR), 92, 122
Insulin sensitizers, 20–21, 29

Insulin-like growth factor 1 (IGF-1), 77, 202–203, 211
  receptor, 8
Insulin-like growth factor receptor (IGFIR), 259
Insulin-like growth factor-binding protein-1 (IGFBP-1), 77
Insulin-sensitizing agents, 25, 79–80, 94
Interleukin-6 (IL-6), 24–25, 84, 203–204, 268
International Society for the Study of Women's Sexual Health (ISSWSH), 313
  sexual disorders, 314*t*
Interpersonal violence (IPV), 302
Interventions. *See also* Surgical interventions
  dietary, 144
  lifestyle, 19, 87–88, 143–144
  medical, 143
Interventions to address obesity in reproductive health
  medical treatment, 260
  physical activity, 260
  surgical treatment, 260–261
Interventions to improve outcomes in pregnancy
  bariatric surgery, 260–261
  weight management, 261
Intraabdominal pressure (IAP), 226
Intraabdominal visceral fat, 230
Intracavernosal arteries (IC arteries), 111
Intracytoplasmic sperm injection (ICSI), 119, 300
Intrauterine contraceptive devices in obese women, 51
Intrauterine devices (IUDs), 44, 67, 230–231, 246–247
Intrauterine insemination (IUI), 127, 129
IPV. *See* Interpersonal violence (IPV)
IR. *See* Insulin resistance (IR)
Irisin, 26
Irregular bleeding, 69
Irritable bowel syndrome, 287
ISSWSH. *See* International Society for the Study of Women's Sexual Health (ISSWSH)
IUDs. *See* Intrauterine devices (IUDs)
IUI. *See* Intrauterine insemination (IUI)
IVF. *See* In vitro fertilization (IVF)
IVF/ICSI. *See* In vitro fertilization/intracytoplasmic sperm injection (IVF/ICSI)

J

Janus kinase 2/signal transducer and the activator of transcription 3 (JAK2/STAT3), 203
Jejunoileal bypass, 57

K

3-Keto-desogestrel (DSG), 60
Kisspeptin, 26

L

Laparoscopic ovarian drilling (LOD), 26, 28–29
Laparoscopic surgery, 219–220, 227–228

Laurence–Moon syndrome, 122
LBRs. *See* Live birth rates (LBRs)
LC–MS. *See* Liquid chromatography and mass spectrometry (LC–MS)
LDL. *See* Low-density lipoprotein (LDL)
Leptin, 25, 84, 93, 122, 203, 268
Letrozole, 28, 206
Levitra. *See* Vardenafil
Levonorgestrel (LNG), 49–51, 246
Levonorgestrel-containing intrauterine systems (LNG-IUS), 11, 69–70, 173–174
  efficacy, 69
  health benefits, 70
  health risks, 69–70
  safe prescribing, 70
  side effects, 69–70
Levonorgestrel-IUD device, 58–61
Levonorgestrel-releasing intrauterine system. *See* Levonorgestrel-containing intrauterine systems (LNG-IUS)
Leydig cell function, 136–138
LH. *See* Luteinizing hormone (LH)
Lifestyle, 26–27
Lifestyle interventions, 19, 87–88, 143–144
Lipid metabolism
  in nonpregnant obese women, 257
  in pregnant obese women, 257–258
Lipotoxicity, 86, 226
  theory, 128
Liquid chromatography and mass spectrometry (LC–MS), 112
Liraglutide, 145, 147–148
Lithotomy, 235
Live birth rates (LBRs), 87–88, 100–101
LMWH. *See* Low-molecular-weight heparins (LMWH)
LNG. *See* Levonorgestrel (LNG)
LNG-IUD system. *See* Levonorgestrel-IUD system (LNG-IUD system)
LNG-IUS. *See* Levonorgestrel-containing intrauterine systems (LNG-IUS)
LOD. *See* Laparoscopic ovarian drilling (LOD)
Long acting contraceptives
  copper intrauterine device, 58–59
  levonorgestrel containing intrauterine device, 59–61
  progesterone-only injection, 61–62
Long-form receptor (LRb), 203
Long-term contraceptive care
  CHC, 67
  copper intrauterine device, 68–69
  etonogestrel-releasing implant, 70–72
  LNG-IUS, 69–70
Low-density lipoprotein (LDL), 110, 255–256
Low-molecular-weight heparins (LMWH), 239
LRb. *See* Long-form receptor (LRb)
Luteal support, 100
Luteinizing hormone (LH), 20, 23–24, 84, 92, 109, 121, 136
  hypersecretion, 24

M

Magnetic resonance imaging (MRI), 285
Malabsorption, 57

Male fertility, 120
Male infertility, 119
Male obesity
  etiological theories, 121–123
  impact on semen quality, 119–120
  and recurrent miscarriage, 93
  treatment of infertility, 123–124
Malignancy, 224
Malignancy and hyperplasia, 172
Malignant gynecological tumors, 305
MAPK. See Mitogen-activated protein kinase (MAPK)
Massachusetts Male Aging Study (MMAS), 110
Maternal obesity, 43
  and in utero programming for cardio vascular disease, 259–260
MC4R. See Melanocortin receptor 4 (MC4R)
Mean difference (MD), 146
Mechanisms relating obesity to female malignancies, 211–212
Mechanism and effect of obstructive sleep apnoea, 315
Medical interventions, 143
  barriers to weight loss, 148
  diet, 144
  dietary interventions, 144
  exercise, 144–145
  high body mass index on fertility and pregnancy, 143
  lifestyle interventions to improve outcomes, 143–144
  liraglutide, 147–148
  metformin, 145–147
  orlistat, 147
  sibutramine, 147
  weight-loss medications and fertility outcomes, 145
Medically reasonable clinical management, 320
Medicine as profession, ethical concept of, 319
Medicines and Healthcare products Regulatory Agency (MHRA), 246
Mediterranean diet, 88
Melanocortin receptor 4 (MC4R), 283
Men sex with men (MSM), 39–40
Menarche, 7–8, 10
Menometrorrhagia, 296
Menopausal HT, 210
Menopausal obesity, 273–274
  emerging menopausal therapies, 277–278
  obesity, lifestyle intervention, and HRT, 274–277
Menopausal status, 166, 201
Menopause, 296
Menorrhagia, 296
Menstrual abnormalities, 296
Menstrual disorders
  endometrium, 173–174
  malignancy and hyperplasia, 172
  obesity and abnormal uterine bleeding, 171
  ovulatory dysfunction, 172–173
  PALM-COEIN classification, 171–172
  polyps, 171–172
  PCOS, 174

obesity in the absence of, 174–175
Menstrual irregularities, clinical psychosomatic angles to, 297–298
Menstrual problems, 171, 173–174, 296
Metabolic bone disease, 270
Metabolic syndrome, 17, 314–315
Metformin, 20–21, 29, 79–80, 89, 145–147, 206
MHRA. See Medicines and Healthcare products Regulatory Agency (MHRA)
Micro RNAs, 197
Microbial alterations, 204
Microbiome, 204
Migrants, obesity in, 302–305
  characteristics of hospital catchment area, 303–305
  clinical psychosomatic aspect of gender-related violence, 302–303, 303t
Minimally invasive surgery (MIS), 229
  indications, 229
  and obesity, 229–230
  technological features, 227–229
MIS. See Minimally invasive surgery (MIS)
Miscarriage, 91
  obesity and, 91–92
  rate, 100
Mitogen-activated protein kinase (MAPK), 203
MMAS. See Massachusetts Male Aging Study (MMAS)
Morbid obesity, 223
MRI. See Magnetic resonance imaging (MRI)
MSM. See Men sex with men (MSM)
Multidisciplinary approach to treatment, 113–116
Myoinositol, 79–80

N

NAION. See Nonarteritic anterior ischemic optic neuropathy (NAION)
Narrow band imaging (NBI), 228
National Cancer Institute of Cancer Clinical Trials Group (NCIC CTG), 206
National Child Measurement Programme (NCMP), 15–16
National Institute for Health Care and Excellence (NICE), 260
National Institutes of Health (NIH), 23
National Patient Safety Agency (NPSA), 219
Natural killer cell (NK cell), 196
NBI. See Narrow band imaging (NBI)
NCIC CTG. See National Cancer Institute of Cancer Clinical Trials Group (NCIC CTG)
NCMP. See National Child Measurement Programme (NCMP)
NET. See Norethisterone (NET)
Neuropeptide Y (NPY), 268, 281–283
NHS Litigation Authority (NHSLA), 221
NICE. See National Institute for Health Care and Excellence (NICE)
NIH. See National Institutes of Health (NIH)
Nitric oxide synthase within nerve endings (nNOS), 108
NK cell. See Natural killer cell (NK cell)

nNOS. See Nitric oxide synthase within nerve endings (nNOS)
Non-ST segment elevation myocardial infarction (NSTEMI), 257
Nonarteritic anterior ischemic optic neuropathy (NAION), 114
Nongynecologic cancers, 196
Nongynecological causes of chronic pelvic pain, 287
Nonhormonal therapies, 277
Nonpregnant obese women, lipid metabolism in, 257
Nonsurgical management of obesity, 152
Norethisterone (NET), 50–51, 61
Norgestrel, 49
Normal bladder function and causes of incontinence in women, 179–180
Norplant, 70
Norplant II, 60, 70
NPSA. See National Patient Safety Agency (NPSA)
NPY. See Neuropeptide Y (NPY)
NSTEMI. See Non-ST segment elevation myocardial infarction (NSTEMI)
Nutritional energy balance, 294

O

OAB. See Overactive bladder (OAB)
Obese gene (ob gene), 25
Obese women
  anal incontinence, 189–190
  epidemiological data in obese populations, 190
  incontinence symptoms following weight loss, 192
    anal incontinence, 192
    urinary incontinence, 192
  outcomes of incontinence procedures in
    anal incontinence, 191–192
    urinary incontinence, 191
  pathophysiology of incontinence in obese population, 190–191
  urinary incontinence, 189
Obesity, 67, 91, 159, 203, 217, 281
  in absence of polycystic ovary, 174–175
  in adolescence, 3, 15
    clinical manifestation, 9
    counseling, 9–11
    etiology, 4–7
    incidence, 3–4
    management, 11
  and androgen production, 77
  and breast cancer, 210
  and cervical cancer, 210–211
  in childhood, 3
    clinical manifestation, 9
    counseling, 9–11
    etiology, 4–7
    incidence, 3–4
    management, 11
  clinical applications
    cervical cancer, 168
    endometrial hyperplasia and endometrial cancer, 165–166

miscarriage, 164
ovarian cancer, 166–167
pelvic floor dysfunction, 168
polycystic ovarian syndrome and infertility, 164–165
and endometrial cancer, 209–210
and erectile dysfunction, 139
ergonomic considerations, 163–164
and FI, 182
and frozen-thawed embryo transfer, 129
and in vitro fertilization, 128–129
and intrauterine insemination, 129
in male, 155
and miscarriage, 91–92
nonsurgical management, 152
obesity-associated dysregulation of adipokines, 213
obesity-linked diseases, 195
obesity-related hyperinsulinemia, 258
and ovarian cancer, 210
and ovarian function, 77
paradox, 218
pelvic ultrasound, 159–163
and polycystic ovarian syndrome, 78
and prolapse, 182–183
and pubertal development, 8–9
and pubertal transition, 16
and puberty, 7–8
and recurrent miscarriage, 92
and Sertoli cell function, 139
and sexual behavior, 37–38
and sexual function, 38–39
and sexual health outcomes, 39–40
and spermatogenesis, 138–139
ultrasound settings, 163
and urinary incontinence, 181–182
Obesity and infertility, possible mechanisms
ghrelin, 26
hyperandrogenism, 23–24
hypersecretion of luteinizing hormone, 24
hyperinsulinaemia, 24
interleukin-6, 25
leptin, 25
resistin, 25
tumour necrosis factor-α, 26
Obesity-related sexual dysfunction, 109–110
Obesogens, 7
Obstetric(s)
outcomes in obese adolescents, 18–19
professional ethics in, 319–320
clinical ethical concept of medically reasonable clinical management, 320
ethical concept of medicine as profession, 319
ethical principle of beneficence, 319–320
ethical principle of respect for autonomy, 320
professional virtue of compassion, 320
professional virtue of self-effacement, 320
Obstructive sleep apnea (OSA), 137, 218–219, 226–227, 255, 283, 315
OCP. *See* Oral contraceptive pill (OCP)
Odds ratio (OR), 43, 85

OHSS. *See* Ovarian hyperstimulation syndrome (OHSS)
OI. *See* Ovulation induction (OI)
Oncogenesis, potential mechanisms for, 196–197, 197*t*
Oocyte, effects on, 86
Open abdominal surgery, 220
OR. *See* Odds ratio (OR)
Oral contraceptive pill (OCP), 144, 285
Oral hormonal contraception, 62
Orgasm, 106–107
Orlistat, 79–80, 89, 145, 147
OSA. *See* Obstructive sleep apnea (OSA)
Osteoporosis, 265
adipocyte hormones, 267–269
epidemiology, 266
female obesity and
clinical and diagnostic implication of concept-obesity of bone, 269
normal bone metabolism, 265
obesity of bone, 269
relationship between fat and bone, 266–267
treatment implications of concept-obesity of bone, 269–270
Ovarian cancer, 166–167
obesity and, 210
Ovarian dysfunction, 92–93
Ovarian function, obesity and, 77
Ovarian hyperstimulation syndrome (OHSS), 28, 154–155, 251
Overactive bladder (OAB), 179
Overweight, 83, 209
Ovulation induction (OI), 143
Ovulatory dysfunction, 172–173

P

PAI-1. *See* Plasminogen activator inhibitor-1 (PAI-1)
Pain physiology, obesity and, 281–283
PALM-COEIN classification of abnormal uterine bleeding, 171, 172*f*
Pancreatic hormones, 267
Parathyroid hormone (PTH), 265, 269
Paroxetine, 277–278
PCO. *See* Polycystic ovaries (PCO)
PCOS. *See* Polycystic ovary syndrome (PCOS)
5-PDE inhibitors, 114
PE. *See* Pulmonary embolism (PE)
Peak systolic velocity (PSV), 113
Pearl index (PI), 58, 68
Pelvic floor dysfunction (PFD), 168, 179
Pelvic inflammatory disease (PID), 68
Pelvic organ prolapse (POP), 183
Pelvic ultrasound, 159–163
transabdominal approach, 159–160
transperineal approach, 163
transrectal approach, 162–163
transvaginal approach, 160–162
Penile Doppler ultrasound, 113
Penile intracavernosal injection therapy, 115–116
Penile prosthesis, 116
Peri-pubertal obesity, 16
Peripheral vascular disease, 111

PFD. *See* Pelvic floor dysfunction (PFD)
PFTs. *See* Pulmonary function tests (PFTs)
Pharmacokinetic studies, 39, 67
Pharmacological agents, 145
Pharmacological therapy, 114–116
Physical activity, 88
Physiology of sexual function, 105–110
PI. *See* Pearl index (PI)
PID. *See* Pelvic inflammatory disease (PID)
Plasminogen activator inhibitor-1 (PAI-1), 24–25, 84–85, 256–257
PMDD. *See* Premenstrual dysphoric disorder (PMDD)
PMS. *See* Premenstrual syndrome (PMS)
Pneumoperitoneum, 227
POC. *See* Progesterone-only contraception (POC)
Polycystic ovaries (PCO), 174
Polycystic ovary syndrome (PCOS), 9, 17, 23, 84, 92, 128, 143, 151, 159, 164–165, 174, 298, 315
in adolescence, 17–18, 20–21
bariatric surgery and, 154
clinical psychosomatic angles to, 297–298
hirsutism and, 78
and infertility, 164–165
and obesity, 298–300, 298*t*
obesity in the absence of, 174–175
recurrent miscarriage and, 92
Polyps, 171–172
POMC. *See* Proopiomelanocortin (POMC)
POP. *See* Pelvic organ prolapse (POP); Progestogen-only pill (POP)
Postoperative hypoxemia, 220–221
Prader–Willi syndrome, 122
Pragmatic approach for obese women, 270
Preconception counseling, preventive ethics in, 321
Preconceptional folic acid, 43
Pregnancy
after bariatric surgery, 154
high body mass index on, 143
lifestyle interventions to improve outcomes, 143–144
pregnant obese women, lipid metabolism in, 257–258
rate, 100
risks of obesity in, 43
Premenstrual dysphoric disorder (PMDD), 297
Premenstrual molimina, 297
Premenstrual syndrome (PMS), 296
clinical psychosomatic angles to, 297–298
Prevention programs, 19
Preventive ethics in preconception counseling, 321
Professional virtue
of compassion, 320
of self-effacement, 320
Professionally responsible clinical management of obese patients
professional ethics
in clinical practice with obese patients, 320–322
in obstetrics, 319–320

Progesterone, 276–277
Progesterone-only contraception (POC), 46, 248
Progesterone-only implant, 60–61
Progesterone-only injection, 61–62
Progestins, 20, 59
Progestogen implants, 70
Progestogen-only pill (POP), 62
Proinflammatory cytokines, 268
Prolapse
  obesity and, 182–183
    weight loss and effects upon continence and, 183–184
Proopiomelanocortin (POMC), 268
PSV. See Peak systolic velocity (PSV)
Psychological morbidity, 21
Psychosexual disorders
  biological and psychological mechanism, 313
    effect of cardiovascular disease on sexual function, 315
    mechanism and effect of obstructive sleep apnea, 315
    normal sexual function, 313–315
  classification of sexual dysfunctions, 314t
  impaired body image, 316
  ISSWSH sexual disorders, 314t
  polycystic ovarian syndrome, 315
Psychosocial distress, 300
"Psychosomatic" stems, 293
PTH. See Parathyroid hormone (PTH)
Pubertal development, obesity and, 8–9
Pubertal timing, 8
Pubertal transition, obesity and, 16
Puberty, obesity and, 7–8
Public Health England, 224
Pulmonary assessment, 233
Pulmonary embolism (PE), 245
Pulmonary function tests (PFTs), 226–227
Pulmonary hypertension, 226

Q

Quality of life (QoL), 189, 297–298

R

Radiotherapy, 205
Rancho Bernardo Study, 110
Randomized controlled trial (RCT), 87–88, 144
RANZCOG. See Royal Australian and New Zealand College of Obstetricians and Gynaecologists (RANZCOG)
RAS. See Robotic-assisted laparoscopic surgery (RAS)
RCOG. See Royal College of Obstetricians and Gynaecologists (RCOG)
RCT. See Randomized controlled trial (RCT)
Reactive oxygen species (ROS), 86, 120–121
  theory, 123
Recombinant LH (rLH), 129
Recurrent miscarriage (RM), 91
  endometrial changes in obesity, 93
  immunological factors, 93
  management, 93–95
  obesity and, 92–93
  ovarian dysfunction, 92–93
  PCOS and, 92
Recurrent pregnancy loss (RPL), 86–87, 93
Renal function, 227
Reproductive functions, 83–84
Resistin, 25, 85, 268
Respiratory function, 226–227
Retinol-binding protein 4, 26
Risk assessment in obese individuals, 255–256
Riociguat, 115
Risk ratio (RR), 100
rLH. See Recombinant LH (rLH)
RM. See Recurrent miscarriage (RM)
Robotic-assisted laparoscopic surgery (RAS), 228
ROS. See Reactive oxygen species (ROS)
Rotterdam criteria, 18, 298
Roux-en-Y gastric bypass (RYGB), 57, 152–153, 270
  and biliopancreatic diversion, 51
  management after, 270
  surgery, 123
Royal Australian and New Zealand College of Obstetricians and Gynaecologists (RANZCOG), 281
Royal College of Obstetricians and Gynaecologists (RCOG), 281
RPL. See Recurrent pregnancy loss (RPL)
RR. See Risk ratio (RR)
RYGB. See Roux-en-Y gastric bypass (RYGB)

S

SAA. See Serum amyloid A (SAA)
SART. See Society of Assisted Reproduction (SART)
Safety and health benefits, 62–63
SC application. See Subcutaneous application (SC application)
Scanning technique, 162–163
Scottish Intercollegiate Guideline Network (SIGN), 3
SCSA. See Sperm chromatin structure assay (SCSA)
Selective ER modulator (SERM), 277
Self-effacement, professional virtue of, 320
Semen quality, impact on, 119–120
  sperm concentration and count, 119–120
  sperm morphology, 120
  sperm motility, 120
Sentinel node mapping, 205
SERM. See Selective ER modulator (SERM)
Sertoli cell function, obesity and, 139
Serum amyloid A (SAA), 257
Severe pelvic/perineal dysfunction, obesity and, 305
Sex drive, 107–108
Sex hormone-binding globulin (SHBG), 8, 16, 24, 77, 119–120, 137–138, 143, 153, 211
Sex hormones, 202
  progesterone, 276–277
  testosterone, 77–78, 109

Sexual behavior, obesity and, 37–38
Sexual dysfunction and obesity-related comorbidities
  coronary artery disease, 111
  diabetes mellitus, 111–113
  hypertension, 111
  peripheral vascular disease, 111
Sexual dysfunction in men, 106–107
  multidisciplinary approach to treatment, 113–116
  physiology of sexual function, 105–110
Sexual dysfunction theory, 122–123
Sexual function, obesity and, 38–39
Sexual health, 37
  sexual health outcomes, obesity and, 39–40
Sexually transmitted infections (STIs), 37, 39–40, 68
SGA. See Small for gestational age (SGA)
Shared decision-making, 321
SHBG. See Sex hormone-binding globulin (SHBG)
Sibutramine, 26, 89, 145, 147
SIGN. See Scottish Intercollegiate Guideline Network (SIGN)
Sildenafil, 114
Simvastatin, 110
Single-incision
  laparoscopic procedures, 229
  robotic-assisted procedures, 229
Sitagliptin, 93
Sleep disorders, 283
Sleeve gastrectomy, 135–136, 152
Small for gestational age (SGA), 152
Society of Assisted Reproduction (SART), 100
Society of Obstetricians and Gynaecologists of Canada (SOGC), 281
SPC. See Summary of Product Characteristics (SPC)
Sperm. See also Fertility; Infertility
  concentration and count, 119–120
  DNA damage, 120–121
  morphology, 120
  motility, 120
Sperm chromatin structure assay (SCSA), 120–121
Spermatogenesis, obesity and, 138–139
Sphincteroplasty, 191–192
Spironolactone, 80
Staxyn, 115
Stendra. See Avanafil
Step-down protocol, 28
Step-up protocol, 28
Sterilization procedures in obese women, 51–52
Steroid hormones, 211
STIs. See Sexually transmitted infections (STIs)
Stress incontinence, 184
Stress urinary incontinence (SUI), 179–181
Strontium ranelate, 269
Subcutaneous application (SC application), 60
Subfertility, 143, 300–302
SUI. See Stress urinary incontinence (SUI)
Summary of Product Characteristics (SPC), 71

Super-Trimix, 115–116
Surgery, 20, 184
　alternatives for class III obese patients, 230–231
　benefits of minimally invasive surgery, 227–230
　in breast cancer, 204–205
　complications, 231–232
　cost-effectiveness, 231
　intraoperative considerations
　　entry techniques, 236–237
　　medication, 236
　　patient positioning, 234–235, 235b
　　surgical technical diligence, 236–238
　　theatre staff, operating room, and equipment, 234
　physiological changes in (obese) surgical patient, 226–227, 226t
　　cardiovascular alterations, 226
　　drug pharmacology, 227
　　effects of CO_2 pneumoperitoneum, 227
　　gastrointestinal changes, 227
　　immune system, 227
　　renal function, 227
　　respiratory function, 226–227
　　venous stasis, 227
　postoperative considerations, 238–239
　preoperative preparation, 232–234, 234b
　　additional considerations, 233–234
　　pulmonary and cardiovascular assessment, 233
Surgical interventions
　assisted reproduction after bariatric surgery, 154–155
　bariatric surgery and polycystic ovarian syndrome, 154
　bariatric surgery as weight loss measure, 152
　impact of bariatric surgery on fertility, 153–154
　nonsurgical management of obesity, 152
　obesity and fertility, 151
　obesity in male, 155
　potential of bariatric surgery for negative impact on fertility, 154
　pregnancy after bariatric surgery, 154
Surplus abdominal visceral adipose tissue, 256

T

Tadalafil, 110, 115
Teenage pregnancy, 18
Tension-free vaginal tape (TVT), 184
Terminal hair, 77–78
Testicular hyperthermia theory, 123
Testosterone, 77–78, 109
Testosterone replacement therapy (TRT), 116
TGs. *See* Triglycerides (TGs)
The Obesity Society (TOS), 233
Thiazolidinediones (TZD), 20–21, 267
3D high-definition video technology, 228
Thromboprophylaxis, 219
Tibolone, 277
Tilt test, 235
TNF-α. *See* Tumor necrosis factor-α (TNF-α)
TOS. *See* The Obesity Society (TOS)
Total testosterone (TT), 108
Traffic light system, 294
Transabdominal approach, 159–160, 161f
　ultrasound in obese women, 162t
Transgenerational epigenetic effects, 139
Transgenerational inheritance, 83
Transperineal approach, 163
Transrectal approach, 162–163
Transvaginal approach, 160–162
　ultrasound in obese women, 162t
Transvaginal natural orifice transluminal endoscopy, 229
Triglycerides (TGs), 49, 256
Trimix, 115–116
TRT. *See* Testosterone replacement therapy (TRT)
Truncal adiposity, 226
TT. *See* Total testosterone (TT)
Tumor necrosis factor-α (TNF-α), 24, 26, 84–85, 203, 255
Tumorigenesis, 306
TVT. *See* Tension-free vaginal tape (TVT)
TZD. *See* Thiazolidinediones (TZD)

U

UI. *See* Urinary incontinence (UI)
UK Faculty for Sexual and Reproductive Health (FSRH), 58, 249
UK Medical Eligibility Criteria for Contraceptive Use (UKMEC), 58, 249, 249t
Ultrasound, 159, 284
　settings, 163
Umbilical stalk elevation (USE), 236
Unintended pregnancy, 39
uNK cells. *See* Uterine natural killer cells (uNK cells)
Unscheduled bleeding, 71
Urethral sphincter mechanism, 180
Urinary incontinence (UI), 179, 189. *See also* Anal incontinence (AI); Fecal incontinence (FI)
　incontinence symptoms following weight loss, 192
　obesity and, 181–182
　outcomes of incontinence procedures in obese women, 191
　pathophysiology of incontinence in obese population, 190–191
　selected symptoms and signs of, 180t
Urinary tract injuries, 232
Urodynamic stress incontinence (USI), 180–181
USE. *See* Umbilical stalk elevation (USE)
USI. *See* Urodynamic stress incontinence (USI)
Uterine natural killer cells (uNK cells), 93

V

Vardenafil, 114
Vasomotor symptoms (VMS), 273
Vaspin, 26
Vellus hair, 77–78
Venous stasis, 227
Venous thromboembolism (VTE), 44, 49, 231–232, 245, 285
　assisted conception, 251
　contraception and, 49–50
　guideline recommendations, 248–249
　gynecological cancer, 252
　gynecological surgery, 251–252
　hormonal contraception, 246–248
　HRT, 249–251
　interplay between obesity and VTE risk, 245–246
Very low–calorie diets, 144
Very low–energy diet (VLED), 137–138
Viagra. *See* Sildenafil
Visceral adipose tissue, 256–257
Visceral fat, 257–258
Visceral obesity, 202
　pathogenesis of, 256–257
Visfatin, 26, 85
Vitamin D, 270
VLED. *See* Very low–energy diet (VLED)
VMS. *See* Vasomotor symptoms (VMS)
VTE. *See* Venous thromboembolism (VTE)

W

Waist circumference (WC), 223
Waist-to-hip ratio (WHR), 201
Weight gain, 45–46. *See also* Obesity
Weight loss, 87–88
　barriers to, 88–89, 148
　incontinence symptoms following, 192
　medications, 145
　treatment, and contraception, 248
Weight management, 113–114, 261
WHI. *See* Women's Health Initiative (WHI)
WHO. *See* World Health Organization (WHO)
Whole System Obesity (WSO), 224
Whole Systems Approach (WSA), 224
WHR. *See* Waist-to-hip ratio (WHR)
Women's Health Initiative (WHI), 206
World Health Organization (WHO), 3, 15, 67, 83, 119, 223, 281, 293
WSA. *See* Whole Systems Approach (WSA)
WSO. *See* Whole System Obesity (WSO)

Y

Yasmin, 80

Z

Zonulin, 26

Made in the USA
Columbia, SC
01 November 2023